KNOWLEDGE, LANGUAGE AND LOGIC:
QUESTIONS FOR QUINE

BOSTON STUDIES IN THE PHILOSOPHY OF SCIENCE

VOLUME 210

W.V. Quine
[Photo: Harvard News Office]

KNOWLEDGE, LANGUAGE AND LOGIC: QUESTIONS FOR QUINE

Edited by:

ALEX ORENSTEIN

Queens College and the Graduate Center,
City University of New York

and

PETR KOTATKO

The Philosophical Institute of the Czech Academy of Sciences

KLUWER ACADEMIC PUBLISHERS
DORDRECHT / BOSTON / LONDON

A C.I.P. Catalogue record for this book is available from the Library of Congress.

ISBN 1-4020-0253-X
Transferred to Digital Print 2001

Published by Kluwer Academic Publishers BV,
PO Box 17, 3300 AA Dordrecht, The Netherlands.

Sold and distributed in North, Central and South America
by Kluwer Academic Publishers, PO Box 358,
Accord Station, Hingham, MA 02018-0358, USA

In all other countries, sold and distributed
by Kluwer Academic Publishers, Distribution Centre,
PO Box 322, 3300 AH Dordrecht, The Netherlands

Printed on acid-free paper

For Professor Quine
in celebration of his 90th birthday

TABLE OF CONTENTS

PREFACE

This collection originated at a conference organized by the department of the philosophy of language and philosophy of science of the Philosophical Institute of the Czech Academy of Sciences held in Karlovy Vary in August 1995. Most of the papers in this volume were presented on that occasion. Professor Quine presented the paper included here and also commented on those given at that time. The questions for Professor Quine put forward in the present volume are arranged into three sections. They bear on issues in the theory of knowledge, the philosophy of language and the philosophy of logic.

To a some extent the organization parallels the treatment Quine himself gives to his views in some recent works. Quine's naturalizing of questions of epistemology involves a naturalistic account of the language employed in recording our knowledge claims and this in turn leads to questions about language and meaning, and logic and reference. The initial essays on naturalized epistemology deal with questions concerning the role of observational factors, which Quine addresses in his paper on Triangulation and which Szubka and George then take up. Grayling and Lehrer question the very project of naturalizing epistemology. Gibson considers Quine's relationship to Wittgenstein, Bergstrom an empiricist definition of truth, Miscevic the status of the a priori and Gjelsvik offers a naturalistic account of decision making.

The section on the philosophy of language concentrates on Quine's well known conjecture of the indeterminacy of translation. Segal, Antony, Horwich, Pagin and Stoutland examine the conjecture from varying vantage points: Segal suggests that a mentalistic semantics can mitigate against the indeterminacy and Antony that data concerning language acquisition argue that same result, Horwich focuses on whether translation provides the way to approach meaning, Pagin on the public nature of our knowledge of meanings, and Stoutland on the normative nature of rules of language.

The last part on logic and reference begins with an examination of the problem concerning vacuous names which Quine dubbed Plato's beard. Parsons and Woodruff then take up matters concerning vagueness. Neale, Ray, and Recanati raise questions concerning issues surrounding opacity: Neale and Ray on modality and Recanati on propositional attitudes.

We, and most particularly I, Alex Orenstein, would like to express indebtedness to Queens College of the City University of New York for a presidential research award which provided time to work on this project, to the Faculty Research Award Program of the City University of New York for a grant which aided us in preparing the manuscript for publication, to Wolfson, to St. Anne's and to Exeter Colleges, Oxford for the use of their facilities during much of the preparation of this work and to Roger Gibson and Douglas Quine for their help with the proofs. Last of all, we would like to record our gratitude to Marx Wartofsky, whose untimely death prevented our thanking him in person. His help made it possible for this collection to appear in the *Boston Studies in the Philosophy of Science* series.

Alex Orenstein and Peter Kotatko

W.V. QUINE

I, YOU, AND IT: AN EPISTEMOLOGICAL TRIANGLE

The triangle I have in mind is one that Donald Davidson has occasionally invoked. It has you at one vertex, some one of you. I am at another vertex, and at the third vertex there is some object. The object is reflecting light at both of us, but in somewhat different patterns owing to our difference in perspective. The light rays trigger our nerve endings, setting off one train of events in your nervous system and another in mine.

The object at the third vertex is a creature, unfamiliar to me. You tell me a name for it: aardvark. What went on in your nervous system and mine when we observed the aardvark differed in perspective and probably in more. We are differently wired. Our sensations may have differed too, if that makes sense. All we clearly shared was the distal cause of our neural events, the aardvark. Still I end up associating this same word with my stimulus, my neural intake, as you did with your different intake – numerically different certainly, and somewhat different in further ways. We thus differ in the proximal causes of our concordant use of the word, but we share the distal cause, the reference, farther out on our causal chains.

This is something worth wondering about. The causal chains from the aardvark to you and to me part company already at the aardvark itself and make their divergent ways into our unlike nerve nets, but we both end up calling the creature an aardvark. We shall find that what forges our link in the triangle, linking us, is a preestablished harmony.

I have now learned the word "aardvark", thanks to you, and I proceed to apply it to similar creatures as they turn up. You do likewise, independently, and we find ourselves in agreement in our continuing use of the term. This suggests that we are both good judges of similarity, in some objective sense of the term. But wait: I know no objective sense of the term, apart from geometry.

Subjective similarity, yes. We can make good psychological sense, experimentally testable, of perceptual similarity of neural intakes for an individual. We can test it by reinforcing and extinguishing his responses. Stimulus A is perceptually more similar to stimulus B than to C if A evokes a response that was rewarded on the occasion of B but penalized on the occasion of C.

Back now to the aardvarks. The next time one turns up and you and I are there to observe it, I say "There's an aardvark" and you agree. My neural intakes from the two aardvarks were perceptually similar by my lights, and your two were perceptually similar by yours. There is no presumption of perceptual similarity between your intakes and mine; intersubjective perceptual similarity is not even defined.

Alex Orenstein and Petr Kotatko (eds.), Knowledge, Language and Logic, 1–6.
© 2000 Kluwer Academic Publishers. Printed in Great Britain.

What we have rather is a parallelism, a preestablished harmony, between your standards of perceptual similarity and mine. In general, if external events out there at the third vertex on two occasions produce neural intakes in both of us, and yours are perceptually similar for you, mine are apt to be perceptually similar for me.

Without this harmony, my perceptual similars might not have paired off so nicely with yours. My later intake might not have qualified as perceptually similar for me to my earlier one, though your later one did so qualify for you. Our scales of similarity might have diverged. Thanks to the harmony, they did not. Aardvarks look pretty much alike to me, and they look pretty much alike to you.

The preestablished harmony is needed to account for our meeting of minds not only on aardvarks, but also on what to call them: on the mellifluous Dutch disyllable "aardvark" itself. The phonetic constancy of a word, from one utterance of it to another, is itself a product of the speakers' subjective standards of perceptual similarity. Thanks to the harmony, communication proceeds apace. Oh, we sound alike. Oh, who says so? Each of us, by his own standards of perceptual similarity, all of which are in harmony.

Perceptual similarity is vital not only for language, but for all learning, all habit formation, all expectation. It was its role in learning, or habit formation, that underlay the experimental test of perceptual similarity itself. Since it is presupposed in learning, moreover, perceptual similarity cannot itself have been learned, not all of it; some had to be innate, though it gets overlaid and changed as learning progresses.

So perceptual similarity standards are in part innate and are in preestablished harmony. Natural selection accounts nicely for both traits. The relevance of similarity standards to survival becomes evident when we consider the role they play in expectation. We tend innately to expect similar events to have sequels that are in turn similar to each other. This is primitive induction. There is survival value in successful induction, successful expectation: it expedites our elusion of predators and our pursuit of prey. Natural selection, then, has favored similarity standards that mesh relatively well with the succession of natural events. This explains why our expectations have run a better than random score of success. It also explains the preestablished harmony: the standards are largely fixed in the genes of the race, the species.

What I call an observation sentence, for a given speaker, is any expression, such as "aardvark", to which he has learned to assent in response to all neural intakes within some range of perceptually similar intakes. The expression may be a single word, as here, or it may be a grammatical sentence: "It's cold", "It's raining". They are "about" the external world, not about sense data or stimulation. They are directed at the "it" vertex of the triangle, where reference lies.

An observation sentence is an *occasion* sentence, true on some occasions and false on others. An occasion sentence qualifies as an observation sentence for a speaker if he is disposed to assent to it outright when appropriately stimulated. It is an observation sentence for a society of speakers if it is an observation sentence for each member and if, further, all members who witness an occasion are disposed to agree in assenting or in not assenting.

The child's first cognitive utterances are observation sentences – thus "Mama", "Milk", "Dog" – though it will be a while before the child gets a grip on the schematism of space, time, and recurrent enduring objects, to which these observation sentences as we know them are committed. These adult refinements make sense of sameness of a recurrent object and distinctness of an exactly similar object. For an infant the observation sentence "Ball" would merely signal a recurrent state of affairs, at first, like "Thunder" or "It's cold". Only later does it make sense to wonder whether it is the same ball next time around or another one like it. Only then is reification in full flower. But the reference is out in the world early and late, at the "it" vertex of the triangle. The child merely has to learn more about it. Meanwhile the range of neural intakes prompting the child's assent to the observation sentence "Ball" can remain unchanged.

At the primitive or infantile stage the observation sentences are the human analogues of the apes' cries, the signals by which an ape alerts his kin to the approach of a predator or the prospect of some fruit at a treetop. Language is, itself perhaps a luxuriant outgrowth of the apes' cries. In any event the child promptly outstrips the apes, amassing new observation sentences hand over fist and learning ways of combining them with help of prepositions and conjunctions to form further observation sentences.

He then somehow caps this achievement with what I picture as his next great acquisition, the *observation categorical*. It is a generalized expression of expectation, the fruit of induction. It is a way of joining two observation sentences to express the general expectation that whenever the one observation sentence holds, the other will be fulfilled as well. Examples: "When it snows, it's cold"; "Where there's smoke, there's fire"; "When the sun rises, the birds sing"; "When lightning, thunder". They are our first faltering scientific laws.

The survival value of the apes' cries, and of our ordinary observation sentences, lay in vicarious observation. One learns from one's kinsman, or kinsape, about something that one couldn't see from where one was sitting at the time. Now the observation categoricals bring us much more: they bring us vicarious habituation, vicarious induction. We get the benefit of generalized expectations built up over the years by some veteran observer or even by his informant, long dead. Observation categoricals can be handed down.

There is more to be said to the glory of the observation categoricals: they are the empirical checkpoints of scientific theory. If a theory implies an observation categorical, we can subject the theory to an experimental test by bringing about fulfillment of the observation sentence that is the protasis of the categorical and then watching for fulfillment of the apodasis. I see in this the distilled essence of the experimental method, and the only empirical check to which science is susceptible. Failure of the test brands a theory as false. Any appeal must rest on challenging the observations themselves, whereof more anon.

Language and science go on developing hand in hand, in the child and in the race. I already remarked on the big ontological leap achieved in conceiving of an enduring body, a body intermittently observed between unseen trajectories. Another big ontological leap was the reification of numbers and other abstract objects. It became clearly recognizable only with the advent of relative pronouns, or variables.

Observation sentences were at first acquired at mother's knee, and affirmed on appropriate occasions without peradventure. The growth of theory, however, calls upon us to think twice. "That's a rabbit", affirmed under appropriate stimulation, had seemed final. But at a later stage, when we allow for the identity over time of an intermittently observed body, we may come to suspect that yesterday's supposed rabbit was just this toy that we now find in the garden.

We continue to accrete observation sentences throughout life. Some, like the child's early ones, are learned outright as wholes by direct conditioning to appropriate ranges of neural intake. Mostly such acquisitions will subsequently become infected with theory in varying degrees. Other observation sentences may be assembled from words learned in earlier contexts. Speakers vary in how they arrive at the same usage of an observation sentence.

There are sentences that are learned only through theory, but become observational later in the specialist's career; thus "There was copper in it", said by the chemist after a glance at the solution, or "There goes a hyperthyroid", said by a physician after a glimpse of a stranger's face.

What I have said thus far will have been pretty familiar to those of you who have been reading my writings. But there now comes a change, the effect of correspondence with Professor Bergström. He propounded what he called an empiricist theory of truth, which I found both appealing and disconcerting. In the end I did not adopt the theory, but I came to this decision only by sharpening my own views in a quarter to which I had hitherto given too little attention: the *recanting* of observation sentences.

The speaker assents outright to "Rabbit" and retracts his assent later on discovering that what he had seen was a toy. In *Word and Object* (p. 44) I treated such intrusions of error as showing that the retracted sentence was not purely observational; theory had intruded. Theory had indeed, but I now see a better way of sorting matters out.

Immediacy of assent is still my criterion of observationality, but emphatically and unequivocally so. The assent may be recanted, but its susceptibility to recantation becomes an independent dimension, *theoreticity*, for separate consideration.

The range of neural intakes to which the speaker's assent to the sentence is keyed will of course be vague along the edges. The speaker may hesitate over "It's raining" in a fine mist, and over "That's a swan" in the startling presence of a black one. Vagueness of boundaries invests language at every turn, and I shall continue to take it in stride. But assent must be immediate when the stimulation is in the clean-cut range between vaguenesses. Immediate but fallible.

As always, the subject matter of the observation sentence is the real world, the "it" of the triangle, and it is the domain of fallible scientific theory. "It's raining", affirmed in full view of a rain-drenched window, is recanted on spotting the garden hose, and "Rabbit" on spotting the toy. Such recantation reflects conflict of observations through theoretical connections between observation sentences. The degree of susceptibility to recantation measures how theoretic the observation sentence is. It is its *degree of theoreticity*.

An observation sentence that is perhaps minimally theoretic is "This looks blue". I say "looks" here, rather than "is", to allow for the possibility that reflected light or environmental contrast may be affecting the color that the object would otherwise show. The reference of the sentence is still to an external patch or body, or is to become so with the flowering of reification.

This move brings me to terms with Kuhn, Hanson, and Feyerabend, who have objected to observation sentences that they are suffused with theory. They are indeed, I now agree, in all degrees, but their observational role remains. The immediacy of assent shows that in one way or another they have become conditioned as wholes to ranges of perceptually similar neural intakes and so serve as observation points in the experimental testing of hypotheses. The experimenter is still free, indeed advised, to examine the circumstances of his observation for possible experimental error, or even to try again.

Sophisticated observation sentences, such as the ones about copper and the hyperthyroid, are apt to be reducible to more primitive ones, delineating more directly sensory evidence. These will tend to be less theoretic by the stated criterion, that is, less susceptible to recantation. Some sophisticated ones, however, are not thus reducible. I think of the subtle traits that the wine expert learns to detect.

Such reduction, where possible, bolsters scientific theory; for the increased resistance to recantation of observation sentences increases their dependability in checking scientific theory.

A thoroughgoing reduction project of the kind, however, is surely a forlorn hope. It is utterly alien to what goes on and went on in the development of language and science in the child and in the race. Observation sentences, already theoretic in varying degrees, are learned outright and helter-skelter by direct holophrastic conditioning. Further ones are synthetized along the way from bits of those at hand. It is not in general a matter of basing more theoretical ones on less theoretical ones. They vie with one another in a surging equilibrium of evidential claims. Such, in my sometime collaborator Joe Ullian's vivid phrase, is the web of belief.

Only limited reduction of overall theoreticity, then, is to be hoped for by reduction of some observation sentences to others. Some reduction may also be sought, however, in another way: by comparing several observation categoricals that are deducible from a theory as checkpoints. One of these may contain less theoretic observation sentences than others, and we could choose it for testing.

The last traces of analogy between my naturalized epistemology and the phenomenalist's dream of a reduction to sense data have now, it seems, been dissipated. Observationality becomes a simple behavioral matter of prompt response to stimulation, and overall reduction of the more theoretic to the less theoretic ceases to figure as an ideal.

My observation categoricals were always a schematic caricature of what goes on in testing scientific theory. They represented an adverse pair of observations as refuting a theory beyond redemption, whereas real life is less clean-cut. Now my recognition of theoreticity as a second dimension for observation sentences enhances the realism of my caricature by accommodating the inconclusiveness of an experimental verdict. The surging equilibrium becomes part of the picture.

Settling then for a modicum of irreducible theoreticity in the observation sentences, let us assess the consequent debilitation of the experimental evidence for science. A theory, let us suppose, has been found to imply an observation categorical. The experimenter has fulfilled the observation sentence that is the protasis of the categorical, to his satisfaction, and the predicted observation does not ensue. He concludes that the theory has failed the test. There are two places where the test itself can have gone wrong. Perhaps the experimental condition, the protasis, only seemed to have been fulfilled; something intruded to create the illusion. Or perhaps the predicted result did occur, but something intruded to obscure it.

At this point the morals to draw are two: a word of caution and a consoling reflection. The word of caution is "Repeat your experiment". It is a superfluous caution, for the scientist would be eager to do just that – the more so since he was probably hoping to save his theory rather than refute it. The consoling reflection is that at worst the error is rejection of a tenable theory and not the retention of a false one. Ignorance is less pernicious than outright error.

Harvard University

NOTE

I am indebted to Burton Dreben, as usual, for a critical reading of earlier drafts and helpful suggestions.

TADEUSZ SZUBKA

QUINE AND DAVIDSON ON PERCEPTUAL KNOWLEDGE

One of the main differences between Quine's and Davidson's theories of knowledge and mind lies in their accounts of the content of perception and the way in
which it contributes to our knowledge of the external world. Both thinkers are very
sensitive to these differences and it has been the subject of discussion between
them in recent publications. To put it very roughly, Quine holds firmly to the position that although we finally manage to get veridical knowledge of the external
world, the content of our perceptions are just the triggerings of our sense receptors
that give us reliable clues about the objects and happenings in our environment.
Davidson considers this view to be a naturalized successor of an older defective
empiricism which should be abandoned. In its place he proposes an externalist
theory of perceptual content, according to which content is fully determined or constituted by the objects and events in the external world. This move, among other
things, bypasses many of the troubles that Quine's approach faces and gives a solid
ground for our intersubjective communication. In other words, if the central
concern of Quine's epistemological project is epistemology naturalized, so the
central concern of Davidson's corresponding project is epistemology externalised.[1]

In the present essay I shall outline, in the first part, Quine's view of perceptual
knowledge, and subsequently, in the second part, Davidson's arguments against it
and his own positive account of perceptual content. In the third and final part I shall
try to say why I find both of these options unacceptable, and then make some suggestions towards what I take to be a more satisfactory solution.

1. NEURAL INTAKE AND OBSERVATION SENTENCES

Quine believes that someone who has given up the idea of philosophy as independent of and prior to science, and who accepts its continuity with science, has only
one available option in epistemology: she must endorse a non-mentalistic version
of empiricism; one transformed into the physics of stimulus and response. Of
course, this does not mean that she is restricted, while doing epistemology, to rely
exclusively on the theories and hypotheses of contemporary physics; by no means –
she can draw upon such disciplines as neurology, psycholinguistics and evolutionary genetics. That is, the task of a scientific or naturalized epistemologist is to
adhere to the ancient slogan *nihil in mente quod non prius in sensu* but to give it a
distinctive and broadly understood physicalist turn.

The assumptions from which the naturalist epistemologist starts would pass
today as hardly disputable truisms or platitudes. We are highly complicated living

7

Alex Orenstein and Petr Kotatko (eds.), Knowledge, Language and Logic, 7–19.
© *2000 Kluwer Academic Publishers. Printed in Great Britain.*

organisms, animals of some sort, interacting with the environment in various ways. How do we get information about this environment and the external world? Quine writes:

Science itself teaches that there is no clairvoyance; that the only information that can reach our sensory surfaces from external objects must be limited to two-dimensional optical projections and various impacts of air waves on the eardrums and some gaseous reactions in the nasal passages and a few kindred odds and ends.[2]

And more recently:

Our avenue of continuing information about the world is the bombardment of our sensory surfaces by rays and particles, plus some negligible kinaesthetic clues to the ups and downs of our footpath.[3]

The result of these causal impacts or bombardments by rays and particles are irritations of our surface or, more precisely, the triggerings of our exteroceptors. Confining our attention to a single subject triggered on a given occasion we can define the subject's global stimulation, or neural intake, as the temporary ordered set of all firing of her exteroceptors on that occasion.[4] These neural stimulations or intakes prompt the subject to assent to or dissent from various observation sentences, sentences such as "It is raining", "That is a rabbit", or simply "Mama", "Milk", "Cold" etc. These sentences plays a central role both in learning a language and in testing our scientific theories about the world.[5]

When we turn to the issue of observation sentences and assent to them, we seem to leave the safe domain of commonplace scientific truths and enter a rather murky area, the field of real philosophical difficulties. One of the problems here is to get an account of observation sentences which ensures that they successfully meet the following requirements: (i) each such sentence should be associated affirmatively with some range of one's stimulations and negatively with some other range; (ii) each such sentence should command the subject's assent or dissent outright, on the occasion of an appropriate stimulation, without further investigation and independently of her interests, etc. (iii) each observation sentence *must* command the same verdict from all linguistically competent witnesses of the occasion.[6] The first two conditions appear to be not very difficult to meet, provided that we keep in mind the adjective "appropriate" in the phrase "appropriate stimulation" (there can, of course, be cases when the subject confronted with a black swan cannot, without further investigation, say "That is a swan" or "That is not a swan") and are able to distinguish from the wide class of observation sentences for various groups of speakers, with various scientific backgrounds, etc. a subclass of very simple and unsophisticated observation sentences common for all speakers of a given linguistic community. But the third condition, i.e. the condition of intersubjectivity, is not very easy to meet. This is because Quine has to get intersubjectivity out of the subjective and idiosyncratic stimulations of each particular member of the community. The quick reply would be to say that these stimulations are not subjective in the Cartesian sense: they are successfully investigated by the intersubjective methods of neuroscience. But this reply is scarcely satisfactory if the observation sentences are supposed to be "the entering wedge in the learning of language". The very idea of a mother making recourse to complicated neurological tests in order to settle whether her child associates with a given observation sentence a stimulation of the right kind, strikes us as totally absurd. So the resolution to this problem certainly does not rely on taking the subjectivity of individual stimulations as merely apparent.

Quine has become increasingly aware of this difficulty. In *The Roots of Reference* he tries to remove it by the following account of observation sentences:

A sentence is observational insofar as its truth value, on any occasion, would be agreed to by just about any member of the speech community witnessing the occasion. This definition depends ... on the idea of membership in the speech community, but that presents no problem; we can recognize membership in the speech community by mere fluency of dialogue, something we can witness even without knowing the language.[7]

The account, as it stands, suffers from the ambiguity of the word "occasion". But further clarifications and the context of his whole epistemological project clearly suggest what Quine has in mind here: the occasions are simply sensory impingements, surface irritations, neural intakes, etc. Another defect of this account is more troublesome, namely the essential dependence of the definition of an observation sentence on the idea of membership in a speech community. Can fluency of dialogue function as a criterion or even a reliable indicator of belonging to the same speech community? Hardly. For example, Poles visiting the Czech Republic often engage in fairly fluent conversations (about matters of daily life) with the people living there, but this does not prove that they belong to the same speech community. There is only some affinity between their languages. So further refinements of the fluency criterion are needed. However, doing this membership in the same speech community comes to be based on agreement about which observation sentences are assented to or dissented from on any given occasion. If so, the above account becomes uninstructive and circular.[8]

In his more recent writings (from the eighties and nineties) Quine explicitly recognizes that this way of defining observation sentences is faulty beyond repair. There is no other choice but simply to be satisfied with an account that says which sentences are observational for each individual speaker. Here is the passage applying this strategy:

An observation sentence is an occasion sentence that the speaker will consistently assent to when his sensory receptors are stimulated in certain ways, and consistently dissent from when they are stimulated in certain other ways. If querying the sentence elicits assent from the given speaker on one occasion, it will elicit assent likewise on any other occasion when the same total set of receptors is triggered; and similarly for dissent.[9]

When, however, a given sentence is observational not just for a single speaker but for a whole community (that is, for each and all its members)? The answer, at first sight, is simple: when it is observational for each individual member of the community. And maybe this answer is *really* simple and easy to apply in deciding which sentence is observational, and which not, for someone who has extraordinary cognitive powers and capacities that enable her to look in the brains of all the speakers concerned and see whether they assent to a given sentence only when their sensory receptors are stimulated in a certain determinate way. But for ordinary speakers with ordinary cognitive powers and capacities this answer seems to be rather useless.

But Quine would reply that ordinary speakers are not in such an impoverished situation. They have empathy, a "gift of human nature", that enable them to correlate, however fallibly, their observation sentences with the observation sentences of other members of the speech community. Empathy plays, for instance, an indispensable role in learning a language. The mother assesses the appropriateness of her child's observation sentences by taking into account the child's orientation and how the environment looks from there. And this use of empathy extends to various sorts of interaction between humans. To put it in Quine's words: "We all have an uncanny knack for empathizing another's perceptual situation, however ignorant of the physiological or optical mechanism of his perception".[10]

So far I have been considering only one problem of Quine's naturalized epistemology: the problem of defining observation sentences without abandoning objectivity. The other, which I merely mention now and to which I shall return briefly in the third part, concerns the nature of the relation of the triggerings of sense receptors or neural intakes to observation sentences. Is this a brute non-rational causal connection that makes us just happen to assent to various observation sentences? Or, is this not just a causal but *also* an epistemic relation where neural intake is evidence for or against a given observation sentence? Quine's writings seems to be rather vague on that matter.

I turn now to arguments Donald Davidson has put forward against Quine's views of content, perception and observation sentences, and his own positive account of these epistemological issues.

2. EXTERNAL CONTENT AND OBJECTIVE MIND

Davidson presents two arguments against Quine's epistemological views: (i) from the nature of justification, and (ii) from the threat of scepticism. Here is the first argument.

Let us suppose, says Davidson, that we have sensations, in the form of neural intake or stimulation, which justify our observation sentences, or more generally, our beliefs about the external world. According to that picture, having the sensation of seeing a green light flashing, or having the neural intake which amounts to that, may justify the belief or observation sentence that a green light is flashing. But now, Davidson asks, does the sensation or neural intake justify the belief? And he gives the following answer:

Of course, if someone has the sensation of seeing a green light flashing, it is likely, under certain circumstances, that a green light is flashing. *We* can say this, since we know of his sensation, but *he* can't say it, since we are supposing he is justified without having to depend on believing he has the sensation. Suppose he believed he didn't have the sensation. Would the sensation still justify him in the belief in an objective flashing green light?[11]

The point of this passage can be put as follows: sensations or neural intakes as such do not play any role in the justification of our beliefs or sentences; only beliefs or sentences can justify or support other beliefs or sentences. We can refer to sensations or neural intakes while giving a descriptive and causal account of how people acquire their knowledge, but an individual thinker cannot put them at the bottom of justificatory relationships holding among her beliefs and sentences. Sensations or stimulations, claims Davidson, "cause some beliefs and in *this* sense are the basis or ground of those beliefs. But a causal explanation of a belief does not show how or why the belief is justified".[12] There is no way of transmuting a cause into a reason, and even positing various epistemological deliverances of the senses (e.g. sensations, sense data, the given, neural intakes) won't accomplish this feat. If these deliverances are to stand in logical or justificatory relations to one's beliefs or sentences, they must already be beliefs or sentences. If they are not full-fledged beliefs or sentences, they cannot serve as justificatory reasons for other beliefs or sentences. There are, of course, causal intermediaries between our beliefs or sentences and the external world, but there are no epistemic intermediaries.

Moreover, and this is the second argument, Quine's naturalized epistemology with its reliance on neural intakes leads to scepticism in much the same way as traditional epistemologies invoking such mental entities as sense data, impressions, and the like, did. This is because it shares with them the idea that empirical knowledge requires an epistemological bridge between the external world and our beliefs, sentences or theories about it. Davidson shows how exceptionally easy it is to generate for Quine's conception the old problem of scepticism concerning the senses:

[L]et us imagine someone who, when a warthog trots by, has just the patterns of stimulation I have when there's a rabbit in view. Let us suppose the one-word sentence the warthog inspires him to assent to is "Gavagai!" Going by stimulus meaning, I translate his "Gavagai!" by my "Lo, a rabbit" though I see only a warthog and no rabbit when he says and believes ... that there is a rabbit. The supposition that leads to this conclusion is not absurd; simply a rearranged sensorium. Mere astigmatism will yield examples, deafness others; little green men and women from Mars who locate objects by sonar, like bats, present a more extreme case, and brains in vats controlled by mad scientists can provide any world you or they please.[13]

So scepticism arises because we have to start with the sensory stimulations of our receptors, that also give a meaning to our observation sentences, and then infer out of this how the external world is like or what is really the case out there. And since there are no rigid connections between what happens in the world and our neural intakes (i.e. it is not the case that the same kind of neural intake can be caused only by one particular kind of events in the world), these inferences may result in false claims about the external world.

Given the function which sensory stimulations or neural intakes have in Quine's account of observation sentences, Davidson ascribes to him (with some qualifications) a proximal theory of meaning and evidence.[14] Meaning, according to this theory, depends on, or better, is determined by, what is going on in our head, what happens within our skin. Davidson contrasts it with an alternative account, a distal theory, which he himself favours. The distal theory holds that "the events and objects that determine the meaning of observation sentences and yield a theory of evidence are the very events and objects that the sentences are naturally and correctly interpreted as being about".[15] Only such a theory preserves the interpersonal sameness of meaning which is required for language and communication.

To show that clearly, Davidson considers in detail a primitive learning situation.[16] Let us suppose that a child babbles, and is rewarded when it produces a sound "table" in the presence of a table; the process is repeated and the connection between the presence of tables and emitting the sound "table" becomes firmly established. Of course, generalization and perceived similarity play a crucial role in such a learning situation: the child's responses depend heavily upon grouping simultaneous or successive objects into classes of the same or similar items, and we can group the child's stimuli by the similarity of responses which they elicit in the child. But why, an advocate of Quine can wonder, go so far for common stimuli? Why not say that the child's responses are not to tables in the external world, but rather to patterns of stimulation at its sensory surfaces, to patterns of its neural intakes? After all, these patterns always produce the relevant verbal behaviour, whereas tables produce it only under favourable circumstances (e.g. when the child clearly sees a table, classifies it correctly as a table, and the like). Moreover, as has been noticed by Quine, when we accept a distal theory of meaning, we will immediately face the problem of giving a satisfactory specification of those common external stimuli, and this problem does not seem to arise in the case of the proximal theory. For we have to decide whether the child responds verbally to the presence of a table, or to a certain feature of it, or to a much wider situation, of which a table is only a part. Of course, we can talk here more vaguely simply about a repeated situation to which the child responds, but this is not theoretically very helpful and introduces a suspicious ontology of situations.

Nevertheless, claims Davidson, it still seems natural to say that the child is responding to a table, rather than to a specific neural intake. "It seems natural to us because it *is* natural – to us".[17] We find certain similar sounds made by the child ("table"), and we find that the natural (to us) class in the world corresponding to those sounds is a class of tables. In opposition to that we cannot easily observe and classify the visual patterns moving between the table and the child's eyes, nor observe and classify the stimulation of its nerve endings. And even if we could observe such items, we would probably discover their similarity only by grouping them in accordance with their respective causes in the external world (i.e. we would treat certain visual patterns or neural intakes as similar because they were caused by objects or events of the same kind).

From these considerations, Davidson holds, we eventually get a picture in which there are three patterns for spotting similarity:

The child finds tables similar; we find tables similar; and we find the child's responses in the presence of tables similar. It now makes sense for us to call the responses of the child responses to tables. Given these three patterns of response we can assign a location to the stimuli that elicit the child's responses. The relevant stimuli are the objects or events we naturally find similar (tables) which are correlated with responses of the child we find similar. It is a form of triangulation: one line goes from the child in the direction of the table, one line goes from us in the direction of the table, and the third line goes from us to the child. Where the lines from child to table and us to table converge "the" stimulus is located.[18]

To express this idea of the distal common stimulus we don't need, Davidson holds, any suspicious ontology of situations.[19] The situations are made similar by the similarity of the responses of the teacher or parent and the similarity of the child's responses to the world. If we press further the question of what makes the situations in question similar apart from the similarity of responses, the only answer that can be given is: they are situations in which a table is seen. The cases where a table appears to be seen, and there is in fact none to be seen, are relatively rare.

Now supposing that there is a close connection between language, thought and knowledge, the above considerations have far-reaching consequences for our theories of mind and of knowledge. To justify this supposition it is enough to point out that the content of our minds is constituted, for the most part, by such states as beliefs, intentions, and desires. According to the usual account, these states are prepositional states, that is their objects are the meanings of sentences, or, to put it differently, propositions. Thus what is true about the meanings or propositions must also be true about the mental states which are prepositional attitudes. However Davidson is even prepared to hold that we can establish the connection and justify the supposition without committing oneself to any particular theory of prepositional attitudes. He believes that "from the fact that speakers are in general capable of expressing their thoughts in language, it follows that to the extent that the subjectivity of meaning is in doubt, so is that of thought generally".[20] The further connection with knowledge is also quite easy to establish: knowledge is true belief of some kind and has to be expressed in language. In that way the distal theory of meaning leads us to the externalist theory of mind and knowledge.

If we look at our mental states in that externalist way, we will no longer view mind as an isolated container of essentially subjective or private thoughts, some of which may correctly represent the objective or public world. Here are two characteristic passages from Davidson's writings expressing this idea:

Our thoughts are "inner" and "subjective" in that we know what they are in a way no one else can. But though *possession* of a thought is necessarily individual, what gives it content is not. The thoughts we form and entertain are located conceptually in the world we inhabit, and know we inhabit, with others. Even our thoughts about our own mental states occupy the same conceptual space and are located on the same public map.[21]

Beliefs are true or false, but they represent nothing. It is good to be rid of representations, and with them the correspondence theory of truth.[22]

The interpretation of these remarks in the light of Davidson's other views seems to be simple: beliefs, intentions and desires are relational states, and their content is determined by objects and events in the common public world, or, to put it even more straightforwardly – these objects and events are their content. Thus individual thinkers do not determine or constitute their own thoughts; they only entertain them. This explains the seemingly paradoxical, and – *pace* to Rorty – by no means "postphilosophical" remark that although beliefs are true or false, they do represent nothing. The beliefs do not represent how objects and events are out there in the world, because these objects and events simply constitute them, as their parts.

This sheds new light on our knowledge, and especially on the relationships which holds between its three varieties: knowledge of the external world, knowledge of other minds, and knowledge of my own mind (that is, self-knowledge). None of them, according to Davidson, is basic in the sense that the other two varieties are reducible to it. We have then three irreducible kinds of broadly empirical knowledge, which are, moreover, mutually dependent. To see the latter it is enough to reconsider the triangulation model involved in the exposition of the distal theory of meaning. It follows from that model that in order to assign meaning to a simple word or sentence the two interacting speakers must know and agree on the common cause of the sounds produced by them, and be able to identify, both in her own case, and in the case of the other speaker, to what sequence of sounds they give assent, or, otherwise, what beliefs they hold. These three sorts of knowledge, Davidson writes, "form a tripod: if any leg were lost, no part would stand".[23]

This has been a very general sketch of Davidson's views relevant to his account of mental content and knowledge. It by no means does justice to the subtlety of Davidson's actual position and arguments. But it seems to me that what has been said constitutes a good background against which my worries about the cogency and tenability of both Quine's and Davidson's epistemological views would perhaps make sense.

3. A PLEA FOR SOME ADJUSTMENTS

It seems, at the first glance, that in this disagreement Davidson has had the better of the argument. Why then does Quine remain so firmly attached to his form of naturalized empiricism? Here is his short answer:

> my interest is epistemological, however naturalized. I am interested in the flow of evidence from the triggering of the senses to the pronouncements of science. My naturalism does allow *me* free reference to nerve endings, rabbits, and other physical objects, but my epistemology permits the subject no such starting point.[24]

I think that, contrary to what may appear at first glance, Quine is right in his insistence on continuing a more traditional epistemological project. Let me elaborate on this.

It is hard to conceive of epistemology otherwise than as an attempt to answer the question of how each of us from her own particular perspective is able to attain knowledge about the objective world, and what the conditions are for possessing such knowledge. So we have here, on the one hand, an objective common world, and, on the other, a number of individual subjects, constrained by their local perspectives or points of view. In fact Davidson agrees with this, considering it even a truism. He writes: "Minds are many; nature is one. Each of us has one's own position in the world, and hence one's own perspective on it".[25] But it seems to me that if we carefully put together all the details of Davidson's externalism, not much is left of that truism. It eventually appears that our thoughts and beliefs are, as far as their content is concerned, part of the common public world. Only their possession and entertaining is subjective and private. Davidson maintains that this feat is achieved by overriding traditional first person Cartesian epistemology by a "resolutely third person approach to epistemology".[26] It is an approach not from the point of view of a single thinker or speaker trying to figure out from the limited evidence which is at her disposal the way the world is, but from the point of view of an interpreter ascribing to her various (mostly veridical) beliefs. Davidson explicitly says:

It is not the *speaker* who must perform the impossible feat of comparing his beliefs with reality; it is the *interpreter* who must take into account the causal interaction between world and speaker in order to find out what the speaker means, and hence what he believes.[27]

But how, we may wonder, does the interpreter know all of this? Either she is in an epistemically privileged position in the relation to the interpreted speaker and knows the objective content of her own thoughts straightaway, or she has to wait on the verdict of another interpreter who in turn ascribes veridical beliefs to her. I don't find either of these alternatives very convincing. Choosing the first we have to explain why the interpreter is in such a privileged position, and choosing the second we commit ourselves to the idea of an infinite regress of interpreters.[28]

So if we want to stay with the intuitive idea of many individual subjects being as it were epistemically on a par we cannot completely eliminate the first person point of view in epistemology. Perhaps we can make epistemology less Cartesian and subjective by making it more naturalistic, but it is unlikely that it will be completely externalized and we will stop asking the question of how each individual is in principle able to construct an objective conception of the world, beginning from her meagre and often idiosyncratic evidence.

But where, Davidson might reply, is a place for the first person point of view in the justification of our knowledge claims? We simply acquire some beliefs, whose content is externally constituted, and justify them by other beliefs having the same character. Nowhere in this process is reference made to the way the world looks or appears to this or that subject. The justification of beliefs by the subject's perceptual states is a myth which should be abandoned. Let us suppose for the moment that this doxastic assumption, that is the idea that only beliefs can justify other beliefs, is true. In what sense then will the whole system of our beliefs be justified? There is not much problem with the justification of the beliefs located somewhere in the middle of such a system; there will always be several other beliefs which entail or support them. But what about the simple beliefs located on the edge of the system? How is the belief that there is a tree before me justified? If I reply: I am justified in holding it because I accept the belief that it just looks like as if there is a tree before me, one faces the problem of how to justify this last belief. The only plausible and non-regressive answer of which I can think in this context is: beliefs which we entertain are not voluntary states; they are the result of various causal processes and are, for the most part, veridical. Indeed this is what Davidson says at some point.[29] But be that as it may we have no alternative but to terminate justifying our beliefs with something that is forced on us, rather than with something that we have accepted because of adequate reasons. John McDowell, who discusses this issue at length, put Davidson's predicament in the following way:

> Davidson's picture depicts our empirical thinking as engaged in with no rational constraint, but only causal influence, from outside. This just raises a worry as to whether the picture can accommodate the sort of bearing on reality that empirical content amounts to, and this is just the kind of worry that can make an appeal to the Given seem necessary. And Davidson does nothing to allay the worry. I think we should be suspicious of his bland confidence that empirical content can be intelligibly in our picture even though we carefully stipulate that the world's impacts on our senses have nothing to do with justification.[30]

But in what way, we may wonder, can the world's impacts on our senses have anything to do with justification? McDowell himself proposes a conception according to which the impact on our senses is already somehow conceptual, that is, where experience is "receptivity" in operation and conceptual capacities that belong to "spontaneity" are already at work there. Clearly this is one possible way of dealing with Davidson's predicament. But there is another possibility, namely putting in doubt the presumably last dichotomy underlying both Quine's and Davidson's philosophy: the dichotomy between the causal and the epistemic or logical (broadly construed). This opposition is quite evident in Davidson's views – we have here logical or epistemic relations holding between beliefs or other prepositional attitudes, and causal relations obtaining outside that domain and to some extent shaping it (recall the case of simple observational beliefs caused by physical events and processes). Quine seems to be less decisive on that point. There are passages where he almost explicitly says that epistemic, or logical, or evidential relations don't reach beyond the level of simple observation sentences. But in other places he seems to suggest that these relations also embrace neural intakes or stimulations (for instance, in the already quoted passage he says that he is interested "in the flow of evidence from the triggering of the senses to the pronouncements of science").[31]

However why not simply say that there are relations that are both causal and epistemic? In order to make sense out of this idea let us suppose that there are many different kinds of causation. Some of them – purely mechanical or physical – operate only on the level of elementary physical features of objects and events. But there are other which work on more complicated and higher levels of organization, and are not adequately explicable in the terms of a purely physical model of causation. Perception furnishes a good example. It seems that in this case beliefs are caused not only by processes occurring on the physical micro level, but through causation working by way of content of perceptual states as well. If so, then one of the conditions for producing a given belief in a subject is her seeing, or having a perceptual experience, that something is this or that way. Such a condition would be both causal and epistemic. If it did not obtain, the subject's belief could perhaps be caused in some other way, but then it would not be justified. Of course, holding such a view presupposes that seeing or having a perceptual experience that something is the case is not always believing that something is the case. Davidson questions this presupposition but it seems to me that there are good reasons for it. There are plenty of cases where I see that something is a certain way but I do not believe that it is that way. A Müller-Lyer illusion of two seemingly unequal lines is one example, perceiving a large object which from a long distance looks very small, much smaller than the window through which I look at it, is another.[32]

So far I have mainly been discussing Davidson's views. Now I want to return to Quine's position and ask, how to modify his account of the relationship between neural intakes and observation sentences in order to reconcile it with the above suggested view of perception. The modification, I believe, should go in the direction of assigning to the subject's perceptual states a more important role than Quine does. For although he talks from time to time about perception, perceptual similarity, and the like, their role in our verbal reactions, that is in assenting to or dissenting from observation sentences, appears to be almost non-existent. Let me quote two passages from Quine's writings illustrating that point:

Each perception that it is raining is a fleeting neural event. Two perceptions by Tom that it is raining are apt to differ, moreover, not only in time of occurrence but neurally, because there are varied indicators of rain. Tom's perceptions of its raining constitute a class of events that is perhaps too complex and heterogeneous neurally to be practically describable in neurological terms even given full knowledge of the facts. Yet there is also, we may be sure, some neural trait that unites these neural events as a class; for it was by stimulus generalization, or subjective similarity, that Tom eventually learned to make the observation sentence "It's raining" do for all of them.[33]

What does perceptual similarity relate? What are perceptually similar? Bombardments, I said: raw inputs which, however unlike one another intrinsically, are similar in their perceptual effects. It is the overall bombardment of the subject on one occasion that is perceptually similar, for him, to the overall bombardment on another occasion, and their perceptual similarity for him can be know to us by his responses.[34]

So what is really important for perceptual similarity is not simply the similarity of perceptual states but rather some common neural trait or the common pattern of bombardment. This is so, even though individual subjects differ enormously in their neural structure and the same verbal reactions can be caused by highly dissimilar surface irritations or bombardments. Of course, it is hard not to agree with Quine that when we observe or interpret the verbal behaviour of others, we can identify perceptual similarity only by their responses and by the stimuli or situations to which they react. But in the subject's own case what really matters and what determines her verbal responses is the intrinsic similarity of the perceptual states. The fact that these states are in different ways realized or constituted by neural intakes, surface irritations, and the like, is largely irrelevant for her.

The obvious reply to this line of argumentation would be to say that in order to assign to perceptual states some important epistemological role it is not enough to show their indispensability in the first person account of perception and verbal behaviour. It has to be shown that they also play a significant role in other cognitive procedures. I believe that this can be shown. Let us take, for example, the method of empathy on which Quine puts so much emphasis in his most recent writings.[35]

When I empathize with another person I, by no means, try to imagine what kind of bombardments or neural intakes she gets while being in that particular position. I try rather to imagine what perceptual states I would have being in her position, looking on the world from her angle, etc. Ontological claims in Quine's philosophy have been guided by the principle that we can accept the existence of something only if it has a firm place in our best theories. It is plausible to suppose that perceptual or experiential states (which need not necessarily be some ghostly sense data or impressions) have such a place.[36]

Catholic University of Lublin, Poland
The University of Queensland, Australia

NOTES

[1]   This does not mean, however, that Davidson thinks his epistemology externalized is *not* a version of naturalized epistemology. Quite the opposite. See e.g. D. Davidson, "Epistemology Externalized", *Dialectica* 45(1991), p. 193, and R.F. Gibson, "Quine and Davidson: Two Naturalized Epistemologists", *Inquiry* 37(1994), pp. 449–463.
[2]   *The Roots of Reference* (La Salle, Ill.: Open Court, 1974), p. 2.
[3]   "Reactions", in: P. Leonardi & M. Santambrogio (eds.), *On Quine: New Essays* (Cambridge: Cambridge University Press, 1995), p. 348. See also a similar account in his "In Praise of Observation Sentences", *The Journal of Philosophy* 90(1993), p. 108.
[4]   "Reactions", p. 349. Quine says there that he now tends to use the phrase "neural intake" rather than "stimulation" in order not to cause misunderstandings on the part of readers who use the latter word differently.
[5]   These are not the only roles played by observation sentences. In his "In Praise of Observation Sentences" Quine describes their five other functions (pp. 110–113).
[6]   See W.V. Quine, *Pursuit of Truth*, rev. ed. (Cambridge, Mass.: Harvard University Press, 1992), p. 3.
[7]   *The Roots of Reference*, p. 39.
[8]   The similar objections have been made by Donald Davidson. See his "Meaning, Truth and Evidence", in: R.B. Barrett & R.F. Gibson (eds.), *Perspectives on Quine* (Oxford: Basil Blackwell, 1990), pp. 70–71.
[9]   W.V. Quine, "Empirical Content", in his *Theories and Things* (Cambridge, Mass.: Harvard University Press, 1981), p. 25. See also *Pursuit of Truth*, p. 40.
[10]   *Pursuit of Truth*, p. 42.
[11]   D. Davidson, "A Coherence Theory of Truth and Knowledge", in: A.R. Malachowski (ed.), *Reading Rorty* (Oxford: Blackwell, 1990), pp. 124–125.
[12]   *Ibid.*, p. 125.
[13]   D. Davidson, "Meaning, Truth and Evidence", p. 74.
[14]   Quine himself thinks that this ascription is rather misleading, especially as far as his semantic views are concerned. See "In Praise of Observation Sentences", p. 114, n. 1.
[15]   "Meaning, Truth and Evidence", p. 72.
[16]   D. Davidson, "The Second Person", *Midwest Studies in Philosophy* 17(1992), pp. 262 ff.
[17]   *Ibid.*, p. 262.
[18]   *Ibid.*, p. 263.
[19]   D. Davidson, "Pursuit of the Concept of Truth", in: P. Leonardi & M. Santambrogio (eds.), *On Quine*, pp. 19 f.
[20]   D. Davidson, "The Myth of the Subjective", in: M. Krausz (ed.), *Relativism: Interpretation and Confrontation* (Notre Dame, In.: University of Notre Dame Press, 1989), p. 164.
[21]   D. Davidson, "Three Varieties of Knowledge", in: A.P. Griffiths (ed.), *A.J. Ayer: Memorial Essays* (Cambridge: Cambridge University Press, 1991), p. 165.
[22]   "The Myth of the Subjective", p. 165.
[23]   "Three Varieties of Knowledge", p. 166.

[24] *Pursuit of Truth*, p. 41–42.

[25] "The Myth of the Subjective", p. 159.

[26] "Epistemology Externalized", p. 193.

[27] D. Davidson, "Empirical Content", in: E. LePore (ed.), *Truth and Interpretation: Perspectives on the Philosophy of Donald Davidson* (Oxford: Blackwell, 1986), p. 332.

[28] One can argue that this way of putting the matter overlooks one plausible possibility, namely, that even though the interpreter herself is not in a particularly privileged epistemic position, the whole procedure of interpretation has nevertheless a special epistemic status in virtue of the interaction of the interpreted speaker and the interpreter. But the viability of this possibility depends crucially upon showing what in particular – from the epistemic point of view – the interaction of two or more subjects adds to their cognitive abilities; and I do not know any successful attempt of showing that.

[29] "I do not invent by beliefs; most of them are not voluntary" ("Empirical Content", p. 331).

[30] J. McDowell, *Mind and World* (Cambridge, Mass.: Harvard University Press, 1994), p. 14–15.

[31] In one of his recent essays Quine attempts to remove this unclarity of his views in the following way:
Some of my readers have wondered how expressions that are merely keyed to our neural intake, by conditioning or in less direct ways, could be said to convey evidence about the world. This is the wrong picture. We are not aware of our neural intake, nor do we deduce anything from it. What we *have* learned to do is to assert or assent to some observation sentences in *reaction* to certain ranges of neural intake. ("In Praise of Observation Sentences", p. 110–111).
It seems to me that this clarification is still far from satisfactory. In particular, more should be said about what the "keying to our neural intake" consists in, and in what circumstances we are entitled to take its results at face value. Of course, Quine is obviously right that we do not have introspective access to our neurophysiology and do not deduce anything from our neural intakes, but such absurd claims are not entailed by the thesis that our responses which result in acceptance of observation sentences have an epistemic facet.

[32] These include cases such as when we acquire beliefs other than those *prima facie* suggested by the content of our perceptual experiences, and cases when we suspend beliefs about what appears to us in perception. An interesting attempt to show that Davidson is unable to give a satisfactory explanation of those cases, especially the latter ones, has been undertaken by Michael Martin in his paper "The Rational Role of Experience" (*Proceedings of the Aristotelian Society* 93(1922/3), pp. 71–88). To put it very roughly, Martin argues that in order to explain them we must admit that the relationships holding between the way in which the things appear to the subject and her former beliefs are not only causal but also logical or rational. The gist of the explanation is that when what appears to the subject is inconsistent with her former beliefs, her perceptual experiences fail to produce corresponding beliefs. Although I find Martin's arguments convincing, my suggestions hint in a rather different direction – in the direction of insisting on the variety of causal mechanisms, that on a certain level of complexity acquire an epistemic facet, and not in the direction of showing that in addition to causal relationships holding between perceptual states and beliefs, there are also epistemic or rational relationships holding between them.

[33] *Pursuit of Truth*, p. 62.

[34] "Reactions", p. 348.

[35] Roger F. Gibson has pointed out to me in discussion that the procedures constitutive of the method of empathy have always played an essential role in Quine's views. In his later writings they have merely acquired a common name.

[36] For helpful comments on an earlier version of this paper and stimulating discussions I am indebted to Alexander George, Roger F. Gibson, Alex Orenstein, Barry C. Smith, Tim Williamson, and the participants of the staff-students seminar at the Philosophy Department, University of Queensland.

ALEXANDER GEORGE

QUINE AND OBSERVATION*

> [T]here is more to the equating of stimulations than
> meets the eye, or indeed perhaps rather less than
> seems to do so.
>
> –W.V. Quine, *The Roots of Reference*

Observation is central to empiricism, whose leading idea is that all knowledge is ultimately conveyed to us through our perceptual experiences. Analytic philosophy is in part distinguished by its transformation of traditional philosophical questions into ones about language, and by its emphasis on the primacy of the sentence as linguistic vehicle. So it is no wonder that the notion of an *observation sentence* is central to W.V. Quine's philosophy, which in many ways represents the zenith of empiricism as pursued in the analytic tradition. A proper understanding and assessment of his philosophy depends on a satisfactory construal of that notion, one that answers to all the roles he calls upon it to play. And yet, Quine has found "the notion of observation ... awkward to analyze" and has returned to its characterization again and again.[1]

In the following, I shall sketch the history of changes rung upon this notion, the internal pressures that have led to them, and the problems they raise. I shall also offer a reconstruction of Quine's most recent conception of observation sentence and assess its consequences for the joint tenability of his views on the objectivity of science and on the indeterminacy of translation. In doing so, I hope to illuminate not only Quine's doctrines, but also something about the movement of which his philosophy is in many respects the culmination.

1. ACCOUNTS OF THE OBSERVATION SENTENCE

Quine's early metaphor that likened language, or scientific theory, to a man-made fabric that touches experience only along its edges soon came to be spelled out more fully.[2] By 1960, in *Word and Object*, the periphery gave way to observation sentences, and the tribunal of experience to excitation of our sensory receptors, to "the triggering of ... nerve endings."[3] Observation sentences were there defined as those (1) assent to which and dissent from which are closely tied to current stimulation and (2) whose stimulus meanings across speakers are equivalent. (The stimulus meaning of a sentence for a speaker is the ordered pair of stimulations of the speaker's sensory receptors that, upon being queried with the sentence, would respectively prompt her assent or dissent.)

21

Alex Orenstein and Petr Kotatko (eds.), Knowledge, Language and Logic, 21–45.
© *2000 Kluwer Academic Publishers. Printed in Great Britain.*

Intuitively, the identification of stimulation with the triggering of one's receptors was meant to secure the shareability of experience, for it was assumed that the surface neural structures of different individuals are more or less homologous. And, while (1) was intended to guarantee that a speaker's response to an observation sentence is closely tied to his experience, (2) was designed to ensure that that response be the same for all speakers.[4]

Both these features answer to roles that observation sentences play in Quine's over-all view. First, observation sentences are the points at which one's theory of the world makes contact with one's experience; without such contact, empiricism is untenable. Secondly, observation sentences are the *lingua franca* of theories; without them, resolution of intertheoretic disagreement about the nature of the world would be impossible.[5]

Because Quine conceives of language learning as a kind of theory testing,[6] these two epistemological roles have language-acquisition counterparts. First, observation sentences are those learned without the need for "scholarship,"[7] that is without the benefit of any prior language; if some sentences could not be learned this way, he holds, language learning would be impossible. Secondly, observation sentences, or the experiences which they tag, are the *lingua franca* of learning; without them, language acquisition as Quine conceives it, in which learner and teacher exploit their common experience, would founder.

In the light of these roles of the observation sentence, its analysis in terms of stimulus meaning can be seen to express Quine's view that the ultimate conduit for justification and meaning is experience: "The two cardinal tenets of empiricism" remain "unassailable. ... One is that whatever evidence there *is* for science *is* sensory evidence. The other ... is that all inculcation of meanings of words must rest ultimately on sensory evidence."[8]

On Quine's view, while translation-relevant stimulation, be it by a tack or a tachistoscope, must be shareable, we are unlikely, because of what he calls "interference from within," to find even an approximate homology between the inner physiological states of speakers who are intuitively having similar experiences.[9] But stimulation must also be such that it should not be describable without reference to the relevant speakers: a rabbit running in the distance is not a candidate for our shared stimulation.[10] Experience, insofar as our detection of it in others constrains translation, should be understood neither as an internal event, however physicalistically acceptable, nor as an event involving objects beyond the speaker. Quine split the difference, as it were, and located experience midway between the two, at the surface of the observing subject's body.

Over the years, no doubt because of the centrality of the notion of observation sentence for his views, Quine has tried to address problems with his account of experience. The first expression of worry came just a few years after publication of *Word and Object*: in 1965, Quine fretted about whether his construal of stimulation as patterns of sensory triggerings had, ironically, the effect of making stimulation unshareably personal: "If we construe stimulation patterns my way, we cannot equate them without supposing homology of receptors; and this is absurd, not only because full homology is implausible, but because it surely ought not to matter."[11]

In fact, Quine's above account of experience entails that there are no observation sentences, for no two speakers could associate equivalent stimulus meanings with a sentence. And even if this worry could somehow be allayed with regard to humans, it can be made vivid again for extraterrestrials, what with their alien sensory apparatus. It would be surprising were Quine to deny the possibility of linguistic communication between humans and any being outfitted with differently structured surface receptors, and this reveals that something has already misfired in his account about us.[12]

This matter can be put more generally. A translation, viewed abstractly, is a preserving map from one set to a second. We have not given this notion any substance, however, until we have specified what precisely is to be preserved. Without such a specification, no constraints have been placed on an adequate translation; indeed, we might say that no substantive construal of the notion of translation has yet been given. What Quine has noted is that his previous account of translation, as a map from one language to another that preserves whatever stimulus meaning can be assigned to sentences, fails this test, for equivalent stimulus meanings cannot be attached to sentences by different speakers, or even by the same speaker at sufficiently different times.[13,14]

Quine's response to this unsatisfactory situation, in *Theories and Things* (1981), was to define observation sentence for an individual speaker and then count as observational *tout court* any sentence that was observational for all speakers of the language.[15] Unlike the account presented in *Word and Object*, this does not have the undesired consequence that there are no observation sentences. It still fails, however, because it counts as observational any sentence which, though observational for each speaker, is such that speakers are disposed to assent to it in intuitively very different stimulatory conditions.[16] On this conception, the important roles of the observation sentence that were intended to be guaranteed by (2) of the earlier account cannot be secured.

Quine's skepticism about the "homology of receptors" across individuals is not in conflict with his analysis of what he calls "perceptual similarity," for this is a relation that holds between an individual's "global stimuli – ordered sets of [triggered] receptors."[17] Nor is it in conflict with his claim that there exists (perhaps for reasons of natural selection) a harmony between the perceptual similarity spaces of different individuals. In *From Stimulus to Science* (1995), Quine says something that might be read as an attempt to use these notions to provide a constraint on translation of observation sentences: "Within the individual the observation sentence is keyed to a range of perceptually fairly similar global stimuli .... It is thanks to the preestablished harmony, again, that they qualify as observation sentences across the community."[18] The proposal might seem to be this: S (a sentence of Speaker$_1$'s language) can legitimately be translated as T (a sentence of Speaker$_2$'s language) only if (i) S is conditioned to a range of global stimuli that are perceptually similar for Speaker$_1$, and (ii) T is conditioned to a range of global stimuli that are perceptually similar for Speaker$_2$, and (iii) the perceptual similarity spaces of Speaker$_1$ and of Speaker$_2$ are in harmony with one another. S and T are linked by virtue of the fact that they are respectively associated with ranges of global stimuli that are respectively located in two perceptual spaces that are in harmony.

This would not do, however. To say that Speaker$_1$'s and Speaker$_2$'s perceptual similarity spaces are in harmony is just to say that "If two scenes trigger perceptually similar global stimuli in one witness, they are apt to do likewise in another."[19] Therefore, from (i)–(iii) we can infer that if two scenes trigger in Speaker$_1$ perceptually similar global stimuli to which S is conditioned, then they will also trigger perceptually similar global stimuli for Speaker$_2$; and that if two scenes trigger in Speaker$_2$ perceptually similar global stimuli to which T is conditioned, then they will also trigger perceptually similar global stimuli for Speaker$_1$. But all this is consistent with S's being conditioned to perceptually similar global stimuli for Speaker$_1$ that are caused by scenes that cause in Speaker$_2$ perceptually similar global stimuli to which T is *not* conditioned. Under these circumstances, it would be intuitively inappropriate to translate the one by the other. (The problem here is reminiscent of the one that plagued Quine's account in *Theories and Things*.)

One might suggest adding an extra clause to (i)–(iii), namely (iv): the scenes that cause the range of global stimuli to which S is conditioned also cause the range of global stimuli to which T is conditioned. What this amounts to is just that S and T are to be translated only if the scenes that would lead Speaker$_1$ to assent to (dissent from) the first sentence would also lead Speaker$_2$ to assent to (dissent from) the second.[20]

Quine, as I understand him, should not find this proposal acceptable as an analysis of what translation preserves. We saw earlier that he will not accept external causes or events as candidates for stimulation. Consequently, the present account of translation is not one that depends on anything that Quine would recognize as sensory experience. And for just this reason, the account fails to safeguard the "cardinal tenets of empiricism" that Quine holds "unassailable": for they call for an analysis of how words acquire their communal meanings, of what translation is to be faithful to, that crucially employs a notion of stimulation, or "sensory evidence." Although empiricism demands that experience be substantively implicated in the acquisition of meaning, this account of translation gives it no significant role to play.

By contrast, Quine's *Pursuit of Truth* (1992) can be read as articulating yet another approach, which is simply to grant the privacy of stimulus meanings, and to count as observation sentences those which are intimately tied to perceptual experience and which each individual would agree in assenting to, or dissenting from, "on witnessing the occasion of utterance."[21] Instead of talking of congruence of assent/dissent behavior in the face of like stimulation, Quine there speaks of such congruence in the face of like perceptions of the relevant occasion. What Quine might mean by this will be taken up in section 4 below.

2. INDETERMINACY OF TRANSLATION

Before considering whether translation of observation sentences is indeterminate, we should say a few words about Quine's thesis of the indeterminacy of translation.

The argument for indeterminacy that pervades most of Quine's work proceeds as follows. There is a range of acceptable evidence that the translator may appeal to in choosing a manual of translation; this constitutes "the ultimate data for the identification of meanings."[22] This evidence is restricted to that which Quine believes one has, in principle, at one's disposal when learning a first language, that which one has to go on when "rating [another] as a master of the language."[23] All such evidence, Quine insists, is thoroughly public. The plight of the radical translator is meant to render vivid these everyday circumstances and the publicity of the relevant evidence: "All the objective data he has to go on are the forces that he sees impinging on the native's surfaces and the observable behavior, vocal and otherwise, of the native."[24] Furthermore, this publicly observable evidence cannot be brought to bear on the translation of sentences taken one at a time, but instead is relevant to accepting an entire package of translation. Finally, Quine argues that the totality of "objective data" is consistent with mutually incompatible manuals of translation. He concludes from this that the choice of manual of translation is indeterminate, that is, that there is no correct choice of manual, that there is no fact of Nature that this choice must reflect.

Some (for example, Michael Dummett) have parted company with Quine on account of his assumption that we cannot in general evaluate hypotheses about the translation of individual sentences independently of hypotheses about the translation of other sentences. Others (for example, Noam Chomsky) have dissented from Quine's conclusion because they question an important presupposition of the above inference. Evidence, they claim, is not being treated here as it is in the natural sciences. There, one standardly assumes that evidence provides one with information about how it is in some determinate corner of reality. Quine, by contrast, treats the totality of "objective data" as if it were *constitutive* of the domain of facts for which, one might have thought, it is merely evidence. Facts about meaning are exhausted by the accessible evidence, on his view, for "there is only the natives' verbal behavior for the manuals of translation to be right or wrong about."[25] "It is the very facts about meaning," Quine holds, "that must be construed in terms of behavior."[26] Quine explicitly draws the contrast he sees between the nature of evidence in semantics and in naturalistic inquiry: "Dispositions to observable behavior are all there is for semantics to be right or wrong about .... In the case of systems of the world, on the other hand, one is prepared to believe that reality exceeds the scope of the human apparatus in unspecifiable ways."[27] While physical reality is typically assumed to extend beyond the results of measurement, there is nothing to semantic reality beyond the verbal behavior of speakers. Putting it yet another way, we might say that while physical facts are not usually taken to supervene on facts about measurement, Quine holds that all truths about meaning supervene on truths about verbal behavior.

This constitutivity of evidence is what Quine means when he says that verificationism is an "attractive" doctrine that supports one route to indeterminacy.[28] Another, more usual, expression of this is his avowal of behaviorism: "In psychology one may or may not be a behaviorist, but in linguistics one has no choice."[29] In part because of his *penchant* for this latter way of expressing himself,

Quine is often dismissed as someone whose arguments are discredited by a stubborn attachment to a superseded psychological approach to language.

This misplaces the root of Quine's adherence to the view that truths about meaning do not outrun the observable evidence. It stems, rather, from the conviction that one's grasp of language is thoroughly public. This presupposition may, in turn, be supported by certain empirical assumptions about the learning process or communication. And there may also be more general philosophical views leading to its adoption. At any rate, the presupposition serves to delimit the conception of meaning operative in his argument: meaning that is publicly observable or manifestable, in some sense of these terms. It is no accident that the first sentence of *Word and Object* reads: "Language is a social art."[30]

The publicity of meaning is one of the most central assumptions operative in the analytic tradition, at least in work that has dominated the second half of this century. The supposition, arguably a departure from early work in this tradition, is generally unsustainable and has distorted much recent thought about language and mind.[31] I shall return to it below in the final section.

3. TRANSLATION OF OBSERVATION SENTENCES

We may now ask: are observation sentences determinately translatable? Usually, Quine insists that indeterminacy does not affect their translation. "Observation sentences can be translated," he writes in *Word and Object*, "translated without recourse to analytical hypotheses." "There is uncertainty," he says, "but the situation is the normal inductive one." Over three decades later, he still writes that observation sentences "are free of the indeterminacies that beset translation of theoretical sentences."[32]

Yet, one can also find declarations that translation of observation sentences, like that of theoretical ones, is indeterminate, likewise dependent upon a choice of analytical hypotheses. The reason is that identification of a speaker's observation sentences depends on identification of stimulus meanings, which in turn depends on identification of the speaker's signs for assent and dissent. But this latter is allegedly indeterminate, hence translation of observation sentences is as well:

> ... let us imagine the fantastic case of two independent radical translators who differ in their identifications of assent and end up with unlike manuals of translation, both of which provide successful training in the language as judged by smoothness of dialogue and influence on behavior. Is the native really performing an act of assent when he uses the word that was hit upon by the one translator, and some mental act other than assent when he uses the word that was hit upon by the other translator? I say there is no fact of the matter. Both manuals are successful, and the native's two mental acts, however unlike, are equally deserving of the name of assent.[33]

Likewise, the two sets of observation sentences corresponding to each manual are equally deserving of the name; hence, there is no fact of the matter about the translation of the native's observation sentences.[34]

These positions appear mutually exclusive: it seems that either there is a fact of the matter concerning the identification and translation of a speaker's observation sentences, and so they should be distinguished as regards translational determinacy from the theoretical sentences, or there is not.

Perhaps a way of resolving the contradiction is to attend to Quine's suggestion that indeterminacy is not a condition that either holds or fails to hold absolutely; rather, "the indeterminacy of translation comes in degrees."[35] Translation of theoretical sentences is worse off than that of observation sentences, which is, in turn, worse off than translation of assent and dissent. The alleged inconsistency can now be seen to be only apparent: relative to theoretical sentences, observation sentences are determinately translatable, but relative to, say, assent and dissent, they are not.

Or can it? What does it mean, we need to know, to say that indeterminacy comes in degrees? Quine has written that we also best speak of "degrees of observationality"[36] – can this help us make sense of degrees of indeterminacy? One might view gradations of observationality as reflecting the extent of the involvement of collateral information, as opposed to experience, in a speaker's response. The two gradations could then be related, for the more collateral information shapes a speaker's verbal behavior with regard to a particular sentence, the more play there is in making compensatory adjustments between the speaker's meanings and beliefs: the lower a sentence's degree of observationality, the higher its degree of translational indeterminacy.

Of course, this will not illuminate the concept of graduated indeterminacy unless the notion of degree of observationality is itself on a firm footing. Yet the intuitive notion of collateral information that it can be seen as approximating has no empirical substance, according to Quine, and so we cannot lean on it for an analysis.[37]

Quine's early attempt to cultivate a "strictly vegetarian imitation"[38] of collateral information was to talk instead of the variation of a sentence's stimulus meaning across members of the community: "We have defined observationality for occasion sentences somewhat vaguely, as degree of constancy of stimulus meaning from speaker to speaker."[39] This will no longer do, however, for we saw that this account of observationality has lately been repudiated: "The view I have come to, regarding intersubjective likeness of stimulation," Quine now writes, "is rather that we can simply do without it."[40]

In spite of this, Quine continues to hold that "We must recognize degrees of observationality," and the extent of "infection of observation by theory" remains an important factor.[41] No doubt Quine's intention is to rely on the idea briefly introduced at the end of section 1, and to hold that the more observational a sentence, the more it will "command the same verdict from all linguistically competent witnesses of the occasion."[42] What it is for two speakers to witness the same occasion is something to which I shall return below. Let us assume, then, that Quine can make sense of a graduated observationality (by appealing to "degree of constancy" across speakers in the witnessings that would lead, respectively, to assent or dissent) and, in terms of it, of indeterminacy by degrees as well.

Now, Quine must hold that there are some sentences whose degree of observationality (indeterminacy) is sufficiently high (low) that we can speak of them as having essentially the same empirical content for different speakers. If such sentences failed to exist, then Quine's belief in the objectivity of science could not be sustained. This is because, according to Quine, "The requirement of intersubjectivity is what makes science objective."[43] Intersubjective agreement between two speakers is possible only if there exist pairs of sentences <S, T>, where S belongs to the one speaker's language, T to the other's, such that S and T elicit just about the same response from the two speakers when they both witness the same occasion. But this is precisely what it is for S and T to be observational to a high degree for the two speakers.

In sum, the objectivity of science requires, for Quine, that there be sentences whose translation is determinate to all but a trivial degree.

4. PERCEPTION, EMPATHY, AND "ULTIMATE DATA"

With this conclusion before us, let us now return to Quine's revised account of the observation statement. His basic idea, again, is to secure the intersubjectivity of experience without the implausible and irrelevant assumption of a homology of "triggered exteroceptors."[44] And he does this by reference to what individuals would agree in assenting to, or dissenting from, "on witnessing the occasion of utterance."

Quine offers no analysis of what such communal witnessings consist in. Because he takes a perception to be a "fleeting neural event,"[45] there is a temptation to say that for two speakers both to perceive some situation is for there to be some common neural state they momentarily pass through. And Quine does believe that, for any given speaker, all witnessings that p will be united "as a class [... by] some neural trait."[46] Yet he does not think there will generally be any such "neural trait" in common to all speakers' perceptions that p.[47] In accordance with his commitment to anomalous monism, he acknowledges that "there are irreducibly mental ways of grouping physical states and events."[48] Furthermore, we saw that Quine finds speaker neurophysiology as irrelevant to an empirical analysis of shared linguistic meaning as are mentalistic notions. Even if type-physicalism were sustainable, it would not be germane: communication, we noted earlier, does not require access to another's neurophysiology any more than it requires extrasensory perception of another's mental state. As Quine puts it, all this "surely ought not to matter."

What does matter to an analysis of shared and public meaning is not commonality at the neurophysiological level, but rather a similarity capturable only through the use of the mentalistic idiom; what matters, in particular, is a commonality at the level of perceptual state. These states, though token-identical to neurophysiological states, are, in the course of radical translation, perforce picked out using mentalistic vocabulary. It is the mentalistic descriptions of such states that figure in the observable regularities (e.g., that the speaker responds in this particular way when he perceives that so-and-so) to which translation must be faithful.

As far as communication is concerned, there is no criterion regarding the identity of a speaker's perceptual state beyond the judgments of attentive observers. (I shall turn momentarily to Quine's view of the basis of such judgments.) The practice of making such judgments is part of our "mentalistic heritage," Quine says, "a mentalistic strain" in our reflective life.[49] In the context of radical translation, the only criterion for the truth of the claim that two speakers both witness the same occasion is that appropriate observers so judge, i.e., that the one speaker's perceptual state is likened to the other's by observers competent to wield the relevant mentalistic predicates (e.g., "perceives that a lemur is in the tree"). This is of a piece with Quine's view that "What is utterly factual is just the fluency of conversation": if a mindful observer would judge that Robin now perceives that a lemur is in the tree, then it does not matter, from the point of view of communication, what is going on inside Robin's brain, at the neural extremities of his body, or in his mind.[50]

This shift, as far as translation is concerned, from stimulation construed neurophysiologically to its mentalistic characterization, induces a corresponding change in the analysis of the observation sentence. Now, in order for a sentence to be (very) observational for a community, its members' assent/dissent response with regard to it must be closely tied to their perceptual state and, in addition, they must all respond (very) similarly to it on those occasions in which they are having similar perceptions.[51]

We concluded in the previous section that there are such sentences, according to Quine, that is, sentences whose translation is determinate to all but a negligible degree. Their existence entails that truths about the perceptual states of speakers be reckoned part of the evidence available to the radical translator. For should such truths not be so reckoned, there would be no basis for translating any sentences of a language. What Quine calls the "objective data" for translation must include information about speakers' perceptual states, about what they are witnessing, for otherwise we could not even articulate the regularities that serve to ground translation. Another way of putting the point would be to note that without appeal to facts about the witnessings of speakers, about the content of their perceptual states, no substance can be given to degrees of observationality. And if there is no substance to the distinction, in point of degree of observationality, between sentences of a language, then a requirement for the objectivity of science cannot be met.

There is a natural objection to my claim that, in the attempt to define the notion of observation sentence so that it can play all its assigned roles, Quine takes truths about the perceptual states of speakers to be within the bounds of the "theoretically accessible evidence"[52] available to the radical translator. For it might be argued that substantive hypotheses about translation are not in fact confirmed or disconfirmed by data that make reference to perceptual states. Rather, the objection continues, talk of perceptual states of speakers is best taken as a heuristic that can help us discover the real "objective data," namely, speakers' behavior under stimulation, which are in turn further refined and rendered manageable by the "solvent" of "Query and assent, query and dissent."[53] It is against such data, the objection concludes, that hypotheses about translation must ultimately be tested. "Discovering where stimulus synonymy holds is one thing," Quine says, while "defining what it is that one thus discovers is another."[54] The notion of perceptual state, this objection maintains, should be treated as at best an aid for discovering the empirical content of, say, the native's *Gavagai*. But what this empirical content consists in is to be understood in terms of stimulus meaning.

This objection is not sustainable in the light of Quine's abandonment of the inter-subjectivity of stimulus meaning.[55] In order to say what the empirical meaning of *Gavagai* is, insofar as this meaning is to constrain translation, we must now make reference to regularities in the native's responses when queried with the sentence upon witnessing different occasions. What it is for *Gavagai* to be correctly trans-lated as *There goes a rabbit* is for there to be a similarity between our response (when queried with *There goes a rabbit*) and the native's (when queried with *Gavagai*) upon our both witnessing the same occasion of utterance. The notion of perceptual experience is therefore not a mere device for discovery: claims about perceptual states are centrally involved in the assertion that this sentence in the native's language is determinately translated by that sentence in ours.

This might still be resisted. For it might be tempting to hold that a speaker's being in a particular perceptual state is not actually a datum for translation, but is rather the content of a hypothesis designed to organize what *really* are the "objec-tive data," namely facts about the speaker's observable behavioral responses in observable stimulatory conditions. Perceptual states, one might claim, are not observed; at best, conjectures about them are supported by what can be observed, that is, patterns of stimulus and response.[56]

Again, this cannot be Quine's present position. Given his commitment to anom-alous monism,[57] he cannot take attribution of a propositional attitude, such as "Robin perceives that a lemur is in the tree," to organize data described in neuro-physiological terms. There are no laws linking perceptual states (states described using mentalistic vocabulary) to physiological ones (states picked out using physi-calistic vocabulary), hence we can infer little regarding the neural states of individ-uals or their motor activity from judgments about what they are perceiving. Such an inference would require a general connection between mental states (so described) and neurological ones (so described) of the kind that anomalous monism denies exists. Therefore, while a perceptual state is a neural state, a judgment about an individual's perceptions, insofar as it is relevant to communication, is not (as this proposal suggests) justified by information about his or her "triggered exterocep-tors." Quine cannot accept, then, that truths about the perceptions of others are inferred from non-mentalistically characterized data, which are for that reason more basic; rather, Quine must now grant, it is such truths that are (partly) constitutive of the "ultimate data for the identification of meanings."

We can summarize the role for mentalistically described perceptual states in this way: if (a) there is no room for a conception of experience as perceptual state, then (b) experience remains the ultimate conduit for meaning (and justification) only if (c) meanings are not shareable (and science is not intersubjective). We have seen that Quine takes (b) to be an "unassailable" principle of empiricism and strongly denies (c). He is therefore committed to rejecting (a), that is, to affirming that there is a place for perceptual states in characterizing the evidence for meaning.

If one denies (c) and wishes not to be under any pressure to reject (a), then one must reject (b) as well. Someone who holds this view might well urge that Quine likewise jettison (b), at least in so far as it is supposed to have any content beyond the causal claim that our sensory apparatus is needed for the acquisition of language and knowledge.[58] And for Quine, its content does go beyond this. For him, there is a notion of "sensory evidence" which plays a part in "the intersubjective business of semantics," but which is not to be identified with neural stimulation.[59] So if the "cardinal tenets of empiricism" are to hold, then the conception of experience that they enshrine cannot be described by adverting solely to the neural machinery of the speaker and its workings. Furthermore, such a description should not become any more acceptable to Quine if one were to add that what look after the "intersubjective business of semantics" are just causes in the environment that are salient for both speaker and translator: for Quine, we saw, will not construe these causes as candidates for shared experience. In the context of Quine's overall views, then, empiricism cannot be confined to the above causal claim. Thus, if he is to maintain his commitment to empiricism, (a) must be rejected and a place found for a conception of experience as perceptual state.[60]

We noted that, as far as communication is concerned, truths about a speaker's perceptual state are intimately linked to the judgments an attentive observer would make. On what basis are these judgments made, according to Quine, if they are not based on facts about the speaker's response to stimulation, non-mentalistically characterized? We should expect Quine's answer to be that we base our judgments about the perceptual states of others on our own perceptual experience: in general, one *sees* that another perceives that something is the case. Why should we expect Quine to hold this? Recall that Quine must, if the intersubjectivity of science is to be maintained, take truths about a speaker's perceptual states to be part of the "ultimate data" for meaning. Because Quine's focus is on a thoroughly public notion of meaning, semantic facts, which are just the facts that constrain translation, cannot outrun the observable. Indeed, Quine is emphatic that "There is nothing in linguistic meaning beyond what is to be gleaned from overt behavior in observable circumstances."[61] Now, when conjoined with the claim that truths about the perceptions of speakers are part of the "ultimate data," this publicity requirement entails that a speaker's perceptual states constitute "observable circumstances" of his responses.

This expectation is indeed borne out. For instance, we find Quine offering the following schematic characterization of the beginnings of language learning: "Martha's business is to encourage Tom in uttering the sentence, or in assenting to it, when she *sees* that he is noticing appropriate phenomena, and to discourage him otherwise."[62] Likewise, the child learning language, he claims, "In his as yet inarticulate way *perceives* that the speaker perceives the object or event."[63] One's basis for judging that a speaker perceives that *p* is itself a perception whose content is that the speaker perceives that *p*.

To claim that one can perceive that another is witnessing that such and such is the case is not to claim that reasoning plays no role in determining the content of one's perception. For it is not implausible that the content of a subject's perceptual experience, what is perceived to be the case, can be affected by the subject's inferences. Unconscious inference about the properties of objects, can lead one to have a perception whose content is, say, that one object is in front of another. Similarly, unconscious inference might also partially fix the content of one's person-involving perceptions. Such reasoning about persons, might well avail itself of assumptions and techniques that differ from those appealed to in reasoning about trees and ponds.

Quine has emphasized one particular kind of inference that is involved in determining the perception one perceives another to be having. "When we ascribe a perception, in the idiom 'x perceives that p'," he says, "our evidence consists in observing the percipient's orientation and behavior and appreciating that we in his place would feel prompted to volunteer the content clause ourselves."[64] In the setting of radical translation, the deliverances of the linguist's faculty of projection will play a role in determining the data for translation.[65] This is meant to capture the fact that in the course of communication, of "rating [another] as a master of the language," one must make use of truths about what speakers are witnessing, truths that are seen to obtain through perceptual experiences whose content is shaped by the deliverances of one's empathic reasoning. "We judge," Quine writes, "what counts as witnessing the occasion, as in the translation case, by projecting ourselves into the witness's position."[66] In sum, a kind of reasoning, or "appreciating," informs one's perception of, and hence in general one's judgment about, what another is witnessing. Because such judgments, insofar as our concern is public meaning, in effect settle part of the "ultimate data" that confront hypotheses about translation and render them substantive, we see that empathic reasoning plays a role in determining the "theoretically accessible evidence" for meaning.

This role that Quine now has empathy play might well seem surprising in the light of his own past pains to warn against the uncritical projection of our familiar mentalistic ways onto the native, thereby obscuring from ourselves the indeterminacy of translation. The illusion of determinacy is enhanced, Quine has long cautioned, because "the lexicographer comes to depend increasingly on a projection of himself, with his Indo-European *Weltanschauung*, into the sandals of his Kalaba informant."[67] And he has repeatedly warned against tainting the thought experiment of radical translation by uncritically "reading our own provincial modes into the native's speech."[68] Is this not inconsistent with the present interpretation of Quine's view?

To see that it is not, it will be useful to draw a distinction between two different kinds of projection. In the context of radical translation, a projection is *illicit* if it makes assumptions about semantic features of the native's language, and it is merely *innocent* if it imputes psychological states that do not involve such semantic assumptions. Quine's prohibitions all concern illicit projection, an instance of which is precisely *not* the empathy that now figures in the translation of observation sentences. Quine alerts us to "outright projection of prior *linguistic* habits,"[69] to imputation of "our sense of *linguistic* analogy unverifiably to the native mind."[70] By contrast, the empathic projections operative in the identification and translation of observation sentences do not involve assumptions about the speaker's language. In fact, they do not even presuppose that the observed individual has a language, for we likewise project ourselves into the perspective of pre-linguistic infants and even of language-less creatures: "The language is that of the ascriber of the attitude, though he projects it empathetically to the creature in the attitude."[71] In this sense, the projections are indeed semantically innocent.

This conclusion, accurate as far as it goes, should be tempered by the comment that there is an obvious sense in which even innocent projections fail to be linguistically neutral. This can be seen by considering even a schematic rendering of the empathic reasoning that might inform our perception that a speaker perceives that a lemur is in the tree. It involves our "appreciating," at some level, that, were we in the position we see the speaker to be in, we should assent to "A lemur is in the tree." This thought experiment is one that is conditioned by the language of the experimenter. So, while it is true that, in the context of radical translation, the linguist does not directly project features of her language onto the native's language, she must nevertheless rely on her language in determining the content of the native's perceptual states, which, as we saw above, is an essential preliminary to translating his words. The shadow of the translator's language does fall indirectly upon the language being translated.[72]

In connection with this last point, there are two features of this conception of empathic projection worth mentioning here. First, the projection avails itself of mentalistic characterizations of the observed speaker. A thought experiment couched in, say, physiological terms would not only be unlike any we are accustomed to make, but it would also be one from which, given anomalous monism, no conclusions regarding perceptual state could follow. Secondly, and relatedly, it is important to recognize that empathic reasoning regarding what one "would feel prompted to volunteer" does not issue in a bare prediction concerning one's response, but rather in a conclusion about how one *ought* to respond; indeed, usually the only way we have of determining what our response would be is *via* a determination of what it should be, of how it would be reasonable to respond. Furthermore, what makes this determination applicable to the case of another is imputation to the latter of a like-minded rationality.[73] In short, both mentalistic characterizations and norms will be involved in thought experiments about how we would respond were we in someone else's position, as well as in the application of this reasoning to the speaker whose cognitive position we are simulating.

In spite of all this, the empathic projection that eventuates in a judgment about another's perceptions remains semantically innocent. Of course, this innocence should not be confused with irrelevance. In this section, we have seen that deliverances of such projection shape what we perceive others to be perceiving, and hence form a basis for our judgments about the perceptual states of others. And these judgments, insofar as communication is concerned, provide primary access to facts which partly constitute the constraints on translation, facts without reference to which we could not even articulate the substance of the claim that a particular translation is correct.

5. EXPERIENCE AND NATURE

We have seen that Quine, on the present rational reconstruction, now insists that translation does not depend on information about the muscular or neural activity of speakers. For the most part, these goings-on are idiosyncratic to each speaker, and information about them is not generally available. Rather, the public enterprise of translation is grounded in evidence about a speaker's perceptions (and responses), in the acquisition of which empathic projection by attentive observers plays a central role.

Quine need not always appeal to such a notion of perception. Some projects to which he is committed can be carried out with the aid of his individualistic notion of stimulus meaning. "I remain," he writes, "unswerved in locating stimulation at the neural input, for my interest is epistemological, however naturalized. I am interested in the flow of evidence from the triggering of the senses to the pronouncements of science."[74] If one's goal is to understand how "neural input" causes the development in *Homo sapiens* of a complex of dispositions to verbal behavior – that is, why, after "light rays strike [our] retinas [and] molecules bombard [our] eardrums and fingertips," we eventually "strike back, emanating concentric air waves"[75] – then we will have to attend carefully to the nature of stimulation so construed.[76]

By contrast, stimulation in this sense "is no part of the intersubjective business of semantics." "My identification of stimulus with neural intake," Quine says, "is irrelevant to that."[77] For projects that require a notion of *shareable, justifying*, and *public* experience (for example, accounts of communication and public meaning, or of the justificatory support experience provides for claims about the world), Quine must instead appeal to the perceptual states of agents. We can see why these requirements are met if we reflect on the role deliverances of projection play in settling the nature of perceptual states. Thus, such states are shareable, for an observer can have the same empathic response upon projection into different perspectives. Furthermore, that such experiential states are also capable of entering into justificatory relations follows from the fact that the deliverances of projection characterize the perceptual experience as content-bearing: because such a state is described as a *perception that p*, we can see how it could justify a belief. Finally, we already saw that these perceptual states are public in that empathic reasoning can lead an observer to see that another is perceiving that something is the case.

We have noted that Quine is also committed to this second kind of project. Science's objectivity is, for him, bound up with its intersubjectivity, in particular with the existence of observation sentences whose translation is free of all but trivial indeterminacy. Accounting for such "shared checkpoints"[78] between speakers is part of "the intersubjective business of semantics," a business to be discharged with the aid of a notion of perceptual experience that is shareable, capable of entering into justificatory relations, and public.[79]

While some have suggested that only the first conception of experience is made use of by Quine, both are required for the above distinct projects.[80] Furthermore, if we cannot find room for the two conceptions within Quine's view, we will be unable to resolve what will otherwise appear to be conflicting remarks by him about the role of experience. Thus, sometimes he seems to say that stimulation cannot function as evidence, cannot be viewed as a justification of anything: "Some of my readers have wondered how expressions that are merely keyed to our neural intake, by conditioning or in less direct ways, could be said to convey evidence about the world. This is the wrong picture. We are not aware of our neural intake, nor do we deduce anything from it."[81] Yet elsewhere, Quine appears to claim that it can, as when he writes that "the proper role of experience or surface irritation is as a basis not for truth but for warranted belief."[82]

One can see how such remarks might be reconciled with one another and a place found for both conceptions of experience by appreciating Quine's embrace of anomalous monism. Rational linkages are mediated by laws or inferential rules, which are linguistic, and so obtain between items insofar as they are described in a particular way. Furthermore, there are no laws linking items described as content-bearing and items not so described; this is one way of characterizing the core of anomalous monism.[83] As a consequence, an item cannot participate in a justification, it cannot be rationally linked to another item described as content-bearing, except insofar as it is likewise so described. If we refer to the states of speakers using neurological terminology, then experience so described cannot enter into justificatory relations. However, if we refer to these (very same) states employing irreducible mentalistic vocabulary, then experience so described is content-bearing and hence capable of entering into such relations.

John McDowell rejects anomalous monism as incapable of providing a satisfactory account of experience.[84] He does so in the context of a discussion of Donald Davidson's views, not those of Quine, whom he mistakenly (from the perspective of the present interpretation) takes to operate only with a notion of experience devoid of conceptual content.[85] But his objection to Davidson naturally extends to Quine, as here interpreted, and its consideration will highlight a surprising and significant development of Quine's thought.

McDowell's argument is that philosophical worries about the gap between mind and world will not dissolve until we recognize that experience is contentful, even when viewed as a natural phenomenon: "experience," he says, "has its conceptual content precisely as whatever natural phenomenon it is." Anomalous monism does not help us to the required recognition because it holds that "it cannot be as the natural phenomena they are that impressions are characterizable in terms of" content.[86] For example, consider the perception that a lemur is in the tree: though itself a natural event or state, the monist insists, the anomalousness of the mental guarantees that it is not in virtue of the state's natural properties that it is a perception that a lemur is in the tree. According to McDowell, any account that remains blind to the conceptual content of experience, so long as the experience is viewed as something within the natural order, is deficient.

I am not concerned here to assess whether such an account of experience is really unsatisfactory, that is, whether it does make a satisfactory understanding of the relation between mind and world unattainable. Rather, I want to examine whether the allegedly deficient view of experience is Quine's. To establish this, it would have to be shown that, for Quine, to treat something as contentful experience is to treat it as something outside the natural order. We do find Quine writing that mentalistic predicates "complement natural science in their incommensurable way, and are indispensable both to the social sciences and to our everyday dealings."[87] That

Quine finds mentalistic predicates incommensurable with those found in, say, physics does not establish that he treats them as having the extensions they do for other than natural reasons, that is, for other than the way the world is. After all, incommensurability involves a resistance to some kind of reducibility, and Quine believes that scientific theories often resist reducibility, in any clear sense, to one another.[88] The question, rather, is whether at least some mentalistic predicates must be taken to be included in our theory of the world – not merely a "complement" to it, but rather a part of it.

Quine's belief in the objectivity of science commits him to the substantial determinacy of the translation of observation sentences. In order for there to be such determinacy, there must be facts about the content-bearing experiences (in particular, the perceptual states) of speakers. These considerations together with Quine's naturalism (his rejection of a first philosophy, of a perspective outside ongoing rational inquiry), entail that facts about the content-bearing experiences of speakers are natural facts, facts that Quine takes to obtain in the natural order of things. Quine must hold that to view something as a content-bearing experience is to view it as a natural phenomenon. Thus, the allegedly deficient view of experience cannot be Quine's.

It might be useful to spell out this argument more fully. Quine adheres to the following four theses. The first is naturalism, which entails that the study of science is part of science itself: "Epistemology is best looked upon ... as an enterprise within natural science," which poses "scientific questions about a species of primates, and they are open to investigation in natural science, the very science whose acquisition is being investigated."[89] The second is that the study of science will incorporate the study of its objectivity, including the empirical commensurability of scientific theories and their testability, which permit scientists to turn to the world for adjudication of conflicts. By the first thesis, the objectivity of science is itself a fit topic for naturalistic inquiry and the conceptual resources we will find ourselves appealing to in the course of this particular study will *ipso facto* be part of the arsenal of scientific inquiry. The third thesis is the determinacy of translation of observation sentences. Objectivity requires this determinacy, for if there were no manual-independent facts about the translation of observation sentences, then there would be no *lingua franca* within which conflicting theories could find a common ground; indeed, talk of conflict at all would become obscure. Hence, the tools employed in characterizing and explaining this determinacy are to be reckoned part of scientific research. Finally, the fourth thesis is that an account of this determinacy will have to refer to truths about the perceptual states of speakers. If translation of observation sentences is to be determinate, then of course there must be a range of constraining facts to determine it. And these facts, on Quine's current analysis of observation sentences, are in part constituted by truths about what speakers are perceiving; without such truths, there would be nothing to which translation need be faithful.

A consequence of these four theses is that the scope and language of science will have to allow for the perceptual states of speakers, so described, since naturalistic research into the objectivity of science will ultimately appeal to their existence. It is important to see that the conclusion here is not that "Experience ... is a worthy object of philosophical and scientific clarification and analysis." Rather, it is what Quine once denied when, continuing this last thought, he said that experience nevertheless "is ill-suited for use as an instrument of philosophical clarification and analysis."[90] Quine's naturalistic account of the objectivity of science will make eventual reference to facts about the experiential states of speakers, and as a consequence the language of science will have to encompass "irreducibly mental" predicates.[91]

Quine might appear to acknowledge this when he says that the propositional attitudes *de dicto* do not "resist annexation to scientific language."[92] But this has to be understood in the context of his view that propositional attitudes are "relations between people and sentences." For "Sentences differ, if merely in phrasing or spelling, though beliefs [doubts, hopes, regrets, perceptions, etc.] be reckoned identical." And in fact Quine is pessimistic regarding the individuation of beliefs and other objects of the attitudes: "All these," he declares, "are *entia non grata* by my lights." He continues: "Thinking is a bodily activity in good standing, however inadequate our physiological understanding of it. The acts of thinking – the *thoughts* in one sense of the term – are justly reified. I would say the same of perceptions, as events. But thoughts in the other sense, identifiable and distinguishable thought contents, I despair of accommodating."[93]

These views leave no room for the position I am here attributing to Quine. For the reasons discussed earlier, Quine must find it intelligible to say that two speakers are having the same perception: experiences, after all, in so far as they are what observation sentences are pegged to, must be shareable. This identity of perceptual state cannot consist in the speakers' being in the same physiological state, because of anomalous monism: "There is token identity, to give it the jargon," Quine says, "but type diversity."[94] Nor can this identity consist in the two speakers' being related to the same sentence, for an observer of the two may report their perceptual states using differently phrased or spelled subordinate clauses. Nor, the above makes clear, will Quine want to understand identity of perceptions to consist in identity of perceptual content. There is a conflict, then, in Quine's views about objectivity and observation, in particular, between the demands of intersubjectivity and the misgivings about perceptions viewed as the contents of perceptual experience.

This conflict does not of, course, suggest its own resolution: it has as many remedies as ingredients. It should be noted and emphasized that nothing in Quine's most general doctrines forbids him from welcoming propositional attitudes, in particular the attitude of perception. Given his naturalism, his hesitation about such notions must emanate from within ongoing rational empirical exploration. Hence, his doubt must be subject to quelling by consideration of the needs of this naturalistic inquiry. A doubt not in principle amenable to such alleviation would be one that stemmed from the tenets of a first philosophy, which Quine's naturalism forbids him to invoke. If it should happen that naturalistic inquiry into the objectivity of science finds itself appealing to the contents of perceptual states, as Quine's own analysis suggests it might, then perhaps he will have to treat them as *entia grata* after all, and to make whatever adjustments necessary to accommodate them.

Because the source of Quine's resistance to a more full-blooded conception of propositional attitudes is the thesis of the indeterminacy of translation, we should expect that the present conflict, between, on the one hand, the claim of science's objectivity and, on the other, his pared-down notion of propositional attitude, signals a conflict between objectivity and the central forces leading to indeterminacy. This expectation is borne out, as we shall see in the following and final section.

6. OBJECTIVITY AND PUBLICITY

We have suggested that Quine's belief in the objectivity of science is in tension with his thesis of indeterminacy. On the present reconstruction, this belief should push Quine to expand the language of science to accommodate descriptions of the perceptions of speakers. "Ultimate data" for translation include the general linkages between experience and response exhibited across speakers. These connections are the descendants of Quine's notion of stimulus synonymy, now abandoned because of the recognized unshareability and semantic irrelevance of stimulus meaning. Translation is constrained by the need to match these linkages across speakers; it is these linkages that give the notion of translation whatever determinacy it has. What remains to be examined, then, is the effect this expanded conception of the data for translation has on the roots of indeterminacy.

We saw that what drives Quine's argument for indeterminacy of translation is a doctrine of the publicity of meaning: he is emphatic that "There is nothing in linguistic meaning beyond what is to be gleaned from overt behavior in observable circumstances." All the evidence for meaning must be observable. In particular, as we also saw, this could lead him to the view that our access to the experiential component of the above linkages is itself perceptual: the experiences of speakers, insofar as they are relevant to translation, constitute "observable circumstances" of their responses.

What enables one's perception of the perceptual experience of another is empathic projection into the other's perspective. We noted earlier that this involves not so much a direct prediction about one's response in some hypothetical situation as it does an "appreciating" of how one would best respond in that situation. In addition, to render this thought experiment relevant to another's perceptual experience, one must take the latter to share one's own standards of rationality. Consequently, the reasoning that operates to shape one's perception of what another is seeing will inevitably draw on a considerable cognitive stock, which will include a dense tapestry of concepts, beliefs, and norms.[95] While empathy does not require projection of the semantics of one's language onto another's language, it does demand that one employ as a springboard one's own language, the notions it makes available to one, and the standards that govern its use. These will be in play in the reasoning which accompanies empathic projection and which informs the content of one's perceptual state regarding the perceptions of another.

Thus, a rich cognitive stock is assumed by the inferences that shape one's perception of what a speaker is witnessing, which perceptions in turn form the basis for our judgments about another's perceptual experiences. Where public meaning is concerned, these judgments provide our only criterion for one component of the "objective data" pertaining to the translation of a speaker's language. It is natural, then, that that on which these judgments are themselves ultimately based should likewise be reckoned part of the evidence for meaning. This cognitive stock therefore belongs amongst the "objective data" for public meaning by virtue of the fact that, in its absence, there would be no grounds for the reasoning without which we would fail to have a basis for our judgments about what other speakers are perceiving.

But then, again by the publicity requirement, the elements of that stock must likewise be observable. Not all will be, however. For, included within it, we will find all those unobservable concepts, beliefs, and norms that might be appealed to when reasoning about how best to respond to some hypothetical situation, and about the applicability of one's conclusion to another person. For instance, assume that, in the course of some translation, part of the "ultimate data" includes the regularity that whenever Robin sees that a lemur is in a tree he assents to a particular sentence of his language. This means, in effect, that Robin so responds whenever an attentive observer judges that Robin sees that a lemur is in a tree. This judgment will be grounded in perception: an attentive observer will base her judgment that Robin sees that a lemur is in a tree on the fact that she sees he is in that perceptual state. Furthermore, she sees that Robin perceives that a lemur is in a tree by empathically projecting herself into his perspective, "appreciating" how she would respond (for example, how she would assent to the query "Is there a lemur in the tree?") and, lest her reasoning be inapplicable to Robin, by making assumptions about the latter's rationality. Whatever is involved in all this reasoning has a claim to being part of the "ultimate data" for translation, for without it there would be no basis for judgments regarding Robin's perceptual state, which, we have seen, must be counted amongst the "ultimate data." Yet it is not plausible that all that is so involved is observable: in general, deliberations about how it would be reasonable to respond or about another's rationality are not based solely on assumptions that can be observed to obtain.

The problem can be put this way. The publicity requirement demands that the evidence on which translation is ultimately based be acquirable through observation. Quine's adherence to the intersubjectivity of science might require that the perceptual states of speakers constitute part of such evidence. But the observational conveyance of these states calls for reliance on a cognitive stock some of whose elements are not themselves amenable to such delivery, not themselves capable of being *seen* to hold. In sum, if science is to be objective, then Quine (duty reconstructed) may have to grant that the publicity requirement is self-defeating.

***

Naturally, Quine would not recognize, let alone accept, very much of the above. Yes, he would agree, science is objective, and this means that it is intersubjective. Yes, the intersubjectivity of science requires that observation sentences be determinately translatable, and such translation in turn demands that there be real regularities about which it can be right or wrong. Above, we have worried the question about the nature of these regularities. Once upon a time, they involved the notion of stimulus meaning. But, as Tallulah Bankhead long ago said, "There is less in this than meets the eye." Quine concurs, and he has cast about for a substitute that can figure in these regularities. A central constraint on this search is that the substitute be recognizable as sensory experience; otherwise, the "cardinal tenets of empiricism" could not be upheld. It is at this point that Quine and the above reconstruction of Quine part company.

I have suggested on Quine's behalf, leaning heavily on some of his remarks, that the sought substitute might be some notion of perceptual experience, of a "witnessing" of some occasion. I have developed this view and tried to articulate some ways in which (surprisingly) it seems to dovetail with other remarks by Quine. And I have argued that if these ideas are further elaborated in a certain direction, then the indeterminacy of translation – and likewise the publicity of meaning, on which it depends – will be threatened.

Quine, surely, will have none of this and will insist that I have leaned so heavily on some of his remarks as to have deformed them entirely. Perhaps. But the alternative I face is a mystery about how Quine hopes to accommodate the demands of empiricism, as he understands it, and of the objectivity of science, once the privacy and semantic irrelevance of stimulus meaning is acknowledged.

Amherst College

NOTES

*   I am grateful to Amherst College for a grant that facilitated completion of this paper. I am also indebted to G. Lee Bowie, Jay L. Garfield, A.W. Moore, Joseph G. Moore, and W.V. Quine for helpful comments.
[1]   *Pursuit of Truth* (revised edition), Harvard University Press, 1992, p. 2.
[2]   See Quine's "Two Dogmas of Empiricism," reprinted in his *From a Logical Point of View*, Harper & Row, 1961, pp. 20–46.
[3]   This is the way Quine describes what his explicit position has been since *Word and Object*. ("On the Very Idea of a Third Dogma," in *Theories and Things*, Harvard University Press, 1981, pp. 38–42; p. 40.) But what he actually says there is somewhat different: "A visual stimulation," he wrote in 1960, "is perhaps best identified, for present purposes, with the pattern of chromatic irradiation of the eye" (*Word and Object*, MIT Press, 1960, p. 31). The triggering and the irradiation are not identical: the latter is the cause of the former.
[4]   See *Word and Object*, secs. 9 and 10. An additional "parameter," as Quine puts it, is that of linguistic community. A sentence might be an observation sentence within one community, but not within another. (See, e.g., *From Stimulus to Science*, Harvard, 1995, p. 22.) The following discussion is not affected by this relativity.
[5]   *From Stimulus to Science*, pp. 44–45.
[6]   Quine writes of "the partnership between the theory of language learning and the theory of scientific evidence" ("The Nature of Natural Knowledge," in *Mind and Language*, Samuel Guttenplan (ed.), Oxford University Press, 1975, pp. 67–81; p. 74).
[7]   "The Scope and Language of Science," in *The Ways of Paradox and Other Essays* (enlarged edition), Harvard University Press, 1976, pp. 228–245; p. 232.
[8]   "Epistemology Naturalized," in *Ontological Relativity and Other Essays*, Columbia University Press, 1969, pp. 69–90; p. 75.
[9]   *The Roots of Reference*, p. 20.
[10]   *Pursuit of Truth*, p. 41; *From Stimulus to Science*, p. 18. Quine's rejection here is perhaps overdetermined, for he is, in addition, pessimistic about the project of articulating a

respectable theory of events. See *Pursuit of Truth*, p. 42; and his "In Praise of Observation Sentences," *The Journal of Philosophy*, Volume XC, No. 3, March 1993, pp. 107–116; p. 114.

[11] "Propositional Objects," in *Ontological Relativity and Other Essays*, pp. 139–60; p. 157. The observation surfaces again when Quine worries about how to define "receptual similarity" across speakers; see *The Roots of Reference*, Open Court, 1974, pp. 23–4, 41.

[12] These worries carry over to what Quine actually says in *Word and Object* (see note 3 above). If the pattern of irradiation is measured just at the surface of a speaker's body, then the shareability of stimulation is questionable, since each person's contour is unique. However, if the patterns are to be measured along the surface of some containing bubble, then (putting to one side concerns about arbitrarily barring the possibility of communication with creatures that do not fit into this space) there is the concern that stimulation would then be describable without reference to the speaker, something that Quine wants to avoid.

[13] Quine's observation is *not* that he had been mistakenly "crediting the field linguist with a grasp of my concept of stimulus or indeed of any theoretical psychology" ("Response to Gibson," *Inquiry*, Volume 37, pp. 501–2; p. 502). The notion of stimulus meaning is not meant to describe anything the field linguist consciously employs, but instead is meant to be part of an account of what constrains her practices, of what she is trying to get right. Quine's point, rather, is that his notion of stimulus meaning simply cannot be part of such an account. See also Quine's "Reactions," in *On Quine*, Paolo Leonardi and Marco Santambrogio (eds.), Cambridge, 1995, p. 348.

[14] We have, as Quine has, focused on the unshareability of stimulation. But Quine's worries about how to understand identity of stimulation across speakers carry over to the problem of defining equivalence of behavioral response across speakers. In *The Roots of Reference*, he says that "A theoretical definition of behavioral similarity is readily imagined. It might be sought in terms of the total set of fibres of striped muscle that are contracted or released on one occasion and on another, or a more functional approach might be devised" (p. 21). But behavioral similarity across speakers would then involve hypothesizing a homology of striped muscle fibers, which, it seems Quine would want to say, is not only dubious but "surely ought not to matter."

[15] "Empirical Content," in *Theories and Things*, pp. 24–30; p. 25.

[16] The objection was raised by Lars Bergström in "Quine on Underdetermination," in *Perspectives on Quine*, Robert B. Barrett and Roger F. Gibson (eds.), Blackwell, 1990, pp. 38–52; p. 39.

[17] *From Stimulus to Science*, p. 17. Global stimulus a is perceptually more similar to b than to c for speaker S when, roughly, S is disposed to respond to a as he does to b, rather than as he does to c. Quine's analysis of perceptual similarity can be found in *The Roots of Reference*, pp. 17–18.

[18] *From Stimulus to Science*, p. 22. In personal correspondence, Quine says that he was not there intending to articulate a new standard for translation.

[19] *From Stimulus to Science*, p. 21.

[20] Donald Davidson seems to make a similar point (in his "Meaning, Truth and Evidence," in *Perspectives on Quine*, pp. 68–79; pp. 76–77) and urges Quine to abandon a *proximal* theory of meaning (one according to which translation and justification are closely tied to experience) and instead to adopt a *distal* one (according to which an analysis of meaning makes reference rather to causes in the environment). Quine has recently said that he is now fully with Davidson in "treating translation purely in terms of the external objects of reference" ("Progress on Two Fronts," *The Journal of Philosophy*, Volume XCIII, Number 4, April 1996, pp. 159–163; p. 160).

Quine acknowledges an earlier unease with this position and diagnoses the discomfort as a failure to appreciate why, given "the intersubjective diversity of nerve nets and receptors" (*ibid.*), an external cause should ever elicit agreement in response across speakers. That hesitation has lately been dispelled by a recognition that natural selection will lead to "a preestablished *harmony* of standards of perceptual similarity" (*ibid.*).

As I indicate in what immediately follows, I think that there is another reason why Quine should feel unease with the "object-oriented line in describing the procedures of translation" (*ibid.*). (Although this phrase, "procedures of translation," might suggest that Quine is referring to how translation proceeds, it is clear that he is in fact discussing what translation aims to preserve, what it is that the translator strives to get right, regardless of the actual methods employed.)

²¹ In *Pursuit of Truth*, secs. 2, 16, p. 43; and also in "In Praise of Observation Sentences," pp. 108–9. Quine first presented this approach in his "Three Indeterminacies," printed in *Perspectives on Quine*, pp. 1–16, section 2. It makes a formal appearance in *The Roots of Reference* (p. 39) where Quine says that a sentence "is observational insofar as its truth value, on any occasion, would be agreed to by just about any member of the speech community witnessing the occasion." Quine quickly stresses, though, that his talk of "joint witnessing" is a mere *façon de parler*: in "a more precise statement, it would speak of witnesses subject to receptually similar impingements" (p. 41). While he goes on to note that this raises the problem of non-homologous sensory receptors, he does not there take the step of repudiating shared stimulus meanings, and it is only in the context of such a repudiation that talk of "joint witnessing" becomes substantive.

²² *Pursuit of Truth*, p. 46. The passage is lifted whole from his "Indeterminacy of Translation Again," *The Journal of Philosophy*, Vol. LXXXIV, No. 1, January 1987, pp. 5–10; p. 7.

²³ *Pursuit of Truth*, p. 38.

²⁴ *Word and Object*, p. 28.

²⁵ "Indeterminacy of Translation Again," p. 9.

²⁶ "Ontological Relativity," in *Ontological Relativity and Other Essays*, pp. 26–68; p. 27, with an elision.

²⁷ *Pursuit of Truth*, p. 101. Note that sometimes Quine writes of meaning being beholden to facts about behavior, and sometimes to facts about dispositions to behavior. (Indeed, Quine's conception of stimulus meaning makes tacit reference to behavioral dispositions.) A disposition to verbal behavior is, for Quine, a hypothesized "physiological arrangement" with certain "characteristic effects" (*The Roots of Reference*, pp. 13, 10). Success in isolating the physiological mechanism would permit an explanation for the behavioral display by reference to which we initially individuate that physiological structure. When Quine writes that semantics is about behavioral dispositions, there is therefore a temptation to think that he takes facts about meaning to be determined by facts about (presently unknown) physiological states. But this would fly in the face of his insistence that nothing fixes what a speaker means beyond the speaker's observable verbal behavior; in particular, an inquiry into a speaker's physiology has no place in the attempt to translate him. For this reason, when Quine occasionally says that a speaker's meaning is determined by the speaker's dispositions to verbal behavior, I think it best to understand him to be talking about the "characteristic effects" of such dispositions, rather than about the "physiological arrangement[s]" themselves. For Quine, if a manual of translation enables "fluency of conversation and the effectiveness of negotiation" (*Pursuit of Truth*, p. 43), then there are no further facts to which it must be faithful and nothing more to its correctness.

²⁸ "Reply to Roger F. Gibson, Jr.," in Lewis Edwin Hahn and Paul Arthur Schilpp (eds.), *The Philosophy of W. V. Quine*, Open Court, 1986, pp. 155–57; p. 155. By verificationism, he intends the doctrine that "evidence for the truth of a sentence is identical with the meaning of the sentence."

²⁹ "Indeterminacy of Translation Again," p. 5.

³⁰ *Word and Object*, p. ix. For a reconstruction of Quine's argument along the present lines, see Dagfinn Føllesdal's "Indeterminacy and Mental States," in *Perspectives On Quine*, pp. 98–109; pp. 103–104.

³¹ For a discussion of Dummett's claim that the principle can be traced back to Frege, see Alexander George's "Has Dummett Oversalted His Frege?: Remarks on the Conveyability of Thought," in Richard Heck (ed.), *Realism, Thought, and Language: Festschrift for Michael Dummett*, Oxford University Press, 1997, pp. 35–69; for an examination of Dummett's own views, see Alexander George's "How Not to Refute Realism," *The Journal of Philosophy*, Vol. XC, No. 2, February 1993, pp. 53–72.

³² *Word and Object*, p. 68, with a quotation from p. 76 interpolated; "In Praise of Observation Sentences," p. 111.

³³ "Reply to Manley Thompson," in *The Philosophy of W. V. Quine*, p. 566.

³⁴ Elsewhere, Quine says that "The linguist's decision as to what to treat as native signs of assent and dissent is on a par with the analytical hypotheses of translation that he adopts at later stages of the enterprise; they differ from those later ones only in coming first, needed as they are in defining stimulus meaning. This initial indeterminacy, then, carries over into the identification of the stimulus meanings" ("Reply to Hintikka," in *Words and Objections*, D. Davidson and J. Hintikka (eds.), D. Reidel, 1975, pp. 312–315; p. 312).

35 "Reply to Hintikka," in *Words and Objections*, p. 312.

36 *Word and Object*, p. 42. In "Progress on Two Fronts," Quine suggests that we talk instead of an observation sentence's degree of "theoreticity" (p. 162); the slight shift of view that he describes there does not affect the discussion that follows.

37 *Word and Object*, §9.

38 *Word and Object*, p. 67.

39 *Word and Object*, p. 43.

40 *Pursuit of Truth*, p. 42.

41 "In Praise of Observation Sentences," pp. 108, 109.

42 *Pursuit of Truth*, p. 3.

43 *Pursuit of Truth*, p. 5; see also, for example, "In Praise of Observation Sentences," p. 111.

44 *Pursuit of Truth*, p. 3.

45 *Pursuit of Truth*, p. 62.

46 *Pursuit of Truth*, p. 62.

47 Not least because "people's nerve nets differ" (*Pursuit of Truth*, p. 62).

48 *Pursuit of Truth*, p. 72.

49 *Pursuit of Truth*, p. 74.

50 *Pursuit of Truth*, p. 43. Note that this marks a second sense in which observation sentences are "parametrized" by community, now not the community whose language is being translated (see note 4), but the community of those who are translating.

51 As we saw above (see note 14), an intersubjective replacement for stimulus meaning requires not only a shareable notion of stimulation, but also a shareable notion of behavioral response. And not only does talk of contractions of muscle fibers fail to provide this, but it also fails to take seriously that "What is utterly factual is just the fluency of conversation." Here, too, Quine may be led to rely on our "mentalistic heritage," now in the individuation of response by appeal to the mentalistic idiom (e.g., through use of the predicate "chases the lemur"). An effect of this would be to make the notion of behavioral response, as it functions in the intersubjective descendant of stimulus meaning, approximate more closely to that of action, where actions, as Davidson has put it, "are typically described not merely as motions but as motions that can be explained by the reasons an agent has [... and] are individuated along the same lines as propositional attitudes" ("Could There Be a Science of Rationality?," *International Journal of Philosophical Studies*, Vol. 3, No. 1, March 1995, pp. 1–16; p. 5).

52 *Word and Object*, p. 75.

53 "Mind and Verbal Dispositions," *Mind and Language: Wolfson College Lectures 1974*, Samuel Guttenplan (editor), Oxford, 1975, pp. 83–95; p. 88.

54 "Reply to Hintikka," p. 313.

55 Again: "... I still locate the stimulations at the subject's surface, and private stimulus meanings with them. But they may be as idiosyncratic, for all I care, as the subject's internal wiring itself" (*Pursuit of Truth*, p. 44).

56 Quine himself encourages such a view when he says that ascriptions of perceptual states are "firmly testified in behavior" and as a consequence "well under the control of empirical evidence" (*Pursuit of Truth*, p. 67).

57 See *Pursuit of Truth*, section 29.

58 On my reading of Davidson, this is his position and his recommendation to Quine; see his "Meaning, Truth and Evidence," p. 76.

59 "In Praise of Observation Sentences," p. 114, n. 1.

60 See also note 20. The position that I am claiming Quine must occupy is neither quite distal nor quite proximal. It fails to be distal, for translation involves attending to the experience, or perceptual state, of the speaker. But neither is it fully proximal, because, for the purposes of translation, there is no criterion for attributing perceptions to a speaker beyond the judgments of attentive observers. Commenting on Davidson, Quine himself describes his new account as "an intermediate point between Don's distal and my old proximal position" ("Comment on Davidson," in *Perspectives on Quine*, p. 80).

61 *Pursuit of Truth*, p. 38.

62 *Pursuit of Truth*, p. 61, italics added.

63 *From Stimulus to Science*, p. 89; italics added. Quine goes on: "*Perception* of another's unspoken thought, however – up to a point – is older than language" (italics added). See also *The Roots of Reference*, pp. 37–38 and "Mind and Verbal Dispositions," p. 84.

[64]   *Pursuit of Truth*, p. 66.

[65]   Quine writes that "Practical psychology is what sustains our radical translator all along the way, and the method of his psychology is empathy: he imagines himself in the native's situation as best he can." And he concludes: "Considerations of the sort we have been surveying are all that the radical translator has to go on" (*Pursuit of Truth*, pp. 46–47; these sentences also appear in "Indeterminacy of Translation Again," pp. 7–8). Such passages make it clear that empathic reasoning delivers information about the "ultimate data" for translation. In fact, it is striking the number of times Quine now writes of our empathic projection into another's perspective. (See, for example, *Pursuit of Truth*, pp. 42–43, 45–46, 62–63, 68–69; and *From Stimulus to Science*, p. 89.) This is perhaps related to his abandonment of intersubjective stimulus meanings and his reliance instead on perceptual states in the articulation of the regularities that translation must capture. For while anomalous monism guarantees that empathic reasoning regarding the mental states of a speaker will be of no use in reaching conclusions about the state of the speaker's neural extremities, such reasoning will be helpful in inferring to what the speaker is perceiving.

[66]   *Pursuit of Truth*, p. 43.

[67]   "The Problem of Meaning in Linguistics," in *From a Logical Point of View*, pp. 47–64; p. 63. Many years later, Quine would repeat that the translator "wisely depends ... on projecting himself into the native's sandals" ("Comment on Parsons," *Perspectives on Quine*, pp. 291–293; p. 292).

[68]   *Word and Object*, p. 77.

[69]   *Word and Object*, p. 70, italics added.

[70]   *Word and Object*, p. 72, italics added.

[71]   *Pursuit of Truth*, p. 68. See also Quine's "Promoting Extensionality," *Synthèse*, Volume 98, 1994, pp. 143–151; p. 146. Since such projections involve an imputation of rationality to the mind being projected into, this remark assumes that not all forms of rationality require possession of a language.

[72]   Davidson believes that this feature of the enterprise of communication renders it unlike inquiry in the natural sciences (though not, for that reason, any less a form of rational inquiry). For one cannot assess the appropriateness of the language (norms of rationality, etc.) that guides one's interpretation of another without appealing to that very language. Davidson holds that in naturalistic inquiry, by contrast, we make sense of phenomena by appealing to concepts which can be "agreed to by all concerned" ("Could There Be a Science of Rationality?," p. 15).

But it seems that there are notions deeply implicated in naturalistic inquiry disagreements about which are so fundamental that anything one might appeal to in adjudication already presupposes just that conception one is trying to justify. Indeed, the very example Davidson chooses to illustrate his thesis, that of the natural numbers, seems to be rather an instance of the very opposite, of a locus of dispute so profound that all the wells of argumentation are poisoned from the start. For a relevant discussion, see Alexander George and Daniel J. Velleman, "Two Conceptions of Natural Number," in *Truth in Mathematics*, Garth Dales and Gianluigi Oliveri (eds.), Oxford University Press, 1998, pp. 311–327.

[73]   I am here expressing a thought voiced also by Richard Moran in the following:

But normally a person has very little of an inductive, norm-free basis for predicting his own future or hypothetical behaviour. In the ordinary case, involving something more than sheer conjecture, he can answer such first-person questions with confidence only because he can make a here-and-now decision about what course of action to endorse. And without the normative basis he loses the rationale for applying this same prediction to another person, for he cannot assume that whatever inductive basis he had for his first-person prediction would apply to anyone else. ... The kind of interest we do have in the "egocentric" question [about what *I* would do or believe in some situation] is expressed in the fact that we answer it in the same manner as we would answer a deliberative question about what *to* do or believe, rather than as a request for a theoretical prediction about a particular person's future behaviour. ("Interpretation Theory and the First Person," *The Philosophical Quarterly*, Vol. 44, No. 175, pp. 154–173; pp. 163–4.)

[74]   *Pursuit of Truth*, p. 41.

[75]   "The Scope and Language of Science," p. 228.

[76]   Quine tries to do just that in the "Appendix on Neural Intake" to his "In Praise of Observation Sentences."

[77] "In Praise of Observation Sentences," p. 114, n. 1.

[78] "In Praise of Observation Sentences," p. 111.

[79] John McDowell argues that internal demands will necessitate Quine's appeal to a notion of experience that can enter into "rational relations to beliefs or world-views" (*Mind and World*, Harvard, 1994, p. 133). While I have focussed on what Quine, by his own lights, needs to do in order to safeguard the objectivity of science, McDowell arrives at his conclusion by reflecting on what Quine needs in order to be able to hold that a scientific theory "is *about* the empirical world, a stance correctly or incorrectly adopted according to how things are in the empirical world" (*Mind and World*, p. 134). These routes can be related. For, according to Quine, objectivity requires intersubjectivity of observation sentences, and the latter can be viewed as consisting in the existence of shareable information about the world that is conveyed by these sentences.

[80] McDowell, for example, claims that "Quine conceives experiences so that they can only be outside the space of reasons, the order of justification" (*Mind and World*, p. 133).

[81] "In Praise of Observation Sentences," pp. 110–111.

[82] "Empirical Content," p. 39.

[83] See Davidson's "Mental Events," in *Essays on Actions and Events*, Oxford, 1985, pp. 207–227.

[84] See *Mind and World*, pp. 74–76, 145–146.

[85] See note 80 above.

[86] *Mind and World*, p. 76.

[87] *Pursuit of Truth*, p. 73.

[88] See, for example, "Goodman's Ways of Worldmaking," in *Theories and Things*, pp. 96–99; p. 98.

[89] "The Nature of Natural Knowledge," p. 68. Davidson finds that inquiry into meaning is legitimate even if it should turn out not to be included in science, not part of naturalistic inquiry; for it aims to provide answers that "the philosopher wants" ("Could There Be a Science of Rationality?," p. 13). Quine, however, because of his naturalism, is committed to the view that any rational inquiry into reality is naturalistic inquiry. If "what the philosopher wants" is to know how things are, then there is nothing for it but to turn to science.

[90] "Reply to Shuldenfrei," in *Theories and Things*, pp. 184–6; p. 184.

[91] Ironically, the view I have been arguing that Quine is led to bears some resemblance to the position that McDowell urges upon us. He says that nature is not exhausted by what "natural science aims to make comprehensible" (*Mind and World*, p. 71) and that our view of the natural must be expanded so as to include normative or content-bearing notions. I have been claiming that forces in Quine's position should lead him to acknowledge that natural science itself must find a place for the irreducibly mental. These positions differ regarding the scope of natural science, but agree (if not completely) on the scope of the natural world.

[92] *Pursuit of Truth*, p. 71.

[93] *Pursuit of Truth*, p. 93.

[94] *From Stimulus to Science*, p. 87.

[95] The apt phrase "cognitive stock" is from Richard Wollheim's "Art, Interpretation, and Perception," which adopts a perspective that is congenial to the position presented here. See his *The Mind and Its Depths*, Harvard University Press, 1993, pp. 132–143; p. 134.

A.C. GRAYLING

NATURALISTIC ASSUMPTIONS

Naturalized epistemology is epistemology based on accepting the deliverances of our best current theories about the world, and premissing them in the account we give of how we get those theories. One of its principal attractions is that it allows us to make progress with other tasks in philosophy and elsewhere, unhampered by sceptical doubts. The paralysing effect of self-conscious questions about the getting and testing of beliefs – prompted in the tradition of epistemology by an acute sense of the finitary predicament not just of each putative knower taken individually, but of the collective even as it pools the results of its members' best endeavours – is solved in naturalised epistemology by the simple expedient of avoiding those questions altogether.

But naturalised epistemology has a tendency to make one uneasy. Its attractiveness can come to seem a corner-of-the-eye affair, lasting only while one's gaze is fixed elsewhere. Under direct scrutiny it appears to have two serious defects, each individually fatal to it, but interestingly linked. These are that it is circular, and that it seems comprehensively to miss the point – or if not, to duck the demand – of traditional epistemology. Without doubt these complaints, in one or another formulation, are wearisomely familiar to naturalism's proponents, but I have yet to see a satisfactory response to them, and so take this opportunity to seek one.

In Quine's view, epistemology naturalised is epistemology treated as part of science – in effect, as an empirical psychological enquiry into how we get our beliefs about nature. In an often-quoted passage from the eponymous paper that launched the debate, Quine writes: "Epistemology, or something like it, simply falls into place as a chapter of psychology and hence of natural science. It studies a natural phenomenon, viz. a physical human subject. This human subject is accorded a certain experimentally controlled input – certain patterns of irradiation in assorted frequencies, for instance – and in the fullness of time the subject delivers as output a description of the three-dimensional external world and its history. The relation between the meagre input and the torrential output is a relation that we are prompted to study for somewhat the same reasons that have always prompted epistemology: namely, in order see how evidence relates to theory, and in what ways one's theory of nature transcends any available evidence".[1]

On the face of it naturalisation would seem to mark a change of focus as against traditional concerns, bearing out the charge laid by Rorty and others that Quine has substituted a causal enquiry for the justification-seeking one that the label "epistemology" traditionally denotes. If so, of course, there could be a complaint only

Alex Orenstein and Petr Kotatko (eds.), Knowledge, Language and Logic, 47–56.
© *2000 Kluwer Academic Publishers. Printed in Great Britain.*

about bending the name to a different purpose, a very minor injury to those still interested in justificatory questions, for what's in a name? – they could continue their task and call it "Fred". But Quine insists that there has been no change of subject matter; naturalised epistemology is truly epistemology; what it does is to improve on the traditional variety. Like the traditional variety it tackles the question of the relation of evidence to theory; but it is epistemology grown up, explaining the relation of sensory inputs to outputs of theory on the basis of empirically check-able facts about how we learn to speak of the world. It transpires that *learning to speak* is the epistemic crux for Quine, for doing so, as he puts it, "virtually enacts the evidential relation"; and he has a psychogenetic account that sketches how.[2] It follows that epistemology is no longer "first philosophy" in the Cartesian sense, because there is no magisterial justificatory task for it to perform. Insofar as justification is to the point at all – and it is not – it comes tangentially and free, because it comes pragmatically; total theory works, which is all that needs to be said. (I speak here of course of justification as it would be sought by what Quine asperses as a "supra-scientific tribunal", namely, traditional epistemology. There is an internal, naturalised sense of justification, the getting of evidence for theories by observation and hypothetico-deductive reasoning which science itself, in the form of psychology, teaches us: but by "justification" in all of what follows I shall mean the ambitious variety of traditional epistemology.)

One of Quine's main motives for supplanting traditional with naturalised epis-temology is that, in his view, the former attempted to get science from sensings by a reductionist project after the various manners of Hume, Russell in one of his phases, and Carnap – who is Quine's principal target in this connection. Quine famously rejects the reductionism involved, but accepts that epistemology "con-tains" science in the sense that science is indeed projected from data, as the output consequent to sensation. But the containment in that direction is reciprocal to a containment in the other: science contains epistemology, because epistemol-ogy is part of psychology. The reciprocity has the effect that epistemology itself turns out to be, like the rest of science, "our own construction or projection from stimulations".[3]

It is this "reciprocity" that looks squarely circular (so to speak). Quine recognises this, but argues that giving up the dream of deducing science from sense data breaks the circle, or perhaps renders it virtuous. If the enterprise of effecting a reductive translation of physical language into sense datum language cannot be carried through, then the Janus face of psychology begs no questions in looking both ways.

What is the objection to this? Well, it is that Quine has taken it that the one alter-native to naturalised epistemology – that is, psychology – is epistemology which involves reductive translation. His immediate target is Carnap's "rational recon-struction", but this can serve to represent phenomenalisms in general, since it shares with them an ambition to exhaust external-world talk in experiential talk without remainder. But why should such a strategy be thought the only alternative to naturalised epistemology? And why should it be thought that rejecting it dis-poses of any interest in justification?

Let us note how Quine comes to think that the reductivist project is naturalism's only alternative. He starts by stating that epistemology is concerned with the foundations of science, which includes mathematics. In just the way that study of mathematics neatly divides into "conceptual" and "doctrinal" studies – that is, investigations respectively into meaning and truth – so too does the study of natural knowledge. On the conceptual side, the task is to show how natural knowledge is based on sense experience. On the doctrinal side, the task is to justify natural knowledge in sensory terms. In Quine's view the doctrinal side fails; we have not, he says, advanced beyond Hume's despair on this head. But on the conceptual side there has been progress, made possible by the technique of contextual definition.

Carnap's "heroic efforts" to effect a rational reconstruction of natural knowledge in terms of sense experience supplemented by logic and set theory, lie on the conceptual side. If successful his efforts would identify and clarify the sensory evidence for science, even if they failed to show how its possession *justifies* science; and they would deepen our understanding of our discourse about the world. But they do not succeed, because the reductive translation required does not go through: Quine's argument needs no rehearsing here. Quine therefore writes, "If all we hope for is a reconstruction that links science to experience in explicit ways short of translation, then it would seem more sensible to settle for psychology".[4]

Let us grant the point about reductive translation. Quine next assumes that naturalised epistemology and epistemologies that turn on translation between them exhaust the options. But this is just incorrect; there are a number of non-reductive ways for the justificatory enterprise to proceed. Take just one example: a remarkable feature of Quine's claim about the foundering of the doctrinal project in Hume is his neglect of what that failure immediately prompted, namely, Kant's critical enterprise, and what that directly or indirectly prompted in its turn, namely, the possibilities it suggested to some – Strawson and the very late Wittgenstein, in not very different ways, included – for exploiting more forensic conceptions of justification than is envisaged by austere versions of empiricism taken alone.[5] It is not necessary to dilate on these options here; we have merely to be reminded of their existence to see that we are not faced with a simple disjunctive syllogism. If naturalised epistemology is to rest to any degree on ruling out the opposition, it will have to show that all other approaches to the doctrinal task founder too.

Quine resisted the charge of circularity, recall, by saying that so long as investigation of the evidential link between sense data and theory is not regarded as addressing questions about justification, no circle is at stake. He assumed that the attempt to secure justification consists in the attempt to reduce theory to data, and so could argue that abandoning hopes of a translation amounts to abandoning the justificatory enterprise. If so, the circle breaks, or turns virtuous: where there is no attempt at justification, it begs no questions to premiss in epistemology what traditional epistemology took itself as having to justify. But the foregoing remarks show that so far we have not been shown that the justificatory enterprise is bankrupt, because we are not bound to identify it with the reductionism Quine rejects.

The next step in showing that naturalism does not escape circularity is to demonstrate something stronger, namely, that the justificatory question is anyway unavoidable. Quine says that epistemology is concerned with the foundations of science. A concern with the foundations of science – a foundational concern – sounds like a concern with assumptions, aims, ontology, methodology and reasoning, one of the chief points in investigating all which – if not *the* chief point – is to ascertain how to tell good theories from bad. This is much more than just seeking to understand the links between evidence and theory, where such understanding falls short of providing justifications. We wish to know when the links are strong enough to bet one's money or one's life on them. We recognise that there are different kinds of evidential relations, individuated by subject matter. We wish to control what can count as evidence for a given subject matter, and to know when caution (or to call it otherwise: doubt) is appropriate, and what can legitimately prompt it. Most of our concerns – in the physician's consulting room, on the construction site, in the pharmaceutical laboratory, on the battlefield – are austerely practical ones, where reliable means of forming judgements are at a premium. We therefore have to try to advance beyond merely *understanding* evidential links where doing so falls short of showing how they bestow a license to rely on certain of them, for such understanding must be the basis for yielding something more: namely, norms. A conception of "getting it right" in a given area of endeavour matters to us; and it demands tests and a way of recognising when they have been passed or failed. Hence, the justificatory enterprise is a non-negotiable part of epistemology. It is why epistemology got started.

So that there is no chance of misunderstanding what is at stake here, allow me to iterate. Quine describes the enterprise of naturalised epistemology as the endeavour (I quote) "simply to understand the link between observation and science" when he is specifically seeking to avoid the circularity charge. That makes it sound as if the task is a very modest one; the passage in full reads, "scruples against circularity have little point once we have stopped dreaming of deducing science from observations. If we are out simply to understand the link between observation and science, we are well advised to use any available information, including that provided by the very science whose link with observation we are trying to understand".[6] Now as we see, the strongly normative character that we expect to follow from understanding evidential links alters the picture. Getting into possession of epistemic norms neither *has* to, nor *does* amount to, deducing science from observations, but neither is it *just* understanding a non-justificatory connection. So we could entirely eschew the attempt to carry out remainderless translations of talk about the external world into a reducing class of sensation sentences, but the attempt to demonstrate the nature and relative strengths of the support provided to theories by observations is a practical necessity, and for this it will not do to premiss what is to be tested in an account of what tests them. Given that, the circle on which naturalised epistemology bases itself is vicious.

At this juncture, by way of aside, one might register a concern about the non-justificatory notion of "evidence" in play here. Standardly, the notion seems intrinsically to be one that exists precisely to play the traditional epistemological role of confirming and infirming, supporting and weakening given claims by its respective presences or absences. In Quine the notion is tied to sensory stimulation, and left otherwise – and explicitly – unexplained: he writes that in his theory "the term 'evidence' gets no explication and plays no role".[7] This, obviously, is a luxury that can be afforded only when we have given up on doctrinal studies altogether.

The foregoing remarks are based on a *pro tempore* acceptance of Quine's claim that epistemology concerns the foundations of science. But – and here is the second of the two objections to be urged against the naturalising programme – epistemology is surely concerned with a good deal more than this. It is more even than a concern with science (that is, with the superstructure as well as the foundations of science), unless the word "science" is stretched to denote with complete generality everything we hope we know. (It had this meaning in the early modern period.) For there are plenty of things we might hope to know beyond the structure and properties of the material world, which is what the natural sciences deal with. We wish also to know about history, for example, and the motives and feelings of others, and whether there is anything of value in the world, ethically and aesthetically; and we are surely interested in the differences between ways of knowing some of these things and knowing others. And as before, we have our exigent practical concerns, which makes getting things right a vital interest, and poses the demand that traditional epistemology exists precisely as an attempt to satisfy. Even the most hardened non-cognitivist in ethics, for example, must accept that we are good readers of the intentions of our fellows, and although physical entities play their part in this – raised eyebrows, air-waves issuing from mouths – it takes an even more exotic form of reductionism than Quine repudiates in Carnap, namely a belief that in the end everything will be expressible in the language of physics, to think that our other epistemic interests, in history and folk psychology and these other spheres, can be naturalised along with the epistemology of science.

Commitment to physicalist reduction cannot be foisted on Quine, however, because although it is naturalistic, it does not exhaust naturalism. Roger Gibson reminds us that Quine infers from the fallibility of science the view that future science might tell us that there is more to evidence than sensory stimulation, and that therefore adherence to empiricism must be tentative.[8] So naturalism is not to be defined as acceptance of the deliverances of physical theory together with what it tells us about how we get it (that is, by empirical means); it is rather to be defined as acceptance of *current* such theory, for any "current". If the theory changes, acceptance must go with it; naturalism lies, in short, in following the fashion. This is integral to naturalistic epistemology's break with justificatory concerns, as witness these facts: a good naturalist in the 13th century AD, or the 13th century BC, would on his naturalistic principles have been bound to go with the current theory of the day. The dramatic nature of past theory change teaches us that being naturalistic at any point in the history of science is therefore a matter quite independent of whatever justification scientists in different periods have had for their claims, whether about the fifth essence, the pulsative faculty of the heart, phlogiston, or the colour and strangeness of quarks: take your pick.

Naturally we all feel confident that even if quarks go the way of phlogiston, current science has got it much righter than ever before, and in some cases, we suspect, has got it right *simpliciter*, so it at least seems safe to be epistemologically naturalistic *now*. We have our Enlightenment belief in progress and the growth of knowledge – I too find such faith compelling – and some even take this to mean that we are more and more closely approximating the truth of things. But this means we should, in the interests of consistency, brush aside the possibility that our remoter descendants, if we have any, might split their sides over 20th century text-books of physics.

These remarks might independently serve to prompt scepticism about naturalised epistemology, but in connection with the circularity problem they add an emphasis. Naturalism enjoins the premissing of *current* theory in explanations of how we come by that theory. To see again why this is deeply unsatisfactory, take a naturalised stand on behalf of our 13th century AD or BC epistemologist, and note how, from that perspective, flat earth and geocentric theories seemed good science in their day. And the fact is, they *are* good science for their day. But that is surely not the point; we wish to know something different, namely, whether they are good science.

In discussing the independence of naturalism from the particular *content* of given current science, Gibson remarks that even if the ontology of current science were to go the way of phlogiston, "naturalism would persist, for naturalism is not wedded to any particular ontology, for example to physicalism".[9] Adding this to the claim that future science might reject empiricism too – so that theory and method both might go – we get the result that *naturalised epistemology* is immune to the fate of whatever it is itself premissed upon at any point in the history of science. This makes it in Popperian terms irrefutable; which in Popperian terms means that it is vacuous.

To see how naturalism misses the point of traditional epistemology – or at least, ducks its demands – we must remind ourselves of what motivates this latter. It is the thought that someone sometime has to take a hard look at the familiar, the taken for granted, the apparently indubitable among our beliefs – and sometimes even to challenge some of them. A charming Chinese drawing in the possession of the British Library shows a Qing Dynasty literatus giving his books their annual airing in the Spring sunshine; something like this has to be done for our beliefs. Philosophy is where it happens; it is where, among other things, our deepest and most general assumptions come in for their check-ups. In connection with momentous questions about what we believe, and why, and whether we are entitled to, and with what degree of confidence, we cannot assume the very things philosophy should be inspecting. One might put the matter this way: when crossing the street in the path of an oncoming bus, it is advisable to make the usual assumptions about the external world. But the study armchair is not in the path of a bus, and so provides a place apart for thinking about why acceptance of the assumptions is advisable when crossing the street.

The point in epistemology is not to substantiate one or another agniological thesis, which Quine seems to think is threatened by traditional epistemological investigations. Such investigations, familiarly, elicit points of concern: evidence underdetermines theory, theories are defeasible, the ontological status of entities (apparently) referred to by certain theoretical terms is moot, and so forth.

Naturalising epistemology is an attempt to cauterize any haemorrhage of confidence such reflections might provoke. Quine writes that it is a "peculiarly philosophical fallacy" to "question the reality of the external world, or deny that there is evidence of external objects in the testimony of our senses, for to do so is simply to dissociate the terms 'reality' and 'evidence' from the very applications which originally did most to invest those terms with whatever intelligibility they may have for us".[10] Now this by implication is to suggest that traditional epistemology, with its central positioning of justificatory concerns, leads to just such questioning and denial: but who (Peter Unger and one or two other *provocateurs* apart) ever seriously thought this? Consider Descartes: he is not one jot a sceptic: he is indeed far less of one than a self-respecting naturalised epistemologist should allow himself to be, out of respect for the open-mindedness of science. Descartes makes methodological use of certain sceptical moves adapted from the antique canon in order to illustrate a point. But the motivation is the excellent one of subjecting the nature of justification to as rigorous a scrutiny as possible, which involves considering even outré possibilities of defeasibility and difficulty. Traditional epistemology thus explores ways of taking serious account of worst-case epistemic situations in order to make the best case for knowledge-claims: an interesting, legitimate, and consequential enterprise.

The objection always advanced by naturalisers to this characterisation of the epistemological task is that it involves boot-strapping, or an attempt to occupy a theoretical vantage point which is not really available. The traditional task takes itself to be a self-reflexively critical endeavour to stand aside, so to speak, and from that position of notional disengagement to bring under review as many of the facts and possibilities of epistemic life as can be mustered, and to make sense of them. It therefore implicitly rejects the premiss of the Neurath ship analogy, which is that we are *essentially* context-bound, *strictly* limited in the task of critical self-reflection by the framework we occupy. And it rejects it because it holds that there are indeed imaginative, comparative, speculative, abstracting, formal, critical, debated, assumption-testing opportunities of reflection available to us, by whose means we interpret our world and our practices to ourselves, and by whose means we push at boundaries. One of the chief of these opportunities is philosophy itself, but such forms of thought are to be found in many other pursuits too, from creative literature to theories of the quantum field.

The link between the two objections has already been made. It lies in the naturaliser's refusal to see epistemology as a justificatory enterprise. To sustain this refusal would be to escape the circularity charge, and to give epistemology a new point: the non-justificatory understanding of connections between experience and science. But if epistemology is stubbornly a justificatory enterprise, to naturalise it is to go in a circle and to duck its chief demand.

What is lost by abandoning naturalisation? First, one notes that the baby need not go out with the bath water. A refusal to premiss current science in one's epistemology is not a refusal to take current science seriously. Like Descartes, we are not sceptics in any but the healthy sense praised by Russell, and our interest lies in extracting the best lessons science has to offer. Sometimes the lesson is that we do not do well in investigating certain phenomena to employ the techniques that were successful with other phenomena. On the naturalistic view, the psychology of perception – even understood as extendible by the finest technology available at the time – seems to provide a model for too much in the way of enquiry.[11] Still, science is its own recommendation as a success, and invites the liveliest epistemological interest. But that is not the same thing as giving up thinking about it critically.

In the fabric of Quine's philosophy, naturalism is internally connected with holism, realism, and the crucial matter of language acquisition with the empiricist theory of meaning it entails. What effect does giving up naturalism have on these other commitments? Gibson has drawn a clear picture, endorsed by Quine, of how the dependencies lie: holism and realism support naturalism, naturalism supports natural science, and natural science supports empiricism.[12] I shall not here consider holism and empiricist theory of meaning, which are too attractive in principal to argue with, although a corollary of Quine's version of the latter gets some comment below. But there is room for unease over realism. It is not clear to me that realism understood as a thesis about the independence from theory of the things we theorise about is either inherently intelligible or – not surprisingly, if the unintelligibility charge is right – consistent with the other two anyway, although longish tales need to be told why. For present purposes the point need only be that in the version espoused by Quine, realism goes if naturalism does; for the following reason.

On Gibson's endorsed account, Quine's realism is "unregenerate realism", by which he says he means a sort of naive realism glossed and theorised into more sophisticated shape by naturalistically understood natural science. It consists in the "archaic natural philosophy" imbibed with mother's milk, extended in the process of actual and epistemic growing up into something deeper and broader. On this view, common sense is continuous with science.[13] "To disavow the very core of common sense," says Quine, "to require evidence for that which both the physicist and the man in the street accept as platitudinous, is no laudable perfectionism; it is a pompous confusion, a failure to observe the nice distinction between the baby and the bath water".[14] Well: for Russell archaic natural philosophy was the metaphysics of the cave-man, and the metaphysics of modern man – namely, science – in his view shows common sense to be false. For Berkeley, grass is green in the dark, and the views of Locke and the New Philosophers (as the corpuscularians were called) absurdly controvert common sense (of which Berkeley took himself to be the doughtiest of champions) by proposing that grass is no colour unless seen, that solid objects enclose mainly empty space, and so on. Now it just seems to me that common sense has its place, that particle physics offers models of a world that is pretty unlike the world of common sense, that any decent theory has to save the endoxa, as Aristotle required, but it does not have to be the endoxa and indeed perhaps had better not be; and that anyway, "archaic natural philosophy" is not as a whole an historical invariant (remember our 13th century AD and BC predecessors) even if part of it – the medium sized dry goods part, along with its associated ways of talking – is; but it is hardly enough to get science going. In fact "unregenerate realism" could well do with a bit of regeneration. It is a natural companion for naturalised epistemology, of course, but its not clear that either gains by association with the other.

Another important feature of the view that presents some difficulties, in unmodified form at least, is the observation sentence-theoretical sentence considerations that go with Quine's language-acquisition story. The difficulty is very familiar, but bears restatement. It is that for the two kinds of sentence to be relevantly distinguishable – as occupying appropriate logical slots with respect to each other – they have to differ enough in kind. Put schematically, observational sentences have to be relevantly untheoretical, theoretical sentences relevantly unobservational. The latter might be granted, but it is a famous stricture on the former that there are no theory-light sentences. This is to be taken seriously as entailing that any sentence counts as an observation sentence only as a high matter of some theory (and on other occasions might well not so count). Now, observation sentences have a crucial place in the economy of naturalised epistemology; they are "the cornerstone of semantics", fundamental to language learning as what gets it started, and firmest in meaning because they have their own empirical consequences.[15] They therefore play both conceptual and doctrinal roles, in the minimal sense of the latter allowed by naturalism, as stating the (non-justificatory) evidence for theory. If we are attracted by the idea that there are empirical controls on meaning from its roots up, and that this is a fact about language-acquisition which blocks any move to theories of meaning based on transcendent conceptions of truth, then some way has to be found of construing talk of observation sentences that accommodates Neurath's point about them.

I anticipate that what naturalists will most dislike in the foregoing is of course the insistence on the traditional epistemological conception of the justificatory task. And that is as it should be, since it is that – the question that Quine identifies as the doctrinal one – that has been at the centre of the trouble all along. Now I can register one concluding point. In drawing the parallel between conceptual and doctrinal studies in mathematics and those aimed at knowledge of the natural world, Quine exported the meaning-truth pairing from one to the other more or less intact. Early in "Epistemology Naturalised" he talks of Hume's despair as to "the justification of our knowledge of truths about nature". Of *truths*, note; a traditional truth-incorporating notion of knowledge is in play. But this is not as it should be, although it too often was so in the minds of traditional epistemologists. We have no decent notion of truth to serve the turn, and do not need one; indeed we need nothing which even remotely resembles the familiar realist versions of truth on offer. For "knowledge" read "reliable beliefs" and scrap truth; we can get by with *success most of the time*, because that is how we do get by. So the notion of justification in traditional epistemology is not as exigent as it clearly seemed to Quine in his celebrated paper – although it weakens in a different direction from the one he prefers. I offer it as a thought that this misperception of what justification is supposed to yield in traditional epistemology is perhaps what mistakenly made naturalism feel so natural in the first place.

Birkbeck College, University of London

NOTES

[1]   "Epistemology Naturalised" *Ontological Relativity and Other Essays* Columbia UP (New York) 1969 pp. 82–3. Cf also: "Epistemology is best looked upon ... as an enterprise within natural science. Cartesian doubt is not the way to begin. Retaining our present beliefs about nature, we can still ask how we can have arrived at them. Science tells us that our only source of information about the external world is through the impact of light rays and molecules upon our sensory surfaces. Stimulated in these ways, we somehow evolve an elaborate and useful science. How do we do this, and why does the resulting science work so well? These are genuine questions, and no feigning of doubt is needed to appreciate them. They are scientific questions about a species of primates, and they are open to investigation in natural science, the very science whose acquisition is being investigated." – "The Nature of Natural Knowledge" in S. Guttenplan *Mind and Language* Oxford 1975.

[2]   The Nature of Natural Knowledge" pp. 74–5, and see *Roots of Reference* (Open Court 1974) p. 3.

[3]   "Epistemology Naturalised" p. 83.

[4]   *ibid.* p. 78.

[5]   "Forensic": Kant saw the question of justification on analogy with legal justification: in effect, as an establishing of right or entitlement to given beliefs. – It might seem implausible to assign Wittgenstein a place in the Kantian tradition, but it is not too far from the point; he had after all been much influenced by Schopenhauer, and *On Certainty* has its antecedents in the *Tractatus*.

[6]   "Epistemology Naturalised" p. 76.

[7]   "Comment on Davidson" in *Perspectives on Quine* eds. R. Barrett and R. Gibson Oxford: Blackwell 1990, p. 80.

[8]   R. Gibson, "Quine on the Naturalising of Epistemology" *On Quine: New Essays* eds. Paolo Leonardi and Marco Santambrogio CUP 1995, p. 90.

[9]   *ibid.*

[10]   "The Scope and Language of Science" in *The Ways of Paradox* Harvard UP 1979, p. 229.

[11]   It does not for example help in judging the qualities of a poem, say.

[12]   Gibson "Quine on the Naturalising of Epistemology" *op. cit.* p. 93.

[13]   Quine "The Scope and Language of Science" *The Ways of Paradox op. cit.* pp. 229–30; Gibson *ibid.* pp. 94–5.

[14]   Quine *ibid.*

[15]   "Naturalised Epistemology" p. 89.

KEITH LEHRER

JUSTIFICATION, COHERENCE AND QUINE

Quine has influenced epistemology beyond his own project of naturalization. His early work influenced my own work in the coherence theory of knowledge,[1] and I should like to take this occasion to raise some questions about coherence and knowledge. Quine was impressed by the failure of the reductionist program of phenomenalism. I agree with him about the failure of that program. I also agree with him that the consequence of that insight is that it is some system that confronts the world and our sensory experience of it. Knowledge, I concluded, must result from some combination of coherence and truth. I learned that from Quine but am less content to follow him down the path of naturalization. I want to explain why I took another path from his insights, a less revisionary one.

Anthony Grayling has argued that there are different conclusions that might be drawn from the failure of phenomenalism. There were, historically considered, two objectives of phenomenalism. The first was to give an account of the meaning of statements concerning material objects in terms of the language of sense data and, in that way, to reduce the meaning of material object statements to the sound foundation of sense experience. Let us call this the *meaning basing objective*. A second objective of phenomenalism was concerned, not with meaning, but with the justification of inference of material object statements from statements of sensory experience. Let us call this the *inference basing objective*. These two objectives of phenomenalism were run together historically, but they are distinct.

I have argued that success of phenomenalism in meeting the meaning basing objective would not have met the justification basing objective. The simple explanation is that those material object statements analyzed phenomenalistically would affirm more about sensory experience than any one person and, indeed, than all people would experience. It would tell us about future sensory experiences as well as present ones and leave ample room for the skeptic to drive a Humean wedge in the argument. Thus, the success of phenomenalism would not have been sufficient to answer a skeptic about the material world.

A similar point is made by Quine himself who remarked, assuming the success of phenomenalistic translation,

"But the mere fact that a sentence is *couched* in terms of observation, logic, and set theory does not mean that it can be *proved* from observation sentences by logic and set theory. The most modest generalizations about observable traits will cover more cases than its utterer can have had occasion to observe."[2]

Alex Orenstein and Petr Kotatko (eds.), Knowledge, Language and Logic, 57–61.
© *2000 Kluwer Academic Publishers. Printed in Great Britain.*

Phenomenalism fails, and if it were to succeed it would not suffice for obtaining the justification basing objective. It is at this point that Quine decides that we should settle for psychology to provide us with an account of the relationship between sensory stimulation and the external world. He says,

"The stimulation of his receptors is all the evidence anybody has had to go on, ultimately, in arriving at his picture of the world. Why not just see how this construction proceeds? Why not settle for psychology?"[3]

Against the objection that using psychology to validate our claims about the external world is circular, Quine replies,

"However, such scruples against circularity have little point once we have stopped dreaming of deducing science from observation. If we are out to understand the link between observation and science, we are well advised to use any available information, including that provided by the very science whose link with observation we are seeking to understand."[4]

Thus, Quine recognized that the success of phenomenalism was also not necessary to obtaining the inference basing objective. He might have gone on to argue that once we combine falliblism with the assumption that it is always some scientific system of statements rather than a single statement with which we confront our experience of the world, we can use what finds it way into that system to justify our inferences. What we need to justify the inference from sensory experience to claims that exceed what we have so far experienced is some additional premise. But as we confront experience with our description of it, we do so with a background of statements that may tell us when our descriptions are justified and when they are not. It is, after all, not only perceptual statements that comprise our background system but statements about when we should trust our senses and when we should demure.

That is what I made of Quine and went on to define justification as coherence with a background system. But Quine himself took a more radical turn toward psychologism. It is only psychology, he proposes, that we should use to obtain our understanding of the relationship between sentences about sensory stimulation and sentences about the external world. Can psychology satisfy the justification basing objective? Can psychology tell us when our inferences from sensory experience to the external world are justified? It appears not, though this depends on how strictly one takes the notion of psychology. For, it appears that psychology would offer us an account how sensory experience influences the formation of beliefs about the external world while remaining silent about whether the beliefs so formed are justified. That is how I understand the point that Anthony Grayling has raised. After all, part of psychology, abnormal psychology, might tell us how the beliefs of psychotics are formed from sensory experience without commenting on whether they are justified. In fact, it appears that psychotics reason in quite normal ways, as well if not better than the rest of us, which suggests that they may suffer sensory distortion. The psychotic may, for example, believe that he observes that President

Clinton is speaking to him personally when Clinton appears on television and is communicating a secret message that only he and Clinton can understand. Now we think that these beliefs are not justified, but that conclusion is a step beyond anything that psychology tells us. Psychology tells us how beliefs originate but about whether those beliefs are justified she remains silent as a stone.

One naturalistic reply is that science can tell us when the inference will lead to a scientific conclusion that is true and that we have no better guide than science itself to tell us when our conclusions from sense experience are true and when they are not true, when, for example, the objects we think exist really do exist and when we are deluded. That is suggested by the passage above and others as well. The point is that once we give up the idea that we can validate conclusions, that is, prove them, from sensory experience, we see that we have nothing better that science to tell us whether those conclusions are true or not. And why, after all, should we want anything better than science? There is, of course, a circle in this, but that is the consequence of giving up the theory of foundational validation and is not to be deplored. In short, science cannot only tell us how our beliefs are formed from sense experience but can also use that information to sort the true from the false, or, at least, to do the best we can.

I think that there is a question remaining, however, which concerns what we are justified in believing. Quine may avoid the question and, in past discussions, has done so. Why concern oneself with the question of what one is justified in believing? Either one has an interest in truth or one does not, and if one does and thinks that science is the best way to get to it, then one follows the path of science. And so Quine does, and admirably so. Is there a question left over? There is. We may have an interest in truth, an interest in the objectives of science, and yet raise the question of whether we are justified in accepting what we do in the interest of achieving our objectives. We may appeal to other things that we accept, the contents of a background system, to answer this question. But a question remains. Are we justified in accepting any of things we do? Obviously we are justified. But why? How does justification get into the system?

This question is not a quest for a foundation but for an explanation. A philosopher who affirms that we just are justified in what we accept and that is all there is to it leaves us with kind of epistemic surd. What I want to suggest is that there is no need to leave ourselves with the surd once we have followed Quine into the perspective of allowing us to use our background system to explain why we are justified in accepting what we do to meet our scientific objectives. We need only ask ourselves what it is that we accept that is supported by other things we accept that enables us to explain why we are justified in accepting what we do.

The answer to the question is not difficult to find. We trust ourselves in accepting what we do to obtain our objectives. Moreover, we trust ourselves in preferring the objectives we do. We also consider ourselves worthy of our trust in what we thus accept and prefer. To put the matter another way, we consider ourselves trustworthy in what we accept, though acknowledging our fallibility. We think that we can be both trustworthy and fallible. Now suppose that we are trustworthy as we think we are in accepting, among other things, that we are trustworthy. That, I suggest, explains why we are justified in accepting what we do, including that we are trust-worthy. If we are trustworthy in what we accept and in our preferred objectives, then we are justified in accepting what we do for those our objectives. That does not mean that we automatically attain our objectives, that what we accept is always true, for example, for fallibilism is our cognitive burden. What it means is that there is at least a minimal level of justification in what we accept if we are trustworthy in accepting what we do to obtain our objective. And, of course, we accept that we are trustworthy.

Is our acceptance that we are trustworthy in what we accept a kind of epistemic surd? If asked why we are justified in accepting that, must we say simply that we are justified and that is all there is to it? We can do better. We can answer that we are trustworthy in accepting that we are trustworthy. The advantage of this answer is that if it is correct, it does explain why we are justified in accepting it. If we are trustworthy in accepting what we do, including that we are trustworthy, why, then, that is why we are justified. We are justified because our acceptance of what we thus accept is a consequence of our trustworthiness. If so, then our obtaining our objectives by accepting what we do is not unexplained or just a matter of luck but is instead a result of our being trustworthy as we suppose ourselves to be.

Is our acceptance of our trustworthiness a foundation? No. It is, in fact, sup-ported by other things we accept rather than being the foundation for them. Of course, we may appeal to our acceptance of our trustworthiness to justify those other things we accept. The metaphor I prefer is that of a keystone. The acceptance of our trustworthiness is a keystone in the system of things we accept. It is sup-ported by the other things we accept but the vaulted arch of science collapses when it is removed. It does, in a way, support itself, but only as a result of being sup-ported by the other stones in the arch.[5]

Once we suppose that the background system of justification contains the key-stone, knowledge results from coherence with the system combined with truth. But the initial justification of the contents of system is essential or the system counts for naught. Given the initial justification of what we accept, we can proceed to accept justified conclusions even though, as Quine says, we cannot validate or prove them. Why is the justification not a validation or a proof? Because, of course, circular rea-soning cannot prove or validate anything. No skeptic can be expected to accept the reasoning above concerning trustworthiness if she does not concede that we are trustworthy in what we accept or in our preference for our objectives. But, though she cannot thus be proven wrong, we may be trustworthy as we think we are nonetheless. We are then justified in accepting that the skeptic is wrong, though we cannot prove or validate this claim.

Quine has argued at the end of the symposium that immediacy is something we should trust, that what we immediately accept, admitting this to admit of degrees, is a fallible guide to the truth of the matter. I suggest that he accepts that immediacy is worthy of our trust, though we are fallible. He also accepts that we are worthy of our trust in what we accept. And that, I suggest, is the keystone of our acceptance system. Our trustworthiness results, as I have noted, from our trustworthiness in the way in which we change what we accept and change our ways of changing. Is trustworthiness just a matter of being correct with a certain frequency? All we can say about how often we must be right in order to be trustworthy is that we must be correct a trustworthy amount of the time. That shows that trustworthiness is not reducible to anything naturalistic which, I suspect, will lead to Quine to eschew such discourse. For all that, Quine considers what he accepts to be worthy of his trust, and he has proven brilliantly justified in accepting his trustworthiness.

University of Arizona

NOTES

[1]   Keith Lehrer, *Theory of Knowledge* (London and Boulder: Routledge and Westview Press, 1990).
[2]   Willard Van Orman Quine, "Epistemology Naturalized," in *Ontological Relativity and other essays* (New York: Columbia University Press, 1969), p. 74.
[3]   *Op. cit.*, p. 74.
[4]   *Ibid.*
[5]   Keith Lehrer, *Self-Trust: A Study of Reason, Knowledge and Autonomy* (Oxford: Clarendon Press, 1997), Chap. 1.

LARS BERGSTRÖM

QUINE, EMPIRICISM, AND TRUTH

In an earlier paper[1] I have argued that certain passages in W.V. Quine's work indicate that he should accept what I call an empiricist conception of truth. According to such a conception, a sentence is true, roughly speaking, if it is entailed by an empirically adequate theory. Quine is unwilling to accept my proposal, and he has indicated certain problems to which it gives rise. In this paper, I shall discuss Quine's response, and I shall also try to develop a more satisfactory version of the empiricist conception of truth.

1. TRUTH AND OBSERVATION

I shall begin by briefly presenting the case for an empiricist conception of truth within a Quinean perspective. Quine is an empiricist. His empiricism can be summarized in the two theses that "whatever evidence there is for science is sensory evidence," and that "all inculcation of meanings of words must rest ultimately on sensory evidence."[2] Empiricists may differ as to how the fundamental notion of "sensory evidence" should be understood. For Quine, sensory evidence is identified with the triggering of sensory receptors.[3]

According to Quine's naturalism, empiricism is itself a scientific hypothesis, and as such it might conceivably turn out to be mistaken. So far, however, current science tells us that information about the external world can reach us through, and only through, our sensory receptors.[4] Personally, I find this very reasonable.

In order to study the relation between science and observation, Quine focuses on the corresponding linguistic formulations. Thus, he takes a *theory* to be a conjunction of sentences,[5] and an *observation sentence* for a given speech community is a sentence which is directly and firmly associated with sensory stimulations (i.e. the triggering of sensory receptors) for every member of the community, and on which all members give the same verdict when witnessing the same situation.[6] Observation sentences are the link between theory and evidence. They are also the first sentences we master when we learn our first language as children.

Observation sentences are occasion sentences: they are true on some occasions and false on others. Under what conditions are they true? Clearly, an observation sentence such as "It's raining" is true on a given occasion if and only if it is raining on that occasion. But it seems that we can also say that "It's raining" is true on that occasion if the sentence would be assented to (on that occasion) by competent members of the linguistic community witnessing the occasion. This is so because

63

Alex Orenstein and Petr Kotatko (eds.), Knowledge, Language and Logic, 63–79.
© 2000 Kluwer Academic Publishers. Printed in Great Britain.

you would not qualify as a competent member of the linguistic community unless you were disposed to assent to "It's raining" on rainy occasions. Conversely, it's raining on a given occasion if a competent speaker witnessing the occasion would assent to the sentence "It's raining." This might be surprising, but it is a consequence of Quine's claim that a sentence is observational for a group "if it is observational for each member *and* if each would agree in assenting to it, or dissenting, on witnessing the occasion of utterance."[7] Given that "It's raining" is an observation sentence for the group, and that someone assents to it on a certain occasion, everyone would assent to it on witnessing that occasion. And if everyone uses the sentence this way, then its meaning is thereby such that it is true on that occasion. Since the speakers are competent, they have learnt to use the sentence on the right occasions, and the right occasions are the occasions on which it is true. In general, then, an observation sentence S is true on occasion O if and only if every competent speaker of the language to which S belongs would assent to S if he or she were witnessing O.

So far we have only defined truth for observation sentences. But we can take a further step. Observation sentences are related to our theories, but they are not implied by theories. Rather, the observational part of a theory – if it has any – consists of so-called observation categoricals.[8] These are general sentences of the form "Whenever X, Y", where X and Y are observation sentences. Observation categoricals can be implied by sufficiently comprehensive (conjunctions of) theories. We can now say that an observation categorical "Whenever X, Y" is true if, and only if, for every occasion O, if X is true on O, then Y is true on O.

This, or something like this, is a basic claim of the empiricist conception of truth. Some qualification or further explanation will be needed, however. Surely, we must recognize the possibility that competent speakers of the language are in fact wrong on certain occasions. I shall try to accommodate this point in sections 3 and 4 below, but I shall first consider Quine's response to the basic idea.

2. QUINE'S RESPONSE

Quine says that my empiricist definition of truth for observation categoricals "has its evident appeal, but it has an alarmingly Protagorean ring, making man the measure of all things."[9] So he is not quite happy about it. He continues:

Perhaps the dilemma can be resolved by taking account of the subjunctive conditional, or the disposition that is implicit in "would assent." Dispositions are theoretical, positing appropriate mechanisms. The appeal to human propensities to assent in unrealized cases is thus fallible theory over and above the observation categoricals themselves. It can thus be argued that truth even of all observation categoricals does not reduce utterly to observation.[10]

I quite agree that dispositions are theoretical. But, as far as I can see, this is no objection to my proposal. Of course, the question of whether a given observation categorical is true cannot be *directly settled* by observation. This much already follows from the fact that observation categoricals are general statements, without spatial or temporal restriction. Nevertheless, Quine seems to accept my criterion of truth for observation categoricals. He makes no objection to it, and he says that it has "evident appeal." What he offers is a consideration which might soften its "alarmingly Protagorean ring."

The Protagorean ring is still there, however. It may be alarming or surprising, but is it fatal? I think not. In this respect, observation categoricals are in the same boat as simple observation sentences. Surely, such a sentence has to be true on a given occasion if all, or nearly all, competent users of the language would assent to the sentence on witnessing that occasion. For the truth conditions of an observation sentence are determined precisely by the actual use of that sentence within the linguistic community. If the community assents to the sentence on certain occasions, then the truth conditions of the sentence obtain on those occasions – rather than the other way round. Of course, linguistic usage does not cause the conditions to obtain, but linguistic usage makes the conditions acquire the status of truth conditions for certain sentences. Consider the sentence "It's raining." Our use of this sentence does not produce rain, but it makes the occurrence of rain (by our lights) constitute the truth condition for the sentence. Therefore, it is really not surprising that our linguistic usage determines the truth of the sentence.

One might object that although usage determines truth conditions, the latter then take on a life of their own and determine truth independently of usage. But this seems wrong. For if usage were to change, it would thereby take away the status of truth conditions from the original truth conditions. Therefore, there is indeed a sense in which "man is the measure of all things." As far as I can see, there is nothing objectionable about this.

It may be easier to accept the claim that "man is the measure of all things" if this is divided into two parts. The first part would say that "man is the creator of language", and the second part would add that "language is the measure of all things." I am willing to accept this.

3. OBSERVATIONAL MISTAKES

Let me now return to the objection that the empiricist conception of truth, as it has been formulated above, does not seem to allow for observational mistakes. This objection is serious, but on the other hand, there is less room than one might think for observational mistakes on a large scale.

Suppose, for example, that I am looking through the window and think I see rain outside. My sensory receptors are triggered in a way which disposes me to assent to the observation sentence "It's raining." But, in fact, there is no rain. Instead, someone is watering the lawn outside the window.[11] Clearly, I am mistaken. Moreover, there may be a lot of competent speakers in the room, all of which are disposed to assent to "It's raining." This makes no difference. We are all mistaken. Doesn't this show that the empiricist criterion of truth for observation sentences must be rejected?

No. There is no reason to suppose that the empiricist criterion would be satisfied in this and similar cases. Even if everyone who happens to witness the occasion in fact assents to "It's raining," we have no reason to conclude that every competent speaker *would* assent to it if he or she *were* witnessing the occasion. It is very easy to imagine that some competent speakers would know that it is not raining, because they have collateral information[12] concerning the weather and the watering of the lawn; they may have come into the house just recently and they may have noticed the gardener watering the lawn. In cases like this, observation sentences can be falsified by collateral information, and on such occasions they are of course false. We learn that such cases exist when we learn the language. But such cases are the exception rather than the rule. For suppose all or nearly all occasions on which we use and learn to use "It's raining" were in fact occasions on which someone is sprinkling water on a window-pane. I submit that "It's raining" would then mean just that, and the sentence would then be true. If all of us always used the sentence on such occasions, we would be using it correctly. It would be raining.

Let us consider a slightly different example. Suppose we all learn "It's raining" in the usual ostensive way, and then some natural catastrophe occurs which radically changes the causal chains which lead to "rainy" stimulations of our sensory organs. For example, from a certain time on all our "rainy" stimulations are caused, not by rain, but by some strange kind of air pollution.[13] If we continue to use the sentence "It's raining" as a response to "rainy" stimulations, I would claim that what we say is still true. From an external point of view, one might claim that the meaning of what we say has changed, but that is another matter. From a more internal point of view, we may find that we have to change our theories about rain and related phenomena in order to accommodate our more recent observations. If this turns out to be very difficult, we may instead change our observational language in such a way that the conditioning of our disposition to assent to "It's raining" to "rainy" stimulations is dissolved. In the transition period, we may prefer to suspend judgment concerning the truth-values of utterances of "It's raining."

Again, suppose I am a brain in a vat and that a computer built by some mad scientist feeds stimuli to my nerve endings, thereby making me have qualitatively the same experiences I have in this world. I can also interact with what I take to be my environment. It seems to me that I can make things happen. The computer responds to my efforts by causing me to have perceptions of the relevant effects. In particular, I can learn to use language in the vat. For example, I learn to use the sentence "It's raining." Of course, the linguistic community to which I belong does not include people outside the vat, and the situations I perceive and to which I apply the sentence "It's raining" are not the rainy situations to which people outside the vat apply it. Still, I can learn to apply the sentence correctly in the language I am being taught. After a period of training, the mad scientist can report to his colleagues that I have finally managed to do so. (The computer has been programmed to report to him.) And then, when I use the sentence on a given occasion, what I say is (normally) true. When I say "It's raining", it is raining – not, of course, in the sense that it is raining outside the vat, but in the sense that my utterance is true. This is what matters to me. I do not care whether the same sentence shape that I use is also used by some alien tribe (outside the vat) with a different meaning, so that my utterance of it is true at the same time as their utterance of it is false.

In general, then, I suggest that there is much less room for observational mistakes than one might think. Still, such mistakes must of course be possible.

4. APPEARANCE AND REALITY

In learning observation sentences, sooner or later we reach a stage where we are introduced to the distinction between appearance and reality. We learn that what looks like a rabbit may really be something else, a hare perhaps, or a small dog seen at a distance, or even a counterfeit rabbit.[14] Similarly, we understand that water on a window-pane may look just like rain. At this stage we become more careful in reporting our observations; we know that we run a risk of being wrong. On some occasions, we can see that the risk is quite high – as e.g. when we get a quick glimpse of what we take to be a herring in the sea. On other occasions, we feel confident that the risk is low or negligible – as when I hold up my hand and say "This is a hand."

Quine has often pointed out that observationality is a matter of degree.[15] The degree has to do with our readiness to assent to a sentence under given stimulation; there may be more or less delay or hesitation. As I understand this, the point is not merely that different sentences have different degrees of observationality. Rather, degrees of observationality are determined by three factors: a sentence, a speaker, and an occasion. Some sentences may be more observational than others (for most speakers and most occasions), but a given sentence can be more observational for one speaker than for another, and more observational on one occasion than on another. For example, "That's a herring" is less observational for me than for an experienced fisherman, and it's degree of observationality is lower the greater the distance, and the muddier the water, between the speaker and the supposed herring. In general, the greater I believe the risk is that I am wrong in assenting to a given sentence on a given occasion, the lower is the degree of observationality of that sentence for me on that occasion.

The empiricist conception of truth employs the notion of a competent speaker of the language. It may be asked whether a competent speaker of a language must also be a competent speaker of every sentence of the language. The natural answer to this is, I think, negative. For example, someone may be a competent speaker of English even though he is not a very competent speaker of the sentence "That's a herring." He may know that a herring is a fairly small fish, but "That's a herring" may not be observational for him or it's degree of observationality for him may be extremely low. Such a speaker should of course not be allowed to decide whether the sentence "That's a herring" is true on a given occasion.

Quine says that a sentence is "observational for a group if it is observational for each member *and* if each would agree in assenting to it, or dissenting, on witnessing the occasion of utterance."[16] Presumably, however, very few, if any, sentences are observational in this sense for the group of competent speakers of a language. The problem is that competent speakers of the language – even if they are also competent speakers of the sentence itself – may not always agree in assenting or dissenting to a given observation sentence.[17] For example, two experienced fishermen may disagree on "That's a herring" when they get just a quick glimpse of a small fish in muddy and unknown waters. And even a competent speaker of "It's raining" may mistakenly assent to the sentence when someone is sprinkling water on his window-pane.

This shows, I think, that the notion of an observation sentence for a group should not contain the requirement that each member of the group would agree in assenting or dissenting to the sentence in question on each occasion. That would make the set of observation sentences almost empty. Rather, we should say that a sentence is observational for a group when it is observational for each member.[18] We may then assume that *competent* speakers of the sentence would agree in assenting to it, or dissenting from it, on each *favourable* occasion – where a "favourable" occasion[19] is one where the sentence's degree of observationality, for competent speakers of it, is high.

We have seen that competent speakers of a language can be mistaken in their assent to observation sentences. Consequently, the empiricist conception of truth must be reformulated so as to contain some protection against risky assent and dissent, or against assent and dissent where the degree of observationality is low. An observation sentence can be true on an occasion even if some competent speakers of the language would dissent from it on witnessing the occasion. A better formulation of our criterion is this:

(O) An observation sentence S is true on occasion O if O is a favourable occasion and every competent speaker of the language to which S belongs, which is also a competent speaker of S, would assent to S on witnessing O.

This criterion contains a restriction to favourable occasions and to competent speakers of the sentence in question. However, it should be possible for S to be true on O even if O is not a favourable occasion. Therefore, the truth condition should be sufficient but not necessary. The "if and only if" of our first proposal must give way to a simple "if."

So much for the truth of observation sentences. Let us now see how an empiricist conception of truth for other sentences can be developed from that basis.

5. THE EMPIRICIST CONCEPTION OF TRUTH

As far as observation categoricals are concerned, we can stick to our earlier proposal: an observation categorical "Whenever X, Y" is true if, and only if, for every occasion O, if X is true on O, then Y is true on O. But how should the truth of more theoretical sentences be accounted for? One suggestion, which naturally comes to mind, is the following:

(E) A theoretical sentence S is true if and only if S is entailed by a tight theory which entails every true observation categorical and no false observation categorical.[20]

A theory is *tight*, roughly speaking, if it contains no superfluous assumptions – where an assumption is superfluous, in a given theory, if it adds nothing to the empirical content or explanatory value of the theory.[21] Something like this requirement of tightness seems necessary to avoid inconsistency, but it is not quite clear how it should be formulated to do its job properly. I'll come back to this question.

Let us now consider Quine's reaction to (E). If I understand him correctly, he is unwilling to accept (E). He writes:

> I see truth in general as far exceeding ... anything that can be checked in observation categoricals. This answers [Bergström's] next question: whether I would declare a theoretical sentence true "if and only if entailed by a theory which entails every true observation categorical and no others." Contrary to the positivist spirit, I do not repudiate sentences for lack of empirical content.[22]

Of course, most theoretical sentences have no empirical content (of their own), and they can nevertheless be true. This is a well-known Quinean thesis, and in this sense truth exceeds "anything that can be checked in observation categoricals." Still, this does not answer the question of whether every true sentence has to be part of some theory which in turn can be checked in observation categoricals.

In particular, consider what Quine often refers to as "our total, or overall, theory of the world." We can take this total theory to be (equivalent to) the conjunction of all theoretical sentences we hold true. Now Quine's empiricism leads him to claim that "[o]ur overall scientific theory demands of the world only that it be so structured as to assure the sequences of stimulation that our theory gives us to expect."[23] This is metaphorical, of course, but as far as I can see the idea is that whatever our total theory says of the world, its truth does not presuppose anything over and above empirical adequacy. As long as the world is so structured that the observation categoricals entailed by our theory are true, nothing more can be asked for in the way of positive feed-back from the world. If the world *is* "so structured as to assure the sequences of stimulations that our theory gives us to expect", then the world is as our theory says it is – and then, presumably, the theory is true. In that case, it is not only that we *hold* the theory true; it *is* true. And if the theory as a whole is true, then surely everything entailed by it is true. This is the idea that (E) tries to exploit.

Quine insists that a theory says "incomparably more" about the world than is said by its observational part.[24] Yes, I agree. I was mistaken on this point in my earlier paper. But what the theory says in its theoretical part has to fit in with what it says in its observational part, and it is doubtful whether Quine has any further (empirical) requirement on theoretical sentences than this. Quine writes:

> [Bergström] rightly quotes me as saying that if a theory conforms to every possible observation, "then the world cannot be said to deviate from what the theory claims", but this only requires truth to be compatible with observation, not determined by it.[25]

Yes, but two kinds of determination should be distinguished here. The observational part of a theory does not determine the *content* of its theoretical part. This is Quine's point about the underdetermination of theory by observational data.[26]

However, it might still be held that the *truth-value* of the observational part determines the *truth-value* of the theoretical part. Clearly, if the observational part is false, so is the theoretical part (since it entails the observational part). But, from an empiricist point of view, one may also say that if the observational part is true, so again is the theoretical part. Indeed, this is precisely the so-called *ecumenical* position, which, sometimes at least, Quine himself finds appealing.[27]

This is also the only way I can interpret Quine's statement that "the world cannot be said to deviate from what the theory claims" if the observational part of the theory is true (i.e. if the theory conforms to every possible observation). To say that "the world deviates from what the theory claims" is, I suppose, equivalent to saying that the theory is false. Moreover, in this context I take it that a theory is true if it is not false. My conclusion is that Quine might agree that a theory is true if its observational part is true.

We may think of a theory as the deductive closure of its axioms. This includes the observation categoricals implied by the theory, which constitute its empirical content. The rest is the theoretical superstructure. Quine's underdetermination thesis says that drastically different theoretical superstructures can imply the same empirical content. In one place, Quine says that "[w]hat the empirical under-determination of global science shows is that there are various defensible ways of conceiving the world."[28] Clearly, for an empiricist a theory does not constitute a defensible way of conceiving the world unless it is compatible with observation. But Quine seems to hold that different theoretical superstructures can all be defensible. The difference between two alternative, underdetermined, theories "is a measure of man's contribution to scientific truth."[29] As far as I can see, this is in the spirit of (E).

Metaphysical realists may object that our theories are true only if they correspond to objective facts which exist independently of our theories. However, this way of thinking is certainly not Quine's way. For Quine, there is no external perspective from which our theories can be said to correspond to objective facts. Our perspective is always internal. Facts are immanent, just as truth is immanent. Quine says, for example:

I can attribute reality and truth only within the terms and standards of the scientific system of the world that I now accept; only immanently.[30]

As I read this, it is not (only) a statement about our abilities. It is not merely epistemological. It is also ontological. There *are* no facts independently of our theories. Quine says:

Truth, for me, is immanent. Factuality, or matterhood of fact, is likewise immanent.[31]

Therefore, there is a sense in which truth is invented rather than discovered. What goes beyond observation is "man's contribution to scientific truth." Truth is built up by us on the basis of observation and linguistic invention. What is built up is "defensible," and thereby true, as long as it conforms to every possible observation. This is the way I read Quine.[32]

6. TIGHTNESS

So far, it seems that Quine's response has not given us any ground to give up the empiricist conception of truth, in the form of something like (E). However, Quine now adds a further consideration.

Much that is accepted as true or plausible even in the hard sciences, I expect, is accepted without thought of its joining forces with other plausible hypotheses ... to form a testable set. Such acceptations may be prompted by symmetries and analogies, or as welcome unifying links in the structure of the theory. ... Having reasonable grounds is one thing, and implying an observation categorical is another.[33]

I myself find this very plausible, and it could perhaps be taken to create a problem for (E). However, if it does, surely it is only the "only if" part of (E) that is threatened. Nothing has yet been said in criticism of the more interesting 'if' part. Besides, it seems to me that (E) may still be compatible with the quoted passage if the tightness requirement in (E) is given a suitable interpretation.

The point of the tightness requirement is to restrict the support, which a theory accumulates by being empirically successful, to those sentences in the theory which really contribute to its success or to its other theoretical virtues. Empirical success is just one of the theoretical virtues. Other virtues are simplicity, explanatory value, coherence, precision, elegance, and comprehensiveness. Exactly how these virtues should be defined and how they are related to each other is a difficult question which may be answered in somewhat different ways by different people. However, the answers are relevant to the process of theory construction, and since they are thereby related to the rest of our theories they need to be included in our total theory of the world. These philosophical or metascientific answers may not contribute anything to the empirical adequacy of our total theory, but they do contribute something to its coherence and comprehensiveness. Therefore, they should not be ruled out by the tightness requirement. The role of the tightness requirement is to rule out the possibility that sentences are being made true simply in virtue of being *gratuitously added* to an otherwise defensible theory.[34]

What is included in our total theory of the world is what we believe about the world (or, rather, the sentences by which we express these beliefs). Our total theory is itself part of the world, so what we believe about our total theory and its parts is part of the theory. If we believe that some parts of the theory constitute good grounds for believing some further sentence to be true, then this latter sentence may be added to our total theory – even if the addition of this sentence does not influence the empirical content of our theory in any way at all. I suppose this is the way many philosophical claims are incorporated into our total theory, but the same mechanism may also be at work within the various sciences themselves. As Quine puts it, "having reasonable grounds is one thing, and implying an observation categorical is another." As long as we have reasonable grounds for our beliefs, they belong in our total theory, even if they do not contribute to its empirical content. The tightness requirement should be interpreted in such a way that this is not ruled out.

7. IDEALIZATION

As we have seen, Quine claims that truth is "immanent." He has explained this notion in somewhat different ways in different places. For example, in his response to my earlier paper he says the immanence of truth means that "true" is a predicate within science.[35] In an earlier paper, he puts the point as follows:

Whatever we affirm, after all, we affirm as a statement within our aggregate theory of nature as we now see it; and to call a statement true is just to reaffirm it. Perhaps it is not true, and perhaps we shall find that out; but in any event there is no extra-theoretic truth, no higher truth than the truth we are claiming or aspiring to as we continue to tinker with our system of the world from within.[36]

It is clear from this passage that truth is not "immanent" in the (intolerable) sense of being identical with "what we believe." This is welcome, for most of us strongly believe that some of the things we believe may be false and that some of the things we do not (yet) believe may be true. We do not conclude from this that truth may change over time. Rather, we believe that the truth may be different from what we affirm at any given time, even though "to call a statement true is just to reaffirm it." In a recent response to Davidson, Quine writes:

There is a remarkable feature of our use of the truth predicate that lends truth a dignity beyond disquota-tion. When a scientific tenet is dislodged by further research, we do not say that it had been true but became false. We say that it was false, unbeknownst, all along. Such is the idiom of realism, integral to the semantics of "true."[37]

However, this is quite compatible with the empiricist conception of truth. Our current theories may not be true according to (E), for our total theory may not entail all true observation categoricals and no others, and it may not be completely embeddable in some hitherto unknown theory which does. As long as our total theory of the world can still be improved upon in its empirical content, it does not yet contain the whole truth and it may even contain parts that are false. And even if its empirical content cannot be improved upon, there may still be theoretical truths that we have not yet incorporated. Consequently, truth may be thought of as an idealization of our current theory of the world. In this sense, as well as in the sense indicated by Quine's remarks, truth is immanent.

The idea that truth is an idealization of our current theory should not be taken to involve the idea that there is one unique ideal version of our theory. As Quine has pointed out,

we have no reason to suppose that man's surface irritations even unto eternity admit of any one system-atization that is scientifically better or simpler than all possible others.[38]

But perhaps Quine also holds that there cannot be any ideal versions of our theory at all. He writes:

[Bergström] asks, rhetorically I suspect, whether I hold that "a sentence is true if and only if it follows logically from our theory or an idealized version of it." No. Such an ideal theory is impossible by Gödel's theorem.[39]

If I understand him correctly, the point is that for every idealized version of our theory, there must be true sentences not logically implied by it. I accept that, and it seems to me that there are also other reasons for being sceptical about the idea of a theory that includes every truth and which can nevertheless be specified without the notion of truth. For example, our language may be constantly growing. Consequently, the best strategy for the empiricist may be to replace (E) by the following one-way conditional:

(E*) A theoretical sentence S is true if S is entailed by some tight theory which entails every true observation categorical and no false observation categorical.

In my opinion, one definite advantage of (E*), as compared to (E), is that (E*) leaves open the possibility that some true sentences can never be known to be true, even in principle. Such sentences are not entailed by *any* tight theory which entails every true observation categorical. For if they were, it would be possible in principle to have good grounds for assenting to them, and hence to know them to be true. In this particular sense, truth is transcendent rather than immanent, but surely this is compatible with Quine's claim that truth is immanent in the sense that "true" is a predicate within science.

It may be asked whether there really are true sentences which cannot be known. Well, this realist thesis is true (i.e., there are such sentences) if it is entailed by some tight theory which entails every true observation categorical and no others. The thesis is part of my own total theory of the world, and I would support it by the usual arguments. I may of course be mistaken, but if the thesis satisfies (E*), I am not mistaken. In any case, (E*) suggests that provability implies truth, but not conversely. I like that. We have all learnt to make a distinction between truth and provability, and I see no a priori reason to assume that every truth is provable "in principle." (Our capacity to understand sentences which are unprovable derives from our understanding of the words that occur in them together with our mastery of their syntactic structure.)

8. LOGICAL IMPLICATION AND TRUTH

It may be objected to the empiricist conception of truth that neither (E) nor (E*) can explain the notion of truth, since both presuppose the notion of entailment, or logical implication, which itself can only be explained in terms of the notion of truth.[40] In other words, it might be held that my explanation is circular.

I think this objection is not fatal. Maybe (E*) does not provide us with a non-circular explanation of truth, but the question I am interested in is whether it provides us with a criterion of truth (for theoretical sentences). I think it does, for it states a sufficient condition for truth, which we can understand whether or not we accept it.

We do not need (E*) in order to understand the word "true", but it seems plausible to assume that we need to understand "true" in order to understand "entail" (or "logically implies") – and, therefore, that we need to understand "true" in order to understand (E*).

I believe that we learn the word "true" already in connection with observation sentences. We learn sentences such as "It's raining" and "That's a dog", and after a while we are also introduced to the idea that some utterances of such sentences are true whereas others are not true (false). For example, our parents say "That's a dog" while pointing to a picture of a cat, just in order to test our mastery of the sentence, and they then comment on their own utterance by saying "No, that was not true. It's a cat." Similarly, when the child says "That's a dog" while pointing to a pig, we say "No, that's a pig" or "No, that's wrong" or "No, that's not true." And so on. In this way we teach the child that "This sentence (or utterance) is true" is applicable on the very occasions where the sentence referred to by "This sentence" is applicable. To understand this is part of understanding the language. And to understand this is to understand a great deal about truth.

What is taught in this way is, I think, the idea of truth as correspondence. The child will eventually distinguish between its utterance and what it perceives (what is pointed to by its parent) on the occasion of utterance, and it will come to appreciate the requirement that the former has to match or correspond to the latter in order for the former to be true. This correspondence idea of truth is then extrapolated by analogy to theoretical sentences, where the appropriate perceptual contents are lacking. Thus, we come to think of truth conditions existing independently of ourselves, even if we cannot perceive them in the way we are used to in the case of observation sentences. In the case of theoretical sentences, truth-conditions are certain weird entities which are somehow similar or analogous to ordinary perceptual contents – except in the crucial respect that they can never be the content of any human perception. (I suggest that when we conceive of gravitational fields and electrons, for example, we conceive of them in such a way that it would be possible, in some science fiction world, for non-human beings with sensory organs different from ours, to perceive them just as we perceive corn-fields and billiard-balls.)

Thus, what we learn to associate with the truth predicate are the very conditions we associate with the sentences to which the truth predicate is applied. This means that truth-predications have no substantial and invariant content apart from the contents of the sentences of which truth is predicated. In particular, truth-predications are neither normative ("I have the right to assert ...") or modal ("It is possible to verify ..."). Of course, in the case of observation sentences, such normative and modal conclusions can be immediately drawn by a person who witnesses an occasion on which an observation sentence is true. But with theoretical sentences, the connection is less close. A person may believe that a theoretical sentence is true, without believing that he has the right to assert it or that it is possible, even in principle, to verify it. There is nothing linguistically illegitimate or strange about this.

When the child has learnt to handle the truth predicate, we may also introduce it to the idea of entailment or logical consequence. We can say that a sentence is a logical consequence of other sentences just in case it has to be true if the other sentences are true. Moreover, and more importantly, we can illustrate this with simple examples involving observation sentences. We can draw the child's attention to the fact that if it's raining and it's cold, then it's raining. And so on.

The point of all this is that we should have no problem of understanding (E*), even if this presupposes that we understand the notion of entailment, which in turn presupposes that we understand of the notion of truth. In short, we can understand the notion of truth long before we understand (E*).

9. UNDERDETERMINATION, INCONSISTENCY, AND TRUTH-FUNCTIONS

One natural objection to the empiricist conception of truth, in the form of (E) or (E*), starts from the Quinean thesis of underdetermination. I discussed this objection in my earlier paper,[41] but maybe it should be mentioned here as well. In the following quotation, Quine appears to use the underdetermination thesis as an argument against an empiricist conception of truth:

we have no reason to suppose that man's surface irritations even unto eternity admit of any one systematization that is scientifically better or simpler than all possible others. It seems likelier ... that countless alternative theories would be tied for first place. Scientific method is the way to truth, but it affords even in principle no unique definition of truth.[42]

But what Quine says here is compatible with an empiricist conception of truth. (E*) does not presuppose the existence of a unique best theory. The requirement is that *some* ideal theory entails the sentence under consideration.

Things would look worse, of course, if some sentence as well as its negation would satisfy (E*). Suppose sentence S is entailed by one ideal theory and $\neg S$ is entailed by another ideal theory. It would seem that S and $\neg S$ cannot both be true. Well, so it may seem, but in Quine's philosophy most sentences have a meaning only relative to a theory. Meaning is immanent. Quine writes:

Unless pretty firmly and directly conditioned to sensory stimulation, a sentence S is meaningless except relative to its own theory; meaningless intertheoretically.[43]

Therefore, S and $\neg S$ in our example are not incompatible. They can both be true, since they belong to different theories. If both belong to theories which entail all true observation categoricals and no others, both are indeed true. (Of course, this means that the syntactic shape "S" is not the same sentence in the two theories.)

There are other possible objections, closely related to the one just mentioned. For example, suppose S is entailed by one ideal theory and T is entailed by another ideal theory. Then both are true by (E*), but it is not clear that the conjunction

S & T is also true by (E*), for S & T may not be entailed by any ideal theory.[44] This seems to violate ordinary logic.

I think there are two ways to respond to this. First, it may be held that truth-functional operators can only operate on sentences which belong to one and the same language, and the sentences S and T do not belong to the same language, since they are entailed by different theories. Second, it may be pointed out that (E*) does not rule out the possibility that some sentence, such as S & T, is true even though it is not entailed by any ideal theory. For (E*) merely states a sufficient condition for truth.

The first of these responses is perhaps a bit awkward. I believe we should opt for the second one. Notice, moreover, that Quine sometimes seems to accept the poss-ibility – pointed out to him by Donald Davidson in 1986 – of two radically different theories expressed in the one and the same language. The idea seems to be that the languages of the two theories are combined into one comprehensive language, to which our truth predicate can apply.

Meaningful application of the truth predicate ... extends to the whole language and is not limited to any particular theory formulation. Empirically equivalent and logically compatible theories can be accepted as true descriptions of the world even if one of them uses terms irreducibly alien to the other. There is no call to fuse them into a single redundant theory. Our language can embrace the full vocabularies of both theories, and our truth predicate can then apply to each on its separate merits.[45]

What we can do, then, is this. We include the ordinary logical laws into our theory of the world, and if S and T but not S & T satisfy (E*), we conclude that S & T is also true, in virtue of the logical laws, even though this conjunction does not belong to any tight theory.

In any case, it seems to me that we need to resist the idea that there is a strict one-one correspondence between theory and language. A theory, for Quine, is something fairly precise – a particular conjunction of axioms – but surely a lan-guage is something more indefinite. It is hard to believe, for example, that our total theory of the world corresponds to a language which is different from every lan-guage corresponding to some of its various subtheories. Similarly, it is hard to believe that the language of physics, say, is replaced by a new language every time a new discovery is made in physics. And the meaning of a term like "democracy", for example, does not change when new theories are established in biology or hydrology. So meaning can hardly be immanent in the strict sense that it changes as soon as our total theory changes.

I would suggest, rather, that a language is somewhat like a family. A family has no clear boundaries, and it can remain "the same" even when some of its members are born and others die. Similarly, a person's language can remain the same when certain components of the person's total theory are dropped and other components are added. When the changes are drastic enough, however, the language itself also changes and its old terms acquire new meanings.

In fact, the capacity of our language to remain the same when our theories change seems essential to the empiricist conception of truth. For (E*) seems to pre-suppose that a sentence S in our current theory has the same meaning (or is the same sentence) as S in some ideal revision or extension of our theory. Meaning immanence must not be interpreted so as to prevent this.

10. TIGHTNESS AND TRUTH

One advantage of the empiricist conception of truth is that it appears to solve a classical problem in the philosophy of science. The problem I have in mind concerns the status of the various "theoretical virtues" that philosophers of science have discussed – factors such as empirical accuracy, simplicity, broad scope, fruitfulness, etc. Should these be thought of as guides to truth or are they independent criteria of value that our theories should satisfy? According to the former position,[46] the virtues are fallible inductive indicators of goodness in theories. They can be formulated as methodological rules, which can be used for finding good theories, and good theories can in turn be identified with theories which are true or which have a high degree of verisimilitude. According to the latter position,[47] the virtues are themselves constitutive of goodness in theories. The aim of science, on this view, is not truth or verisimilitude, but rather virtuous theories – i.e. theories with a high degree of accuracy, simplicity, scope, and so on.

The problem with the first position is that there seems to be very little ground for supposing that the theoretical virtues really are indicators of truth or verisimilitude. For example, it seems quite possible that the truth is far from simple. Similarly, it has been argued that inference to the best explanation cannot be used to show that our best theories are true. The problem with the second position is that it may be wondered why we should care about theoretical virtues at all if they are not indicators of truth or of approximate truth. In this case, theories which have the virtues to the highest degree may nevertheless be completely false. If this is so, the value of simplicity and other virtues seems quite dubious.

However, given the empiricist conception of truth, we can see why simplicity and the other virtues play an important role. For they determine the tightness of theories, and tightness is in turn a "means" to truth in the sense that the main criterion for truth of theoretical sentences is entailment by a certain kind of tight theories. Or so I have argued in this paper.

Stockholm University

NOTES

1   "Quine's Truth."
2   *Ontological Relativity and Other Essays* , p. 75 (italics have been removed here).
3   See *Pursuit of Truth*, p. 2.
4   *Ibid.*, pp. 19–21.
5   The conjunction may have only one conjunct. According to Quine, "[t]here will be no need to decide what a theory is or when to regard two sets of sentences as formulations of the same theory; we can just talk of the theory formulations as such", *Theories and Things*, p. 24. Moreover, a theory formulation "is simply a sentence – typically a conjunctive sentence comprising the so-called axioms of the theory", "On Empirically Equivalent Systems of the World," p. 318.

6   *Pursuit of Truth*, p. 3.
7   *Ibid.*, p. 43.
8   See *ibid.*, pp. 9–11.
9   "Response to Bergström," p. 496.
10 *Ibid.*, pp. 496–7.
11 This particular example has been suggested to me by Burton Dreben.
12 For the notion of collateral information, see e.g. *Word and Object*, p. 37f.
13 An example of this kind was suggested to me by Peter Pagin.
14 For the counterfeit rabbit, see *Word and Object*, p. 42.
15 See e.g. *Word and Object*, p. 42, *Pursuit of Truth*, p. 3, and "In Praise of Observation Sentences," p. 108f.
16 *Pursuit of Truth*, p. 43. This quotation might create the impression that observation sentences for Quine are sentence *tokens*, sentences which are uttered just once, on a particular occasion. But I take it that this is not Quine's intention. For he often says that observation sentences are occasion sentences, and occasion sentences are true on some occasions and false on others (see e.g. *ibid.* p. 3).
17 Folke Tersman has pointed out to me that this might require some revision of Quine's definition of "observation sentence." We wish to classify a sentence like "It's raining" as an observation sentence, but we can hardly claim that all competent speakers of the language would agree on its truth-value on all occasions. See also my paper "Quine on Underdetermination," p. 39.
18 This is what Quine himself said in 1981. See also *Pursuit of Truth*, p. 40. In 1975 Quine said that the "really distinctive trait of observation terms and sentences is to be sought not in concurrence of witnesses but in ways of learning. Observational expressions are expressions that can be learned ostensively"; see "On Empirically Equivalent Systems of the World," p. 316.
19 Notice that an *occasion* in this context, or an "occasion of utterance" in Quine's words, is simply a spatio-temporal location or region – a "time and place" as Quine puts it in his "Response to Gibson," p. 502 – and that it is the location of the *utterance* that is referred to, not the location of the *content* or subject matter of the utterance. Thus, if I say "That's a rabbit" when I see a rabbit 50 meters in front of me, the relevant occasion includes me and my utterance, but not necessarily the rabbit or the place 50 meters in front of me. When other speakers are supposed to agree in assenting to the sentence "on witnessing the occasion of utterance," the idea is not that they should witness the rabbit, or the place 50 meters in front of me. Rather, they should be supposed to be in *my* place. "We judge what counts as witnessing the occasion, as in the translation case, by projecting ourselves into the witness's position"; *Pursuit of Truth*, p. 43. In particular, an "occasion of utterance" in Quine's sense should not be identified with what Donald Davidson calls the "distal stimulus" of an utterance, e.g. in "Meaning, Truth and Evidence," p. 72–3. See also Roger F. Gibson, "Quine and Davidson: Two Naturalized Epistemologists," p. 461.
20 This is, roughly, my proposal in "Quine's Truth."
21 The requirement of tightness goes back to Quine's paper "On Empirically Equivalent Systems of the World," p. 351. See also my paper "Quine on Underdetermination," pp. 41–2.
22 "Response to Bergström," p. 497.
23 *Theories and Things*, p. 22.
24 "Response to Bergström," p. 497.
25 *Ibid.*
26 For references, see my "Quine on Underdetermination."
27 For example, in "Three Indeterminacies." In this paper, he says: "A total coherent system of the world remains the scientific and philosophical ideal, but, again, multiple ways of achieving it can all be accounted true" (p. 15).
28 *Pursuit of Truth*, p. 102.
29 "Reply to Harold N. Lee," p. 316.
30 *Ibid.*
31 "Reply to Robert Nozick," in *The Philosophy of W.V. Quine*, p. 367.
32 It must be admitted, however, that there are also certain passages in Quine's work that seem to point in another direction. In one place, for example, he says that "surface irritation is a basis not for truth but for warranted belief," *Theories and Things*, p. 39.
33 "Response to Bergström," p. 497.
34 See e.g. my paper "Underdetermination and Realism," p. 351.

35 "Response to Bergström," p. 497.
36 "On Empirically Equivalent Systems of the World," p. 327.
37 "Response to Davidson," p. 500.
38 *Word and Object*, p. 23.
39 "Response to Bergström," p. 496.
40 This objection has been brought to my attention by Dag Prawitz.
41 "Quine's Truth," p. 430.
42 *Word and Object*, p. 23.
43 *Word and Object*, p. 24
44 This, or a very similar, objection was brought to my attention by Sten Lindström.
45 "Three Indeterminacies," pp. 14–5. See also *Pursuit of Truth*, pp. 99–100.
46 This position is held e.g. by W.H. Newton-Smith. He writes: "The factors relevant to theory choice in science are not constitutive of a good theory. The goodness of theories is constituted by their degree of verisimilitude. The factors are fallible inductive indicators of that"; see *The Rationality of Science*, p. 225.
47 This position is held e.g. by Carl Hempel. He writes: "The imposition of desiderata may be regarded, at least schematically, as the use of a set of means aimed at the improvement of scientific knowledge. But instead of viewing such improvement as a research goal that must be characterizable independently of the desiderata, we might plausibly conceive the goal of scientific inquiry to *be* the development of theories that ever better satisfy the desiderata. On this construal, the desiderata are different constituents of the goal of science rather than conceptually independent means for its attainment"; see "Turns in the Evolution of the Problem of Induction," p. 404. For further discussion of this distinction, see my paper "Scientific value."

BIBLIOGRAPHY

R.B. Barrett and R.F. Gibson, eds., *Perspectives on Quine* (Cambridge: Blackwell, 1990).
Lars Bergström, "Underdetermination and Realism," *Erkenntnis* 21 (1984): 349–65.
Lars Bergström, "Quine on Underdetermination" in Barrett and Gibson, eds., *Perspectives on Quine*, pp. 38–52.
Lars Bergström, "Quine's Truth," *Inquiry*, 37, 4 (December 1994): 421–35.
Lars Bergström, "Scientific value," *International Studies in the Philosophy of Science*, 10, No. 3 (1996): 189–202.
Donald Davidson, "Meaning, Truth and Evidence," in Barrett and Gibson, eds., *Perspectives on Quine*, pp. 68–79.
Roger F. Gibson, "Quine and Davidson: Two Naturalized Epistemologists," *Inquiry*, 37, No. 4 (December 1994): 449–63.
L.E. Hahn and P.A. Schilpp, eds., *The Philosophy of W.V. Quine* (La Salle: Open Court, 1986).
Carl Hempel, "Turns in the Evolution of the Problem of Induction," *Synthese*, 46 (1981): 389–404.
W.H. Newton-Smith, *The Rationality of Science* (London: Routledge&Kegan Paul, 1981).
W.V. Quine, *Word and Object* (Cambridge: The M.I.T. Press, 1960).
W.V. Quine, *Ontological Relativity and Other Essays* (New York: Columbia University Press, 1969).
W.V. Quine, "On Empirically Equivalent Systems of the World," *Erkenntnis*, 9 (1975): 313–28.
W.V. Quine, *Theories and Things* (Cambridge, Mass.: Belknap Press, 1981).
W.V. Quine, "Reply to Harold N. Lee," in Hahn and Schilpp, eds., *The Philosophy of W.V. Quine*.
W.V. Quine, "Reply to Robert Nozick," in Hahn and Schilpp, eds., *The Philosophy of W.V. Quine*.
W.V. Quine, "Three Indeterminacies," in Barrett and Gibson, eds., *Perspectives on Quine*, pp. 1–16.
W.V. Quine, *Pursuit of Truth*, revised edition (Cambridge: Harvard University Press, 1992).
W.V. Quine, "In Praise of Observation Sentences," *Journal of Philosophy*, XC, No. 3 (March 1993): 107–16.
W.V. Quine, "Response to Bergström," *Inquiry*, 37, 4 (December 1994): 496–8.
W.V. Quine, "Response to Davidson," *Inquiry*, 37, 4 (December 1994): 498–500.
W.V. Quine, "Response to Gibson," *Inquiry*, 37, No. 4 (December 1994): 501–2.

ROGER F. GIBSON

QUINE, WITTGENSTEIN, AND HOLISM*

1. INTRODUCTION

"Holism" has become a "buzz-word" of contemporary philosophy. It figures prominently in current discussions in philosophy of language, philosophy of mind, philosophy of science, and epistemology. However, as is frequently the case with
"buzz-words," its meaning rarely remains fixed from context to context or even
within a single context.

Two prominent philosophers whose writings have contributed significantly to the
recent "'holism' phenomenon" are W.V. Quine and Ludwig Wittgenstein. In particular, the Quine of "Two Dogmas of Empiricism" (1951) and later, and the
Wittgenstein of *On Certainty* (1969) both evince holistic tendencies, but are their
holistic tendencies comparable? One might doubt that they are. After all, in the
sources just cited Quine's holism emerges largely in reaction to Rudolf Carnap's
philosophy, while Wittgenstein's holism emerges largely in reaction to
G.E. Moore's. I will address this question of sameness and difference after first
explaining Quine's holistic tendencies and then Wittgenstein's.

2. QUINE'S HOLISM

"The primary reference for my holism is 'Two Dogmas'." W.V. Quine

In "Two Dogmas of Empiricism" (hereafter TDE), Quine repudiates the ana-
lytic/synthetic distinction (dogma 1) and reductionism (dogma 2). Dogma 1 purports to distinguish those statements that are true by virtue solely of their meanings,
independently of how the world is (the analytic ones), from those statements that
are true by virtue of their meanings together with how the world is (the synthetic
ones). Dogma 2 purports "that each statement, taken in isolation from its fellows,
can admit of confirmation or infirmation ... ."[1]

In TDE Quine's repudiation of both dogmas, but especially his repudiation of the
dogma of reductionism, relies on his advocacy of an *extreme holism*: "My countersuggestion [to reductionism], issuing essentially from Carnap's doctrine of the
physical world in the *Aufbau*," Quine explains, "is that our statements about the
external world face the tribunal of sense experience not individually but only as a
corporate body."[2] What makes this holism *extreme* is Quine's taking "corporate
body" to mean *the whole of science*: "The unit of empirical significance," he writes,
"is the whole of science."[3] However, by the time Quine published *Word and Object*

81

(1960) he had come to see that a *moderate holism* is both more accurate of scientific practice and still sufficient for undercutting reductionism (and the A/S distinction).[4] Let's look more closely at Quine's moderate holism:

It is holism that has rightly been called the Duhem thesis and also, rather generously, the Duhem-Quine thesis. It says that scientific statements are not separately vulnerable to adverse observations, because it is only jointly as theory that they imply their observable consequences. Any one of the statements can be adhered to in the face of adverse observations, by revising others of the statements.[5]

Quine emends, and thus moderates, this formulation of holism, or the Duhem thesis, by adding the following two reservations:

One reservation has to do with the fact that some statements are closely linked to observation, by the process of language learning. These statements are indeed separately susceptible to tests of observation; and at the same time they do not stand free of theory, for they share much of the vocabulary of the more remotely theoretical statements. They are what link theory to observation, affording theory its empirical content. Now the Duhem thesis still holds, in a somewhat literalistic way, even for these observation statements. For the scientist does occasionally revoke even an observation statement, when it conflicts with a well attested body of theory and when he has tried in vain to reproduce the experiment. But the Duhem thesis would be wrong if understood as imposing an equal status on all the statements in a scientific theory and thus denying the strong presumption in favor of the observation statements. It is this bias that makes science empirical.[6]

So, Quine's *first reservation* regarding holism is that a given statement's suscept-ibility to tests of observation is a matter of degree, with observation statements rep-resenting a limiting case. Thus, holophrastically construed, observation statements are indeed separately susceptible to tests of observation because they are learned (or could be learned) by being conditioned to fixed ranges of confirming and infirming patterns of sensory stimulation. In time, however, these same observation statements become linked to theoretical statements (statements which are remote from sensory stimulation) by virtue of their sharing some vocabulary. For example, the holophras-tic observation statement "This+is+water" can become linked to the theoretical statement "Water is H_2O" in a person's web of belief just as soon as that person learns both to parse the holophrastic statement "This+is+water" into the analyzed statement "This is water" and, of course, some chemical theory.[7]

Thus, observation statements enjoy a double life: Holophrastically construed, they are conditioned to patterns of proximal stimuli; analytically construed, they are linked to other statements, including theoretical ones by virtue of a shared vocabulary. The former connection accounts for observation statements' susceptibility to being confirmed or infirmed individually; the latter connection accounts for how considera-tion of systematic efficacy for theory can sometimes override the former connection.

Quine's *second reservation* regarding holism "has to do with breadth. If it is only jointly as a theory that the scientific statements imply their observable conse-quences, how inclusive does that theory have to be? Does it have to be the whole of science, taken as a comprehensive theory of the world,"[8] as Quine maintained in TDE? Quine now thinks that it does not:

Science is neither discontinuous nor monolithic. It is variously jointed, and loose in the joints in varying degrees. In the face of a recalcitrant observation we are free to choose what statements to revise and what ones to hold fast, and these alternatives will disrupt various stretches of scientific theory in various ways, varying in severity. Little is gained by saying that the unit is in principle the whole of science, however defensible this claim my be in a legalistic way.[9]

So, Quine's moderate holism recognizes (1) that, in general, a statement's susceptibility to tests of observation is a matter of degree and that some statements (observation statements) are individually susceptible to such tests, and (2) that it is more accurate of current scientific practice to think of *significant stretches of science*, rather than the whole of science, as having observable consequences.

3. QUINE'S GROUNDS FOR HOLISM

As we have seen, in TDE Quine proffered holism as a "countersuggestion" to the dogma of reductionism. However, apart from the plausible story that Quine tells there about his countersuggestion there is but one meager argument for holism to be found in TDE. That argument is the following reductio: if reductionism were true, then we ought to be able to come up with an explicit theory of confirmation, but as Quine notes, "apart from prefabricated examples of black and white balls in an urn"[10] this endeavor has not met with success. Thus, it is *likely* that reductionism is false (and, therefore, that holism is true).

However, searching beyond the pages of TDE for further sources of support for holism, we find that Quine relies on two further arguments. One is what one might call the *language learning argument*.[11] This argument is extracted from some of Quine's speculations regarding how theoretical (i.e., non-observational) language is learned. The crucial idea is that while a person can learn the observational part of language by extrapolating along lines of observed (subjective) similarities, theoretical language cannot be learned that way. Rather, the learning of theoretical language requires irreducible leaps of analogy on the part of the learner. These analogies forge multifarious and somewhat tenuous links among a person's repertoire of statements. And since a person's language is learned from other people, many of those links in an individual's web of belief come perforce to resemble those of other people, thereby making communication possible. More to the present point, however, if some cluster of a person's statements which includes theoretical ones implies a particular observation statement which subsequent observation shows to be false, then (because of the aforementioned multifarious and tenuous links) there is some latitude as to which statement(s) in the implying cluster to cull so as to block the false implication, i.e., there is holism. This language learning argument for holism also goes some way toward explaining why moderate holism occurs. If human language consisted entirely of observation statements, each one learnable by extrapolating over observed similarities, then each would have its own unique sets of confirmining and infirming patterns of sensory stimulation. Holism, then, would not occur – but then neither would theoretical science; as Quine says,

"... I see no hope of a science comparable in power to our own that would not be subject to holism, at least of my moderate sort. Holism sets in when simple induction develops into the full hypothetico-deductive method."[12]

Another of Quine's arguments for moderate holism is what one might call the *scientific practices argument.* This argument maintains that as a matter of empirical fact scientists involved in testing some hypothesis H must assume the truth of various axillary assumptions A, and that H can always be saved by making drastic enough adjustments to A. Suppose, for example, that the conjunction of H and A entails the observation statement O. Suppose also that upon inspection O turns out to be false. Quine's claim is that H could always be saved from refutation by replacing A with A' such that the conjunction of H and A' would not longer entail (the false) O. (Notice that this claim is much weaker than the dubious claim that H could always be saved by replacing A with A' such that the conjunction of H and A' entails not-O. Quine disavows this stronger claim.)[13] Of course, it is also true that H could be saved, without altering A, were one to refuse to accept the falsification of O. If one's giving up the truth of O portends cataclysmic consequences for one's web of belief, one might choose to hold fast to the truth of O in spite of a seemingly recalcitrant observation. A person might even go to the extreme of pleading hallucination in order to maintain O's truth.[14]

In sum, there is a dialectic of epistemic values at work in Quine's conception of moderate holism; these values include observation, on the one hand, and considerations of conservatism, simplicity, and generality of theory on the other. Moreover, conservatism, simplicity, and generality are themselves competitors in the dialectic. For example, simplicity of theory can give way to complexity, if great gains in generality are to be achieved; but generality can occasionally bow to simplicity, if complexity makes the theory unwieldy. It is important to recognize that for Quine there is no recipe, no algorithm, for adjudicating conflicts within this dialectic of values; he would say that the values are incommensurable. Finally, as we noted previously, observation statements are holophrastically conditioned to the ranges of proximal stimuli, which tend to confirm/infirm them. However, these very same observation statements are connected in a piecemeal fashion to various theoretical statements by sharing vocabulary with them. Thus are observation statements pulled in opposite epistemic directions: toward sensory stimulation, on the one hand, and toward considerations of systematic efficacy for theory, i.e., toward conservatism, simplicity, and generality of theory, on the other.

4. HOW QUINE USES HOLISM

The Duhem thesis, moderate holism, plays a major part in Quine's systematic philosophy. As we have noted, he relies on it in arguing against the two dogmas of empiricism, viz., the A/S distinction and reductionism, but he also relies on it in accounting for mathematical truth, in supporting his thesis of indeterminacy of translation, and in responding to global skepticism.

4.1 The two dogmas

If, in light of the considerations canvassed in the preceding section we conclude that moderate holism is true, then not only is reductionism false, it is also very unlikely that there are analytic statements, statements that are true by virtue solely of their meanings, independently of how the world is . As Quine has argued, any statement can be held true independently of how the world is, if we make drastic enough revisions to others of our statements. According to moderate holism, then, the statements most likely to count as analytic are those that are extremely remote from sensory stimulation, including statements like "There have been black dogs". But surely advocates of analyticity do not want such statements to count as analytic. In advocating moderate holism, has Quine therefore proved that there are no bonafide analytic statements? I think not, but what he has done is to supplant a less adequate theory of the relation between scientific theory and the world (reductionism) with a more adequate theory (moderate holism): "Holism in this moderate sense is an obvious but vital correction of the naive conception of scientific ... [statements] as endowed each with its own separable empirical content."[15]

4.2 Mathematics

According to the Logical Positivists, notably Carnap and A.J. Ayer, mathematical truths lack empirical content and are necessary. These philosophers argue that both of these traits of mathematical truths are explicable in terms of analyticity: mathematical truths are devoid of empirical content because they are analytic, i.e, they make no claims about the world. And they are necessary because they are analytic, i.e., they are true solely in virtue of the meanings of their terms. Thus, by relying on analyticity, and without abandoning their empiricistic scruples, these philosophers can cheerfully admit that some truths are indeed necessary.

But how is an empiricist like Quine, one who shuns analyticity, to respond to these two problems?

I answer both ["Quine writes"] with my moderate holism. Take the first problem: lack of content. Insofar as mathematics gets applied in natural sciences, I see it as sharing empirical content. Sentences of pure arithmetic and differential calculus contribute indispensably to the critical semantic mass of various clusters of scientific hypotheses, and so partake of the empirical content imbibed from the implied observation categoricals.[16]

What of the second problem, the necessity of mathematical truths?

This again is nicely cleared up by moderate holism, without the help of analyticity. For ... when a cluster of sentences with critical semantic mass is refuted by an experiment, the crisis can be resolved by revoking one *or* another sentence of the cluster. We hope to choose in such a way as to optimize future progress. If one of the sentences is purely mathematical, we will not choose to revoke it; such a move would reverberate excessively through the rest of science. We are restrained by a maxim of minimum mutilation. It is simply in this, I hold, that the necessity of mathematics lies: our determination to make revisions elsewhere instead. I make no deeper sense of necessity anywhere. Metaphysical necessity has no place in my naturalistic view of things, and analyticity hasn't much.[17]

So, by relying on moderate holism, and without abandoning his empiricistic scruples, Quine believes that he can account for both the empirical content and apparent necessity of mathematical truth.

4.3 Indeterminacy of translation

One of Quine's more contentious philosophical claims is that two linguists working independently of one another on translating some hitherto unknown language could end up constructing manuals of translation which "might be indistinguishable in terms of any native behavior that they give reason to expect, and yet each manual might prescribe some translations that the other translator would reject. Such in the thesis of indeterminacy of translation."[18] Quine summarizes a central argument supporting his thesis in the following passage: "If we recognize with Peirce that the meaning of a sentence turns purely on what would count as evidence for its truth, and if we recognize with Duhem that theoretical sentences have their evidence not as single sentences but only as larger blocks of theory, then the indeterminacy of translation of theoretical sentences is the natural conclusion."[19] Quine has other arguments for indeterminacy of translation, but this one clearly rests upon his commitment to Duhem's thesis, moderate holism.

4.4 Global skepticism

One of Quine's finest essays, one that is frequently overlooked by his critics and commentators, is "The Scope and Language of Science" (1954). This essay is important because it contains an early statement of Quine's reciprocal containment thesis. This thesis says, in effect, that ontology (the theory of what there is) and epistemology (the theory of method and evidence) contain one another but in different ways. This notion of reciprocal containment plays an important role in Quine's response to global skepticism, so let us examine it is some detail.

Quine is emphatically a naturalist. A naturalist of his kind rejects first philosophy and accepts the view that it is up to science to tell us what exists (ontology) as well as how we know what exists (epistemology). As a naturalist, Quine accepts a physicalist ontology (including sets) and an empiricistic epistemology. He does so because he believes that physicalism and empiricism are themselves empirical hypotheses championed by our best current (if tentative) scientific theories. Also as a naturalist, Quine believes that ontology contains epistemology in the sense that empiricism is to be articulated in physicalistic terms, e.g., in terms of physical forces impinging on nerve endings. On the other hand, he believes that epistemology contains ontology in the sense that physicalism is our own construction and projection from those very same empiricistic resources.

Some of Quine's readers have thought that this talk of ontology being a construction and projection from some meager empiricistic input inexorably leads to global skepticism (or, perhaps, to instrumentalism or to idealism). For example, one might reason as follows:

It is thus our very understanding of the physical world, fragmentary though that understanding be, that enables us to see how limited the evidence is on which that understanding is predicated [i.e., *ontology contains epistemology*]. It is our understanding, such as it is, of what lies beyond our surfaces, that shows our evidence for that understanding to be limited to our surfaces [i.e., *epistemology contains ontology*]. But this reflection arouses certain logical misgivings: for is not our very talk of light rays, molecules, and men then only sound and fury, induced by irritation of our sensory surfaces and signifying nothing? The world view which lent plausibility to this modest account of our knowledge is, according to this very account of our knowledge, a groundless fabrication [i.e., *global skepticism*].[20]

However, to so reason, Quine explains, is to succumb to fallacy,

a peculiarly philosophical fallacy, and one whereof philosophers are increasingly aware. *We cannot significantly question the reality of the external world*, or deny that there is evidence of external objects in the testimony of our senses; for, to do so is simply to dissociate the terms "reality" and "evidence" from the very applications which originally did most to invest those terms with whatever intelligibility they may have for us.[21]

Beyond this sort of paradigm case argument against global skepticism, Quine explains why we should "accept physical reality, whether in the manner of unspoiled men in the street or with one or another degree of scientific sophistication":[22]

We imbibe an archaic natural philosophy with our mother's milk. In the fullness of time, what with catching up on current literature and making some supplementary observations of our own, we become clearer on things. But the process is one of growth and gradual change: we do not break with the past, nor do we attain to standards of evidence and reality different in kind from the vague standards of children and laymen. Science is not a substitute for common sense, but an extension of it. The quest for knowledge is properly an effort simply to broaden and deepen the knowledge which the man in the street already enjoys, in moderation, in relation to the common-place things around him. *To disavow the very core of common sense, to require evidence for that which both the physicist and the man in the street accept as platitudinous, is no laudable perfectionism; it is a pompous confusion*, a failure to observe the nice distinction between the baby and the bath water.[23]

Thus, given Quine's naturalistic stance, the fact that the best scientific theory of method and evidence (empiricism) underdetermines the best scientific theory of what there is (physicalism) is not a reason for repudiating the latter together with common sense.

Before concluding my discussion of Quine's holism, I would like to point out two further extremely important points regarding the passage last quoted. *First*, it endorses both coherentist and foundationalist elements of knowledge. There can be no doubt that both elements are found throughout Quine's writings on the nature of natural knowledge. He sounds like a coherentist when he talks about theoretical statements and considerations of systematic efficacy for theory, but he sounds like a foundationalist when he talks about holophrastic observation statements and evidence. This is just what one would expect from an empiricist advocate of moderate holism. *Second*, in the last quotation Quine accords a special status to common sense: to disavow the very core of common sense, to require evidence for that which both the physicist and the man in the street accept as platitudinous is no laudable perfectionism; it is a pompous confusion. Quine's view of the status of core common sense beliefs is similar to those of Moore and Wittgenstein, but, as we shall see, each of the three gives a different explanation of the grounds of their view.

5. WITTGENSTEIN'S HOLISM

"141. When we first begin to *believe* anything, what we believe is not a single proposition, it is a whole system of propositions. (Light dawns gradually over the whole)." Ludwig Wittgenstein

Wittgenstein died on April 29, 1951 – three months after the publication date of "Two Dogmas of Empiricism" in *The Philosophical Review*, and four months after Quine read the paper at the Eastern Division of the American Philosophical Association in Toronto. For four separate periods during his final eighteen months Wittgenstein concerned himself with certainty and related topics. In fact, the last entry in his notes on these topics was made just two days before he died. In 1969 these notes were published in their entirety in book form under the title *On Certainty*.

The notes that comprise *On Certainty* were largely precipitated by three papers that G.E. Moore published between 1925 and 1941, papers in which Wittgenstein took a keen and lasting interest. The three are: "A Defense of Common Sense" (1925), "Proof of the External World" (1939) and "Certainty" (1941). In *On Certainty* Wittgenstein is concerned with some of the same topics that Moore addressed in these three papers. In particular, Wittgenstein agrees with Moore's view articulated in "A Defense of Common Sense" (hereafter ADCS) that there is a core of common sense beliefs which can neither be justified nor doubted, though Wittgenstein rejects Moore's account of why this is so. Wittgenstein also rejects Moore's view, articulated in "Proof of the External World" (hereafter PEW), that a proof of the external world is needed and can be given.

In ADCS, Moore articulated a great number of beliefs belonging to what he called the Common Sense view of the world, beliefs he claimed to know with certainty to be true, but which could not be justified, beliefs such as:

There exists at present a living human body, which is *my* body. This body was born at a certain time in the past, and has existed continuously ever since, though not without undergoing changes ... . Ever since it was born, it has been either in contact with or not far from the surface of the earth; and, at every moment since it was born, there have also existed many other things, having shape and size in three dimensions ... .[24]

Two further beliefs that Moore said that he knows with certainty to be true are (1) that he has two hands, and (2) that there is an external world. Indeed, in PEW he argues that he can prove (2) by appealing to (1), though he admits that he cannot prove (1) since he cannot prove that he is not dreaming.

As I understand *On Certainty*, Wittgenstein agrees with Moore's view found in ADCS that there is a core of common sense beliefs that are certain (i.e., cannot be doubted), but he denies that Moore *knows* such beliefs. For Wittgenstein Moore's utterance of "I know I have two hands", or "I know there is an external world", or the like, involves a misuse of the idiom "I know". According to Wittgenstein, "I know" is used correctly only when it is possible to muster evidence for or against the relevant claim, and mustering evidence is a public activity. However, it is not possible to muster evidence for or against those core common sense beliefs that both Moore and Wittgenstein regard as certain. Thus, Wittgenstein is driving a logical or grammatical wedge between certainty and knowing. It is correct to say "I am certain there is an external world", but not "I know there is an external world". It is correct to use "certain" in contexts where giving evidence or doubting are inappropriate. Moore's tendency to conflate certainty and knowledge might be due to his assuming that both certainty and knowledge are mental states, accessible to introspection. Wittgenstein, of course, denies that knowledge is a menatal state. Finally, Wittgenstein rejects Moore's assumption that philosophers' utterances such as "This is a hand", or "There is an external world", or the like, express sensible propositions at all. For Wittgenstein, such utterances are *without sense* (senseless), but *not nonsense*.[25]

In sum, then, Wittgenstein rejects the following three of Moore's assumptions: (1) that "I know" is being used correctly by a philosopher who says things like "I know there is an external world"; (2) that knowing is a mental state, accessible to introspection; and (3) that philosophers' utterances like "This is a hand" are not senseless, i.e., they express sensible propositions. It follows that Wittgenstein also rejects the proof of the external world that Moore proffers in PEW, since he regards both the "premisses" and the "conclusion" of Moore's "proof" as senseless.

Wittgenstein does a great deal more in *On Certainty* that criticize Moore. In particular, he provides a positive account of the grounds for the insight that he and Moore share, viz., that there is a core of common sense beliefs that are certain and, therefore, impervious to doubt. In an excellent new book entitled *Moore and Wittgenstein on Certainty*, its author, Avrum Stroll, argues persuasively that Wittgenstein provides not one but two logically distinct accounts of the ground of such certainty. Following Stroll, let's refer to these two accounts as relative foundationalism and absolute foundationalism.[26]

5.1 Relative foundationalism

Here are a few quotations from *On Certainty* which indicate the nature of Wittgenstein's relative foundationalism:

144. The child learns to believe a host of things. I.e. it learns to act according to these beliefs. Bit by bit there forms a system of what is believed, and in that system some things stand unshakeably fast and some are more or less liable to shift. What stands fast does so, not because it is intrinsically obvious or convincing; it is rather held fast by what lies around it.[27]

152. I do not explicitly learn the propositions that stand fast for me. I can *discover* them subsequently like the axis around which a body rotates. This axis is not fixed in the sense that anything holds it fast, but the movement around it determines its immobility.[28]

225. What I hold fast to is not *one* proposition but a nest of propositions.[29]

96. It might be imagined that some propositions, of the form of empirical propositions, were hardened and functioned as channels for such empirical propositions as were not hardened but fluid; and that this relation altered with time, in that fluid propositions hardened, and hardened ones became fluid.[30]

97. The mythology may change back into a state of flux, the river-bed of thoughts may shift. But I distinguish between the movement of the waters on the river-bed and the sharp shift of the bed itself; though there is not a division of the one from the other.[31]

98. But if someone were to say "So logic too is an empirical science" he would be wrong. Yet this is right: the same proposition may get treated at one time as something to test by experience, at another as a rule of testing.[32]

99. And the bank of that river consists partly of hard rock, subject to no alteration or only to an imperceptible one, partly of sand, which now in one place now in another gets washed away, or deposited.[33]

So, as children we learn a system of beliefs, some of which are certain and indubitable, while others are more or less susceptible to doubt. Those beliefs that stand fast do so by virtue of those that shift: "The game of doubting itself presupposes certainty."[34] However, we do not explicitly learn the propositions that stand fast, but we can subsequently discover that we have acquired them. Also, what is held fast is not one proposition but a nest of propositions. Such propositions have the *form* of empirical propositions, but they are not empirical propositions for they are not functioning as empirically testable propositions but, rather, as rules of such testing. The point of the river-bed analogy is that there is a difference between empirical propositions and propositions merely of the form of empirical propositions and, further, that in different situations one type of proposition may take on the role of the other. Hence the aptness of Stroll's referring to this position as *relative* foundationalism. However, one might quibble with the aptness of calling Wittgenstein's position *foundationalism*, insofar as that term is often opposed to holism, for there certainly are holisitc tendencies in the passages just quoted.

There are similarities and difference between Wittgenstein's relative foundationalism and Quine's moderate holism. One similarity is that virtually every proposition is up for revision but not equally so. Revising certain propositions might be avoided because revising them would be too disruptive for the system. On the other hand, though Wittgenstein admits that there is no sharp division between the propositions making up the river-bed and those comprising the waters, he tends to think of them as different in kind and not merely different in degree. For example, he sometimes refers to the river-bed type propositions as "rules of testing", "grammatical rules", "world pictures", "scaffolding for our thoughts", and so on. Moreover, Wittgenstein maintains that such propositions are outside of the language games which they make possible. Even so, I am reluctant to saddle Wittgenstein with anything as severe as an A/S distinction. I would be more inclined to say that he embraces something like the internal-question/external-question dichotomy. However one characterizes Wittgenstein's view, it nevertheless seems at odds with

Quine's insofar as Wittgestein thinks there is a difference in kind betwen those (senseless) propositions that stand fast and regular empirical propositions.

5.2 Absolute foundationalism

Stroll claims that as Wittgenstein's thought progressed in *On Certainty* Wittgenstein gradually came to favor absolute (non-propositional) foundationalism over relative (propositional) foundationalism; Stroll explains:

> We have seen that one metaphor Wittgenstein uses for certainty is "standing fast". I believe this concept is ambiguous as he employs it, that it denotes two different notions. On the one hand, it is hinge propositions that are said to stand fast; on the other, each in a set of non-propositional features is said to stand fast.[35]

The hinge propositions that Stroll here refers to derive from Wittgenstein's claim that "the *questions* that we raise and our *doubts* depend on the fact that some propositions are exempt from doubt, are as it were like hinges on which those turn."[36] Stroll continues with his explanation of Wittgenstein's propositional and non-propositions accounts of what is said to stand fast:

> We shall begin with the propositional account. It is marked by three characteristics: (i) that foundational propositions form a system and (ii) that some hinge propositions do not stand absolutely but only relatively fast, and (iii) that some hinge propositions – "that the earth exists," for example stand absolutely fast. The emphasis he gives to the propositional theory stresses it relativistic character; the absolutist version is more hinted at than explicitly stated. In holding this propositional account, Wittgenstein thus differs from Descartes, who thinks of the *cogito* as the sole foundational item and from Moore, whose common sense propositions do not form a system, and from both Descartes and Moore, who think all foundational propositions hold absolutely. In his later view Wittgenstein's foundationalism abandons principles (i) and (ii) of the propositional account. Since the new view is non-propositional, it cannot be a system of propositions, and the foundations it describes are absolutist in character.[37]

According to Stroll, the new view, absolute foundationalism, is developed by Wittgenstein along three lines: "(1) that certainty is something primitive, instinctual, or animal, (2) that it is acting, and (3) that it derives from rote training in communal practices."[38]

I am convinced by Stroll and by my own reading of *On Certainty* that the view that Stroll calls absolute foundationalism is present in *On Certainty*. However, I am unconvinced by the main thrust of the final chapter of Stroll's book where he pits Wittgenstein's absolute foundationalism against Quine's holism (or, better, Quine's fallibilism). "The central issue is," Stroll writes, "whether there is something that stands fast in the sense that it is neither eliminable nor revisable."[39] His view is, of course, that nothing stands fast for Quine, while something does stand fast for Wittgenstein – and Stroll sides with Wittgenstein, for certainty and against global skepticism.

I believe that Stroll might have come closer to the mark if he had not based his construal of Quine's position entirely on a few passages from TDE. One must remember that in TDE Quine overstates his holism, and when he proffers the claim that any statement can be held true come what may, his target is the doctrine of analyticity. I believe that a more balanced construal of Quine's position can be achieved by recalling two points: (1) Quine's formulation of the holism thesis refers explicitly to scientific theories, not to common sense, and (2) in "The Scope and Language of Science", Quine said that to "disavow the very core of common sense, to require evidence for that which both the physicist and the man in the street accept as platitudinous, is no laudable perfectionism; it is a pompous confusion."[40] The sentiment is surely one that both Wittgenstein and Moore would have found congenial.

Finally, while I do think that Quine would find Wittgenstein's account of relative foundationalism uncongenial (because it turns on something like an A/S distinction), I do think that he would find absolute foundationalism congenial. After all, there is nothing non-naturalistic about that position, nothing about the community and its practices that is not susceptible to scientific study.

In the final analysis, I believe that it is fair to modestly claim that Quine's and Wittgenstein's holistic tendencies are not as dissimilar as either their respective historical precursors (Carnap and Moore) or their different philosophical methods might at first suggest – proving once again that great (original) minds (sometimes) think alike!

Washington University, St. Louis

NOTES

*   This essay was first published in *Wittgenstein and Quine*, edited by Robert L. Arrington and Hans-Johann Glock (London: Routledge, 1996), pp. 80–96, and is reprinted here with the kind permission of the editors and the publisher.
1   W.V. Quine, "Two Dogmas of Empiricism." In Quine *From Logical Point of View* (Cambridge: Harvard University Press, 2d ed. rev. 1980), p. 41.
2   *Ibid.*, my emphasis.
3   *Ibid.*
4   See W.V. Quine, *Word and Object* (Cambridge: M.I.T. Press, 1960), p. 13n.
5   W.V. Quine, "On Empirically Equivalent Systems of the World," *Erkenntnis* 9 (1975), p. 313.
6   *Ibid.*, p. 314.
7   See W.V. Quine, "In Praise of Observation Sentences," *The Journal of Philosophy*, Vol. XC, No. 3 (1993), pp. 107–116.
8   Quine, "On Empirically Equivalent Systems of the World," p. 314.
9   *Ibid.*, pp. 314–315.
10   Quine, "Two Dogmas of Empiricism," p. 41.
11   See Roger F. Gibson, Jr., *Enlightened Empiricism* (Tampa: University of South Florida Press, 1988), pp. 33–42.
12   W.V. Quine, "Reply to Robert Nozick." In *The Philosophy of W.V. Quine*, L.E. Hahn and P.A. Schilpp, eds. (La Salle, Ill.: Open Court Press, 1986), p. 364.
13   See W.V. Quine, Pursuit of Truth (Cambridge: Harvard University Press, 1992, rev. ed.), p. 16.
14   See W.V. Quine, *Methods of Logic* (Cambridge: Harvard University Press, 1982, 4th ed.), p. 2. Also, see "Two Dogmas of Empiricism," p. 43.
15   Quine, *Pursuit of Truth*, 16.
16   W.V. Quine, "Two Dogmas in Retrospect," *Canadian Journal of Philosophy*, Vol. 21, No. 3 (September, 1991), p. 269. Observation categoricals are standing sentences composed

of two holophrastic observation sentences of the form "Whenever this, that": "Whenever it's raining, it's wet."

[17] *Ibid.*, pp., 269–270.
[18] Quine, *Pursuit of Truth*, pp. 47–48.
[19] W.V. Quine, "Epistemology Naturalized." In Quine *Ontological Relativity and Other Essays* (New York: Columbia University Press, 1969), pp. 80–81.
[20] W.V. Quine, "The Scope and Langauge of Science." In Quine *The Ways of Paradox and Other Essays* (Cambridge: Harvard University Press, 1976 rev. and enlarged ed.), p. 229.
[21] *Ibid.*, my emphasis.
[22] *Ibid.*, pp. 230.
[23] *Ibid.*, pp. 229–230, my emphasis.
[24] G.E. Moore, "A Defense of Common Sense." In Moore *Philosophical Papers* (London: George Allen and Unwin Ltd., 1959), p. 33.
[25] See Avrum Stroll, *Moore and Wittgenstein on Certainty* (Oxford: University Press, 1994), p. 114.
[26] *Ibid.*, pp. 138ff.
[27] Ludwig Wittgenstein, *On Certainty*, G.E.M. Anscombe and G.H. von Wright, eds. (New York: Harper and Row, Publishers, 1969), p. 21e.
[28] *Ibid.*, p. 22e.
[29] *Ibid.*, p. 30e.
[30] *Ibid.*, p. 15e.
[31] *Ibid.*
[32] *Ibid.*
[33] *Ibid.*
[34] *Ibid.*, p. 18e.
[35] Stroll, *Moore and Wittgenstein on Certainty*, pp. 155–156.
[36] Wittgestein, *On Certainty*, p. 44e
[37] Stroll, *Moore and Wittgenstein on Certainty*, p. 156.
[38] *Ibid.*, p. 157.
[39] *Ibid.*, p. 166.
[40] Quine, "The Scope and Language of Science," 230.

NENAD MISCEVIC[1]

QUINING THE APRIORI

Empiricism abhors mysteries, and insists on the natural, causal origin of human knowledge, its experiential justification and its fallible and defeasible character. The classical notion of *a priori* knowledge, and the view that our intuitions provide such knowledge, assumes that some knowledge is, for its justification, independent from experience. Since this view sins against the demands listed, Quine was very much right in having rejected it. In this paper I want to defend a version of empiricism about justification that I take to capture the gist of Quine's message. Some defense might be needed in view of the recent revival of interest in apriorism accompanied by criticism of the empiricist program (Bealer, 1992; Bonjour, 1992). Moreover, some prominent empiricists like E. Sober (in his paper (1993)) have come to question the fundamental tenets of Quine's view. In these paper I wish to defend the tenability of empiricism against some of the charges, focusing upon Sober's criticism, but mentioning others along the way. Sober discusses the universal truths of mathematics. There are four kinds of reasons why one should demand empirical test (warrant) for logical and mathematical beliefs:

1) The intended empirical scope of relevant beliefs invites empirical test (if arithmetic is applicable to apples and to cabbages, why not see if it yields the right prediction). Quine himself has always been a friend of universal methods and strategies of research, exemplified prominently by mathematics and logic. He praises their ubiquitousness, their "versatile yield" and points out their indispensability and their contribution to the success of science. The pointing out has been developed into a doctrine – most notably by Putnam – and has earned a special name, Indispensability thesis: mathematics is justified by being indispensable to science.

2) The sophisticated mathematics and logic rest partly on simple mathematical beliefs (like "two plus three equals five") as well as on simple logical rules (some mathematicians and philosophers – from Descartes on – have also assumed that humans have intuitions concerning the very axioms of mathematics and logic). The fallibility – even straightforward dubiousness – of particular intuitions invites empirical testing. Also, the lack of a straightforward naturalistic explanation of their origin demands a tougher control of their reliability.

3) The practice of science involves testing and reinterpreting – if not literally revising – some parts of mathematics and logic. Finally, if one wants to be pragmatic, there is also some good news:

4) The empirical justification is available: the massive empirical success of everyday knowledge and of science in which such beliefs are essentially used does

Alex Orenstein and Petr Kotatko (eds.), Knowledge, Language and Logic, 95–107.
© *2000 Kluwer Academic Publishers. Printed in Great Britain.*

vindicate elementary mathematical intuitions. This is just the Indispensability thesis used to boost empiricism.

This demand for ultimate empirical warrant – to supplement the intuitive obviousness – falls into line with Quine's epistemology naturalized. The appeal to literal truth about the physical world accounts for precisely those phenomena that have plagued empiricist epistemology, e.g. the high reliability of elementary pretheoretical arithmetic. It explains why people have stayed by their mathematical and logical intuitions long before those have become part of the developed science, and shows the reasonableness of their attitude. It is perfectly in accord with Quine's view that logical and mathematical truths capture the actual make-up of the world: if the world is such as logic depicts it to be, than the fact could have been learned through causal interaction with the world, and could have been recorded in our evolutionary epistemic heritage. Indeed, the development of cognitive science has taught us about the importance of innate cognitive structures, possibly even innate knowledge. Maybe our intuitions stem from our cognitive make-up, due to evolutionary learning, which is purely causal and utterly non-mysterious, so its products do not merit mistrust. They are however "prior" to individual experience, and result in compelling intuitive beliefs. In the case of some intuitions, most prominently mathematical ones, we have reason to believe that they are rather reliable. Still, this descriptive news does not threaten the essence of Quinean empiricism. For one thing, on the descriptive level itself, the evolutionary learning is empirical, a series of trial-and-error interactions with environment. Further, on the normative side, any *prima facie* justification intuitions might have should be supplemented with an ulterior, Quinean coherentist justification from the whole of science.

Some critics might object at this point that the talk of warrant or justification – be it empirical or not – is not available to a Quinean. Indeed, Quine is often taken as prohibiting any normative epistemological inquiry, or, at best, restricting it to extremely modest, rather insignificant scope.[2] Some of his remarks on epistemology naturalized support such reading, especially on a traditional reading of the term "normative". However, Quine does offer a clear alternative, which is in keeping with central themes in Western epistemology. Cognition is the pursuit of prediction and truth, and the epistemic terms are technological terms: the cognizer values cognitive strategies as means for finding out the truth. Although his statements are sometimes ambiguous between a narrow pragmatist reading on which the technology of knowledge is geared merely to prediction and to empirical adequacy, and a veritistic reading, taking truth as the ultimate value, one should take into account Quine's recent pronouncement – in his dialogue with Lars Bergstrom, including the ones at this very conference – and disambiguate it veritistically, taking the view that epistemology is the technology of truth-acquisition. I propose that its advice are hypothetical imperatives: if you want to reach truth and avoid falsity use such and such strategy of inquiry! Its norms of appraisal are technological norms: consider how close does such-and-such strategy bring the inquirer to truth in relevant circumstances (where this last, environmental, parameter is on purpose left indeterminate).

Here is the plan of the paper: In the main part of the paper I wish to develop the point (4) above: that empirical success does vindicate mathematics. I first sketch the general Quinean view. Shorn of exaggerations it looks like commonsense and I interpret it as such. Then, in the rest of the paper, I defend the Quinean view against the most important contemporary attack on Indispensability thesis, the one by Sober (1993).

Let me first summarize and briefly illustrate the empiricist line and its Quinean variant. Empiricism in general claims that the ultimate justificatory evidence for a person's beliefs are the person's experiences or observations. As far as intuitions are concerned, they have only a provisional value: intuitions taken in isolation from empirical evidence are justified only in a transitory manner – the cognizer may trust them in the absence of disproof and in a heuristic spirit. They may receive their ultimate justification only from the empirical epistemic success of the whole in which they figure.

The characteristically Quinean brand of empiricism pictures mathematical and logical beliefs as intertwined with the rest of human knowledge, of "theory" as Quine would say. The degree of mutual dependence varies; Quine is careful to note that science "is variously jointed, and loose in the joints in varying degrees" (Quine, 1981; 67). However, one cannot test the empirical part in isolation from the mathematical and logical one, nor can one isolate the latter part and test it alone. Of course, some procedures seem to be more sensitive to errors in logic, others, to errors in observation, but no strand can be conclusively isolated and tested in pure form. This basic and simple idea is also rather general. Every successful activity can be analyzed into components conspiring to produce success: the quality of a good piano concert depends on many components and bears testimony to the quality of all of them. Similarly with cognitive activities. When an inquirer success-fully applies a theory – a whole including a more observational, and a more logico-mathematical part – several things happen.

First, the success of the whole justifies all the epistemic components. The idea is complete commonsense. Take Sherlock Holmes and his "deductions" – in fact, strategies comprising logic, some calculation and reasoning to the best explanation, applied on observational and commonsense experiential material. Both Watson and the reader take the truth of Holmes's predictions to be a testimony in favor of the validity of Holmes's methods, including his logical acumen and arithmetical skills if needed. What is valid for Holmes, should be valid for science too. The success of a theory confirms it as a whole; all its components participate in the final triumph.

Second, within the whole one can apportion praise and blame by varying the contribution of single components. In the case of Holmes's success, the reader praises him primarily for his reasoning capacities; precisely because the observation he starts from, and the experiential material he works on is banal, its validity is not in question; what was tested are Holmes's logical capacities. One can put the point in terms of the right "contrast class": when testing a component C of a theory consisting of C and the rest R, keep R fixed, find the best alternative to C, say C* and test both R-cum-C and R-cum-C*. If the former fares better than the later, then C is confirmed. Applied to science, and to non-elementary portions of mathematics the initial idea is that the scientific community sometimes concentrates on the con-tribution of mathematical statements to the success of science, and tries to judge it on this account. A famous historical episode is the early acceptance of calculus on the grounds of its contribution to physics. The general idea is that every success of science adds a little to the confirmation of its formal, mathematical and logical part. Moreover, in order to gauge the amount of confirmation, we should compare the predictive powers of science using the standard mathematical apparatus to the predictive powers of any alternative methods within our ken. This is in fact sometimes done, as a part of scientific methodology: for instance, researchers are taught to prefer statistical methods to the intuitive rules of thumb because these methods yield better predictions.

The issue is now whether such empiricism can hold water. Can mathematics be really ultimately confirmed by its success? Does Indispensability thesis hold? Although the vindication by success is very much commonsense, and is assumed for all kinds of practices, including intellectual ones, philosophers have famously argued that it is not available precisely for mathematics and logic.

THE INDISPENSABILITY THESIS

I) Sober's argument

E. Sober has in his (1993), offered an extended argument against the Quinean view that mathematics can be confirmed by its empirical success. This is the most sophisticated attempt at refuting the Indispensability thesis to be found in the literature, so it may render the Quineans a service: if the attempt can be foiled, the rather obvious right to use success as vindication may be freely restored to mathematics. This is what I will strive to achieve in the rest of the paper.

Sober argues that it is precisely its ubiquity – Quine's versatile yield – that makes mathematics *not confirmable* by empirical means.[3] In doing this, he gives an original twist to a line of thought that that been extremely popular both with logical positivists (Hempel, 1983) and within Wittgensteinian tradition, up to our days (see Wright, 1980, and surprisingly, Field, 1996). His main argument purports to show that mathematical propositions are not and cannot be confirmed by the empirical tests for theories that contain them. A Quinean can agree – and I do agree – that mathematical propositions are often not tested by empirical tests, but disagree that they are never so tested, and *a fortiori*, that they cannot be so tested. I shall call Sober's conclusion that they cannot be tested the Anti-Quinean Conclusion. To repeat, my disagreement concerns this more radical conclusion.

Sober's basic idea is simple and elegant: since mathematics is a part of both successful and unsuccessful science it is not differentially confirmed by success of science. Sober calls his brand of empiricism contrastive empiricism, since it stresses the importance of the contrast class in testing theories. Theories are always confirmed only relatively to other theories; a theory, or a hypothesis fares better or worse than some other theories (hypotheses), and there is no absolute, isolated success of a theory. The Likelihood Principle he proposes – following Edwards – enjoins the epistemologist to rank confirmation received by hypotheses by comparing likelihoods of observation: A hypothesis A is confirmed relative to some other hypothesis B by observation O if O has greater likelihood on A than on B and O actually takes place. Now, if only one given hypothesis G predicts O with great likelihood, whereas all other give O a very small likelihood, then G is – in the given context – indispensable for predicting and explaining O. (Sober, 1993: 36). This is the link between indispensability and the principles of contrastive empiricism. One more Sober's idea will be useful in the sequel. We sometimes accept theories that apparently do not compete with alternatives. Sober suggests that in such cases "we *can* find competitors if only we set our mind to it" (*ibid.* p. 59), i.e. "there are alternatives constructable" from the material at hand. I find the whole idea of contrastive empiricism very attractive, and I shall not question it here except for some minor points. We now come to the main issue: does the indispensability of mathematics for the empirical science bestow empirical confirmation upon it?

Sober reasonably starts from the usual situation in the developed contemporary science: several physical hypotheses $(H_1, H_2,...H_n)$ sharing the same set of mathematical assumptions M is being tested, and one, say H_1 wins. Does the differential success of H_1 over its rivals confirm M? Suppose an inquirer – call her Challenger – raises the question of justification challenging the validity of M. Can one appeal to the success of H_1 in order to persuade the Challenger that M is true. Sober answers in the negative. He points out that M is part of all n hypotheses, and is therefore neither confirmed nor disconfirmed by the success of one. Sober devotes a lot of attention to the development of the argument, since it is his main weapon against the indispensability thesis. Here is Sober's statement of its central premise:

"If the mathematical statements M are part of *every* competing hypothesis, then, no matter which hypothesis comes out best in the light of the observations, M will be part of that best hypothesis. M is not tested by the exercise, but is simply a background assumption common to all hypotheses under test" (Sober, 1993: 45).

The premise is a conditional. Let me call it *The Contrastive Conditional*: *If* M is a part of every competing hypothesis, *then* M is not confirmed by the success of any hypothesis.

To put it graphically, the Challenger is free to reject the success of H_1 as confirming M in the situation described above, on the ground that the sequence of hypotheses that all contain M is not the right contrast class for testing M. It seems then, by Sober's own advice that the Challenger should propose a relevant contrast class to be compared with hypotheses based on standard mathematics. She should compare the best hypothesis containing M, i.e. H_1, to one of the two kinds of alternatives. Since she wants to know whether M is better than some other alternative, say M^*, or the empty alternative E, she is free to propose testing M it against M^*.

Obviously, Sober has to deny the very possibility of such test. In order to reach his Anti-Quinean conclusion he needs the antecedent of the Conditional, to which we shall give a name:

No Alternative: M is part of every hypothesis ever tested or testable, since there is no alternative M^* to it.

The *Contrastive Conditional* and the *No Alternative* together lead to the:

Anti-Quinean Conclusion: M cannot be tested by empirical success of hypothesis that contain it.

The *No Alternative* forbids any appeal to some alternative mathematics M* on the part of the Challenger. The issue is whether *No Alternative* holds. If it does the argument establishes the Anti-Quinean Conclusion, that *M* cannot be confirmed or justified by the empirical success, of a theory (in our example, of H_1). (Let me note that Michael Resnik (in print) has tried to defend the Quinean line against Sober. His approach differs from ours, in that he takes holism as the principal issue, so that he does not even address the *No Alternative* thesis. I fail to see how his defense of holism helps against Sober's attack on Indispensability thesis.)

Consider the possibilities open to us and to the Challenger. Since she wants to test M she is free to compare the best hypothesis containing *M*, i.e. H_1, to one of the two competitors. Either she avails herself of the next best hypothesis H*, similar with regard to empirical data but containing assumptions from alternative mathematics *M** instead of *M*, or she can rest satisfied with some rather primitive hypothesis *HE* containing no mathematical part at all (i.e. "containing" the empty alternative *E*).

Consider the last option first. The hypothesis *HE* yields no quantitative prediction at all, only a claim to the effect that "certainly such-and-such will happen". (This is something one would reasonably expect from a theory that contains no mathematics.). Such-and-such does not happen; instead the outcome predicted by H_1 occurs. Then H_1 is definitely to be preferred to *HE*, since it gives precise quantitative prediction having come out true. By Sober's own criterion, *HE* loses to any rival that comes even close to predicting precisely an actual observation. The idea might sound far-fetched to those philosophers of science that are accustomed to work with quantitative science only. However, it is close to what has actually happened at the birth of Galilean science: the scholastic physics, devoid of mathematical sophistication, was offering no quantitative predictions, whereas the nascent Galilean physics was offering at least some: this gave it the needed edge, and the verdict of the intellectual community turned out exactly as predicted by our sketch.

Consider now the other possibility. Suppose that the Challenger is offering some alternative mathematical assumptions, *M** embedded within H*, and we are supposed to test them. The reasonable way is to compare H* not with some other hypothesis also containing *M** but with our H_1, that differs from H* in the right place, namely by its mathematical part. If observation then favors our hypothesis over the Challengers, then *M* is vindicated as against *M**. The account fits nicely into Sober's "contrastive empiricism", and should therefore be congenial to him. Remember his suggestion that in problematic cases "we *can* find competitors if only we set our mind to it" and his license to construct an alternative if no one is readily available.

However, Sober is bound to disagree since he needs *No Alternative* to reach his Anti-Quinean Conclusion. Here is his reaction to the possibility of considering alternatives:

"Formulating the indispensability argument in the format specified by the Likelihood Principle shows how unrealistic that argument is. For example, do we really have alternative hypotheses to the hypotheses of arithmetic? If we could make sense of such alternatives, could they be said to confer probabilities on observations that differ from the probabilities entailed by the propositions of arithmetic themselves? I suggest that both these questions deserve negative answer." (*Ibid.*: 45,46).

Notice first that the reasoning attack the very idea that indispensability confirms mathematics, for any set M of mathematical statements. It puts the issue as an all-or-nothing affair – either indispensability confirms mathematics or it doesn't, period, and it squarely denies that indispensability confirms mathematics at all. (It is not geared to the modest conclusion that there are other ways of confirmation besides the appeal to indispensability.)

I shall here discuss extensively only Sober's first and crucial claim with which he supports his *No Alternative*: we really don't have alternative hypotheses to the hypothesis of arithmetic. The claim is meant to establish the absence of contrast class and thereby the impossibility to test mathematics. However, if he means business, he should extend the claim at least to analysis, the crucial part of mathematics relevant to science. The more restricted actual claim is:

(A) Elementary arithmetic has no imaginable alternatives

Sober seems to have suddenly narrowed his focus upon arithmetic only, whereas in the title and in the paper as a whole it is mathematics that is talked about. However, (A) is not sufficient, since it leaves open the possibility that e.g. alternatives to analysis are being imagined and tested. But what is to be shown is that mathematics in general does not get confirmed by application. So, what Sober needs is a stronger claim:

(M) No part of mathematics used in science has imaginable alternatives

Now, Sober has no direct argument for (A) (nor for that matter for (M), which he does not explicitly formulate at all). He argues indirectly from the assumption that people anyway don't take the idea of empirical testing of arithmetic seriously. (The argument is in line with the older and famous one given by Hempel (1983)). Here is then Sober's *one and only argument for his No Alternative premise*:

"In the real world, we frequently encounter quantities that fail to combine additively. Pour two gallons of salt into two gallons of water; you will not obtain a volume of four gallons. Place two chickens together with two foxes; this will not produce four organisms, but just two foxes and a pile of feathers. If we interpret *unadditive* cases in this way, we can hardly claim that observed examples of *additivity* offer genuine confirmation of our arithmetical beliefs" (Sober, 1993: 50).

Sober places this examples in the context of answering an objection to the effect that success does confirm elementary mathematical prediction. In spite of this modest location, the example is central: it is the only example of our actual practice with arithmetic that Sober offers at all, and it is supposed to show that cognizers normally *never* consider disconfirmation of arithmetic.

Let me now summarize (my reconstruction of) Sober's anti-Quinean strategy. His main argument has the form of a *modus ponens*: the *Contrastive Conditional* states that if mathematics M is part of every competing hypothesis it is not confirmed by the success of any, the *No Alternative* claims that M in fact is part of every competing hypothesis. The *No Alternative* is indirectly argued for by appeal to imaginability of alternatives to arithmetic – thesis (A) – and by examples allegedly illustrating people's unwillingness to consider the possibility of an arithmetical judgement being falsified. For his general conclusion Sober obviously needs one more claim: the claim (M) that the accepted mathematics has no imaginable alternative.

II) Discussion

It is now time to assess Sober's argument against Quine. Defending Quine I shall argue that it is multiply flawed, so that its conclusion is not warranted.

Let me first briefly note my uneasiness with the Conditional. I find it rather counterintuitive. Why does indispensability of mathematics disqualify it, instead of qualifying it for confirmation? If some component of a successful activity is indispensable to that activity – so that the activity cannot even start without it – why would such indispensability make it epistemically less praiseworthy than mere usefulness? Why are necessary conditions of success to be deprived of their share on account of their necessity? Why not rather say that M is confirmed – not by its role within the particular hypothesis – but by its very indispensability? Consider the following example. The presence of the oxygen in the sports hall is indispensable for the match, but does not favor John over Jim. It does not enter into explanation of John's beating Jim, but does enter the explanation of John's being able to compete at all, and should get it share of credit on that account.

However, I shall not insist upon this objection. I want to convince those who accept the Conditional and think that no test can confirm M since all alternative hypotheses listed contain M. Therefore, I concentrate upon the *No Alternative*, and argue that it does not hold. The only reason to accept the *No Alternative* would be the truth of (A) and (M), the claims that respectively, arithmetics and mathematics in general have no imaginable alternatives. However, that would be a bad reason. Even if (A) and (M) were true and their use were unproblematic one could still use the contrast class containing hypotheses of non-quantitative "science" of the pre-Galilean kind. Some contrast class is available even if the use of (A) and (M) were unproblematic. However there is worse: (i) the use of (A) and (M) in proving the Antecedent is not legitimate, and (ii) they are not true. This is what I intend to show.

Let me start with (i) the legitimacy of the appeal to (A) and (M). The appeal is not legitimate for very simple reason:

The appeal to both (A) and (M) relies upon (and tacitly presupposes) the following principle: if the best competitor of a theory T is a non-starter, then T is not confirmed by the evidence that apparently supports the very T itself. (Applied to the case at hand: if alternative arithmetic is a non-starter, then standard arithmetic is not confirmed by its own success). The principle is extremely implausible for any competitive activity: if my best opponent has no chance against me, then I am the winner, not the looser of the game. Why should epistemic competition be an exception and why is the poor status of alternative arithmetic deleterious for the confirmation of the standard one? I don't think a good answer could be given. The standard arithmetic is more empirically successful than its best competitor, and this bestows empirical confirmation upon it (in addition to any other sort of justification it might have in virtue of its obviousness).

Besides this general reason why (A) and (W) are useless, there are several particular reasons why one should resist using them.

Consider first (A). It is not true. The counterarithmetical outcome of say, addition, is easily imaginable and conceivable; only the details of how it comes about (the mechanism) are not. Although this kind of unimaginability and the feeling that counterarithmetic is absurd are important in themselves, that is irrelevant for the issue of testing, since the one can test a theory one believes to be absurd (or a theory one is certain cannot come out true). Remember Sober's general optimism that "we *can* find competitors if only we set our mind to it". Why not in case of arithmetic?

Further, testing has little do with our capacity intellectually to justify the theory tested in yet another, more concrete fashion. Consider simple and banal arithmetical mistakes, sloppy calculations or careless use of mathematics that lead to errors in predictions, and then get discarded and carefully avoided in future. It doesn't matter for the present purpose how the mistakes get detected. For purely epistemological purpose these mistaken pieces of mathematical reasoning form a foil, a legitimate alternative that produces unsuccessful hypotheses, which get disconfirmed. Take hypothesis H endowed with the correct mathematics M, and its actual rival H*, burdened with error-ridden M*. If H wins the contest, M is empirically shown to be better than M*, exactly the way Sober's own contrastive empiricism requires.

Let us next consider Sober's only argument for (A) namely, that for arithmetics people never even think of alternatives, but simply discount any apparently counter-arithmetical experience. Remember, it concerns the quantities that fail to combine additively – two gallons of salt into poured into two gallons of water; two chickens placed together with two foxes. Ironically, Quine has a simple, and to my mind perfectly satisfying answer to this kind of examples:

The "plus" of addition is likewise interpreted from start to finish but has nothing to do with piling things together. Five plus twelve is how many apples there are in two separate piles of five and twelve, without their being piled together (Quine, 1981: 149).

What the example alleges is simply not true, namely, that people just interpret away all cases of non-additive combination of quantities. The rules of addition do not concern physical combinations, but counting together, and foxes and chickens in that respect behave exactly like apples or cabbages do. The distinction between the so-called additive cases (e.g. piling apples together) in which the resulting whole retains the number of elements of two initial piles counted together, from non-additive ones (foxes and chickens) in which bringing piles together triggers physical processes that change the number of elements from the two initial piles is a physical distinction. It concerns the ulterior processes that happen *after* the item have been physically combined, and has nothing to do with the validity of arithmetics.

Sober might try a last-ditch defense: But who would doubt arithmetic, he might rhetorically ask, suggesting that its very indubitability makes it unsuitable for empirical confirmation. Answer: Remember, this is a philosophical discussion. Of course, it is inappropriate in many ordinary situations to inquire after empirical credentials of elementary arithmetical beliefs; they are so obvious. However, inappropriateness entails does not entail the lack of truth, as Grice has taught analytic philosophers long ago.

In order to assess the truth – as opposed appropriateness – we need a suitable context. So let me make the situation more vivid and the very question more appropriate by using an imaginary example. The Emperor of a distant kingdom is attained by acalculia, the defect that selectively impairs the capacity to perform particular mathematical operations. He has lost his ability to add, and his memory of having learned to do it has faded away. He is, however, still able to count loud. Instigated by intriguers, he summons the court mathematician and accuses him of being a fraud, who earns his money by senseless drooling. The mathematician has to convince him that his mathematics is an honest occupation, that brings palpable result. So, he lets the Emperor perform experiments, adding apples, pebbles, diamonds. For example, having piled together two piles of diamonds, a dozen each, the Emperor predicts out of blue that there will be fifty altogether, the mathematician bets on twenty-four. And lo and behold, most predictions of the mathematician come out true. For those that don't, a reasonable explanation is quickly found. The Emperor is satisfied, the mathematician gets a rich reward is back in his business. Of course, the Emperor is convinced that the mathematician's statements – e.g. the one about the number of diamonds – have been confirmed by his own final counting.

How would Sober judge the situation? Well, the final counting has not additionally boosted our confidence in adding – it was as high as it can be already – but it did make a huge difference for the Emperor. Is the Emperor wrong or irrational in placing his trust in empirical confirmation? If not, then claiming that elementary arithmetical beliefs are empirical confirmed is only socially inappropriate – since most of interlocutors are not in Emperor's predicament – but is perfectly true. The moral of the story is that elementary arithmetical intuitions are so obvious that they do not need mentioning any additional confirmation by success; nevertheless, should the need arise, the confirmation is still there, available to be appealed to.

Now, the Emperor is a fictional character, but there are real people, moreover our professional colleagues, who are more than willing to suspend their belief in elementary arithmetic. Sociologists of science, social relativists like Bloor and Barnes – in their more radical moments – take seriously the idea of alternative arithmetic. The single biggest problem their idea encounters is that no imagined alternative works in science. Now, is the anti-relativist who points out to them that this is actually a big problem using an absurd strategy of argumentation? Suppose the anti-relativist challenges the relativist to produce a piece of alternative arithmetic, and the relativist does produce a sketch of such a thing. Would not one proper response be simply to test the sketch on some applied problem, and see which alternative works better? Remember that the social relativists claim that our inability to persuade ourselves that the alternative arithmetical theories are viable is simply a result of social inculcation, so they will not accept an appeal to intuition in arithmetics. In fact the only way to show that such opponent is wrong is precisely to point out that alternative proposals have no systematic application. To return to the framework of the debate, a philosopher Challenger wishing seriously to impress upon her audience the need for justification of mathematics does not typically question the fact that we cannot imagine alternatives to fundamental arithmetical truths, but does question the reliability of our imagination, asking us to step back from our certainties. Her question concerns the ground or warrant of these certainties, so she does not take it for granted that it is absolutely not in our power to consider alternatives. To quote Wittgenstein "we are not so naif as to make the self-evidence count in place of experiment." (1967:III, par. 3) Since the Challenger does not trust her intuition, she is prone to presenting mathematics as a possibly empty play with signs. The only argument she must listen to by her own assumptions is the argument from indispensability. That argument at least shows that mathematics is *prima facie* to be taken as true if not as necessarily true. This gives us than the wedge we need even to start the polemics. It would indeed be crippling to deprive oneself of that wedge. Sober's Anti-Quinean conclusion risks to land the philosopher of mathematics precisely into this kind of predicament.

Let me finally briefly discuss (M), the claim that alternatives to accepted mathematical theories in general are unimaginable. Here are a few reasons to doubt it.

First, the fate of the notorious historical claims that there are no imaginable alternatives to Euclidean geometry, so one should not take is as a physically confirmable theory, should teach us that commonsense is a poor guide to what is confirmable and what is not. This point alone should persuade us to reject (M), since (M) would make every applicable part of mathematics immune to doubt, and the example shows that such dogmatism is unreasonable.

Second, the situation akin to that which Sober depicts as impossible for arithmetic has been historically real for analysis. In Berkeley's times the best argument for the calculus was that it works in physics, better than any alternative could. Remember how messy the foundations of calculus were in the eighteen century, and imagine a proponent of calculus arguing with its learned opponent, say Berkeley himself. Both agree that the conceptual foundations of calculus is shaky. Is not the strongest weapon of the proponent an appeal to actual contribution of calculus to mechanics: if it works better than any rival method based on mere approximations, it should be more valid in an objective sense than its rivals. By Sober's lights such a dialogue would be absurd, and our proponent utterly unreasonable. Let me also briefly mention M. Wilson's story of tinkering with math to preserve physics; how mathematicians and physicists have been tinkering with their hunches (intuitions) concerning algebra and complex analysis, trying to bring them in line with the demands of physics and technology. He presents this story as part of a larger picture in which scientists are pictured as adjusting their formal, intuition to the substantial, physical and technical needs. Let me quote his brief summary: "Consider this history. 1790: Lagrange hopes to make the calculus rigorous, so he articulates formal rules of syntactic manipulation. 1830: Cauchy, realizing that blind adherence to Lagrange's rules leads to inferential disaster in celestial mechanics, tries to sift away the bad by analyzing the semantic bases of differential equations.Notice that it is mechanics that prompts correction of calculus. The story continues, with Heaviside's discoveries prompted by needs of telegraphy: "1890: Heaviside, however, uncovers profitable inferential principles ungrounded in Cauchy's semantics. 1950: Laurent Schwartz discovers a reinterpretation of differential equations in terms of 'distribution' which supports the Heaviside calculus and seems possibly more appropriate to physical reality. And on it goes." (1994: 543).

Wilson stresses the interplay of syntactic intuitions and semantical needs: depending upon circumstances, we may cite our semantic theory to correct our syntactic practice and vice versa.

Only a symbiosis of this kind, he claims, suits the pattern of oscillation characteristic of many attempts to install "rigor" within a science. In the cases mentioned it was the success of physics that has directly supported certain mathematical intuitions and has dictated the rejection of others. This is what is to be expected given the enormous empirical success of logic and mathematics, both elementary and advanced.

Let me conclude. Sober's attack on Quine's Indispensability thesis is the most sophisticated and up-to-date criticism of it available. I hope to have shown that it fails, and that the thesis can and should be upheld. Of course, if Quine's thesis holds, as I have argued, the empiricists should use it as pointing to the ultimate warrant of mathematical beliefs. This does not deprive mathematical – and other – intuitions of their *prima facie* credibility; on the contrary, the empirical success enhances and grounds their credibility in the face of the skeptical challenge. There is no reason to tie such justification to an untenable strong holism, nor to undercut it by depriving naturalistic epistemology of its normative force. Quine is not committed to any such kind of holism, and his remarks on pursuit of truth offer a perspective of a truth-centered account of epistemic norms that presents them as hypothetical norms guiding us to discovery of truths. The resulting neo-Quinean picture seems to be coherent; I hope I have rendered it sufficiently attractive to deserve attention.

CONCLUSION

Empiricism is coherent, and the Quinean variant of it can be made plausible. The charge of incoherence does not touch its core, and works only against rather exaggerated and caricatured versions of it. It is therefore important for the empiricist not to give to her opponent any opportunity for such misidentification. First, one should make clear that empiricism is a normative stance, but in the moderate sense of technology of truth seeking. Second, one should accord some *prima facie* justification to intuitions, but insist that ultimate justification demands empirical testing. Third, one should update the descriptive ground of empiricist norm, by taking into account the results of cognitive and biological sciences and their potential to accommodate innate knowledge in a basically empiricist fashion, through an account of evolutionary learning.

University of Maribor

NOTES

[1] I wish to thank Professors Willard Quine, Elliot Sober, Lars Bergstrom, Suzan Carey, and my colleague Boran Bercic for discussions on the issues involved and Elizabeth White for stylistic help.
[2] The list includes Bealer (1992), Stroud (1984), Kim (1993). For a balanced discussion see Gibson (1987), (1995).
[3] Sober seems to pursue two related but distinct goals. The modest goal is to show that mathematics is not justified *purely* by the empirical success of theories that apply it. Part of this modest goal is also to show that it is not the case that "successful prediction provides a *general grounding* for the mathematics used in empirical science" (Sober, 1993: 51). I agree with this modest claim. Unfortunately, Sober argues for the modest claim *a fortiori*, by arguing for a much stronger claim, namely that application as such does not confirm mathematics.

REFERENCES

Bealer, George (1992), "The Incoherence of Empiricism", *Proceedings of the Aristotelian Society*, suppl. vol. LXVI.
Bonjour , L., (1992) "A Rationalist Manifesto" (*Canadian Journal of Philsophy*, Suppl. vol. 19.
Craig, E.J., (1975), "The problem of necessary truth", in Blackburn, S. (ed.), *Meaning, Reference and Necessity*, Cambridge University Press, Cambridge, England.
Field, H. (1996), "The Aprioricity of Logic", Proceedings of Aristotelian Society.
Fodor, J. and LePore, E. (1992), *Holism*, Blackwell, Oxford.
Gibson, R.F. (1987), "Quine on naturalism and epistemology", *Erkenntnis*, v. 27. 57–78.
Gibson R.F. (1995), "Quine on Naturalizing Epistemology" in Leonardi, P. and Santambrogio, M. (eds), *On Quine*, Cambridge University Press.
Hempel, C.G. (1983), "On the nature of mathematical truth" in Benacerraf, P. and Putnam, H. (eds), *Philosophy of Mathematics*, Cambridge University Press, Second Edition.
Hookway, Christopher, (1994), "Naturalized Epistemology and Epistemic Evaluation", *Inquiry*, Vol. 37., No. 4.
Kim, J. (1993) "What is 'naturalized epistemology'?", in *Supervenience and Mind*, Cambridge University Press.
Quine, W.V. (1986), *Philosophy of Logic*, Prentice Hall, Enelwood Cliffs, N.J.
Quine, W.V. (1981), *Theories and things*, Harvard University Press, Cambridge, Mass.
Quine, W.V. (1990), *Pursuit of truth*, Harvard University Press, Cambridge, Mass.
Resnik, M.D. "Holistic Mathematics" (in print, paper presented at a conference at the University of Munich, 1993.)
Skorupsky, J., (1989), *John Stuart Mill*, Routledge, London.
Sober, E. (1993) "Mathematics and Indispensability", *Philosophical Review*, vol. 102. No. 1.
Stroud, B. (1984) *The Significance of Philosophical Scepticism*, Oxford Univeristy Press, Oxford.
Wagner, J. (1993), "Truth, Physicalism and the Ultimate Theory" , in Howard Robinson /ed/:*Objections to Physicalism*, Clarendon Press, Oxford.
Wittgenstein, L. (1967), *Remarks on the Philosophy of Mathematics*, The MIT Press, Cambridge, Mass.
Wright, C. (1980), *Wittgenstein on the Foundations of Mathematics*, Duckworth, London.

OLAV GJELSVIK

THE EPISTEMOLOGY OF DECISION-MAKING "NATURALISED"*

1. INTRODUCTION

This paper does not discuss Quine's important contributions to philosophy. I shall explore a topic Quine has not discussed much, namely decision-making. I hope Quine will appreciate my naturalism; he has done more than anyone to put naturalism at centre stage.

Decision-making is of course of great importance if one were to approach radical interpretation along the lines of Davidson. My focus in this paper, however, is deciding to act contrary to what one thinks is the best way of acting. I hold that such phenomena constitute an important clue for the general philosophical picture of human decision-making. I shall suggest that they motivate an explanatory naturalism, sketch a paradigm naturalist mechanism, and try to bring out parts of a general philosophical picture of the decision-making agent.

The paper has three basic parts, and a natural dialectics. The first part presents the philosophical problem of acting against one's better judgement (acrasia), and argues that acrasia remains a puzzle for traditional philosophy. In the second part I describe a naturalist approach to the explanation of behaviours which support our positive beliefs in acrasia. To naturalise the explanatory issue has a significantly liberating effect upon our further thoughts about acrasia, or so I claim in the third part. It makes us able to see the problem in a wholly new light; where what is experienced as acrasia is located at the interface between the naturalised explanatory story and the system of concepts within which we conceive of ourselves as rational agents. The source of the puzzle of acrasia is our own inability to see, from the inside, the proper limits of the system of concepts within which we see ourselves as performing full blown intentional actions.

I am a naturalist, as Quine is, but I resist a thoroughgoing philosophical naturalism: I resist replacing this system of concepts for intentional agency with the concepts used in the basic explanatory decision-theoretic story. The explanatory task, the task of explaining each individual physical action, or patterns thereof, is not all we need the intentional concepts for. If it were, we could replace them with the concepts used in the explanatory story. We need the intentional concepts in order to account for our own reactive attitudes directed towards ourselves and others. These reactive attitudes are fundamental for moral thought. Since we need both systems, we approach a dual or double aspect picture of the human decision making agent. The two aspects might be seen as reflecting a distinction between the descriptive

Alex Orenstein and Petr Kotatko (eds.), Knowledge, Language and Logic, 109–129.
© *2000 Kluwer Academic Publishers. Printed in Great Britain.*

and the normative, but both system of concepts are seen as having descriptive substance. This dual aspect view is in my judgement the lesson of acrasia. The implications may be quite significant, both for moral epistemology and for epistemology in general.

Traditionally, philosophical accounts of human decision-making give a central place to weakness of the will. Aristotle remarks that an action is a conclusion of a piece of practical reasoning. Given this view, (which may not be Aristotle's considered view), it seems impossible to act freely and deliberately contrary to what we conclude we should do or what we hold best. What seems to be impossible, however, also seems to be real: It is what we do when we act weak-willedly. That something judged impossible is real constitutes a puzzle. The perceived necessity to account for the possibility of acrasia and the real processes in acratic action has since Aristotle structured many an account of practical reasoning and intentional action.

It seems to me that acrasia exists. But what is it I believe in? The examples which support a positive belief in acrasia are all cases of motivational conflict in humans. Acrasia is, furthermore, seen as an exceptional event against the background of the otherwise normally functioning rational agent. The task of accounting for the possibility of acrasia is part of the task of giving the right account of the conceptual apparatus we use to describe this normal decision-making agent and his or her behaviour. I shall soon explain the difficulties, which are rooted in our conception of free, intentional action and the role of reason in the production of intention. First I give an example of acratic behaviour.

Occasionally I watch something really silly and boring on television even if I know that I have much better reasons for going for a run. At times I dislike my own telly-watching strongly while I continue to watch. A friend asks me whether running would not be better, and I say, absolutely sincerely: "Yes, of course, I could not agree more". Still I just sit there, perhaps consoling myself by thinking that I shall indeed run tomorrow. This is weak will on my part, or so it seems. Examples with cigarettes, wines, foods, sweets, and chocolate flourish in the writings of philosophers who believe there is weak will. A weak-willed person in real life is typically someone who gives in to temptations while knowing or believing he is better off not giving in.[2] What is weak about this person is not his or her reasoning ability, but the ability to make decisions and execute them in concordance with what he or she concludes is best to do.

Method in the study of acrasia

a) The roles of candidate conceptual truths

My discussion here will focus on method. The philosophical problem of acrasia has two sides. It is a traditional problem about conceptual possibilities, a conceptual problem. It is also a substantial explanatory problem. Let us start with some candidate conceptual truths about intentional agency and practical reasoning. I acknowledge a general scepticism about conceptual truths, in the spirit of Quine, and I do not believe in a sharp distinction between factual and conceptual truths. Let us for the time being just think of them as very general truths, supported by counterfactuals, truths of a similar standing as Aristotle's purported view that an action is a conclusion of some practical reasoning.[3]

Such truths have two basic functions on my view:

First function: The candidate conceptual truths contribute towards fixing the extensions of the concepts we use to give an account of what intentional action is. This fixing of extensions is constrained by the fact that the puzzle of acrasia is situated in this web: These candidate conceptual truths about intentional agency seem to make acrasia impossible. That is why we have to deal with acrasia if we are to provide an account of intentional action: we cannot take the concept of intentional action for granted.

Aristotle clearly saw the danger of contradictions in the applications of the important concepts if he were to allow for acratic action. When fixing the extensions of these concepts one must on the one hand make contradictions impossible, on the other hand one must maximise truths of ordinary intuitions about intentional action and acrasia.

Second function: These conceptual truths provide the basic conceptual resources we use to describe and explain intentional action, including the acratic ones (if they exist).

Of course these two functions are not fully separate or distinct, but they should be kept apart when we approach the question of method in our approach to acrasia. We need to separate the two functions as we need to separate the two questions, the question about the conceptual possibility of acrasia and the substantial question about the description and explanation of acratic actions.

There are two pairs of plausible candidate conceptual truths about action and intention. The first pair concerns the backward connection, backwards from intentional action to what we judge best. The second pair concerns the forward connection from practical reasoning to intentional action.[4] Here is the pair describing the *backward connection*:

(1) When we act freely and intentionally, what we do is what we want most to do among the things we can do and believe we can.
(2) What we want most to do is what we judge best to do.

If we adopt these two principles, we put constraints on *how we might have reasoned given how we act*. The principles contribute towards an out-put (behaviour) based account of what we mean by judging something best by connecting judging best with what we mean by wanting something most and connecting that with what we do when we act freely. The connection runs backwards from what we do to what we judge best.

Here are the candidate conceptual truths describing the *forward connection*, from reasoning to intentional action:

(3) Practical reasoning is aimed at establishing what course of action is best, and the reasoning about this takes place in the setting of settling what to do.

(4) Free intentional action involves the conclusion of a piece of practical reasoning in the sense that the intention in acting is or corresponds to the practical conclusion.

(3) gives the point of practical reasoning, and (4) connects conclusion in practical reasoning and intentional action. Together (3) and (4) put constraints on *what we can do intentionally given how we reason*. Accepting them can be a crucial step towards giving a practical-reason based account of intention and intentional action.

It seems right to take these four principles linking the forward and backward connections as a package: The backwards and forward connection are like the beginnings of a tunnel from each side of a mountain, they must connect in the middle for the thing to be a tunnel in working order. I shall give an argument for the need to see them as a package.

If one were to start one's theorising about acrasia with principles 1) and 2) only, then one needs a conceptual grip on what intentional action is to get started at all, and to have that one needs the forward connection. One might debate how much one specifically needs about the forward connection. One needs, however, to see a clear connection between the agent's reasons and the action, and 3) and 4) provide that together with a principled view on what practical reasoning is all about.

On the other hand, if one were to start with the forward connection (practical reasoning, 3) and 4)), one is not free of conceptual truths about the backward connection: In general, what one does must have implications for what one judges best if the most plausible accounts of what it is to have propositional attitudes are right. Our appetitive attitudes, as all attitudes, must manifest themselves in behaviour one way or other. We might, of course, debate how direct or overt this manifestation must be. What we should not debate is that there must be manifestation one way or other. 1) and 2) serve as a very reasonable manifestation-requirement: what we do on a particular occasion manifests what motivate us most and which available option we hold best on that occasion.

Taking the four principles as a package, we can attempt to see them as jointly fixing the extensions of the involved concepts: Something is a free intentional action when a) what we do is what we are most motivated to do among the things we can do and believe we can do, b) what we do corresponds to a practical conclusion, and c) this practical conclusion corresponds to a judgement about what is best.

I now turn to acrasia. It is easy to see that if we take these four principles as jointly fixing the extensions of these concepts, then there cannot be acratic actions. The problem of acrasia can be seen as a reminder that simply taking these four principles jointly cannot be the end of story about how the extensions of these concepts are fixed. The problem about the possibility of acrasia can now be seen as the problem of how to fix the extensions of these concepts in such a way that all intuitions, the acratic ones included, can be taken care of. Note that there is another methodological error: It is prima facie wrong to take the extensions of some or of all of these concepts as fixed independently of or prior to the problem of acrasia. There is also the further substantial issue of how to understand and possibly explain what happens when we act acratically. That issue is not yet addressed: From the point of view of traditional philosophical method we need to carry out the meaning fixing part in order to raise the substantial issues.

b) Sources of methodological errors about the order of questions

Let us continue the methodological considerations. We can easily introduce an acratic puzzle with each pair of these principles. (Acrasia is an entrance problem at either side of the tunnel.) Remember my example of just sitting there without going for a run, holding that going for a run is best. If we believe in (1), we can reason from the fact that I watch television when I am free to run and believe that I can run, to the conclusion that watching television is what I want most. If we believe in (2), we can infer that I judge watching television to be best. If this is true, I do not act against my own better judgement when I watch television. Perhaps I am plain wrong about my own judgement when I report that I hold running to be best.

If we do not introduce the pair 1) and 2), but instead start with the pair 3) and 4), the problem is this: If I conclude that running is best in my practical reasoning, how can I watch television freely and deliberately? The intention when watching is, or must correspond to, the conclusion of my reasoning, but the intention seems in my case to be contrary to the conclusion I have reached in my reasoning.

Accepting (3) and (4) will therefore in itself seem to suffice to prohibit weakness of the will, and so will acceptance of (1) and (2). Furthermore, acceptance of the pair 3) and 4) corresponds to one way of disputing that there are acratic actions: This way consists in disputing that these pieces of behaviour are cases which conflict with our principles: One may in fact claim that they are not full blown intentional actions, they are compulsions or things like that, pieces of behaviour which fall short of full agency. Acceptance of the pair 1) and 2) is matched by the other main strategy for disputing the reality of acratic action: To dispute that I really hold best what I claim to hold best. What I do shows what I really hold best, this strategy claims.

Therefore, 1) and 2) and also 3) and 4) as pairs seem to rule out acrasia, and each pair is paired with a known strategy for disputing the reality of the acratic cases. This explains why it is easy and natural to conceive of the problem of acrasia with reference to just one of the pairs. If both pairs of conceptual truths have equal weight, however, then principles about both the forward and the backward connections ought to make up the joint starting point for theorising about intentional action and acrasia. This may easily be overlooked.

One explanation of why this is easily overlooked is the simple fact that acrasia is a deep puzzle relative to just one of the pairs. You don't need both pairs to have a real philosophical challenge. Imagine that you start from a favoured pair of principles, a favoured connection, the forward or the backward, and work very hard to give an account of acrasia and finally succeed in finding something plausible. Then it is not only human but very natural indeed to use the attraction and presumed truth of this account together with the plausible belief that there are acratic actions as evidence for the falsity of the principles describing the connection (forward or backward) you did not start out with.

I claim that this is methodologically wrong. One needs principles both about the forward and the backward connection. One must judge the plausibility of each principle prior to any attachment to specific ideas about what acrasia is. Both pairs of principles have equal plausibility prior to such attachments.

I would not have been at pains to make this point about method unless most writers on acrasia were not prone to making exactly this methodological error. Typically they start with one pair of principles when they explain why acrasia is problematic, and then try to show acrasia compatible with this pair of principles. Donald Davidson starts with the first pair, the backward connection, and Michael Bratman starts with the second pair, the forward connection.[5] Implicitly they must judge their favoured pair of principles or their favoured connection as privileged in comparison with the other pair or the other connection. But for a judgement of that sort we do need philosophical reasons as long when our aim is a unified account of the relationship between the crucial concepts in the philosophy of action.

I sum up and repeat the basic point: Taken jointly these 4 principles bring together the crucial concepts in action theory, and relate them to each other. The concepts of evaluation of what is best, conclusion in practical reason, motivation, and full blown (intentional) action can be seen as tied together in a package of mutual constraints by these principles. We can account for one concept by employing the others, start in one end and work our way to the other, etc. We need to have this whole package clearly in view when approaching acrasia.

c) *The methodological lesson tried out in practice*

It seems clear that if we are to account for intentional action against our own better judgement, we cannot accept all 4 principles as they stand. Look at the second pair, principles 3) and 4).[6] The task is to explain how there can be intentional action contrary to one's best judgement. For that to be the case, given 3) and 4), intention has to be distinct from judgement about what is best, while it is still seen as a conclusion of a piece of practical reasoning. It is therefore mandatory to distinguish between evaluative reasoning and practical reasoning. Success in doing that seems to promise an account of acrasia given 3) and 4).

The point I want to make here is simple. Even if the account were to be judged acceptable from the point of view of this second pair of principles, it is easy to see that if 1) and 2) are true, then it cannot work. It cannot be sufficient for an account of acrasia to force a wedge between practical conclusion and judgement about what is best if you accept 1) and 2). If what you do is what you want most, and what you want most what you judge best, then there is under no circumstances room for intentional action contrary to the best judgement. A defence of an account of how weakness of the will is possible has at best just begun by showing acrasia compatible with the truth of principles 3) and 4). If one needs to show it compatible with the acceptance of both pairs of principles, the task is very hard indeed. It seems necessary to deny principle 2), or to introduce two separate senses of judging something best. Both moves are methodologically unsound if they are made in order to save an account of acrasia which has not given principle 2) its due weight at the outset.

If you start with the pair 1) and 2), as Davidson does, then the task is to separate the best judgement contrary to which you act from the best judgement which has to match what you do intentionally. One way of doing that, is to identify the judgement which is contrary to the best judgement with which you act with a prima facie judgement, and think of the judgement which necessarily is in harmony with the action as an all-out judgement.[7] Imagine that we have found acceptable moves to allow for acrasia given 1) and 2). Can we accept this way of accounting for weakness of the will if we accept principles 3) and 4)? It seems clear that we cannot. As long as I judge running best relative to all considerations, how can the unconditional judgement that telly-watching is best be a conclusion of my practical reasoning without there being errors in this reasoning? We can imagine a causal transition to the state where I judge telly-watching best, but this causal transition is hardly a reasoned transition from the prima facie conclusion. A reasoned transition (which does not have to go by deductive reasoning) without error in reasoning can hardly be imagined. An account like this generated from the starting point of 1) and 2), therefore, seems incompatible with seeing the intention as the conclusion of your practical reasoning. Or rather: it is not compatible unless further premises are brought in, but at this stage that is premature and possibly ad hoc.

It seems therefore is quite clear that all 4 principles jointly rule out weakness of the will. The 4 principles seem to have more or less equal plausibility; the forward and the backward connections have the same standing. I have argued in some detail that modern accounts of acrasia which start with one or the other of these pairs, with the forward or the backward connection as the way into the problem of acrasia, clearly fail if the backward and the forward connection are equally important on a conceptual level. If I am right about the equal plausibility of both connections, then most contemporary theories of acrasia are arrived at by unsound philosophical methods.

PART 2: NATURALISING DECISION-MAKING

A. Ainslie's theory

We seem to be unable to account for the possibility of acrasia, and unable to provide explanations of a type of behaviour we believe exists: Acratic behaviour. We must step outside the circle of concepts inside of which we have posed the problem of acrasia. Recall the basic circularity we face when explicating the contents of the crucial concepts describing intentional action, and in relating them to each other. This is not unlike the circularity we face when accounting for meaning and translation, but there is one difference: There is an additional disturbing puzzlement about not allowing for acratic action. Traditional philosophical approaches here face a real problem: they seem forced to deny the possibility of phenomena which are real.

When we step outside the circle of concepts essential for describing intentional action, we must put the conceptual side of things aside, and look at available scientific explanations of the phenomena we think of as acratic actions, temptations in particular. We must look at the assumptions about rationality in these explanations, and at the available concrete ways of explaining the types of behaviour which feed our intuitions about acrasia. Against that background we can take a new look at the conceptual issues surrounding free intentional action. The tables are turned on the traditional philosopher whose approach took us nowhere. We start with the substantial, explanatory story and then approach the conceptual situation: Starting the other way did not take us to the explanatory story. I conceive of this move as naturalism.

Many typical examples of weak-willed behaviour in the philosophical literature involve nicotine, alcohol and other drugs. Some theories of addiction extend to gambling, sexual behaviour, and overeating. George Ainslie's work on motivation and behaviour is extremely interesting for a naturalist theory of motivation and the explanation of irrational and addictive behaviour. His most fundamental contribution is his theory of hyperbolic discounting of reward value, and his use of this mechanism when developing a general theory of motivation. I shall use Ainslie's theory as a paradigm example a philosophical naturalist can learn from and exploit.

Naturalists of the sort I envisage start with observable facts about animal behaviour, and from there onwards carefully and experimentally work their way to the human case. Ainslie does precisely this. The observable fact which is his starting point is temporary preference. This is universally observed when looked for in animals, and plausible evolutionary explanations have been given.

The next point is that a number of experiments have shown a preference for a smaller earlier reward when the delay is short, and a preference for a larger but later reward when the delay is long. There is no direct experimental evidence for seeing the discount-function as hyperbolic. Its exact shape does not really matter, however, the point is that it cannot be exponential. There is theoretical evidence for seeing it as hyperbolic.[8] In cases of repeated choice animals tend to learn to control their own impulsiveness, for instance pigeons have a limited success in this. The basic point generalises to humans. Temporary preference is clearly seen in small children who do not want to go to bed, and the point is brought home in psychological experiments with adults.

Our state of nature, according to Ainslie, is not to have consistent preferences through time. The latter is the exception rather than the rule. Still, at the end of the day, a human adult seems to control her or his impulsiveness. The former exception is now the rule. A naturalist would want an account of how this comes about when the basic picture is that dynamic inconsistencies make up the state of nature. To achieve consistency, or to achieve it in the right type of way, we can see as achieving willpower, i.e. ability to resist temptations. This may not be all there is to willpower, but it is a part. I shall now introduce a new example with the same structure as that of running and television. This is an example discussed by Michael

Bratman.[9] I have a choice between drinking wine in the afternoon and play the piano well at night. I am a piano player who is very fond of wine, and I face this repeated choice. I may take the 30 coming days into consideration when making my present choice of whether I shall accept today's offer of wine. I ignore nice issues about backward induction; the basic case can also be stated without giving it a structure which invites the special issue of backward induction.

I am offered wine at 6 PM. At that time I choose between

a)   drinking the wine
b)   not drinking the wine, and, as a result, play well later

At 6 PM I prefer a), even if I used to prefer b) until 5.30. But at 6 PM we can also compare the following sequences:

c)   My drinking the wine today and on the next 30 occasions
d)   Not drinking the wine today and not drinking on the next 30 occasions

In this case, my preference is clearly d), since after discounting d) has a much larger utility than c). If we were able to simply choose sequences of 30 actions instead of one after the other, it seems as if we would be much better off than when we choose each day at 6 PM. To be able to choose to live by a rule of not drinking wine until after I have played, would solve the present problem. A choice of rule to live by is equivalent to choosing a very long, perhaps infinite, sequence of actions.

Ainslie's working assumption is that the action which is carried out at any point in time is the action which is supported by maximising considerations at that point in time. It is obvious that when we bring in the next 30 occasions, and discount the value of those occasions to present value, then relative to maximising considerations the best strategy *at 6 PM is to drink today and abstain on the next 30 occasions*. So let us include these options as well:

e)   Drink today and abstain on the next 30 occasions
f)   Abstain today and drink on the next 30 occasions

When e) and f) are included, we can sum up: In the morning every day the preference-ordering of sequences is this:

d)   above e) above f) above c)

But after 5.30 PM my preferences reverse to this:

e)   above d) above c) above f)

The situation is that I have to choose whether to drink or not at 6 PM. What do I choose? If I, at 6 PM, see the choice as a choice between single acts, I will drink the wine. If I see the choice as a choice of a sequence, the best sequence involves drinking today and then abstention. If I believe that what I do today has no consequences for my choice tomorrow, I will choose to drink.

If, however, I see what I do today as influencing what I will do on the next occasion, then this might alter my choice. Imagine the belief that if I do not drink today, I will not drink on the next 30 days. Let us call this the belief in precedence, the belief p. P is a simple indicative conditional. If I have this belief, then d) might be my preferred option. It will be my preferred option if I believe that without choosing to act on this belief p, I will as a matter of fact drink on all these 30 occasions. If I believe I can drink today and then abstain, I will drink today even if I believe p.

Can a naturalist see abstentions as precedents? Does p have a rational foundation? It seems clear that if we do not believe p, we will as a matter of fact drink at 6 PM on all days, even if we have a consistent preference for d) above c) through time, and believe that we will drink tomorrow if we drink today. This follows from the pay-off structure. Without a belief in p we will make the strategy c) real, while we have a consistent preference for d) over c). Our drinking we can see as giving in to temptation on the naturalist view, while being able to act upon the consistent preference for d) rather than c), we can see as having willpower.

It is easily seen that if we adopt the rule to abstain in the afternoon and gained willpower, we would be much better off in terms of total welfare in the long run than without such a rule; we would realise our preference for d. If we were able to believe p, that would make us adopt the rule in question. The question whether this rule can be adopted by rational means, boils down to whether we can give a rational foundation for p.

Ainslie argues that a rational foundation for seeing actions as precedents in the required way can be found if we see the situation as a repeated prisoner's dilemma in a non-co-operative game between successive person stages, or successive motivational states within the person. We choose to play a tit for tat or a similar strategy with tomorrow's self. We make a co-operative move. As in the game with two people, tomorrow's self will have reason to co-operate in the same way with the self of the day after tomorrow and so on. The net result is to my overall gain.

Put simply like this, it needs much detailed backing. The hard problem in Ainslie's positive approach is whether the analogy to the two-person game works. There is no other person with which we interact through time. There are two competing interest, but we are facing rows of person stages. Depending on how we conceive of this situation, on whether we think of these new person stages like people or not, we are either conceiving of the situation as facing an endless row of people, or we are facing the same agent through time, namely ourselves. In the first case, we get no parallel to a repeated game, because there is always a person stage interacting with a new future person stage. Alternatively, there is just one player. This disanalogy with the two-person game seems fundamental. In a game we choose a strategy, where moves on our behalf are seen by us as incentives for the other player to play in certain ways. But the choice of such a strategy does not really make much sense in our situation. If I were to give in to the temptation today, should I punish myself by giving in to temptation tomorrow as well in order not to give myself an incentive to not give in to temptation on day #3? What could be an incentive, and for whom is it an incentive for doing what? In both ways of conceiving of the situation, this question does not make much sense. Remember that we are doing non-co-operative bargaining theory, and looking at what is rational, we do not introduce deals, promises, agreements etc.[10] We choose a strategy.

The crucial point for what I, a rational agent will do, is of course the rational foundation for my belief in p. The point just made is that a rational foundation for a belief p can hardly come from this analogy to games. Assume that I have learnt that drinking one day is followed by drinking the next, and that I hold it very likely that if I drink on day #1, then I will drink on day #2, etc. What matters is the probability I assign my abstaining on day #2 if I abstain in day #1. If drinking on day #2 is just as likely if I abstain on day #1 as it is if I drink on day #1 then of course I will drink on day #1. I need to believe that abstaining on day #1 makes a difference for what I go on to do to rationally abstain on day #1. In this situation it all comes down to how probable I hold p.

The point is this: If I hold p sufficiently probable to actually choose not to drink on day # 1, and also not to drink on day #2, then this seems to have a stabilising influence on my confidence in p on day # 3. This seems true because my choice on day #2 is either a confirming instance of p or not. If I am genuinely uncertain about what the probability of p is on day #1, I can conceive of my choices as tests which influence my probability estimates. If I do not drink on day #2 after having abstained on day #1, p has a confirming instance. My confidence in p is strengthened, and therefore the option of not drinking will have higher utility on day #3 than it had on day # 2. This seems to support the conclusion that if I am rational, and choose rationally when I choose to abstain, I will continue to abstain. Ascribing full rationality to my own actions, p can be seen as true.

But I cannot rationally force myself into assigning a high probability to p if I do not hold it very probable. P will be held to have one probability or other. I need to hold p sufficiently probable to rationally abstain to test the probability of p, but then my choice the next day will confer expected value on the options. Continued abstention will influence and increase the expected value of that choice. This seems to give a role to reason in providing a rational foundation for p. It is a limited role; reason can at most provide a better and more solid foundation for p than the foundation p has when it is barely rational to abstain on day #1. We cannot bootstrap ourselves into giving p sufficient probability to abstain on day #1 if we do hold p less probable than what is needed to abstain rationally on day #1. There is, of a course, a further consideration as well. If I get very confident in my ability to abstain tomorrow, I might choose to make an exception and drink today because I can start abstaining again tomorrow, and tomorrow I make another exception by similar reasoning, and soon I might be drinking every day even if I used to be an abstainer. My ability to live by a principle has then eroded. The danger of erosion is ever-present, and may subside in many types of mechanism. (The lines I have drawn might fade due to new circumstances etc.)

The basic conclusion is that belief-desire based reasoning alone has severely limited power as a pre-commitment device if one starts out with dynamic inconsistencies due to hyperbolic discounting. The result is at best a securing of already existing dynamic consistency. The ability to resist temptation in the first place is, however, what needs to be accounted for. Of course there are other more securely based ways of pre-commitment, but they are both costly and not always available. Given the resources of this naturalist, it seems as if she or he cannot achieve dynamic consistency without relying on these other methods. A claim to the effect that one can identify a pre-commitment device by this appeal to a succession of choices and the parallel pay-off structure with two-person games, seems wrong or exaggerated as an account of dynamic consistency.

We think of ourselves as having a pervasive dynamic consistency. This consistency is an essential part of the substantial rationality we ascribe to ourselves and others when we understand each other as intentional agents. Ainslie sees us as living by personal rules, and the question has been whether reason can install such rules where there is none. This is the question I answer negatively.

B. Bratman's alternative to Ainslie

Perhaps we can view ourselves as having the ability to choose between series of actions of the same type instead of choosing between single actions. A non-reductive account of intention, or a theory of intention which is based in the concepts applied in the 4 principles or closely related ones, can, perhaps, be seen as making room for this possibility. This can in turn be taken as a point in favour of a non-reductive account of intention. The upshot is that the basic choice is a choice between c) and d), and is not seen as a choice between a) and b) or e) or f). We do have a consistent preference for d) over c) through time. We can pre-commit ourselves to abstaining by forming the intention to abstain on all occasions.

The starting-point for a non-reductive approach to intention can for instance be a primitive ability to plan our lives. Michael Bratman recommends that, and opposes Ainslie's view which steers clear of primitive concepts like planning or intention, and sticks to beliefs and desires in the explanatory story. For Bratman it might be rational to stick to one's plans even if one's "preference", or belief/desire based evaluation, says otherwise. This is rational if the intention was rational when formed, and one's reasonable habits of reconsideration do not recommend a reconsideration of the intention. What is it then for an intention to do A at a later time to be rational when formed? A first shot is that a rational intention to do A at a later time t cannot conflict with our known preferences at t. If we now prefer A at t, and we anticipate a reversal prior to t due to hyperbolic discounting in favour of the smaller good B, can we then pre-commit ourselves by forming a rational intention to A?

Bratman holds that it might still be rational to form an intention now to A as long as our reasonable habits of reconsideration will not dictate a reconsideration of this intention. As long as that is so, it will not be rational to change the intention. As long as the intention is not changed, the preference reversal in favour of B is prevented. Forming an intention is then seen a pre-commitment device which can counteract changes in "preference" due to hyperbolic discounting, changes which might seem inevitable on a naturalist perspective if we have no access to more drastic pre-commitment techniques.

In my judgement the starting-points of Ainslie and Bratman are too different for Bratman's alternative to be an alternative available for a naturalist like Ainslie. Ainslie's is trying to give an account of how we can come to be dynamically consistent from a starting point which does not presuppose dynamic consistency. Hyperbolic discounting in its most primitive form can perhaps be understood as saying that even if one does not give up consistency-requirements in one's theory of rational preferences, one gives up the general requirement that options preserve their identities from one choice situation to another. As we move through time, we are consistent at all points in time, but we are also facing new options at each moment in time as we approach the rewards in time. As time changes, we face different options, and different rankings result. If we conceive of dynamic inconsistency this way, we can conceive of the change towards our normal situation as a change towards being able to track options through time and think of them as identical. Ainslie's agent is on the way toward our normal situation. Only very primitive animals have no conception of the identities of options through time. To get closer to dynamic consistency brings with it an enormous change seen from this very primitive level.

It is quite clear that Bratman's planning perspective on intention presupposes both the ability to identify and keep track of options through time, and that his theory of rational intentions presupposes dynamic consistencies as a norm. Therefore the planning perspective on intention is really at odds with hyperbolic discounting, because stronger rationality-requirements are built into the former than into the latter. It becomes a serious issue about the extent to which one can use results obtained within the planning perspective on intention to give an account of how to rationally transcend the dynamic inconsistencies. Methodologically the naturalist Ainslie cannot allow himself to employ results which derive from of a system with a basic dynamic consistency when his aim is to give a naturalistically acceptable account of how we can transcend our dynamic inconsistencies, which is taken as our state of nature.[11]

PART 3. TOWARDS A DUAL ASPECT VIEW

A. Where the naturalist stands

Let us remind ourselves of the overall dialectical situation. If we approach acrasia as a problem about conceptual possibilities, as a basic problem for a philosophical account of what full-blown intentional action is, then we seem to be stuck. If we are naturalists of the Ainslie type, we do not so far really face the earlier problem of the possibility of acrasia, the problem which basically involve intention. Again I stress that I treat the Ainslie type mechanism as a paradigm case: I believe there may be many more naturalist mechanisms in addition to time-discounting which works in ways parallel to the way time-discounting works. Ainslie's mechanism is paradigmatic since substantial rationality is conceived of as an achievement beyond our basic natures, an achievement which always remains somewhat fragile, and more fragile than we think it is.[12]

There are resources within this naturalist approach to provide an account of behaviour contrary to our own better judgement. Such behaviour is easily explained given that the preconditions for substantial rationality no longer holds. If we live out the series c) while seeing the series d) as bringing a larger reward, there is a precise sense in which we acting against our better judgement in a repeated choice. This is so as long as we judge d) best in the sense that we judge that the series d) will bring a much larger summed reward discounted to the present moment than the series c) does. We can judge that the d-type of action in that precise sense is better. This is our better judgement. It is also well explained why we act the way we act and choose c) against out better judgement: There is a mechanism, namely the pay-off structure after discounting, which accounts for why this better judgement for behaviour through time and what we are most motivated to do at a time come apart. If we think of the process from beliefs and desires to a favouring of one option as practical reasoning, then there is no logical error in our reasoning when motivation and better judgement thus come apart.

What is needed for dynamic consistency on the naturalist view, is a belief in precedence, p. If we have such beliefs, we are likely to act in ways which confirm them, and the rules we live by may become self-enforcing. Such beliefs can be accepted on trust, but they always transcend their evidence. On my naturalist view, an individual considered by herself can hardly come to adopt such beliefs by rational means. Still we happen to have them. Perhaps we get them in ways similar to what Tversky and Quattrone describes in a well known paper about how people choose actions because of their diagnostic value even when they know that the condition the action is diagnostic of is unaffected by their choice.[13] If this is right, we acquire the necessary beliefs in some non-rational or magical way. Alternatively we are disposed to have them after being initiated into a praxis. This might be typical of the human animal.

The basic point in this is that on the naturalist perspective the beliefs necessary for substantial rationality and dynamic consistency transcend evidence in a dramatic way. These beliefs are constructions, underdetermined by the evidence as also scientific constructions are. These constructions also differ from scientific constructions. They are interesting constructions for a naturalist since they provide us with a dynamic consistency we would not otherwise have but which we most of the time want to have since it brings us more utility.

The naturalist therefore recognises that there is an ever-present significant chance of breakdowns in dynamic consistency, since there is always reasoning in favour of exceptions to rules and principles which uphold dynamic consistency. Making an exception can be the right thing, but it can also be completely undermining for the construction we live by if we lure ourselves into making new exceptions over and over again. Recognising these dangers, and the possible spillover effects from the breakdown of one rule to that of others, we might become unable to make exceptions when we really ought to, like Davidson seems to be when he decides to get up and brush his teeth when he judges that he is much better off doing it in the morning.[14] Compulsions are seen as arising out of the properties of the impulse-controlling devices we employ to achieve dynamic consistency. Rules need bright lines, otherwise they can erode etc. Therefore the devices we have to use, the rules we live by, are far from optimal, and impulse control can be radically overdone. We might have to enforce the rule of no wine in the afternoon even if the optimal rule is to have a wee bit in the afternoon when the wine offered is really good, just enough to enjoy the taste, of course, and less than what is needed to influence one's piano playing ability at night. This optimal rule may simply not work in practice, the afternoon drink might get gradually bigger and bigger if you try to live by that rule. Mark Twain lived by the rule of one cigar a day, and ended up having specially made cigars the size of a crutch. We might have to forbid drinking to get an effective rule, and letting in exceptions when the wine is exceptionally interesting might also be hard or impossible since all wines start to become exceptionally interesting for us etc.

But of course: once in a while there is an exceptionally interesting wine, and playing badly one night does little damage as long as all the other nights have good music, but this time we might be unable to make the exception. We depart from the best choice in two types of ways: giving in to impulse when we should not, and giving in to compulsive-like impulse-control when we should follow the impulse. There is plenty of room for explaining how behaviour contrary to a specific type of better judgement is possible and what happens when this is actual. The picture of the decision-making human agent is quite different from both the traditional philosophical picture and also from standard rational choice theory.

B. The dual aspect view on acrasia

What about acrasia as a philosophical problem and puzzle? In all theorising about acrasia the concept of intention is vital. We are asking about the limits of full-blown intentional action. According to the picture I have been painting, the intuitive concept of full-blown intentional action has its home in a system of interrelated concepts, a system with a number of substantive normative rationality assumptions built in. One such assumption is dynamic consistency. Acrasia is a question about the limits of this way of understanding ourselves, the way by which we do understand ourselves as rational and moral agents.

To raise the question of whether there is acrasia, and the question of what acrasia is, one needs to be very clear about what question one is asking. The concept of intention appears in the philosophical question about acrasia. To answer the question we need to be clear about what concept of intention is employed in our question. I have through this essay maintained that intention in full-blown intentional action is a concept whose home is a system which presupposes dynamic consistency and substantial rationality as general norms. Can the question we are asking about acrasia be answered positively? It seems to me that we still have no good reason for answering yes to this question; nothing has come up which changes the conclusion of the discussion in the first part of the essay, where I discussed whether we could give a satisfactory account of acrasia from the inside of that system of concepts in which the problem is posed.

The liberating point is that this answer no longer leaves us baffled. We have available a perspective where we see the fragility of the presuppositions of the system of concepts in which the problem of acrasia is posed, and where we also seem able to explain behaviour which is out of accord with best judgement, as well as the normal types of behaviour. We have an enormous explanatory gain.

What we can now say about the philosophical issue of acrasia may be something new. We can think of ourselves in both ways, as described by the system presupposing substantial rationality, and by the naturalist theory. Breakdowns in the preconditions for applying the system of concepts in which intention has its home might lead to questions which are extremely puzzling if we try to see the behaviour *from within that system of concepts.* If we think of the acratic behaviour as describable by that system of concepts, then we are trying to stretch those concepts beyond their normal sound applications. Acratic behaviour becomes a deep puzzle, because it seems right to conceive of that behaviour from within that intentional system of concepts: attempts to see the behaviour as not belonging with this system of concepts are simply not convincing.

As naturalists we see acrasia as a case of breakdown in the normal preconditions for application of the core concept of intention. We might even achieve a view on why acrasia turned into such an impossible problem: We were trying to understand pieces of behaviour from within a system of concepts which was not applicable those pieces of behaviour.

Acratic behaviour surely has much in common with full-blown intentional action, so much that the well-known strategies of denying acrasia have extremely limited appeal: we are not that wrong about our own best judgement, and the behaviour in question cannot be classified as a compulsion. The naturalist uses a system of concepts which is very close to the system in which intention has its original home. And "intentional action" is not sharply distinguished from other behaviour in ordinary talk. We do not normally make finer distinctions between various rationality-assumptions. When we do that in order to sharpen up our own concepts, we are in a sense changing that very vocabulary into a vocabulary where questions can be answered which we formerly did not know how to answer. But that is part of an extension-fixing task where we try to preserve as much as we can of ordinary intuition.

The suggested view on acrasia is that acratic behaviour is much like intentional action, so very like it that we confuse it with it, but it can only be explained by moving to a level where not all the pre-conditions for talk about full blown intentional action holds. This move is hard for us to make, since we react to such behaviour on our own part from within the system of concepts in which the concept of full blown intentional actions has its home. Our reactive attitudes towards our own failings, can only be understood from within the system of intention and substantive rationality.[15] Imagine that we actually give in to a temptation of the Ainslie type, and live in the way described by that theory. We will get irritated and angry with ourselves, lose self-esteem and self-respect etc. This is because we keep track of options through time, we remember our own evaluations of options' respective worth, we regret our own suboptimal choice. These reactive attitudes have a basic function in our own self-understanding, and an instrumental function in preventing giving in to temptations. To make sense of the anger and irritation and the whole range of emotional responses we might have, we need to see things from a perspective with substantive rational norms as an internal part.

C. The need for both aspects

The last observation is vital if we raise the issue of whether we should be realist of sorts about both kinds of descriptions, the naturalist and the other. Why not opt for the naturalist perspective and leave the rest behind? Especially if the naturalist can explain all pieces of behaviour, why do we need anything in addition? The naturalism which was the way out to understand types of behaviour becomes imperialist.

Imagine leaving behind the system of concepts with full blown intentional actions. We look at things through the naturalist system of concepts. What are the questions we can answer then? If my example with watching television and running is a case of acrasia, as it seems to me, then this acratic behaviour should not be puzzling at all. It can be explained quite straightforwardly. This generalises to most explanatory issues. This naturalist will have the resources to explain the behaviour in question, but not our reactive attitudes. But what about the philosophical questions, among them the question of acrasia itself? The "diagnostic" answer given above seems to presuppose the reality of the two different aspects. The source of acrasia is our stretching the concept of intentional action too far. If we are pure naturalists, and stick to that perspective only, then we do not have a question to answer which necessarily involve intention. And unless there is a need in decision-science for a concept of intention, and a scientific usage of the concept, there will be no question about acrasia for this naturalist to answer.

The naturalist I have been considering sees a continuum from primitive behaviour to adult human behaviour. If this is the perspective, then there is no deep philosophical need to provide conceptual identifications of full blown intentional agency as opposed to behaviour which fall short of that. Given that, it is not surprising that we are left with a general picture of the problem of acrasia which is very different from the picture of that problem in the branch of philosophy which deals with agency. On the traditional picture in philosophy, one of the most fundamental task is the deep conceptual problem of what intentional agency is. On the naturalist picture I am now exploring, there is silence about such issues. What there is, is an explanatory theory of behaviour with great scope.

If this "reductive" way were to exhaust a philosophical approach to human decision-making, we are left with a picture of the human agent which is very different from the picture we have encountered in traditional philosophical works. Especially challenging is the connection to moral philosophy. We seem to have given a precise sense in which a person might fail to be motivated to perform an action which can be subsumed under a rule which he or she would prefer to be the general rule she or he or anybody lives by. We have been given no basis for regretting such a motivation.[16]

Here, at the point where we start touch the foundations of morals, is also the point where I see profound limits for this naturalism. I see a need to provide an account of full-blown intentional agency, for freedom of the will, etc. This is a need which does not necessarily arise out of the task to explain behaviour. If the task, however, is to provide an account of agency which connects with how to give moral theories, then we need more than the explanatory theory. I think we can have what we need without giving up naturalism.

We need, among other things, an account of moral intuitions to get started in moral theory. It is far from obvious that all intuitions about what we will do in certain circumstances have an equal bearing on the normative issue of what we should do. In the same way it is not obvious that all our tendencies in reasoning should have an equal bearing on the normative issues about how to reason.[17] All in all, it seems to me that our basic model for the justification of norms, be it the inference rules of logic or moral rules, is something like Rawls' method of reflective equilibrium. In that case we might need at the end of the day something like a "demarcation" of what is a valid intuition and not a slip. In the moral case we need to exclude as a provider of relevant moral intuitions for the justification of norms the type of intuition which cannot be matched by intentions for performing full blown intentional actions. To isolate these intuitions we also need an account of what full-blown intentional action is.

I believe Quine might be on my side here. He writes in his paper "On the Nature of Moral Values": "The empirical foothold of scientific theory is the predicted observable event; that of the moral code is in the observable moral act. But whereas we can test a prediction against the independent course of nature, we can judge the morality of an act only by our moral standards themselves."[18]

It therefore seems clear that there are genuine philosophical and normative needs which are different from the explanatory needs, and that the naturalist story in itself cannot satisfy these needs. These needs are not purely abstract needs. They are intimately connected with the need to understand our own emotional and other reactions to our own inconsistencies.

I therefore claim that we need both systems of concepts, the system for full-blown intentional action, and the naturalist explanatory story. The first system makes up the system in which our self-understanding is located, and explanation by these concepts can therefore capture how the agents sees his or her own action. Still the naturalist provides a possibly full explanatory story when explaining particular actions, an explanation which serves a predictive interest. These are two aspects reflected in two basic interests. Why are these aspects aspects of the same thing? Well, in the normal case they reflect different interests in accounting for the same action. An action is intentional, and as long as we have a "reductive" view on intention, a view which denies seeing intention as in any way independent of beliefs and desires and preferences generated from them, then we can see the two systems of concepts as different ways of accounting for the same intention. That helps us stay naturalist. To see what is described by the two systems of concepts as aspects of the same thing in the normal case, aspects of the reasoning process prior to a practical conclusion, we need, in my view, to see an intention in a full-blown intentional action as closely connected to, and possibly to be identified with a maximising judgement. Not all maximising judgements amount to intentions, this is the lesson of acrasia, but many do.[19]

PART 4. CONCLUSION

This conclusion is now quite brief. The view on acrasia I am pushing sees acratic behaviour as falling short of full-blown intentional action. It is behaviour explainable by a maximising judgement within a naturalised decision-theoretic approach to action, and is easily confused with-full blown intentional action.

It is very difficult not to see acratic behaviour as full-blown intentional action. One reason for the difficulty is that we tend to think in too crude dichotomies when we attempt to classify acratic behaviour: Either it is full blown action or it is fully compulsive or it is a simple case of mistaken belief about what we think is best. On my view acrasia is neither of these things: what we are mistaken about is how to classify this behaviour, whether it is full-blown action or not. The behaviour in question is much more like full-blown intentional action than compulsive behaviour is. It is indeed quite like full blown intentional action, apart from the fact that some rationality-assumptions which are essential for intentional actions are lifted. As long as the operating mechanism is of the Ainslie-type, acratic behaviour may either have elements in common with compulsive behaviour (impulse-control overdone) or it may be impulsive behaviour due to temporary preference. We can depart from the best choice in two ways: giving in to impulse when we should not, and giving in to compulsive-like impulse-control when should follow the impulse. The former type might be experienced as a sort of recklessness, while the latter might be experienced as like a compulsion while one clearly knows that it is not really a compulsion of the sort where one simply could not do otherwise.

I know that my thoughts on acrasia can be seen as trading on how to use words. But, as Quine has thought us, there is no clear dividing line between truths about meaning and truths about the world. We have to make meanings as clear as we can in the light of the empirical theories we believe in. If there any merit in this essay's proposals, Quine deserves some of the credit.

University of Oslo

NOTES

* A version of this paper was given at the 1995 Karlovy Vary Meeting, on the topic "Questions from Quine". I am very grateful to my audience at that occasion. It was an honour to have Quine's comments, and I thank him. Versions of the paper have also been read at London School of Economics, Lund University, and the World Congress in Logic, Methodology, and Philosophy of Science, and I thank the audiences. I am very grateful to Jon Elster's interdisciplinary "Addiction" project which made this research possible. I thank the Norwegian Research Council for generous support, and SIFA, the National Institute for Alcohol and Drug Research, for access to a stimulating research environment. I am also much grateful to the following for comments: George Ainslie, Michael Bratman, Robyn Dawes, Jon Elster, Eyjolfur Emilsson, Dagfinn Føllesdal, Peter Gärdenfors, Jennifer Hornsby, Margaret Gilbert, Al Mele, Wlodek Rabinowicz, Bjärn T. Ramberg, and Ole-Jørgen Skog.
[1] The other type of action which is often mentioned in connection with acrasia has things in common with compulsion. I shall return to this point.

²   For good and thorough disucssion of Aristotle's view, see David Charles, *Aristotles'Philosophy of Action*, London, Duckworth, 1984. For a good critical discussion of Charles's view, see Terry Irwin's "Aristotelian Actions", *Phronesis*, 1986.

³   I borrow this terminology from Davids Pears' book *Motivated Irrationality*.

⁴   See Davidson's "How is Weakness of the Will Possible" in Davidson 1980, and Michael Bratman's "Practical reasoning and Weakness of the Will", *Nous*, 1979. I consider Davidson and Bratman as leading representatives of the two different starting points, the backward or the forward connection.

⁵   This is Michael Bratman's point of departure in his 1979 account of weakness of the will.

⁶   This is Donald Davidson's theory of weakness of the will.

⁷   Ainslie derives this theoretical support from Herrnsteins matching law.

⁸   In Bratman's paper "Planning and Temptation". I owe a lot to Bratman's insightful criticism of Ainslie in the following.

⁹   Making promises to oneself are therefore ruled out. Such things are rather examples of Ulysses-type strategies, quite like side-bets which influence what you will do: you promise yourself a big glass of cold beer if you go for a run.

¹⁰   Bratman has given a very illuminating account of the role of plans and other intentions in structuring our decision problems, but his planning account of intention cannot really be employed to solve the sort of problem the naturalist is trying to solve. Bratman is employing the vocabulary of intention, practical reason and action with which we started out when approaching the problem of weakness of the will. He is a champion of a non-reductive account to intention, an account based upon beliefs, desires and evaluations based upon them. The problems about how acrasia can be possible discussed in the first part of this essay therefore returns with full force on Bratman's view.

¹¹   There will probably be other factors in addition to temporal distance which influence motivation in systematic ways. There is for instance spatial proximity. Here is a case: We have made up our minds about not having dessert, and we would have been able to stick to this decision at the time of finishing the meal, unless the waiter had rolled the dessert-trolley with the most fantastic content right up to our table. We knew all the time that he would serve us at the slightest nod from us. Now he is serving the guests at the next table. The trolley's presence is too much for us, and we reverse our preference in favour of having a dessert after all. We know at the time of preference reversal that it will take longer to be served than it would have taken if we had nodded to the waiter while the waiter was not busy, and we know our preference wouldn't have been reversed if the trolley had stayed away. I believe there are many cases of this sort. They have not been experimentally studied to my knowledge.

It seems to me that proximity in general might have role in producing the awareness in our minds which is essential for preference reversal, and that the role of proximity may not be exhausted by the role of temporal or spatial proximity. What matters might be sensory information or imagination. Still, these factors which must be modelled as influences on the maximising judgement itself, not upon actions contrary to maximising judgement.

¹²   See George A. Quattrone and Amos Tversky: "Self-Deception and the Voter's Paradox", in Jon Elster, *The Multiple Self*. Cambridge 1985, pp. 35–58. Drazen Prelec presented a paper at the Oslo Conference on Addiction in May 1995 where he aired the possibility of seeing some such choices as rational, bringing what he called diagnostic utility.

¹³   See Davidson's "How is weakness of the will possible" in Davidson 1980.

¹⁴   The concept of reactive attitudes is from Strawson's essay "Freedom and Resentment.", in the collection with the same name. I am surely indebted to Strawson in the line I am pushing. See also Strawson's book *Scepticism and Naturalism. Some Varieties*.

¹⁵   Samuel Scheffler's great book *Human Morality*, has an intresting discussion of the ways a naturalist view on motivation can account for motivation out of duty, or for a difference between authoritative motivation and desire-based motivation. Scheffler argues that psychoanalytic thought can help us part of the way. I claim that the Ainslie-type cognitive mechanisms for fighting the temptations make up very interesting material for providing a satisfactory account of the two types of motivation, without becoming motivational dualist. In particular the link between authoritative reasons and generalisability across times and people and thinking of choices as repeated choices is interesting. I think there is more promise in this naturalism than in psycho-analytic approaches. As Scheffler himself points

out, such approaches cannot really provide a link between authoritative reason and moral reason.

[16] Stephen Stich's recent important book, *The Fragmentation of Reason*, MIT-Press, Bradford Books, is a piece of work I would like to approach from an angle like this. In my view Stich makes a mistake in being too naturalistic when letting in relevant evidence for inference rules.

[17] WvQuine, *Theories and Things*, Harvard University Press, 1981, p. 63.

[18] In my paper "Intention and Alternatives", *Philosophical Studies*, 1996, pp. 159–177. I investigate the reasons given by Michael Bratman for seeing intentions as basically different from a maximising judgement, reasons arising out of the insight that intentions are agglomerative. I argue that his argument at this point is mistaken. I think it is important for a naturalist to keep the tight Davidsonian connections between beliefs, desires and intentions, to be precise about why the dual aspect view can claim that these two aspects are aspects of the same thing.

REFERENCES

Ainslie, George. *Picoeconomics*, Cambridge 1992.
—— "Beyond Microeconomics", in Jon Elster (ed) *The Multiple Self*, Cambridge, 1986.
—— "Hyberbolic Discounting", (with Nick Haslam) in George Loewenstein and Jon Elster (eds) *Choice over Time*, Russell Sage Foundation, 1992.
"Self-Control" (with Nick Haslam) in Loewenstein and Elster.
—— "Where There's a Will There's a Won't" Unpublished Manuscript 1993, Presentation for the Symposium *Contemporary Perspectives on Self-Control and Drug-Dependence*, Chicago 1993.
Bratman, Michael , Practical Reasoning and Weakness of the Will, *NOUS*, 1979.
—— *Intentions, Plans and Practical Reason,* 1987.
—— "Planning and temptation", in May, Friedman and Clarke (eds) Minds and Morals, Bradford Books, MIT Press, Cambridge, 1995, pp. 293–310.
Charles, David. *Aristotle's Philosophy of Action*. Duckworth, London, 1984.
Davidson, Donald, *Essays on Actions and Events*, Oxford 1980.
—— "Paradoxes of Irrationality", in Richard Wollheim and James Hopkins (eds) *Philosophical Essays on Freud*, Cambridge 1982.
—— "Deception and Division", in Elster (ed) *The Multiple Self*, Cambridge, 1985.
Gjelsvik, Olav "Intention and Alternatives", *Philosophical Studies*, 1996, pp. 159–177.
Irwin, Terence "Aristotelian Actions", *Phronesis*, 1986.
Pears, David. *Motivated Irrationality*, OUP, Oxford, 1984.
Quattrone, George A. and Tversky, Amos: "Self-Deception and the Voter's Paradox", in Jon Elster, (ed) *The Multiple Self*. Cambridge 1985, pp. 35–58.
W.V. Quine. *Theories and Things*. Harvard University Press, Cambridge, 1981.
Scheffler, Samuel, *Human Morality*, OUP, Oxford 1992.
Stich, Stephen *The Fragmentation of Reason*, MIT-Press, Bradford Books, Cambridge, 1993.
Strawson, Peter F. "Freedom and Resentment", in *Freedom and Resentment and Other Essays*. London, Methuen, 1974.
—— *Scepticism and Naturalism. Some Varieties*. OUP, Oxford 1986.

GABRIEL SEGAL

FOUR ARGUMENTS FOR THE INDETERMINACY
OF TRANSLATION

INTRODUCTION

W.V. Quine's thesis of the indeterminacy of translation has attracted a great deal of attention in the philosophical literature from both supporters and critics. It is intriguing, deep and important. To my mind it is also a thesis that presents a bleak prospect. If it is correct, then we must jettison mentalistic semantics, eliminating locutions like "understands" and "knows the meaning of" and replace them with talk of behavioural dispositions and the neural states underlying them. At least we should do this so long as we wish our discourse to reflect the true structure of reality, as described by sound science.

Indeed, even more may be at stake than that. For in giving up on meaning it seems we must give up also on truth and falsity. For if sentences do not have meaning, it would seem to follow that they do not have truth conditions. And if sentences do not have truth conditions then it seems they cannot have truth values.[1]

My aim in this paper is to examine Quine's arguments for the indeterminacy of translation from the standpoint of a friend of mentalistic semantics. I hope that the mentalism need be neither "pernicious" (Quine 1969b, 27) nor "uncritical" (*ibid.* 80). I certainly don't mean to endorse Cartesian dualism, nor any view that places mind and meaning out of the reach of serious science. To a first approximation, the view is that what a speaker's words mean is what he thinks they mean. If a speaker thinks that "gavagai" applies to an object if and only if that object is a rabbit, then "gavagai" in his language is, indeed, true of rabbits and rabbits only. To a second approximation, the view is that the semantic features of a person's language are determined by the complex psychological states that constitute his linguistic competence. The nature of the psychological states involved is to be determined by scientific linguistics.

I will not defend this mentalistic standpoint here, although at the end of the paper I will briefly indicate what I think such a defence could consist in. Rather, I want to consider whether Quine's arguments ought to dislodge the mentalist from his position. I will try to show that they should not. I will suggest, in effect, that Quine's arguments beg the question against someone who is happy with the idea that there are facts of the matter about what people mean by their words and about how the words of one language should be translated into those of another.

In what follows I will present four arguments for the indeterminacy thesis, and respond to each one. Three of the arguments are found in Quine's writings. The

131

Alex Orenstein and Petr Kotatko (eds.), Knowledge, Language and Logic, 131–139.
© 2000 Kluwer Academic Publishers. Printed in Great Britain.

fourth argument is one that he does not explicitly give (as far as I know), but that does contain elements of a Quinean philosophy.

1. THE THESIS

What, exactly, is the thesis of the indeterminacy of translation? In chapter two of Quine (1960, 27) he introduces it with these words:

> The thesis is then this: manuals for translating one language to another can be set up in divergent ways, all compatible with the totality of speech dispositions, yet incompatible with one another.

The point of this claim is that translation is underdetermined by all the possible observations that might be relevant to it. No matter how many observations one has made and might make, one will never have enough evidence to adjudicate between all rival translation manuals. But this claim does not by itself threaten the notion of meaning. A fan of mentalistic semantics might reply that there is one and only one correct translation manual, even if there are many that are equally well supported by all possible evidence. The real thesis of the indeterminacy of translation is precisely the denial of that mentalistic claim. It is the thesis that there is no fact of the matter about which of the rival translation manuals is correct. As Quine puts it in his (1969a, p. 303): "where indeterminacy of translation applies, there is no real question of right choice".

If a single sentence, S, of one language can be translated with equal justice into two sentences of another language that are in no sense equivalent in meaning, then indeed, the whole idea of meaning is threatened. Switching from translation to semantic theory, the conclusion would be that different semantic theories could disagree over the meaning of S, neither theory being right and neither wrong.[2] And this would indicate that there is no fact of the matter about what S means.[3]

But what is the relation between underdetermination and indeterminacy? The first two arguments I will consider relate closely to this question.

2. THE FIRST ARGUMENT

Quine addresses the question of the relation between underdetermination and indeterminacy as follows (1969a, p. 303):

> [T]heory in physics is an ultimate parameter. There is no legitimate first philosophy, higher or firmer than physics to which to appeal over physicists heads. ....
>
> Though linguistics is of course part of the theory of nature, the indeterminacy of translation is not just inherited as a special case of the underdetermination of our theory of nature. It is parallel but additional. Thus, adopt for now my fully realistic attitude towards electrons and muons and gluons and curved space-time, thus falling in with the current theory of the world despite knowing that it is in principle methodologically underdetermined. Consider, from this realistic point of view, the totality of truths of nature, known and unknown, observable and unobservable, past and future. The point about indeterminacy of translation is that it withstands even all this truth, the whole truth about nature. This is what I mean by saying that, where indeterminacy of translation applies, there is no real question of right choice; there is no fact of the matter even to within the acknowledged under-determination of a theory of nature.

The claim here is that the underdetermination of semantic theory is an additional layer of underdetermination, over and above the underdetermination of theory that we had all along. Our theory of nature, with its muons, gluons and so on is underdetermined by the evidence that supports it. Now suppose we take this whole theory of nature to fix the evidence available to constrain semantic theory (which is what I take "linguistics" to mean, in this context). We find that semantic theory is underdetermined by this evidence. And this, as it were, double underdetermination of theory by evidence yields indeterminacy. As Alexander George puts it (George, 1986) "Where there is slack between observation and theory we have underdetermination, but slippage between total theory (all facts, known or unknown) and theory is indeterminacy".

I think Quine's statement quoted above is intended more as an expression of his view than an argument for it. But it does suggest an argument. The argument could be put roughly as follows. (1) There's some theory, T_1, that is underdetermined by the evidence for it, yet acceptable. (2) Any further theory that is underdetermined by T_1 suffers from underdetermination additional to that of T_1 (3) Any such doubly underdetermined theory is indeterminate. (4) Semantic theory is underdetermined by T1. So semantic theory is indeterminate.

But what is T_1? It can't be a total theory of nature that includes semantics, since then the argument would falter at step (4). So we must take it to be some theory of nature that excludes semantics. The mention of muons, gluons etc. suggests that we can call this theory "physics".

The idea is then that semantic theory inherits the underdetermination of physics, but suffers an additional layer as well. One could at least conclude from this that semantics would inevitably be more underdetermined than physics. And this might suggest, although it would not yet entail, that there is better reason to be realist about physics than about semantics.

There is, though, a problem with this argument. It is hard to see why semantics should inherit the underdetermination of physics, so that it ends up being doubly underdetermined. It is not as if the radical translator needs to base her work on a prior theory of physics. Of course, at least from Quine's point of view, the evidence that the translator is working with is physical evidence. And any amount of underdetermination that affects the sentences describing this physical evidence will of course carry over to the translator's theory. However, the observation sentences of a Quinean radical translator will describe such things as correlations between publicly observable goings on and types of utterance, and hence not be significantly more underdetermined than the observation sentences of advanced physical theory. Both physics and translation will be based upon observations of middle-sized, middle-distanced objects and events. Each theory then departs in its own way to infer hidden goings that would explain these observations.

Hence the underdetermination of advanced physical theory is irrelevant to semantics, and the underdetermination of observation sentences about the physical world applies equally to physics and to semantics.

Certainly it is true that semantic theory is underdetermined by physical theory, in the sense that a choice of physical theories does not mandate a choice of semantic theories. But, as Noam Chomsky (1980) points out, the reverse is also true. In that sense, physics is underdetermined by semantics. A choice of semantic theories does not mandate a choice of physical theories. Indeed, this is typically true of any pair of empirical theories. Each of physics and biology, for example, is underdetermined by the other, in that a choice among competitors in one theoretical domain doesn't mandate a specific choice among competitors in the other.

The mentalist thus retains the right to claim that the business of making sense of the utterances of a speaker of an unknown tongue may be no more intractable in principle than the business of constructing a physical theory. This is not to say that physical theories and mentalistic semantics are symmetrical in every respect. One might hold, along with Quine (and Chomsky too, for that matter), that if there are any facts about what speakers mean by their words, then these facts ultimately depend upon the speakers' physical states. We might also demand that theories of meaning should be compatible with physics. In the case of a clash one should give up the theory of meaning, not the physics. (Although I take it that this should not be an item of *a prioristic* dogma: it should depend upon which of the clashing theories was better supported). Or, more strongly, one might demand that a theory of meaning should ultimately be ratified by an account of how the brain (or the brain and its relations to the environment) manages to attain the mental states it does (as Chomsky thinks, see e.g. his 1986, p. 27). But none of this asymmetry helps Quine's argument. For none of it shows that a semantic theory is worse off than a theory of physics in respect of its relations to evidence.

3. THE SECOND ARGUMENT

The second argument for indeterminacy also hinges on a kind of double underdetermination claim. It is given in Quine (1970). The argument there goes as follows. In order to translate a foreign tongue one needs to know about the speakers' theory of the world around about them. And this immediately gives rise to a good argument for underdetermination. One could imagine that there might be a translation scheme, T_A, that makes sense of what the speakers say on the supposition that they hold theory A, and a different and incompatible scheme T_B which makes equally good sense of native utterances on the supposition that they hold theory B. Thus far we have only an argument for underdetermination. But now there is more.

Both our theory of the world, and the natives', are underdetermined. We can suppose that theories A and B are empirically equivalent, and that both we and the natives face the problem of deciding which is right. And now it appears that the underdetermination of physical theory (the underdetermination of A and B) enters into the business of translation twice over. As Quine puts it (p. 181) "But now the same old empirical slack, the old indeterminacy between physical theories, recurs in second intension."[4] For the underdetermination of physical theory occurs once in its own right, and once again when we come to translate the foreigners' language:

Where physical theories A and B are both compatible with all possible data we might adopt A for ourselves and still remain free to translate the foreigner as either as believing A or as believing B (180).

Although this argument is different from the first one, it suffers from a parallel problem. The underdetermination of physical theory does not "recur in second intension" because we do not need to make a choice between A and B ourselves in order to proceed with translation.

Of course, in practice, radically translating an advanced physical theory would require the translator to have an advanced physical theory herself. But this only means that she must be well enough versed in the ways of physics to have the conceptual wherewithal to understand the natives. The underdetermination of her own physical theory does not carry over to her translation, for the truth of the latter does not rest on the truth of the former. Indeed she might understand theories A and B and believe neither, and yet address the question of which of the two is held by the natives.

4. THE THIRD ARGUMENT

The third argument for indeterminacy is rather different from the first two. We find it in Quine 1987.[5] Here, Quine says that the thesis is a consequence of his behaviourism, and continues:

In psychology one may or may not be a behaviourist, but in linguistics one has no choice. Each of us learns language by observing other people's verbal behaviour and having his own faltering verbal behaviour observed and reinforced or corrected by others. We depend strictly on observable behaviour in observable situations. As long as our command of language fits all external checkpoints, where our utterance can be appraised in the light of some shared situation, so long all is well. Our mental life in between checkpoints is indifferent to our rating as a master of language. (1987, 5).

The conclusion here, stated in the final two sentences, is that there is no more to knowing the meaning of a sentence than being disposed to use it, or react to it, in certain ways in certain observable circumstances. It is a short step from this behaviouristic conclusion to the indeterminacy of translation. The step is another underdetermination claim: for any sentence, S, in a given speaker's repertoire, there will be many different assignments of meaning to S that are compatible with all her dispositions to use S or react to it in observable situations.

What is questionable here is not the move from behaviourism to indeterminacy, but the argument for behaviourism. Indeed, I confess to finding it difficult to see what the argument is. However I have two suggestions. The first focuses on the point about learning, and proceeds as follows. There can be nothing more to the meaning of an expression than can be learned by observing others use it and having one's own behaviour in respect of it reinforced or corrected. But all one can learn in this way is to use the expression in ways that match up appropriately with observable goings on. Hence knowing the meaning of an expression consists only in having the right dispositions with respect to it.

The problem with this argument is that one cannot tell what can be learned from a given range of evidence without a theory of the resources available to the learner. This is a quite general point. An academic psychologist, a small child and a cat, equipped as they are with different resources, would learn rather different things from observation of a rat's behaviour in a Skinner box. Equally, what a pre-lingual human infant can learn from observation of adults' verbal behaviour is very different from what might be learned in the same situation by a chimpanzee or a chihuahua. Everyone in this debate allows that the human infant is not a blank slate, but comes to the learning situation already endowed with some innate learning principles or other.

Given this, no behaviouristic conclusions can be derived merely from reflection on the nature of the learning situation, with no attention to what the learner itself contributes. And, indeed, an argument from the constraints on the learning situation to behaviourism in linguistics seems more or less doomed to beg the question. For the mentalist holds that language learning involves making inferences about the meanings of expressions from the data provided by competent speakers. The child is innately endowed with rich and sophisticated resources that allow it to infer considerable conclusions from the scanty observational data. For example, on observing Mama holding up a pink, velour toy rabbit and saying "rabbit" the child might infer that "rabbit" applies to rabbits (not to the colour, the material, the class of toys nor any of the endless other possibilities compatible with data).[6] Having made many inferences of this general sort, the child arrives at a body of beliefs about the meanings of various expressions of the language.

So, on the mentalist picture, what is acquired is not a set of dispositions to verbal behaviour, but a complex mental state; knowledge (or "cognizance" (Chomsky, 1980, 73) if one prefers a technical term) of the semantic properties of a language. The truth or falsity of a theory of meaning for the speaker is then dependent on the contents of this mental state. This mentalistic view might be questioned, but not on *a priori* grounds. Given the coherence of the view, it is hard to see what justifies Quine's strikingly strong claim that in linguistics one has "no choice" but to be a behaviourist.

The second interpretation of Quine's argument focuses on the comment that one's rating as a master of language depends only upon one's dispositions to use sentences. The idea would be that anyone whose behaviour with respect to the language meets the observable criteria of competence thereby shows themselves to be a master of the language. So knowing a language consists only in having the right dispositions, irrespective of the underlying mental states (if any) that explain them. But this, too, begs the question. As far as the mentalist is concerned, if an object manifests all the right dispositions to use sentences, but does so for the wrong reasons (that is, without the right mental states underlying the dispositions) then the object does not understand the language. We have yet to be shown why this mentalist position is wrong, why the behaviourist approach is mandatory.

I conclude that none of the three arguments so far considered makes headway against the mentalist. None of this goes to show that Quine's conclusions are false or that mentalistic semantics is the road to truth. But it does show that the mentalist's position has not been threatened.

5. THE FOURTH ARGUMENT

The final argument I will consider is not explicit anywhere in Quine. But it is an interesting argument and it has a Quinean feel to it.[7]

The problem for Quine is to find some principled reason for being a realist about physics and an anti-realist about mentalistic semantics, given that both are underdetermined by all possible data. One reason is suggested by this remark from Quine (1969a): "theory in physics is an ultimate parameter. There is no legitimate first philosophy, higher or firmer than physics, to which to appeal over physicists' heads". Why should physics be an ultimate parameter? Why is there nothing to which we can appeal over physicists' heads? One reason might be that we simply have no alternative explanation of the observable facts about the physical world. We cannot reject physics in favour of some other theory, because there is no other theory that can do the work that physics does. But, Quine might say, the same is not true of theories of meaning. There we have an alternative: we can explain the data with behaviouristic psychology and neurology. And this course would appeal to Quine's "taste for desert landscapes", his liking for Occam's Razor. Brains and behaviours are there anyway, and we can attain an ontologically economical picture of the world by jettisoning mentalistic semantics and its baggage and adopting the Quinean alternatives.

Quine rejects a crude version of this argument in (1969b, p. 27):

Semantics is vitiated by a pernicious mentalism so long as we regard a man's semantics as somehow determinate in his mind beyond what might be implicit in his overt behaviour. It is the very facts about meanings, not the entities meant, that must be construed in terms of behaviour.

The problem with mentalistic semantics is not (or not merely) one of ontological extravagance. But there might still be an objection to theoretical extravagance. The objection would be that the whole apparatus of mentalistic semantics, whether or not it requires non-physical objects, is explanatorily redundant: don't hold two theories where one would do.

But here, once again, the mentalist will see the argument as begging the question. For it is far from clear that behaviourist psychology or neurology can explain the data that need to be explained. Even if we agree on a neutral description of the data, and conceive of the data as consisting of, say, pairings of utterance types with observable stimulations, it is not obvious that any behaviourist or neurological theory will explain them. It may be that behaviourist psychology fails to unearth behavioural laws that explain language learning and language use. It has, indeed, failed badly hitherto. And although we would expect neurology to explain bodily motions under some description or other, we have as yet no reason to believe that it will explain the rather abstract correlations observed by the field linguist. Hence the hope that we can simply jettison mentalistic semantics and get by with other things is just that; a hope and no more. The apparatus of mentalistic semantics is part of Neurath's ship, along with neurology, particle physics and the rest of our current total theory of nature. If we wish to keep the ship afloat, "we are not in a position to jettison any part of it, except as we have substitute devices ready to hand that will serve the same essential purpose" ((Quine 1960, 124) discussing the ontology of abstract objects.)

Surely the way to find out whether Quine is right or wrong is to look at the work of mentalistic semantic theories. If the enterprise flourishes, then all will be well. If it dissolves in a cloud of disagreement, then Quine will have been vindicated. But the apparent health of both psycholinguistics and semantics at large should promote an optimistic outlook.[8]

King's College London

NOTES

[1]   There seems to be a problem here for Quine. If truth and falsity are consigned along with other semantic phenomena to the realm of what isn't, then the coherence of some of Quine's formulations of the indeterminacy thesis appears to be threatened. Consider, for example, the section from Quine (1969a) quoted below. However, there may well be coherent formulations of the thesis that would ward off this objection, so I shall not pursue the point here. See Paul Boghossian (1990) for discussion.

[2]   Strictly speaking, it is the "neither right" part that ushers in the serious threat. Someone of what Quine (1990) calls an "ecumenical" disposition would hold that all empirically and formally adequate theories would be true. Donald Davidson is ecumenical in this respect. Although I will be assuming that there is one and only one true theory of meaning for each language, most of what I say should be acceptable to the ecumenically inclined.

[3]   Quine is clear that the two acceptable translations of S might have different truth values (see e.g. 1960, p. 27). This would seem to entail that there is no fact of the matter about S's truth value. Hence, as I said above, truth appears to follow meaning down the path to oblivion.

[4]   It appears that Quine intends "underdetermination" by "indeterminacy". I cannot find a way of interpreting this claim of Quine's that does not require this substitution. For a brave attempt see Miriam Solomon (1989).

[5]   The argument also appears in Quine (1969b, 81), (1975) and (1990, 38).

[6]   See Susan Carey (1984) for relevant empirical work.

[7]   I am indebted to Mark Sainsbury for the main idea here.

[8]   Unsurprisingly, Quine thinks not. He says of "the old notion of separate and distinct meanings" that "[i]n later years ... scientific linguists ... have simply found it not technically useful" (1990, 56). It is true that linguists have done much more work on syntax and phonology than on semantics. But that doesn't mean that semantics has withered away. As far as I can tell, many linguistics departments around the world have a semanticist or two among their ranks. Further, syntactic theory itself appears to require something not too unlike "the old notion of separate and distinct meanings", since it frequently deploys semantic notions in its description of the explananda. Syntactic theory explains such things as why a given string is *ambiguous*, or why a string has this *interpretation* but not that one, why a pronoun in a given position can (or cannot) *corefer* with some antecedent noun, and so on. These apparently semantic notions appear to be important rather than technically useless.

[9]   I am grateful to Matthew Carmody, Alex George, Keith Hossack and Mark Sainsbury for comments on earlier drafts. My views on linguistics and the indeterminacy of translation have been greatly influenced by the lectures and writings of Noam Chomsky. Much of what I say above merely organises and synthesises ideas of his.

REFERENCES

Boghossian, P.A, 1990, "The Status of Content", *The Philosophical Review.*

Carey, S., 1984, "Constraints on Semantic Development." In Mehler, J., and Fox, R., (eds.) *Neonate Cognition: Beyond the Blooming, Buzzing Confusion* (381–398) Hillsdale, N.J.: Lawrence Erlbaum.

Chomsky, N., 1980, *Rules and Representations*, New York, Columbia University Press.

Chomsky, N., 1986, *Knowledge of Language*, New York, Praeger.

George. A, 1986, "Whence and Wither the Debate between Quine and Chomsky?", *The Journal of Philosophy.*

Putnam., H., "The Meaning of 'Meaning'", in K. Gunderson, ed., *Minnesota Studies in the Philosophy of Science, Vol. VII, Language, Mind and Knowledge.* Minneapolis, University of Minnesota Press.

Quine, W.V., 1960, *Word and Object,* Cambridge Mass., MIT Press.

Quine, W.V., 1969a, "Reply to Chomsky", in *Words and Objections: Essays on the Work of W.V. Quine,* eds. D. Davidson and J. Hintikka, Dordrecht, Reidel.

Quine, W.V. , 1969b, "Ontological Relativity", in his *Ontological Relativity and Other Essays,* New York, Columbia University press.

Quine, W.V.,1970, "On the Reasons for the Indeterminacy of Translation", *The Journal of Philosophy.*

Quine W.V., 1975, "Mind and Verbal Dispositions", in *Mind and Language,* ed. S. Guttenplan. Oxford, Clarendon Press.

Quine, W.V., 1987, "The Indeterminacy of Translation Again", *The Journal of Philosophy.*

Quine, W.V., 1990, *Pursuit of Truth,* Cambridge MA, Harvard University Press.

Solomon, M., 1989. "Quine's Point of View", *The Journal of Philosophy.*

LOUISE ANTONY

NATURALIZING RADICAL TRANSLATION

Quine is famous (infamous?) for defending the radical doctrine known as the Indeterminacy of Translation. The thesis states that, beyond small empirically grounded fragments, there is "no fact of the matter" as to the correct translation of one language into another. (Graduate students who have been so unfortunate as to have had their foreign language proficiency exams graded by Prof. Quine may question the depth of his commitment to this principle, but let us eschew *ad hominems*.) Quine has offered at least three different, though related arguments for the Indeterminacy Thesis; I shall focus in this essay on the argument from "radical translation," offered in Chapter Two of *Word and Object*.[1] The objection I'm going to make is a familiar one, but with a twist.

Radical translation is a thought-experiment in which we imagine a linguist attempting to interpret a radically foreign language from scratch. All such a theorist has to go on in such a situation, Quine contends, is her observations of the verbal behavior of native speakers. She can correlate utterances with the enviromental circumstances in which they occur, and she can test her tentative translations by querying the natives in their native tongue. Such data will suffice to attach "stimulus meanings" to the "occasion sentences" (sentences that the native will only utter or assent to in the presence of specific stimuli). By clever contrivances, the theorist can also find evidence for translations of the language's logical connectives. She will also be able to identify a class of sentences that are treated by the natives as synonymous with each other, as well as classes of sentences that are treated, respectively, as true no matter what ("stimulus-analytic") and as false no matter what ("stimulus-contradictory").

This, however, is as far as the behavioral evidence will take her, and this is very far from a complete translation. In particular, the theorist has no empirical grounds for any particular *syntactic* analysis of the natives' truth-functionally primitive sentences. Given its stimulus meaning, the native utterance "gavagai" could be parsed indifferently as "There goes a rabbit" or as "There go some undetached rabbit-parts." No matter how extensive, information about the enviromental contingencies of use will always be insufficient to settle questions about the language's division of speech into words, or about the language's mechanisms of quantification and individuation.[2] If the linguist wants to decide such matters, she will be forced to import general principles Quine calls "analytical hypotheses." Although the choice of analytical hypotheses is not arbitrary (Quine allows that there may well be pragmatic considerations that favor one set over another), they are empirically

Alex Orenstein and Petr Kotatko (eds.), Knowledge, Language and Logic, 141–150.
© *2000 Kluwer Academic Publishers. Printed in Great Britain.*

ungrounded. And insofar as the complete translation depends on the analytical hypotheses, so too is it empirically ungrounded.

> Thus, the analytical hypotheses, and the grand synthetic one that they add up to, are only in an incomplete sense hypotheses .... [The translation of "Gavagai" by similarity of stimulus meaning] is a genuine hypothesis from sample observations, though possibly wrong .... On the other hand, no such sense is made of the typical analytical hypothesis. [Quine 1960, 73]

Now the question is, what has this argument established? It appears to have demonstrated at least the following *epistemological* claim: translation is *under*-determined by the totality of behavioral *evidence*. But it's clear that Quine intends to be defending a much stronger, *metaphysical* claim: translation is *un*determined by the totality of *facts*:

> The point is not that we cannot be sure whether the analytical hypothesis is right, but that there is not even, as there was in the case of "Gavagai," an objective matter to be right or wrong about. [Quine 1960, 73]

What is supposed to license the move from the epistemological claim to the metaphysical claim? Let us spot Quine the assumption that the behavioral evidence captures all the behavioral facts; we still need to know why facts *other than* behavioral facts might not settle matters of translation. Why not facts about cognitive psychology, or about neurology?

Quine's answer refers to the circumstances of language acquisition. The young language learner, Quine contends, has nothing to go on but the observable verbal behavior of others, and their observable responses to hers. The correct use of a term could not depend upon either the speaker or the hearer's being in some particular neurological state, for the child is in no position to make such an association. Moreover, the child's tutors cannot observe the child's neurological state, either, and so could never be in a position to correct or applaud the child's usage. If linguistic facts can be learned at all, Quine concludes, it must be because they supervene on interpersonally accessible facts.

> The sort of meaning that is basic to translation, and to the learning of one's own language, is necessarily empirical meaning and nothing more. A child learns his first words and sentences by hearing and using them in the presence of appropriate stimuli. These must be external stimuli, for they must act both on the child and on the speaker from whom he is learning. Language is socially inculcated and controlled; the inculcation and control turn strictly on the keying of sentences to shared stimulation. [Quine 1969a, 81]

A further consideration is the fact that all that matters to success in language learning is conformity to the prevailing community patterns of linguistic usage. As long as two speakers agree in their verbal dispositions, it is irrelevant how those verbal dispositions are supported internally by "private" events, whether neurological or ideational:

Internal factors may vary *ad libitum* without prejudice to communication as long as the keying of language to external stimuli is undisturbed. [Quine 1969a, 81]

In sum, then, we can attribute to Quine the following argument:

1) Linguistic facts must supervene on whatever facts are evidentially available to a language learner.
2) Only behavioral facts are available to the language learner.
3) Linguistic facts must supervene on behavioral facts.

The relevance of radical translation to the metaphysics of meaning thus becomes clear. Radical translation is a device for simulating, as far as possible, the situation of the child learning a first language. Quine agrees that the radical translator can never recover the linguistic innocence of the preverbal infant. But for that very reason, he contends that whatever disanalogies exist accrue entirely to the side of the determinacy of meaning:

Now of course the truth is that [the radical translator] would not have strictly simulated the infantile situation in learning the native language, but would have helped himself with analytical hypotheses all along the way.... [Quine 1960, 71]

Quine thus presumes that there is no *more* empirical determinacy to language than what is established by the process of radical translation.

But is Quine right to presume this? I think not. I think there is actually considerably *more* empirical determinacy to the output of radical translation as Quine describes it than there would be were we to take premise one very seriously. Note that Quine confers upon his linguist a decided advantage over the pre-linguistic child: the translator, unlike the kid, is given access to *as much* behavioral evidence as she wants. Whenever the linguist has a question that can *in principle* be answered by further observation, Quine obligingly posits the relevant datum. The child, however, does *not* have an arbitrarily large body of behavioral evidence available to her, nor is she as active or as successful as the linguist in manipulating the behavior of her informants in order to elicit crucial data. Given premise one, shouldn't the methodology of radical translation be different? Shouldn't Quine be asking how determinate translation would be if the translator were restricted to the types and amounts of evidence a child *actually possesses* during the time in which the child actually does acquire language? Mustn't the objective basis of meaning be just *that*?

I mean to be pointing to a telling inconsistency in Quine's application of the principle in premise one. On the one hand, his rationale for excluding psychological and neurological facts from the body of evidence to which the radical translator can appeal is the presumption that such evidence is unavailable to the infantile language learner. On the other hand, he makes freely available to the translator an *amount* of evidence that goes far beyond the evidence provided to the infant. Perhaps the translator can choose between "There goes a rabbit" and "There are some rabbit-tracks" as a translation for "Gavagai" on the basis of the native's responses to situations cleverly contrived to discriminate between these two hypotheses. But the two-year-old language learner can stage no such experiments. She cannot necessarily manipulate environmental contingencies in order to choose between two competing, but equally adequate hypotheses as to the meanings of sentences; she is more or less dependent, particularly at the early stages of language development, on whatever her environment happens to bring her.

The following rejoinder may perhaps be offered on Quine's behalf: radical translation is not meant to be a psychological theory of language acquisition, any more than it is meant to be an epistemological thesis. It is rather a philosophical inquiry into the metaphysics of meaning. The point of the thought-experiment is to show that semantic facts do not supervene on non-semantic facts, that the intensional floats free of the physical. Because this is his aim, Quine is at pains to show that the indeterminacy of translation is something other than mere inductive uncertainty – he is not arguing merely that, at any given stage of theorizing, the evidence leaves open a variety of incompatible hypotheses. It is to forestall this particular misunderstanding that he allows the radical translator access to *all possible* behavioral evidence – all possible pairings, that is, of native sentence with circumstances of utterance or assent. Since translation can be shown to remain indeterminate even given *all possible* evidence, translation is not merely underdetermined by the available evidence, but genuinely unfixed by the totality of empirical facts.

But although this rejoinder is correct about Quine's philosophical aims, it misses the current point. In order to conclude from his thought-experiment that translation is genuinely indeterminate, we saw that Quine needs first to have it that the indeterminacy exists even relative to *all the physical facts*. But of course the paradigm of radical translation does not bring to bear all the physical facts: the translator is not allowed access to neurological facts, to the findings of cognitive psychology, not even to commensensical assumptions about the beliefs, desires, and general psychological make-up of the native informants.[3] So, as metaphysicians of meaning, we ask how Quine can safely assume that these excluded facts are not part of what fixes semantic facts. The only answer we ever get is that *children learning language do not have access to these facts.*

But this answer does not go far enough. The child's access to evidence is not just restricted as to *kind*; it is also restricted as to *degree*. Consider: Quine tells us, in effect, that "rabbit" cannot be determined to mean RABBIT rather than UNDETACHED RABBIT-PARTS by virtue of the neurological patterns in an adult speaker's brain, because the child learner can't possibly – and *a fortiori* doesn't – know anything about these. But if so, then by the same token, the meaning of "rabbit" cannot be determined by dispositions of the adult tutor that *the child doesn't actually see realized*. It matters little that some ambiguity *could in principle* be resolved by providing the child with some bit of behavioral evidence that she in fact lacks, for the same could be said about the kinds of neurological and psychological evidence forbidden to the radical translator.

This consideration is not fanciful. Lila Gleitman points out that there exist in natural languages like English a large number of pairs of verbs that apply, almost with necessity, to exactly the same situations. The only semantic difference between the members of each pair is the perspective the verb represents. Thus we have "flee" and "chase," "buy" and "sell," "win" and "beat," "give" and "receive," etc. In any circumstance in which the first member of the pair is appropriate, so will be the second. Gleitman argues that such pairs create an in-principle problem for anyone who contends that word meanings are learned by tracking the contingencies of their use:

Whenever the hounds are chasing the fox, the fox is fleeing from the hounds. If some hounds are racing, even with evil intentions, toward a brave fox who holds its ground, they cannot be said to be chasing him.[4]

Steven Pinker, however, counters with the consideration that there *are* in fact circumstances that could be used to discriminate between the members of these pairs: for example, I might speak of my *fleeing* the city, without being willing to say that the city is *chasing* me; and I say that I *bought* a Coke from the machine, but not that the machine *sold* me a Coke. But while these examples defeat the claim that there are no contingencies of use that could *in principle* enable the child to choose between CHASE and FLEE as meanings for a particular verb, they do not explain the semantic competence of a child who has *not happened to have encountered the crucial cases.* And yet children do acquire such verb-pairs with no particular difficulty.

In fact, the epistemic situation for the child is even worse. Given the ways in which the child must deal generally with word-world matches, it's very implausible that the acquisition of verb meanings could depend on the child's encountering a *single* counterexample to a particular hypothesized meaning. Gleitman points out that the child's learning strategy must be reasonably tolerant of counterexamples, just because there are plenty of reasons why the hypothesized referent of any given term might be absent in a given circumstance. Gleitman cites a study by Beckwith, Tinker and Bloom (1989) in which a very large corpus of maternal utterances was analyzed, with a view to determining what in the child's experience enabled verb learning to occur. Their rather startling finding was that, in a large percentage of the cases, the utterance did not describe actually occurring situations or actually present objects. Thus, for example, there was *opening* going on in only 67.5% of the cases in which "open" was uttered. Looking at matters in the other direction, the counterexamples threaten to be even more frequent: what are the chances, on any given opening that the child notices, that someone will bother to remark on the fact, using the word "open?"

In fact, Gleitman notes, once we start empirically investigating the suggestion that children learn their language by tracking (by methods either rationalistic or associationistic) contingencies of word use, we encounter a whole host of difficulties. Postive imperatives ("Go to sleep, now!") and questions ("Where's your blankie?") form a sizeable percentage of the child's early linguistic data. Yet it's *typical* of such utterances that the described events and referred-to objects are not present. We don't tell people to do things that are already done, nor inquire about the location of things that are sitting out in plain sight. There is also the problem of the level of abstraction at which events and objects are specified. Every instance of "duck" is simultaneously an instance of "bird" or "animal", and while discriminating stimuli exist potentially, there's no guarantee that they will occur within a child's early experience, nor that they will occur often enough for the child to take them seriously as countermanding evidence. Then there is the problem of words that denote non-observable, or (equally irksome) omni-present states of affairs – verbs like "think" or "know," for example.

Now what do we get if we restrict the radical translator to the *actual* data that children apparently have available to them? An extraordinary degree of indeterminacy, far more than Quine is willing to countenance. And what conclusions are we to draw from this? If we insist that linguistic fact supervenes on the actual data available to children, then we'd have to say that meaning is virtually unconstrained, and that it's a crapshoot whether anyone ever understands anything anyone else ever says. (Jerry Fodor has said he sees nothing absurd in this conclusion.) But such a conclusion, however philosophically warranted, runs up against the following fact: children do appear to come to resemble their elders, and each other, in respect of verbal dispositions. This is an achievement children routinely attain, and it remains to explain how this happens.

The point I'm urging here is of course quite of a piece with Chomsky's argument from the "poverty of the stimulus." Chomsky and his followers have shown, quite conclusively in my view, that children do not receive the kind of focused and explicit feedback on their syntactical productions that would be necessary for them to learn their languages' grammars by a process of conditioned reinforcement. Brown and Hanlon (1970) found, for example, that there was no correlation between the grammaticality of a child's sentences and expressions of parental approval or disapproval. The children thus appear to lack "negative evidence" – evidence that a particular construction is *not* grammatical. This fact makes even more mysterious (on the assumption that the child learns through operant conditioning) the fact that certain types of linguistic errors are rare or totally absent in children's early corpus. Children rarely make word order mistakes, for example(Hirsch-Pasek,Golinkoff *et al.*, 1985), and have never been observed to extract verbs from relative clauses in the course of sentence formation. They also tend to obey rules that aren't really there – regularizing irregular verbs despite explicit instruction to the contrary (Baker 1979).

Chomsky's rationalistic explanation of these data is that children are born with built-in constraints that simplify the problem of acquiring language by significantly reducing the hypothesis space that a child must search. A constraint that requires all hypothesized rules to be structure-dependent, for example, would explain why children never attempt to form the question "Is the cat who on the mat is fat?" out of the assertion that the cat on the mat is fat. But syntax is one thing; surely semantics is something else. Native constraints cannot account for the American child's coming to treat "cat" as a name for cats, because French children learn an entirely different term for the same animal. *Surely* the learning of semantics must be a matter of learning word-world contingencies – *for what else is there*?

In psychology one may or may not be a behaviorist, but in linguistics one has no choice. Each of us learns his language by observing other people's verbal behavior and having his own faltering verbal behavior observed and reinforced or corrected by otheres. We depend strictly on overt behavior in observable situations....There is nothing in linguistic meaning beyond what is to be gleaned from overt behavior in observable circumstances. [Quine *The Pursuit of Truth*, 37–8]

But what Gleitman's research shows is that this truism – "we depend strictly on overt behavior in observable situations" is, like so many truisms, *false*. Claims about what is available to the language learner are *empirical* claims – they cannot be known to be true *a priori*. The irony is that Quine, despite his eloquent advocacy of naturalized epistemology, and his stern denunciations of philosophical "make-believe," never does make any serious inquiry into the actual conditions of language acquisition.

This still leaves the question of how kids *do* acquire their semantic knowledge – or, not to beg questions – how they manage to bring their verbal dispositions into conformity with those of their linguistic communities. And it's at this point that it's worth separating two distinct strands in Quine's preferred view of psychology. He is, of course, a behaviorist. But behaviorism involves two independent assumptions about psychology: one is anti-mentalism, the assumption that internal states of the organism do not constitute independent variables in the production of behavior, and the other is empiricism, the view that all knowledge is constructed out of sensory primitives via general principles of learning (or association). Behaviorism more or less entails empiricism, but the reverse is not so. One can be a mentalist, and even speak of internalized rules and principles, and yet still seek explanations of learning in terms of environmental contingencies.

Recall premise two in the argument I attributed to Quine: Only behavioral facts are available to the language learner. The justification for this premise is meant to be the necessity of evidence being intersubjectively available. But as Quine construes "behavioral evidence," it includes far less than all the *information* that is present, and objectively so, in the language learning environment. Gleitman's own solution to the problem she outlines, for example, refers to the mass of *syntactic* information that children have access to in the course of learning the semantics of their native words. Typically, Gleitman found (together with Barbara Landau), when there are pairs or clusters of semantically related verbs, the individual members of the group tend to occur in distinct syntactic environments. Compare, for example, the set of syntactic frames in which the verbs "look" and "see" occurred within one corpus:

Look at NP	See NP
Look D	See?
Look!	See? This is NP
Look! This is NP	Come see NP

It turns out, further, that those "counterexample" cases involving "look" and "see" – cases in which the objects or events referred to are not within sight – are all distinguished syntactically from the "canonical" cases outlined above:

Look AP   See S
Look like NP   See ∅
Look how S

Gleitman's theory, in sum, is that the child is engaged in a process of "syntactic bootstrapping" – that the child is able to use information about the syntactic frames in which a given term occurs as *evidence* about its precise meaning. This, of course, reverses the situation the radical translator finds herself in – she settles stimulus meaning, then finds herself with too many choices regarding syntax. But if Gleitman's data are right, the task of settling even stimulus meaning without availing oneself of the – perfectly public – information about syntax may well be impossible.

Well, I say "perfectly public," but that's not quite right. A good empiricist must ask two questions: first, how does the child know what the right syntactic analyses are? and second, how does the child know what to make of the syntactic information? Some of what's known with respect to the first question should hearten empiricists everywhere, though it provides no special cause for concern among nativists. Recent research by Safran, Aslin, and Newport (1996) indicates that there is sufficient information in the typical primary linguistic corpus to make it possible to infer word boundaries from the patterns of phonemic contingencies, and that further, very small infants are able to make both the discriminations and computations necessary to exploit this information. The child, that is, is capable of noticing that, for example, the probability of hearing the syllable "bé" immediately after hearing the syllable "ba" is much higher than is the probability of hearing "iz" after "be," and thus is apt to infer that "ba-be" is one word, while "be-iz" is not. The rest of what's known is less favorable to empiricism – the information in question is perfectly objective, but is only of use to a creature built to know what to do with it. For one thing, there appear to be physical features of the speech stream that are universally indicative of phrasal boundaries (Cooper and Paccia-Cooper 1980), though chimpanzees, unlike children, are unable to exploit this fact in the learning of speech. Furthermore, Gleitman's theory of syntactic bootstrapping only works on the assumption that the rules that map syntactic profiles onto semantic roles are natively available to the child; otherwise, performing the parsings would get the child nowhere.

All this points to one final difficulty with radical translation as a model of language acquisition: radical translation presumes the existence of a *rational* relation between the environmental information the child relies on, and the linguistic conclusions to which the child eventually arrives. Granted, the empirical evidence does not determine the radical translator's conclusions, for the analytical hypotheses must be imposed without empirical warrant. But the two situations look qualitatively different from each other once we find that the human language learner *in fact* relies on a body of evidence that is (a) composed of stuff of which mature speakers are generally unaware (prosodic patterns, syllable contingencies) and (b) useless without additional information. Now it should come as no surprise to a behaviorist that a human can learn without being consciously aware of the environmental contingencies that are in fact shaping her behavior. But then that makes it *more* surprising that Quine limits the radical translator to evidence that the translator is able to consciously note.

Similarly, there is no dictum in the behaviorist creed that says that operant conditioning will track rational relationships – that I'll only come to associate nausea with cream cheese and jelly (as I did in childhood) if nausea is a *reasonable* response to such a sandwich filling. Yet Quine constrains his translator to discover *evidentiary* patterns in support of her semantic hypotheses, ignoring the possibility that the child, who the translator is meant to mimic, needs considerably less "evidence" to settle on linguistic conclusions than would rationally justify them. The child at the peak of her word-learning form – eighteen months to six years – appears to acquire, on average, nine new words a day. (Carey 1978) Where would she find the *time* to get proper evidence for such wanton hypotheses? And how long, in comparison, does it take a dog, under explicit and assiduous training, to learn even one? Whether it's a matter of the child's specialized "innate similarity space" or her innate knowledge of universal semantic structure (in case these are different possibilities), the child is obviously not proceeding as the radical translator does.

The problem with the paradigm of radical translation is not, then, simply that the constraints on the evidence presumed to be linguistically relevant are too *stringent*; it is that they are utterly *arbitrary*. Given the actual facts, there is nothing of any theoretical interest about the particular set of constraints embodied in the paradigm of radical translation: they do not match the constraints operating on the child learning language; they do not even reflect consistently the behaviorist methodology Quine presumably endorses. What they do reflect, I suspect, is a prior commitment to a particular view of language, a view that says at the outset that linguistic facts are not "real" facts. Considered in the context of Quine's metaphysical goals, the idealization involved in permitting the linguist an unlimited amount of behavioral evidence appears concessive to the meaning realist; in fact, it is quite the opposite.

There is much to say about the indeterminacy argument that I have not said. However, many others have. (Chomsky 1969, Friedman 1975, Fodor and LePore 1992) My goal has been to make a point clear in a new way, a point that many of Quine's critics have urged against him in a variety of ways for many years: his epistemology of language is insufficiently naturalized. It was Quine that taught me, as he taught many philosophers of my generation, that our own cognitive achievements are not transparent to us simply through being our own. They are natural phenomena, as potentially obscure to us as the most distant star. Language, the most unique of our human activities, ought surely to provide the clearest object lesson in this regard. As regards radical translation, I say to Quine what Quine said to Carnap: "Why all this make-believe?"

University of North Carolina at Chapel Hill

NOTES

¹ Quine 1960. The argument from the non-derivability of semantic theory from physical theory can be found in Quine 1969b and Quine 1970, and the argument from the inscrutability of reference is given in Quine 1969a and Quine 1970.
² In fact, it has been shown (independently, by Gareth Evans and Jerry Fodor) that the mechanisms of individuation and predication *can* be translated on the basis of behavioral evidence. See Evans 1975 and Fodor 1994.
³ The prohibition against utilizing psychological assumptions derives from the presumed metaphysical interdependence of semantic and mental content, which Quine will argue for later on. In the discussion of radical translation, Quine actually violates this prohibition several times, most noticeably in his discussion of methods for discerning the natives' devices for expressing assent and dissent. (Quine 1960, 29–30). Davidson, who accepts the radical translation paradigm as a general model of the metaphysics of meaning, is forced by the need to rule out certain intuitively bad translation theories to allow the radical translator to speculate about such matters as whether the native population is scientifically sophisticated (Davidson 1976). As far as I know, neither Quine nor Davidson ever comments on the contamination to the thought-experiment introduced by the importation of these psychological hypotheses.
⁴ Gleitman 1990, 185. All subsequent citations of Gleitman are to this work.

REFERENCES

Baker, C.L. (1979) "Syntactic Theory and the Projection Problem." *Linguistic Inquiry*, 10, 533–81.
Beckwith, R., Tinker, E., and Bloom, L. (1989) *The Acquisition of Non-Basic Sentences.* Paper presented at the Boston Child Language Conference, Boston.
Brown, R., and Hanlon, C. (1970) "Derivational Complexity and Order of Acquisition in Child Speech." In J.R. Hayes (ed.), *Cognition and the Development of Language.* New York: Wiley.
Chomsky, N. (1969) "Quine's Empirical Assumptions." In D. Davidson and J. Hintikka (eds.) *Words and Objections*, 53–68. Dordrecht: D. Reidel.
Cooper, W.E., and Paccia-Cooper, J., (1980) *Syntax and Speech.* Cambridge, Ma.: Harvard University Press.
Evans, G. (1975) Identity and Predication. *Journal of Philosophy*, 72, 343–363.
Fodor, J.A. and LePore, E. (1992) *Holism: A Shopper's Guide.* Cambridge, Ma. and Oxford: Basil Blackwell.
Fodor,J.A. (1994) *The Elm and the Expert.* Cambridge, Ma.: MIT Press.
Friedman, M. (1975) "Physicalism and the Indeterminacy of Translation," *Noûs* 9, 353–73. Reprinted in P.K. Moser and J.D. Trout (eds.) *Contemporary Materialism: A Reader.* London and New York: Routledge, 1995.
Gleitman, L. (1990) "The Structural Sources of Verb Meanings." In *Language Acquisition*, 1,1, Potomac, Md.:Lawrence Erlbaum Associates. Reprinted in P. Bloom (ed.) *Language Acquisition: Core Readings.* Cambridge, Ma.: MIT Press.
Hirsch-Pasek, K., Golinkoff, R., Fletcher, A., DeGaspe Beaubien, F., and Cauley, K. 1985) "In the Beginning: One Word Speakers Comprehend Word Order." Paper presented at Boston Language Conference, Boston.
Quine, W.V. (1990) *The Pursuit of Truth*, 37–8. Cambridge, Ma.: Harvard University Press.
Quine, W.V. 1970. On the Reasons for the Indeterminacy of Translation. *Journal of Philosophy* 67: 178–183.
Quine, W.V. 1969a. *Ontological Relativity and Other Essays.* New York: Columbia University Press.
Quine, W.V. 1969b. Reply to Chomsky. *Words and Objections: Essays on the Work of W.V. Quine*, 302–11. Dordrecht: D. Reidel.
Quine, W.V. 1960. *Word and Object.* Cambridge, Ma.: MIT Press.
Saffran, J.R., Aslin, R.N., and Newport, E.L. (1996) "Statistical Learning by 8-Month Old Infants." *Science*, 274, 1926–1928.

PAUL HORWICH

ON THE EXISTENCE OF MEANINGS

How is it possible for a word or a sentence to be meaningful and yet not to have a meaning? That this is not merely possible but generally true is the startling conclusion of the considerations Quine advanced in chapter II of *Word and Object* and which he has refined in subsequent writings, notably "Ontological Relativity" and *Pursuit of Truth*.[1] In the present essay I would like to hazard a reconstruction of Quine's argument – a reconstruction which I hope does justice to it, but which reveals, I believe, a serious defect: namely that although its main assumptions are indeed correct the shocking conclusion cannot really be derived from them. More specifically, I will argue that we should welcome Quine's sceptical scrutiny of the naive conception of meaning, which he calls "the museum myth", and in addition that we should endorse a qualified form of his behaviourism regarding semantics; but I will suggest that these ideas ought to have led him, and ought to lead us, to a use theory of meanings rather than to a denial of meanings.

Quine's line of thought begins – at least in my formulation of it – by his calling to our attention our tendency to take for granted, without justification, a picture of meaning containing the following elements: (1) that there is a set of entities, known as *meanings* (or *concepts*), with which we are immediately acquainted; (2) that each term of a language is associated with ('possesses') one or more of these entities, and thereby becomes meaningful; (3) that we construct our conceptions of sameness-of-meaning – that is, of synonymy within a language and translation between languages – on the basis of the prior conception we have of these meaning-entities; and (4) that we are able thereby to rationalize the procedures we employ for trying to arrive at manuals of translation, and to explain why it is useful to be in possession of manuals that are correct.

Quine's critique of this naive picture may be divided into two parts: a methodological prologue and a constructive analysis. In the first part, the methodological prologue, he objects to the series of conceptual priorities to which the picture is committed: namely, that we derive explanations of the conditions of justification for our manuals of (i.e. beliefs about) translation, and for the utility of such manuals, from some prior conception of their content, which stems in turn from our immediate grasp of the meaning-entities. According to Quine, this order of explanation should be reversed. We must begin with the idea that certain translation manuals are useful, facilitating foreign travel. This is what explains why some are taken to be 'better' than others – they enable more accurate predictions, hence more successful negotiations. Therefore this is what determines the basis on which we select manuals of translation: i.e. it determines what is taken to be evidence for 'a correct translation'. Thus we arrive at our conception of the facts of translation.

151

Alex Orenstein and Petr Kotatko (eds.), Knowledge, Language and Logic, 151–162.
© *2000 Kluwer Academic Publishers. Printed in Great Britain.*

Then, finally, we may obtain our notion of meaning in terms of this conception of translation: meaning is that which is preserved in correct translation. To put it another way, our conception of what it is for a foreigner to have certain beliefs derives from our procedures (which include methods of arriving at translation manuals) for deciding when they have those beliefs; and these procedures are adopted on the basis of the pragmatic, predictive benefits that they engender. Thus the first component of Quine's critique is the claim that our only route to *meaning* is via *translation* which is itself reached from the *pragmatic function of translation manuals*.

In the second component of his critique Quine proceeds to implement this methodological proposal. His constructive analysis has three stages. First, beginning with plausible assumptions about the pragmatic point of translation, he derives a certain conclusion about the properties of an adequate translation manual: namely that a translation scheme is adequate if it preserves all dispositions to assent and dissent (i.e. manual T adequately translates A's language into B's if, whenever A is disposed to assent to, or dissent from, an utterance, then B is disposed to assent to, or dissent from, the translation under T of that utterance, and vice versa). Second, on this basis he argues that there will be many non-equivalent, yet perfectly adequate, translations into our language of just about any expression. And third, he thereby obtains the conclusion that there can be no such entities as meanings. Thus the naive picture – the museum myth – is shown to be not merely unwarranted but false.

This, I take it, is the structure of Quine's critique of meaning. In what follows I plan to go over the three steps is his constructive analysis, filling in some of the details and explaining where I have reservations. Instead of beginning at the beginning I'm going to consider the steps in reverse order, looking progressively deeper into his rationale for meaning-scepticism. So I will start by considering how one gets from the existence of multiple translations to the non-existence of meanings. Next I will examine the main premise of this argument and show how Quine derives it from his thesis about the conditions sufficient for adequate translations: that is, how he gets to the lemma that many non-equivalent translation manuals will exist from the thesis that any good predictor of assent/dissent dispositions will be an adequate translation manual. And then I will consider the first stage in Quine's constructive analysis (which is the one I think most debatable): his derivation of the adequacy of any assertibility preserving manual from the pragmatic raison d'etre of translation. At the end I will attempt to rectify what I believe is the main defect in this reasoning and I will suggest that the proper standards of adequacy will leave us with unique translations and with a use theory of meaning.

To begin then at Quine's final step, let us imagine that we have established an 'indeterminacy' thesis (I):

There are two adequate manuals of translation between language J and English such that according to one of them the translation of foreign word v is "e" and according to the other the translation is "e*" – where "e" and "e*" are not regarded as co-referential by English speakers (i.e. We do not accept, 'Something is e if and only if it is e*')

And let us also assume thesis (M):

> If a word has a meaning then an adequate translation of that word must have the
> same meaning

and thesis (S):

> If two English (context-insensitive) referring expressions have the same meaning
> then they are regarded as co-referential by speakers of English

It follows from these theses that v does not have a meaning. For if it did then, given (M), any adequate translation of v would have the same meaning. Therefore, given (I), both "e" and "e*" would have that meaning, hence the same meaning as one another. And therefore, given (S), "e" and "e*" would be regarded as co-referential, which , given (I), they are not. Therefore v does not have a meaning; so, given (M) and (I), neither does "e" or "e*". Thus if premises (I), (M), and (S) can be shown to be typically correct, then Quine's renunciation of meanings will be quite justified.

Of these premises the most controversial has been thesis (I) concerning the multiplicity of translations, and we shall be considering in a moment how Quine attempts to support it. It is worth noting, however, that even if (I) is accepted it is nonetheless possible to resist the sceptical conclusion, since one well might choose to deny thesis (M). The rationale for so doing would not simply be an unreasonable attachment to the 'museum myth', but rather a quite general and plausible anti-verificationist view about the relationship between truth and justification: namely, that one should be prepared to countenance some space between them – even space that is in principle ineliminable. Therefore, insofar as "adequate" is taken to mean "useful" or "justified", one might hold that even if there are many *adequate* translations, only one of them is *true*, and only that one need preserve meaning; therefore (M) is incorrect, and so the case against meanings collapses. To amplify a little, one might distinguish between *determinate truth*, which requires determination by underlying facts, and *truth* simpliciter, which does not. In that case one could say that although many of the truths of translation are indeterminate (since the underlying facts of verbal behaviour do not fix them), they nonetheless exist [2]. And in this way one would be able to hang on to meanings as those entitites that are preserved in *correct* (not merely *adequate*, and not necessarily *determinate*) translation, correlated with equivalence classes under the translation relation.

The plausibility of this sceptical attitude towards thesis (M) is enhanced by consideration of an analogous thesis (R) regarding reference: namely

> If a word has a referent, then any adequate translation of that word must have the
> same referent

For it is constitutive of what we mean by "refers", to accept instances of the disquotational schema, "'n' refers to N'. Therefore, even in the face of a plurality of non-equivalent, adequate translations of a foreign term v (including one into "e" and another into "e*"), we will nonetheless be inclined to regard it as perfectly determinate that "e" refers to e and "e*" refers to e*. But we cannot suppose that v refers to the same thing as both "e" and "e*", since they are not co-referential. Therefore (R) must be denied. But having taken this somewhat counter-intuitive step, it is not additionaly counterintuitive to deny (M) as well. These principles stand or fall together; and (R) must go; therefore it is natural to give up (M) too, and to apply to translation manuals the familiar distinction between the true and the merely justified.

Thus the step from indeterminacy of translation to meaning-scepticism is questionable. However, let us now put this issue aside and begin to descend deeper into Quine's line of thought by examining the basis for thesis (I). As we have just seen, it is possible to resist his meaning-scepticism even if one concedes that there exists a plurality of adequate translations. But should we make that concession? Why suppose that there are bound to be several adequate translation of virtually any expression?

For the case of *words* (as opposed to sentences) Quine provides us with a simple and ingenious argument, which turns on what he calls "proxy-functions". Let f be any 1–1 function: for example, the function that takes each physical object into its 'cosmic complement', which is everything in the universe outside that object. Thus it takes this table into the scattered entity that consists in everything but this table; and it takes that scattered entity to everything but that entity – namely, back to the original table. Now let us introduce, corresponding to each name "a", an additional singular term "a*", by means of the stipulation

$$a^* = f(a)$$

and let us introduce, for each primitive predicate "G", an additional predicate "G*" such that

$$\{x\}\{G^*[f(x)] <\!\!-\!\!> G(x)\}$$

and similarly for relational predicates. (Thus, in the case where f is the 'cosmic complement' function, we would introduce the term "Socrates*" to name the cosmic complement of Socrates, and the term "red*" to apply to anything that is the cosmic complement of something red, etc.) In that case it is not hard to see that, for every atomic sentence,

$$G(a) <\!\!-\!\!> G^*(a^*)$$

And it follows that every (extensional) sentence "p" is definitionally equivalent to the sentence "p*" which is derived from "p" by replacing every primitive term in it by the corresponding 'star' term. Therefore, if we assume that such equivalences are recognized by speakers of the language, it follows that "p" and "p*" are co-assertible: that in circumstances in which there is a disposition to assent to (or to dissent from) "p", if asked, there will also be a disposition to assent to (or to dissent from) "p*", and vice versa. Now suppose that the holistic totality of assertibility facts for a certain English speaker, s, is

$$Ds("p", "p*", "q", "q*", ...)$$

That is to say, Ds specifies, for each sentence, the circumstances in which s is disposed to assent to it and to dissent from it. Given the co-assertibility of each sentence with its 'star', it must be that

$$Ds("p", "p*", "q", "q*", ...)$$
$$= Ds("p*", "p", "q*", "q", ...)$$

Consequently, if we transform the body of facts regarding the speaker's assent/dissent dispositions, Ds, by means of the function, T*, which takes each term into the star of that term, then what results is the same as the body of facts with which we started.

Now in order to be able to go from this result to thesis (I) regarding the multiplicity of adequate translation manuals it is necessary (and sufficient) to embrace the following assumption (A):

Any manual of translation that preserves assertibility is adequate

Or more precisely: suppose that language H has sentences h1, h2, ..., that language J has sentences j1, j2, ..., that the totality of assertibility-facts for a certain speaker of H is D(h1, h2, ...), and that the totality of assertibility-facts for a certain speaker of J is D(j1, j2, ...) (i.e. the assent/dissent dispositions of the two speakers are the same, modulo substitution of H-sentences for J-sentences). In that case then, according to thesis (A), the translation schema that matches h1 with j1, h2 with j2, ..., and so on, is adequate.

On the basis of this assumption together with the proxy-function considerations we can prove thesis (I): that there are many adequate translations of virtually every term. For let H be my language and J be the language of a person (e.g. my doppelganger on a completely twin earth) who has exactly the same verbal dispositions as I have. Consider any proxy-function f, and the corresponding translation mapping T* which matches the terms in my language with the star (relative to f) of those terms in J. Any such scheme – of which there are many – will preserve all the assertibility facts. Therefore, given thesis (A), any such scheme will provide an adequate manual of translation.

Thus if thesis (A) is correct – if any assertibility preserving mapping is an adequate translation manual – then, given the proxy-function considerations, it follows that thesis (I) is correct – each of our words is adequately translatable in many non-equivalent ways. And from thesis (I) it can then be argued at stage three (pace the reservations I expressed earlier about this move) that there are no such things as meanings. So everything boils down to the plausibility of thesis (A), and hence to the first stage in Quine's constructive analysis where he attempts to show that this thesis may be derived from the pragmatic function of translation manuals. Let us examine this reasoning.

Before proceeding, however, it is important to recognize a severe limitation of the proxy-function considerations – a limitation emphasised by Quine himself. They purport to provide us with alternative, non-equivalent translations of *words*. Hence they threaten the very idea of *word*-meanings. But they do not offer uncontroversially non-equivalent translations of *whole sentences*. So they can pose no threat to *sentence*-meanings. For although they imply that a foreign sentence, j, might be translated equally well as either h or h*, these alternatives are definitionally equivalent to one another and it is far from obvious that they differ in meaning. Therefore one might respond to the proxy-function considerations by conceding that words do not have meanings, but by continuing to maintain that sentences do have them. Thus the thoroughgoing scepticism about meanings to which Quine subscribes requires a thesis stronger than (I); it requires a thesis which would refer to expressions in general, and not merely to words. And the justification for that strengthened version of (I) will not come out of the proxy-function considerations. Now, neither Quine nor anyone else has offered a *demonstration* of the stronger indeterminacy thesis. And although Quine says that he finds it plausible, he admits that he not only has no proof of it, but despairs of finding even a single clear-cut example. As he would put it: whereas the 'inscrutability of reference' (which concerns the translation of of *words*) is proved by proxy-function considerations, the 'indeterminacy of translation' (which concerns *sentences*), though likely to be true, is not demonstrable.

Let us therefore return to our scrutiny of the supposedly easier Quinean case against *word*-meanings. Our critique of this argument will suggest, I believe, that there is in fact no indeterminacy of word-translation – which will imply that there is no indeterminacy of sentence-translation either.

In order to derive conditions of adequacy for translation manuals from the requirement that they be useful one must make some assumptions about how such manuals are used and what makes them valuable. On the first point we may suppose that an H-speaker, interacting with J-speakers, will use manual T as follows: instead of saying what he would have said if he were interacting with fellow H-speakers he substitutes the translation of what he would have said (i.e. he substitutes the sentence T(h) for the sentence h); and when a J-speaker makes an utterence he responds as if it were an H-speaker making the translation of that utterence (i.e. he imagines that T(j) was uttered rather than j). In other words, for an H-speaker to adopt manual T in his interactions with J-speakers is for his expectations regarding a foreign utterance, j, to be the same as the expectations he has, when dealing with H-speakers, regarding the utterence T(j). As for the question of what it is for such a policy to be successful, i.e. for the manual to be useful, the answer is pretty obvious. A manual is pragmatically adequate if it in fact gives the same sort of ability to operate in the foreign context that we normally enjoy within our own linguistic community. Thus, we are often able to predict on the basis of someone's utterence, some feature of the environment, or something about what the speaker will do or say. So an adequate translation scheme would provide us with exactly the same ability in in dealing with foreigners. Similarly, in our own community we are often able to predict the effects on others of what we say; therefore, an adequate translation should also provide that ability in the foreign community.

To put it another way, we use a translation manual T by applying it to the behavioral/psychological theory we implicitly deploy in the midst of our own community, and converting it into a theory to deploy in connection with foreigners. And this is successful if the theory we thereby generate gives as good an account of the foreigners as the original theory gives of ourselves. In other words, we operate at home with an implicit theory $\$(w1, w2, ...)$ (where $w1$, $w2$, ... are the words of our language H); our use of translation manual T in community J consists in our operating there with $T(\$)$, which is $\$(T(w1), T(w2), ...)$; and this is succesful if $T(\$)$ is just as good there as $\$$ is here.

Now let us see what conclusions about the conditions of *adequacy* for translation schemes can be drawn from these observations about their use and utility. In the first place, we can see that the facts about which scheme is being used and whether it is working properly are open to ordinary observation. They are entirely determined by observable relations between what is said, the environment, and other behaviour. Such behavioural facts are the only data relevant to the assessment of a translation scheme, because, as we have seen, an adequate manual is one that converts reliable patterns of expectation in the home context into into reliable expectations abroad. And whether this is so is a matter of behaviour alone. Thus Quine's behaviourism in this domain is vindicated.

In the second place we must consider whether it is possible to justify thesis (A): that any assertibility preserving mapping is adequate. This is the burden of the first stage in Quine's constructive analysis. If (A) does follow from our assumptions about the way manuals are used and about what constitutes successful use, then his proxy-function argument (at the second stage) will show that there are indeed multiple, non-equivalent, adequate translation manuals. And (at the third stage) this lemma will imply (arguably) that word-meanings don't exist. But how can principle (A) be supported?

Actually I don't think it can be. The best that can be done to support it, as far as I can see, would go something like this. Suppose we are employing a translation manual T that preserves all facts of assertibility. More specifically, suppose that the entire complex of assertibility-facts concerning a certain J-speaker consists in $D(j1, j2, ...)$, and that all the facts concerning what I am disposed to assert are given by $D(h1, h2, ...)$, and that T matches hn with jn. Now suppose that I deploy T as a translation manual in my dealings with J speakers. In that case, my assertibility dispositions in those dealings will exactly match those of the J speakers, and my expectations will be just as reliable as they are at home. Therefore any mapping that preserves all assertibility dispositions is functionally adequate, hence correct.

The flaw in this argument, it seems to me, is that there is too much emphasis on *dispositions*, and not enough on what is actually accepted and on how those conclusions are reached. We can grant that a non-standard manual of translation, T*, which preserves all the facts of assertibility, will be adequate as far as *some* expectations are concerned – expectations of what the foreigner will be *disposed* to accept (under questioning) in various circumstances, and expectations, given what he is disposed to say, of certain environmental conditions. But it does not follow – and nor would it seem to be true – that such a manual will be perfectly adequate with respect to *all* the predictions that we might want to make: specifically, predictions about what will be sponntaneously accepted and about how those commitments will be arrived at.

For nobody comes out with a string of obviously equivalent sentences; typically, one of them alone is maintained. Thus in appropriate circumstances I think and say "That's a rabbit"; but I will not think "The cosmic complement of *that* is the cosmic complement of a rabbit" – even though I might well be disposed to work it out and assent to it if queried. Consequently, if I use such facts about *actual* usage, together with a translation scheme, to generate expectations about what foreigners will say under similar circumstances, then only one of the various assertibility-preserving schemes will be right. For remember, in general, how a translation scheme is deployed to generate predictions: we are to have exactly the expectations we would have had if we were dealing with members of our own linguistic community – modulo the substitution of the appropriate foreign sentence for our own. Thus if a foreign speaker makes a certain utterence we are to have the expectations that we would have had if one of our own people had said whatever, according to our translation scheme, translates that utterence. Similarly, if the circumstances are such as to lead us to expect that one of our speakers would make a certain utterence, then we are to expect the foreigner to come out with whatever is the translation of that utterence.

Thus T* is likely to engender incorrect expectations about what the foreigner will say. In addition it will fail to predict the foreigner's *inferential* behaviour. For we infer h* from h, not the other way round. We first observe (or in some other way discover) that "That's a rabbit" is true, and then, given the definition of h* in terms of h, we reason that "The cosmic complemmment of *that* is the cosmic compement of a rabbit" must also be true. Moreover the fact that we reason in this way is a behavioural fact about us – a fact about our behaviour with respect to h and h*. Consequently a fully adequate translation manual should preserve such behavioural relations, and no merely dispositions to assent and dissent. It should translate h into a sentence of J from which J-speakers infer the sentence into which it translates h*. That is, since we infer h* from h, an adequate translation, T, should be such that J-speakers infer T(h*) from T(h). But suppose that T(h*) = j* and T((h) = j, whereas T*(h*) = j and T*(h) = j*; and suppose that J-speakers infer j* from j. Then T* will not correctly predict the inferential relations, even though it may satisfy Quine's adequacy condition (A), preserving all the assertibility facts.

Having argued against the thesis (A), relied on by Quine, that any assertibility preserving translation is adequate, I would like to end this paper by sketching what I believe is a more plausible condition of adequacy (judged by Quine's own pragmatic and behavioural standards) and considering where this would leave us with respect to the issues of multiple translation and the existence of meanings.

Suppose translation manual T functions perfectly well in a certain community as a device of expectation replacement. In other words, if we employ T to transform what we would say at home into what to say in that community, and to determine how to react to what others are saying, we find that everything goes smoothly. It follows that the correlations between environmental circumstances, acceptance of sentences, and other states of mind are the same in the two communities, modulo the substitution of their words for ours – i.e. modulo the replacement of each of our words w with T(w). Now these correlational phenomena, within each community, are each the product of certain deeper general facts – explanatorily basic regularities governing the overall use of the words in relation to one another and to other factors. In so far as the correlations produced by these basic facts are the same, modulo T, within the communities, it must be that the basic facts themselves are the same, modulo T. In other words, if the theory that determines the overall use of *our* words is $(w1, w2, ...)$, then the theory that determines the overall use of the foreign words is T\$ – i.e. $((Tw1), T(w2), ...)$. Thus our conception of a functionally adequate translation schema presupposes that every person's linguistic behaviour is governed by certain universal principles including a set of interconected basic regularities for the use of words; and that a correct translation schema is one that matches words governed by the same basic regularities. To put it another way, the difference between the theory $, which accounts for the use of our words, and theory T$, which accounts for the use of the foreign words, is merely that the theory structure, $(x1, x2, ...)$, is occupied on the one hand by our words and on the other hand by the associated foreign words. Thus the property which our w1 has and which any adequate translation of w1 must also have, is $(\exists x2) (\exists x1) ... \$(x1, x2, ...)$. Therefore an adequate translation manual is one that preserves the explanatory roles of words, i.e. their basic regularities of use.

That this is indeed our conception is borne out in our practice of translation. Consider, for example, the fact that we translate a foreign word as "red" just in case the foreigners are disposed to ascribe it when red things are under observation. What is being supposed in this practice is that the explanatorily fundamental regularity for our word "red" is roughly:

A tendency to assent to "That's red" iff there is something red under observation

and that the explanatorily fundamental regularity governing the foreign word, say "rouge", is

A disposition to assent to "Ça c'est rouge" iff there is something red under observation

and that, in a similar way, the laws governing our "That's ..." match the laws governing "Ça c'est ...".

Suppose that this conception of translation is correct, where does it leave us on the issues of indeterminacy and the existence of meanings? Well, as we have seen,

Quine's sceptical argument depends on the assumption that any assertibility-preserving manual is adequate. So the question arises as to whether this assumption is consistent with the use-theoretic picture of translation that we have just arrived at.

Clearly, the answer is No. In the first place, the theory we are deploying to explain and predict both within our own community and (relative to a translation manual) within the foreign community, is intended to cover, not merely assent/dissent dispositions, but also what is actually held true, and also the causal relations between these states, the inferential processes that engender them. Therefore, for any expression "k", the theoretical roles of "k" and of "k*" are bound to differ. In the second place, an assertibility preserving mapping might translate a single word of one language as a complex expression of the other. But it is not posssible for a single word to have the same theoretical role (i.e. to be governed by the same basic use-regularities) as a complex expression. For complex expressions will participate in regularities which relate the whole expression to its parts; whereas no such regularity can apply to single words. Thus there are at least some assertibility preserving schemes which are not adequate translations: namely, those that involve mapping words onto complexes. Moreover, it is precisely this type of mapping that is involved in Quine's proxy-function argument. Remember that in this argument the non-standard mapping T* takes the term "Socrates" to, for example, "the cosmic complement of Socrates" (or, in general, to "the object that bears relation R to Socrates") and it takes the predicate "red" to "something that is the cosmic complement of a red thing" (or, in general, to "something that bears relation R to a red thing"). Such mappings, let us grant, do preserve assertibility, but they do not preserve meaning in the intuitive sense. More importantly they do not preserve basic regularities of use; hence they do not provide the sort of translation that best serves the needs of prediction; hence, by Quine's own fundamental standards of adequacy, they are inadequate translation manuals.

Finally, let us consider whether there is any scope at all for multiple translation from the perspective of the present conception. To repeat: a translation manual, T, is deployed by converting the theory, $(w1, w2, ...)$, which generates behavioural expectations at home, into a structurally identical theory, $T(\$)$, [i.e. $\$(T(w1), T(w2), ...)$], which applies abroad; and T is adequate just in case $T(\$)$ is as accurate as $. Given the predominantly behavioral character of these theories, what this strongly suggests is that in order for two manuals T and T* to be equally good, the theories $T(\$)$ and $T^*(\$)$ must be identical. For if they were not identical, then they would to some degree make different behavioral predictions, and so there would be the potential for one to be better than the other. So the question becomes whether this condition – that $T(\$) = T^*(\$)$ – leaves any scope for a difference between T and T*, and hence any scope for a multiplicity of correct translation manuals.

The answer is that it does – but barely. For suppose there were two non-synonymous words whose uses in relation to one another, to other words, and to environmental circumstances, were perfectly symmetrical. In that case, a translation of a foreign term into one of these words would be just as good as a translation into the other; and we would have a real case of indeterminacy of translation. One hypothetical example of such a phenomenon would be provided by a physical theory containing terms "A" and "B" for two types of fundamental particle, where "A" and "B" play qualitatively identical roles in the theory-formulation. The postulates might look something like this

$$A + A \rightarrow X$$
$$B + B \rightarrow X$$
$$A + B \rightarrow Y$$
$$A + C \rightarrow Z$$
$$B + C \rightarrow Z]$$

Thus the uses of "A" and "B" would be exactly similar; and if the same physical theory were formulated by foreigners, but using "A*" and "B*" instead of our "A" and "B", then both the translation of "A*" as "A" and the translation of "A*" as "B" would preserve usage, and hence be perfectly adequate [5].

More precisely, suppose:

(a)   our theory treats words w1 and w2 symmetrically:
       i.e. $\$(w1, w2, \ldots) = \$(w2, w1, \ldots)$.
(b) $T(wn) = vn$ (where w1, w2, .., wn, ... are our words and v1, v2, .., vn, ... are words of the foreign language).
(c) $T*(w1) = v2$, $T*(w2) = v1$, and, for $n>2$, $T*(wn) = vn$

In that case. given (b)

   $T[\$(w1, w2, \ldots)] = \$(v1, v2, \ldots)$

and given (c)

   $T*[\$(w1, w2, \ldots) = \$(v2, v1, \ldots)$

Moreover, from (a) it follows that

   $T[\$(w1, w2, \ldots)] = T[\$(w2, w1, \ldots)]$

which, given (b), implies that

   $\$(v1, v2, \ldots) = \$(v2, v1, \ldots)$

Therefore

   $T[\$(w1, w2, \ldots)] = T*[\$(w1, w2, \ldots)]$

Thus even though the translation manuals T and T* are divergent they are equally adequate since they engender exactly the same theory of the foreigners' behaviour.

However, it is not clear that there are any actual examples of this hypothetical situation – any actual English words w1 and w2 that we treat so symmetrically that $\$(w1, w2, \ldots) = \$(w2, w1, \ldots)$. Certainly there arn't going to be many such pairs. And, as I argued earlier (in criticizing the third stage of Quine's constructive analysis), even in this sort of case, one need not conclude that the words don't have meanings, but merely that the assignment of their meanings to foreign terms is indeterminate. For example we would not compelled to admit that "A" and "B" do not possess meanings. It would be enough, rather, to allow that there is no determinate fact as to whether "A"'s meaning is also possessed by "A*".

Thus my conclusion is that although the basis of Quine's treatment of meaning may be perfectly reasonable – specifically, his insistence that our conception of meaning be extracted from the pragmatic function of translation, and his insistence that the correctness of a translation manual be evaluated solely on behavioral grounds – the surprising, meaning-sceptical conclusion which he draws does not follow. The best translations, from the point of view of prediction, are those that preserve the theoretical roles, the basic use-regularities, of words. Such an adequacy condition will not normally be satisfiable by two non-equivalent translation manuals; hence it provides no grounds for the rejection of meanings. On the contrary, what it suggests is a reduction of meaning-properties to basic regularities of use.[3]

University College London

NOTES

[1]   See W.V. Quine, *Word and Object*, MIT Press, Cambridge Mass., 1960; "Ontological Relativity", in *Ontological Relativity and Other Essays*, Columbia University Press, New York and London, 1969; and *Pursuit of Truth*, Harvard University Press, 1990.
[2]   Robert Brandom (in his "The Significance of Complex Numbers for Frege's Philosophy of Mathematics", *Proceedings of the Aristotelian Society*, 1996, pp. 293–315) suggests that "*i*" and "–*i*" (refering to the square-roots of –1) are instances of this phenomena, providing a genuine case of indeterminacy. For the sake of even greater symmetry (hence more plausible indeterminacy) Hartry Field (in "Some Thoughts on Radical Indeterminacy", *The Moninst*, 1998) imagines a community in which the symbols for the square-roots of –1 are "/" and "\" – where "\=-/" and "/=-\" are both accepted, neither being regarded as more definitional than the other.
[3]   For further articulation and defence of the use theory of meaning see my book, *Meaning*, Oxford University Press, Oxford, 1998. The present essay is chapter 9 of that work.

PETER PAGIN

PUBLICNESS AND INDETERMINACY

This paper is concerned with one rather specific question:

Is indeterminacy of translation a consequence of the publicness of meaning?

As I understand professor Quine, he thinks that the answer to this question is *yes*.[1] I shall provide some support for this interpretation. Personally, I believe that the answer is *no*, but I shall not try to establish that answer. I don't know how to do that, or even if it is possible to do it.

Instead, I shall examinine what I take to be Quine's reasoning from the publicness thesis to the indeterminacy thesis. I shall reconstruct the reasoning into an explicit argument, and try to show that this argument cannot be successful. It is not easy to say in advance in what way that argument cannot be successful, since that depends on the structure of the argument, but I shall make it explicit below (section 3).

1. WHAT IS PUBLICNESS OF MEANING?

The thought that linguistic meaning is public has been a main theme for a long time in the 20th century (analytic) philosophy of language. It features in different versions in Wittgenstein, Quine, Donald Davidson, Michael Dummett and others. The common element in these versions is primarily the rejection of a certain picture of linguistic meaning, according to which mental factors, private in the sense of being epistemically inaccessible to others than the speaker himself, determine what his words mean. The rejection of this picture sometimes takes the form of tenets which explicitly concern *knowledge or knowability* of linguistic meaning, i.e. the possibility of knowing what meaning another speaker attaches to her words (as in Davidson's writings). Sometimes it takes the form of a more metaphysical statement about the nature of language, or meaning, or speakers, e.g. that a speaker's understanding of an expression can be made manifest in practical capacities (as in Dummett's writings).

The more metaphysical versions are, however, motivated by epistemological considerations. You can, for instance, reason like this: "People really communicate linguistically. But it wouldn't be real communication unless they know what the others mean by their words. But for such knowledge to be possible, linguistic meaning must be of such and such a nature." The "such-and-such" a nature can be e.g. reducibility to behavior. You might say that linguistic meaning is public in the

163

Alex Orenstein and Petr Kotatko (eds.), Knowledge, Language and Logic, 163–180.
© *2000 Kluwer Academic Publishers. Printed in Great Britain.*

sense that what a speaker means reduces to his linguistic behavior (which is in itself something publicly observable). In that case you have chosen a metaphysical notion of publicness, even if motivated epistemologically.[2] I prefer to tie the notion of publicness directly to the underlying epistemological idea. One reason for this is that I am interested in the various metaphysical conclusions, about the nature of language, that can be drawn, and that have been drawn, from the epistemological premisses.

I want to distinguish between two different theses, which I shall call *Basic Publicness* and *Standard Publicness*, according to whether the possibility, or the actuality, respectively, of knowledge of other speakers' language is what matters. The thesis of Basic Publicness is then the following:

(BP)   What a speaker means by his words *can* be known by others

So the thesis of basic publicness denies that what meaning a speaker attaches to his words can be something essentially private, epistemically inaccessible to other speakers.

The thesis of Standard Publicness is this:

(SP)   What a speaker means by his words is *normally* known by others

The thesis of Standard Publicness states that, at least in ordinary communicative situations, speakers understand each other correctly, and can be credited with the knowledge of understanding correctly. I also include in that thesis that speakers normally know, of the members of their own speech community, what they normally mean by their words. You know e.g. that your barber normally means *car* by "car", etc.

I shall not further dwell on the intepretation of these two tenets, or their relation to each other. Much can be said. For Quine I think both (BP) and (SP) are important, but the second, that of Standard Publicness, is the more important one, as we shall see.

2. QUINE'S WAY FROM PUBLICNESS TO INDETERMINACY

The thesis of indeterminacy of translation is a metaphysical thesis, a thesis concerning the nature of language. So, an argument from publicness to indeterminacy is an argument from an epistemological premiss to a metaphysical conclusion. As I see it, Quine has such an argument, or at least, if not an explicit argument, a line of reasoning of that kind. The following passages are from *Word and Object* and *Pursuit of Truth*, respectively, with more than thirty years in between:

Language is a social art. In acquiring it we have to depend entirely on intersubjectively available cues as to what to say and when. Hence there is no justification for collating linguistic meanings, unless in terms of men's dispositions to respond overtly to socially observable stimulations. An effect of recognizing this limitation is that the enterprise of translation is found to be involved in a certain systematic indeterminacy;[3]

> Critics have said that the thesis [of indeterminacy of translation] is a consequence of my behaviorism. Some have said that it is a *reductio ad absurdum* of my behaviorism. I disagree with this second point, but I agree with the first. In psychology one may or may not be a behaviorist, but in linguistics one has no choice. Each of us learns his language by observing other people's verbal behavior and having his own faltering verbal behavior observed and reinforced or corrected by others. We depend strictly on observable behavior in observable situations. [...] There is nothing in linguistic meaning beyond what is to be gleaned from overt behavior in observable circumstances.[4]

If we break down the statement about learning (through observation of behavior) into components, we get the following structured representation of Quine's reasoning:

1. Speakers learn language from other speakers (PUBLICNESS)
2. Their only source of information is observation (EMPIRICISM)
3. What can be known by ordinary, everyday observation about what others mean, can be known by observation of their behavior in observable circumstances (BEHAVIOR RELEVANCE).
4. Hence, there is nothing in meaning beyond what is determined by observable behavior in observable circumstances (LINGUISTIC BEHAVIORISM, from 1, 2 and 3)
5. Hence, correctness of translations is indeterminate (INDETERMINACY, from 4)

I have labelled each step in this argument according to which thesis I think it expresses or is a fairly immediate consequence of. According to these labels, what I ascribe to Quine is an argument from the two premises of publicness and empiricism to the intermediate conclusion of linguistic behaviorism, and further from that intermediate conclusion to the final conclusion of indeterminacy. I shall now comment on the steps of this argument in order to support my exegetical claims.

Publicness

Quine points to the fact that we learn our language from others, and I have characterized that view as the tenet of Publicness. More precisely, I think that it, more or less, expresses the view of Standard Publicness, in the form that speakers, in learning a language, do learn what other speakers mean by their words. Clearly, this is not an *obviously* correct interpretation of Quine. I think there are two crucial issues. The first is whether language learning is to be characterized at all as leading to knowledge about other speakers' semantics. The second is that Publicness is stated in terms of meaning, and it almost goes without saying that using the term "meaning", in attributing a view to Quine, can only be done with utmost care. Let's start with the first issue.

Is it really Quine's view that as we acquire a language, we learn what other speakers, our teachers, mean by their words, or what semantic properties their words and sentences have. The relevant alternative view of language learning is that of a process of acquiring a set of speech dispositions which in fact largely overlap with the speech dispositions of others, one's teachers, without necessarily involving any *knowledge* of what they mean, or even of the overlap itself. If this were Quine's view, then ascribing Publicness to him would be wrong. I don't think it is.

Although Quine often writes as if language learning does not consist in anything more than development of speech dispositions, there are passages which do include cognitive components, i.e. components of knowledge about other speaker's speech habits:

In the case of occasion sentences [language learning] amounts to learning what occasions warrant assent to the sentences, or dissent.[5]

It is hard not to understand this passage as concerning knowledge of some kind of community standards of assent and dissent. And in another passage Quine writes

The semantic part of learning a word is more complex than the phonetic part, therefore, even in simple cases: we have to see what is stimulating the other speaker.[6]

In this case knowledge of the other speaker, knowledge of a feature of a linguistic disposition, is explicitly included in a description of what language learning involves.

But I think something else to be more important than such scattered pieces of textual evidence: Quine's reasoning from language learning to linguistic behaviorism would not make good sense if language learning did not include learning what others mean. Clearly, if knowing a language *only* consists in having certain speech dispositions, then, trivially, linguistic behaviorism is true. There can be no interesting argument from language learning to linguistic behaviorism. Therefore, since Quine explicitly reasons from language learning to linguistic behaviorism, I don't think Quine can be credited with such a view of language learning.

Furthermore, if knowing a language consists in having certain speech dispositions, *and* something else, like associating meanings with one's words, then linguistic behaviorism doesn't trivially follow, but then it cannot be argued either. On that assumption, it is perfectly alright to say that the meanings of your words are determined by your private mental experiences and intentions, inaccessible to me, and that my language learning consists in this: on the basis of observing your speech behavior I acquire my own speech dispositions, and begin to associate meanings with my words, in a way which is cognitively inaccessible to you. And if this is true, then linguistic behaviorism is clearly false. Again, it would not be reasonable to ascribe this view of language learning to Quine.

The problem is that if language learning does not amount to anything *more* than developing one's own language, then language learning does not offer any *restriction* on what linguistic meaning can be like. It can be as mentalistic as anything. By contrast, if learning a language does involve learning what others mean, then any restriction on what can be known about others will be a restriction on what meaning can be. If your mental events are epistemically inaccessible to me, and I can know what your words mean, then what your words mean cannot depend on your mental events.

So, if there are restrictions on what can be known about others, it is plausible to reason from those restrictions, and from the assumption that we learn what others mean, to conclusions about linguistic meaning. And Quine does place restrictions on what can be known about others: viz. restrictions coming from empiricism itself (see below). In this light, by far the most natural interpretation of Quine is that it is part of his views that language learning does involve learning what others mean by their words.

The second issue is whether you can ascribe to Quine the view that the knowledge a speaker has, in knowing a language, is knowledge of what other speakers *mean*, or what their words mean. As I understand it, Quine alternates between two notions of meaning, one objective and one uncritical, or intuitive. The intuitive notion is more or less the traditional notion, wedded to the traditional notion of synonymy. In fact, reference to *some* intuitive notion is used for the very formulation of the indeterminacy thesis, for it is with respect to intuitive standards of sameness and difference of meaning that the incompatibility of different translation manuals are judged:

In countless places [the translation manuals] will diverge in giving, as their respective translations of a sentence of the one language, sentences of the other language which stand to each other in no plausible sort of equivalence however loose.[7]

Here, in *Word and Object,* there is appeal to an intuitive notion of equivalence of sentences, of the same language, and in *Pursuit of Truth* we have a similar appeal to coherence and to interchangeability:

The thesis of indeterminacy of translation is that these claims on the part of two manuals might both be true and yet the two translation relations might not be usable in alternation, from sentence to sentence, without issuing in incoherent sequences. Or, to put it another way, the English sentences prescribed as translation of a given Jungle sentence by two rival manuals might not be interchangeable in English contexts.[8]

There is a question whether the indeterminacy thesis can be stated at all without appeal to some uncritical notion. I think it cannot.[9]

By contrast, when Quine states that "there is nothing in linguistic meaning beyond what is to be gleaned from overt behavior in observable circumstances" he is, as I understand him, speaking of meaning in an objective sense, viz. meaning *insofar* as can be gleaned from overt behavior in observable circumstances. There would not be much point to the statement if Quine's view was simply that there is nothing in linguistic meaning at all.[10] Rather, the indeterminacy thesis can be seen as amounting to the claim that there is a gap between objective meaning and intuitive meaning, between objective synonymy and intuitive synonymy, between objective and intuitive equivalence: there is no objective basis for many of our intuitive semantic discriminations.

So, use of the term "meaning" in ascribing the publicness view to Quine is perfectly alright. Clearly, it is the objective sense of "meaning" which is intended in the statement of Publicness.

Empiricism

Empiricism is the view that all knowledge derives from sense experience. As a consequence of empiricism, knowledge about other speakers, including knowledge of what they mean, must be based on observation. We do not have any direct epistemic access to their minds, but must base all knowledge about their minds, about what they feel, mean, want, believe and intend, on observation.

Behavior relevance

Strictly speaking, it isn't ruled out by empiricism that I can gain knowledge about what a speaker means by way of an observation which isn't an observation of that speaker's behavior, nor, in any ordinary sense, an observation of any circumstances of behavior, but an observation of something else, from which I infer something about what she means. So publicness and empiricism are not enough for concluding the truth of linguistic behaviorism. We could bridge the gap by declaring that anything observable could be taken as circumstances of behavior, but that would take much substance away from the doctrine.

Rather, I think it is best to include as a separate premiss that behavior, observable behavior in observable circumstances, is just what is *relevant* for observation based knowledge about what others mean. So we add the premiss that everything that we *can* know, on the basis of ordinary (i.e. everyday) observation, about what others mean, can be known on the basis of observation of their behavior. This does not exclude the existence of indirect evidence, which I can get by observing other things, but it assures us that no indirect evidence is necessary.

The premiss of Behavior Relevance does not exclude the possibility of arriving at knowledge of meaning on the basis of *scientific* observation. Neither does it exclude the possibility that this knowledge could not be had by ordinary observation alone. I might infer something about meaning on the basis of meter readings from a machine that monitors the speaker's neurochemistry. However, the premiss of Publicness tells us that ordinary speakers do have the knowledge there is to have. Since ordinary speakers rely only on ordinary observation, scientific observation isn't necessary either.

Linguistic behaviorism

Thesis 4 is simply a statement of the doctrine of linguistic behaviorism. Linguistic behaviorism is not, for instance, a stimulus-response theory about language learning, or anything similar. This, I think, is pretty clear from the quoted passage from *Pursuit of Truth*,[11] but is also made quite explicit elsewhere.[12]

Indeterminacy of translation

The indeterminacy thesis is presented in the following ways in *Word and Object* and *Pursuit of Truth*, respectively.

The thesis is this: manuals for translating one language into another can be set up in divergent ways, all compatible with the totality of speech dispositions, yet incompatible with one another. In countless places they will diverge in giving, as their respective translations of a sentence of the one language, sentences of the other language which stand to each other in no plausible sort of equivalence however loose. [...] It is in this last form, as a principle of indeterminacy of translation, that I will try to make the point plausible in the course of this chapter.[13]

A manual of Jungle-to-English translation constitutes a recursive, or inductive, definition of a translation relation together with a claim that it correlates sentences compatibly with the behavior of all concerned. The thesis of indeterminacy of translation is that these claims on the part of two manuals might both be true and yet the two translation relations might not be usable in alternation, from sentence to sentence, without issuing in incoherent sequences. Or, to put it another way, the English sentences prescribed as translation of a given Jungle sentence by two rival manuals might not be interchangeable in English contexts.[14]

A reasonable reconstruction is the following:

a) A manual of translation between two natural languages is *correct*, or *acceptable*, if it agrees with the speech dispositions of all speakers concerned.

b) Two manuals of translation between the same two languages are *mutually compatible* if they give, for each sentence *s* in the one language, translations *s'* and *s"* which are intuitively equivalent: it must then be possible to interchange *s'* and *s"* without loss of coherence of the discourse. Then the two manuals can be used in alternation.[15]

c) There are, or can be construed, two manuals between the same two languages which are both correct, and yet not mutually compatible.

In order to give the indeterminacy thesis full substance the criteria of correctness of manuals must be fleshed out. I come back to this issue in section 5.

This concludes my presentation of the indeterminacy argument itself.[16]

3. THE STRATEGY OF ATTACK ON QUINE'S ARGUMENT

As the argument is laid out, it is an argument from three premises, Publicness, Empiricism, and Behavior Relevance, to the conclusion, Indeterminacy, via an intermediate position, Linguistic Behaviorism. I am going to assume the truth of Empiricism as well as of Behavior Relevance. Given those assumptions, what remains is an argument from the premise of Publicness. So Quine's argument, as I see it, has two steps:

Publicness $\Rightarrow$ Linguistic Behaviorism
Linguistic Behaviorism $\Rightarrow$ Indeterminacy

This is the argument I shall examine. I am going to treat the first part as a very strong argument, something like a transcendental argument, i.e. an argument of this form: Publicness is true. Linguistic Behaviorism is a condition for the *possibility* of Publicness. Hence Linguistic Behaviorism is true. Similarly, I shall treat the second part as amounting to a claim that Linguistic Behaviorism necessitates, in some loose sense, Indeterminacy. Quine has never presented his consdiderations in this way, and he does view the indeterminacy thesis as an unproven conjecture (by contrast, he seems to view linguistic behaviorism as established). These choices are, of course, to my advantage, since it is easier to attack a stronger claim, but my attack on the argument still has a general interest, since if I am right, Quine's considerations cannot in principle be strengthened into a conclusive argument.

The problem with (what I shall call) Quine's argument is that linguistic behaviorism comes in several varieties, some stronger and some weaker. The formulation above of the thesis covers them all. What I shall do is the following. I shall first try to show that although Publicness, Empiricism and Behavior Relevance together do support Linguistic Behaviorism, they only support a *weak* form of linguistic behaviorism. Then I shall try to show that although linguistic behaviorism does support the indeterminacy thesis (or rather, a version of it), only a *strong* form of linguistic behaviorism supports that thesis. I shall further try to show that the strong form is indeed stronger than the weak form, in the obvious sense that truth of the weak form is compatible with the falsity of the strong form of linguistic beaviorism. Schematically we have:

Publicness $\Rightarrow$ Weak Linguistic Behaviorism
Strong Linguistic Behaviorism $\Rightarrow$ Indeterminacy
Not: Publicness $\Rightarrow$ Strong Linguistic Behaviorism
Not: Weak Linguistic Behaviorism $\Rightarrow$ Indeterminacy

In fact, my conclusion can be stated in a more perspicuous and general form:
    For any version V of Linguistic Behaviorism it holds that

If (Publicness $\Rightarrow$ V), then (not: V $\Rightarrow$ Indeterminacy)

From this it does not follow that Publicness, Empiricism and Behavior Relevance are compatible with the falsity of the indeterminacy thesis, but it does follow that no conclusive argument for indeterminacy can be given from these premisses *via* linguistic behaviorism.

4. FROM PUBLICNESS TO LINGUISTIC BEHAVIORISM

Let us rewrite the first part of the argument somewhat, by fusing the Publicness, Empiricism and Behavior Relevance into one single premiss:

1. What a speaker means by his words is known, or can be known, by others on the basis of observation of his behavior in observable circumstances.[17]
2. Hence, what a speaker means by his words is determined by his observable behavior in observable circumstances.

As far as I can see, this is a valid argument. It is an argument of the following kind:

X knows that *p*. From *p* X concludes that *q*. In virtue of this X knows that *q*. Hence *q* is a consequence of *p*

If *q* were not a consequence of *p*, X's belief in *q* on the basis of *p* would not be knowledge.

So far so good. There is an acceptable argument from Publicness (and the other premisses) to Linguistic Behaviorism. But there are stronger forms of Linguistic Behaviorism which do not follow from the premisses. In order to see that we shall note a few general things that do not follow from premiss 1.

First, it does *not* follow that if two speakers, John and Jill, have identical speech dispositions, or speech dispositions which are as alike as it is possible for two persons to have, then their words mean the same. It may well be that Ralph by observation arrives at sufficient knowledge of the speech dispositions of John and of the speech dispositions of Jill, and despite the fact that these speech dispositions are identical, Ralph interprets John and Jill differently. He may translate one and the same sentence differently when in the mouth of John from when in the mouth of Jill, regarding the translation scheme for John as giving incorrect results if applied to Jill. Moreover, and more importantly, it does *not follow* from 1 that in so doing he would be wrong, or that he would not have *knowledge* both of what John means and of what Jill means. It is compatible with 1 that he interprets them differently and yet knows what both means.

How can this be? It can be that Ralph is simply disposed to interpret them differently. It can be that his disposition for interpreting John is a result of causal influence on Ralph by various background factors. Similarly for his disposition for interpreting Jill. This need not detract us from regarding Ralph as *knowing* what John and Jill mean. For Ralph may be extremely reliable as an interpreter. It may be that whenever, or virtually whenever, Ralph has firm beliefs about what other speakers mean, he is right. His beliefs may be the result of extremely reliable cognitive processes, which involve making conclusions about meaning on the basis of observation of speech behavior.

Thus, if we allow a *reliability* conception of knowledge of meaning, we have no reason to deny that Ralph has the knowledge in question. Should we allow it? Clearly it cannot be rejected on the basis of a claim that there is a stronger connection between meaning and behavior, since this connection is what was supposed to be established. Rather, whether the reliability conception is acceptable turns on general considerations about knowledge of meaning.[18] I return to the question of knowledge in section 6. In this section I shall go on with the assumption that *reliability* is allowed.

Thus, although the inference from 1 to 2 is still valid, 2 must not be interpreted as implying that two speakers mean the same with their words, if their speech dispositions are identical. By analogous reasoning we can conclude also that one speaker need not mean the same at an earlier time as he does at a later time, even if his speech dispositions have remained the same.

This is still compatible with Linguistic Behaviorism, as stated in 2. If meaning is *determined* by the dispositions to speech behavior in observable circumstances, then, we can say, there is a *function* from speech dispositions to meanings. This is fully compatible with the possibility that *which* function it is depends on contextual factors: had the background been different in relevant respects, the function from speech dispositions to meanings would have been different. With the background factors held constant, on the other hand, there can be no difference in meaning without a corresponding difference in speech dispositions.

So, although it follows from 1 that meaning is determined by observable behavior in observable circumstances, it does *not* follow that meaning is determined *only* by observable behavior in observable circumstances. For if that were true, then there would have to be an *invariant* relation between speech dispositions and meaning, invariant between speakers and across time.

I agree that if two speakers mean different things with the same sentence, then then there must be *some* physical difference in the world underlying this difference in meaning (some form of supervenience), and the difference in Ralph's dispositions to interpret the speakers must be connected with that difference in order for Ralph to be reliable. But that underlying physical difference need not be observable. It can still be causally connected with differences in Ralph's dispositions. Even if it is observable, Ralph need not have observed it. And even if Ralph has observed it, it need not be behavior, or circumstances of behavior, in any reasonable sense of "circumstance". Moreover, Ralph may have no reason to conclude anything about interpretation from these observations, despite the fact that they causally affect his general dispositions for interpreting. So, in more than one way, Ralph's disposition for interpreting may be affected by what is not observable behavior in observable circumstances.[19]

Now, for the sake of making my attack on Quine's argument perspicuous, I shall define a notion of what I shall call *Meaning Determining Behavioral* properties, or *MDB* properties:

The property F is a *meaning determining behavioristic property*, or *MDB property*, iff it holds that
i) whether an expression α, as used by a speaker X at a time t, has F, depends on, and only on, X's dispositions, at t, to observable speech behavior in observable circumstances, and
ii) if an expression α, as used by a speaker X at a time t, has F, and an expression β, as used by a speaker Y at time t', has F, then expressions α and β have the same (objective) meaning

In this definition I intend "observable speech behavior in observable circumstances" to be restricted as before.

Now if all expressions used by speaker X, and all expressions used by speaker Y, have MDB properties, and X and Y use the same expressions, and every expression has the same MDB property in the mouth of X as it has in the mouth of Y, then each of these expressions has the same (objective) *meaning* in the mouth of X as it has in the mouth of Y. And if all natural language expressions, by all speakers, have MDB properties, then, indeed, there is an *invariant* general relation between speech dispositions and meaning, invariant between speakers and across time.

Hence, if it does not follow from the premises of Publicness, Empiricism and Behavior Relevance that there is an invariant general relation between speech dispositions and meaning, then neither does it follow from those premises that all natural language expressions have MDB properties. This will be of central importance.

Let us say that

Any version of Linguistic Behaviorism counts as *weak* iff it is implied jointly by Publicness, Empiricism and Behavior Relevance

From this definition it follows that Weak Linguistic Behaviorism does not imply that all natural language expressions have MDB properties. For if it did, that would, by transitivity, be implied by the three joint premisses also, contrary to what has been argued.

5. FROM LINGUISTIC BEHAVIORISM TO INDETERMINACY

For a direct and conclusive proof of the thesis of indeterminacy of translation three things are needed: first, a set of conditions of correctness of translation manuals; secondly, a proof that these conditions are adequate and that the set is complete (contains all adequate conditions); and third, a proof that mutually incompatible manuals satisfy the conditions.

How plausible you find the indeterminacy thesis depends on how plausible you find both the claims about adequacy and completeness of the conditions, and the claim about the correctness of incompatible manuals, given these conditions.

At least the wording, and perhaps the substance, of the criteria of translational correctness have shifted in Quine's writings over the years. In *Pursuit of Truth* Quine says:

A pioneer manual of translation has its utitlity as an aid to negotiation with the native community. Success in communication is judged by smoothness of conversation, by frequent predictability of verbal and non-verbal reactions, and by coherence and plausibility of native testimony. It is a matter of better or worse manuals rather than flatly right or wrong ones.[20]

Our radical translator would put his developing manual of translation continually to use, and go on revising it in the light of his successes and failures of communication. The successes consist – to repeat – in successful negotiation and smooth conversation.[21]

I shall not here attempt a general evaluation of this proposal. I only wish to note that, if the proposed criteria of smoothness of conversation, frequent predictability of verbal and non-verbal reactions, and coherence and plausibility of native testimony, is taken literally, it does not give us very much of a reason for believing in the second part of the claim: that mutually incompatible manuals satisfy the criteria. All sorts of factors, any seemingly insignificant detail, may play a role in making conversation more or less smooth, etc., and in this way creating a preference between manuals in other respects equivalent.

The criterion does not, of course, give any particular reason for *disbelieving* in indeterminacy either, but this does not matter much, since *any* substantial, non-arbitrary criterion opens the possibility that it will be equally well satisfied by several, intuitively incompatible, manuals. That does not even have anything to do with behaviorism. Clearly Quine intended a stronger conclusion. If the general argument from Publicness to Indeterminacy proceeds by way of the *Pursuit of Truth* criteria of translational correctness, then, as far as I can see, the argument fails in the second step, from Linguistic Behaviorism to Indeterminacy.

Instead I shall turn to Quine's treatment of the subject in *Word and Object*. These are based on the notion of stimulus meaning. Quine defined the stimulus meaning of a sentence *s*, as used by a speaker X, to be a pair of the positive and the negative stimulus meaning, respectively, where the positive stimulus meaning is the set of (types of) *stimulations* (triggering of sensory receptors) which makes X disposed to assent to *s*, and the negative stimulus meaning is the corresponding set related to the disposition to dissent.[22]

In terms of stimulus meaning Quine then defined the concepts of *observation sentence*, *stimulus synonymi* , *stimulus analyticity* (the property of being assented to under any stimulation) and *stimulus contradictoriness* (correspondingly).[23] Quine also explained how logical connectives in the alien language can be identified in terms of stimulus meaning. For instance, negation is characterized by the fact that the positive stimulus meaning of a negative sentence is the same as the negative stimulus meaning of the negated one, and *vice versa*.[24]

On top of this Quine introduced the notion of an *analytic hypothesis*. An analytic hypothesis about an alien language is a hypothesis about grammar, about what *words* its sentences contain, and how these are to be paired with words in the home language.[25] From the analytic hypotheses you can then *derive* translations of whole sentences.

In these terms Quine then could state criteria of acceptability, or correctness, of translation manuals. It shall have a not too complicated analytical hypothesis, and further be such that:

1. Observation sentences are translated into observation sentences with the same stimulus meaning.
2. Truth function particles are translated according to their contributions to stimulus meanings, specified by Quine earlier in the chapter.
3. Stimulus-analytic sentences are translated into stimulus-analytic sentences, as far as possible, and similarly for stimulus-contradictory sentences.
4. Pairs of intrasubjective stimulus-synonymous sentences are translated into pairs of intrasubjective stimulus-synonymous sentences.

Quine does not present this as a complete set of conditions for translational correctness, and the conditions are seen as ideal rather than strict. But let us suppose that these conditions in fact are strict, and that the list is complete. Under this assumption, we can almost prove the indeterminacy thesis, and at least make it eminently plausible. What we have to provide is a *method* which, given an acceptable translation manual M, allows us to construct another manual M', such that M' is both acceptable and incompatible with M. Various tricks can be employed to this effect.

For instance this one. There is clearly an upper limit to the length of sentences which the natives have assent/dissent dispositions to at all. Suppose that this limit is a length of 3000 words. And suppose we have a translation function G which satisfies the conditions. Then we define an alternative translation function H such that

$H(s) = G(s)$ for sentences up to 3000 words
$H(s) = $ "It is not the case that"$^\wedge G(s)$ for sentences of more than 3000 words

Clearly the following holds: H is correct, since it agrees with G on all sentences to which there are any speech dispositions, and H is incompatible with G, since for infinitely many sentences their translations contradict each other. Moreover H is recursively definable if G is (just add the two clauses which defines H to the recursive definition of G).

So, with the suggested set of conditions, indeterminacy can more or less be proved. And we have good reasons to believe that if we have something *like* the suggested set of conditions, then some similar trick will produce the required correct and incompatible alternative manual. Quine himself has devised a method for alternative translation of terms[26] (but Quine does not himself regard these alternative manuals as incompatible), and Gerald Massey has devised a method for alternative translation of whole sentences *together with* reassignments of illocutionary force to the natives' utterances (which unfortunately makes them incorrect in the sense of the *Word and Object* conditions, since Masseys alternatives, because of the reassignments, do not preserve stimulus meaning).[27]

There is no guarantee that every set of behavioristic conditions of correctness will allow several correct and mutually incompatible manuals. But on the other hand it is not even possible to give a direct and conclusive argument for indeterminacy *without* having a set of conditions, which is arguably both adequate and complete. For instance, we cannot just happen to "find" an example of indeterminacy: if we have two alternative manuals we still need a set of conditions accepted as adequate and complete for concluding that both are correct, since otherwise we will just be in the situation of not having made up our minds about which to prefer.[28]

So we have seen that at least some sets of behavioristic conditions of correctness of translation allow us to more or less prove the indeterminacy thesis, and that we need an adequate and complete set for such a proof. Let us use this for our definition of Strong Linguistic Behaviorism:

Any version of Linguistic Behaviorism counts as *strong* iff it determines a complete set of specific behavioristic conditions of translational correcntess.

But now we can easily give a definition of a set of MDB properties in terms of a set of correctness criteria. We do this by way of letting the set of correctness conditions, which do determine sameness of objective meaning, also determine sameness of MDB properties. As an example we can use the set of *Word and Object* conditions:

Two expressions have the *same* MDB Property if, and only if, they are paired by a translation relation which satisfies conditions 1. through 4.

This is an acceptable definition, since the conditions of translational correctness themselves are stated in behavioristic terms. Two expressions are correctly inter-translatable iff they have the same objective meaning, and they have the same objective meaning iff they have the same MDB property. Hence, if we accept those conditions, we accept the existence of MDB Properties, or, more precisely, that all natural language expressions have such properties. Similarly, *any* such set of conditions in behavioristic terms provides a corresponding definition of sameness of MDB properties.

Now for the conclusion: As argued here, any version of Linguistic Behaviorism which allows a demonstration of the indeterminacy thesis, i.e. any version of Strong Linguistic Behaviorism, will imply that natural language expressions have MDB properties. But, as I argued in the previous section, it does *not* follow from the joint premises of Publicness, Empiricism and Behavioral Relevance, that natural language expressions have MDB properties. Therefore, no version of Linguistic Behaviorism which is implied by these premises, i.e. no version of Weak Linguistic Behaviorism, does imply the truth of the indeterminacy thesis. For if it did, it would imply that natural language expressions have MDB properties, contrary to what was argued in the previous section.

If this is correct, then no successful argument from Publicness to Indeterminacy, via Linguistic Behaviorism, can be given.

6. THE NATURE OF KNOWLEDGE OF MEANING

It is time to return to the question of knowing what others mean. Above I have appealed to the reliability conception of knowledge, with this result: given two incompatible translation schemes, which satisfy some set of behavioristic correctness conditions equally well, an interpreter, the child, or the linguist, can simply *know* which is the right one. But is this appeal to the reliability conception licit?

If an interpreter settles for one out of two manuals, there must be some relevant difference between them. Quine has discussed the question of relevant and irrelevant differences before. Concerning the reasons for not perceiving the indeterminacy Quine writes in *Word and Object*:

[A fifth cause is] that linguists adhere to implicit supplementary canons that help to limit their choice of analytical hypotheses. For example, if a question were to arise over euqating a short native locution to "rabbit" and a long one to "rabbit part" or vice versa, they would favor the former course, arguing that the more conspicuously segregated wholes are likelier to bear the simpler terms. Such an implicit canon is all very well, unless mistaken for a substantive law of speech behavior.[29]

So some principles of selection are merely practical supplementary canons of no theoretical importance. But what the dividing principle is, is not obvious, and neither is it obvious why such a thing as sentence length should belong to the supplementary side. On Quine's view, stimulus meaning is a behavioristically relevant property, and sentence length is not. But if that division is to have any force in the present context, it must be a consequence of Publicness and the other two premisses, and it is not at all clear why that should be so.

If there is a reason, then, it has to do with the concept of knowledge. You may object that where knowedge is concerned there must be *evidence* for what is known. A mere disposition for preferring one manual could not count as a basis for knowing that it is right. The interpreter must base his choice on *evidence*, and this is why we distinguish between stimulus meaning and sentence length.

I do not think that this answer is good enough. It is not *a priori* clear what to count and what not to count as evidence for the correctness of a manual. Our best shot for deciding the question is to se what interpreters, children and linguists, *actually rely upon* in arriving at their interpretations. At least, this strategy is clearly not in conflict with canons of naturalized epistemology. And if we follow that strategy, sentence length cannot dismissed off hand.

But the objection can be pushed one step further. You can argue that where there is evidence, there must also be *justification*. The interpreter must be able to *justify* his choice of manual. I do not mean that the requirement of justifiability would force us to distinguish between stimulus meaning and sentence length. I don't think it would. But I think that this requirement does imply the existence of an *invariant* connection between speech dispositions and meaning, or between speech dispositions and further observable features, on the one hand, and meaning, on the other. For without stable general principles to appeal to, there would not be any justification. You cannot just appeal to one observable feature in the one case, and another observable feature in the other case. Unless backed up by further justification, such a procedure is simply *ad hoc*.

Again, however, I think that this objection fails. There is no *a priori* reason why knowledge in general, or knoweldge about meaning, should *be* fully justifiable. And again I cannot see that the requirement of justifiability is in any way a consequence of the canons of naturalized epistemology.[30]

So I conclude that my appeal to the reliability conception of knowledge in section 4 is alright. And then, Quine's argument from Publicness, Empiricism and Behavioral Relevance, via Linguistic Behaviorism, to Indeterminacy, interpreted as a strong, transcendental like argument, fails.

But, again, this is not a refutation of the indeterminacy thesis. It is not even a refutation of the claim that indeterminacy is a consequence of publicness, since the possibility remains that this be established by some quite different argument. But that remains to be produced.

Stockholm University

NOTES

[1]   I do not here use the term "consequence" as standing for logical, or semantic consequence. Else it would be wrong to ascribe the view to Quine. There was some terminological
misunderstanding on this point between Professor Quine and myself after my talk in Karlovy
Vary. What I mean to ascribe to Quine is the view that publicness is a *reason* for believing in
the indeterminacy thesis, or, similarly, the belief that the indeterminacy thesis is true *because*
of publicness.
[2]   In Fredrik Stjernberg's *The Public Nature of Meaning*. Stockholm Studies in Philosophy
10, Almqvist & Wiksell International, Stockholm 1991, Stjernberg oscillates between epistemological and metaphysical alternatives, and finally (p. 18) proposes

Meaning is accessible, given suitable capacities, constraints, and (empirical) evidence

This is to say that the meaning of a linguistic expression is knowable by others than the user
of the expression, given that they fulfil certain extra conditions. I do not agree with
Stjernberg about the need for stating those extra conditions, but that is a topic for another
occasion.
[3]   Quine, W.V.O., *Word and Object*, MIT Press, Cambridge, Mass. 1960, p ix, opening
passage of the preface.
[4]   Quine, W.V.O, *Pursuit of Truth*, 2nd edition, Harvard University Press, Cambridge,
Mass. 1992, pp. 37–38.
[5]   Quine, *The Roots of Reference*, Open Court Publishing, La Salle, Ill., 1973, p. 63.
[6]   "Ontological relativity" p. 28.
[7]   *Word and Object*, p. 27.
[8]   *Pursuit of Truth*, p. 48.
[9]   In his reply to Harman in Davidson and Hintikka (eds), *Words and Objections*, Reidel,
Dordrecht 1969, pp. 295–97, Quine regrets the appeal to the uncritical notion of equivalence
in *Word and Object*. But clearly, the notions of coherence and of interchangeability in
English contexts are equally uncritical.

As long as it is essential to the indeterminacy thesis that it is stated in terms of translation
from an alien language to English (The Home Language), it is also essential that the thesis
rely on uncritical semantic notions. If the thesis can equally well be stated in terms of translation between two alien languages, without assuming that either is translated into English,
then use of uncritical notions can be avoided. The reason is that we cannot, from the present
point of view, apply any uncritical semantic notions to another language save as relative to a
manual of translation into English.

I do not myself regard it as a weakness of the thesis that it employs an uncritical notion,
since it is in virtue of this employment that the thesis has a direct bearing on that uncritical
notion. If two non-equivalent English sentences can be correctly translated to the same alien
sentence, and hence have the same objective meaning, we can conclude that the uncritical
distinction in meaning, between the two English sentences, does not have any objective
basis. In fact, I see this as the gist of the indeterminacy thesis.

By contrast, it is hard to see how we can reach a similar conclusion about meaning, in the
uncritical sense, without having to rely on further assumptions, if the thesis itself is stated
without uncritical notions. One suggestion for stating it thus is to use the idea of alternating
between manuals. We can exploit this idea without appealing to coherence, by deeming the
alternation itself as an unacceptable manual. More precisely: Take the two correct manuals
M and M'. Construct a third manual M* by simply taking half of the translations from M and
half from M'. So for every second sentence s, M*(s) = M(s), and for the rest it holds that
M*(s) = M'(s). And now it may turn out that M* is *not* a correct manual. (In similar vein
Andrzej Zabludowski, in 'On Quine's indeterminacy thesis', *The Philosophical Review*,
Vol. XCVIII, 1989, pp. 35–63) suggests that we take the union of two manuals (p. 46)).

The problem with this suggestion is that so far we do *not* have the right to conclude that
correct manuals do not preserve meaning (in the uncritical sense). We could do this, if we
could conclude that M* does not preserve meaning, buth we are not entitled to this either.
Sure, if both M and M' preserve meaning, then so does M*, but the reason why M* is *incorrect* may be that it lacks *additional virtues* which correct manuals have (like having a particular overall structure). It may be that both M and M' have these additional virtues, and that
M* does not, even though it preserves meaning. In order to reach the desired conclusion
about intuitive meaning, we would need the extra assumption that preservation of meaning is

enough for making a manual correct. And surely that assumption is controversial, even from a behaviorist point of view.

[10]   Also, Quine says in *Pursuit of Truth*, p. 37: "The meaning of a sentence of one language is what it shares with its translation in another language".

[11]   Passage referenced in note 4.

[12]   Quine, "Linguistics and philosophy", i *The Ways of Paradox*, Harvard University Press, Cambridge, Mass. 1976.

[13]   *Word and Object*, p. 27.

[14]   *Pursuit of Truth*, p. 48.

[15]   I have here amalgamated Quine's compatibility conditions, in b), from *Word and Object and Pursuit of Truth*. The reader may, as Quine himself does, prefer the latter conditions only.

[16]   To my knowledge this view about Quine's reason for the indeterminacy thesis is not to be found in the literature. On the whole, too little attention has been paid to the theme of publicness in this context. In Robert Kirk's monograph *Translation Determined*, Clarendon Press, Oxford, 1986, Quine's linguistic behaviorism is dismissed as unjustified (pp. 89–90), and isn't even considered important for the indeterminacy thesis itself. On Kirk's behalf it can be said that the book appeared the year before Quine's "Indeterminacy of translation again", *Journal of Philosophy* 84, 1987, containing the same passage quoted above from *Pursuit of Truth*.

Fredrik Stjernberg (*op. cit.* chpt 2) shares Kirk's view: Quine's behaviorism is inessential to the indeterminacy thesis. Stjernberg goes as far as claiming that not even telepathy would be of any relevance to the thesis, but then he seems to view telepathic capacity as a capacity to observe others' mental symbols (inner linguistic behavior, as it were), rather than a capacity to directly perceive thought content.

Dagfinn Føllesdal has stressed the relevance of publicness to indeterminacy, in "Indeterminacy and mental states", in Robert Barrett och Roger Gibson (eds), *Perspectives on Quine*, Basil Blackwell, Oxford 1990, and in 'In what sense is language public', in Paolo Leonardi och Marco Santambrogio (eds), *On Quine*, Cambridge University Press, Cambridge 1995, but without making the publicness thesis precise and without presenting an articulated argument.

Closest to my own view is Dorit Bar-On in 'Semantic Verificationism, linguistic behaviorism, and translation', *Philosophical Studies* 66, 1992, pp. 235–259. Bar-On's semantic verificationism is almost the same as Basic Publicness here. Bar-On, however, discerns roughly this argument for linguistic behaviorism: Semantic psychologism (à la Chomsky) is compatible with the falsity of semantic verificationism. Therefore semantic psychologism must be rejected, and then linguistic behaviorism is a plausible alternative. Here we differ, and even more so in that Bar-On seems to view linguistic behaviorism as a stimulus-response theory about language learning. Fredrik Stjernberg helped me to get clear about these differences.

[17]   We reason from the three premises like this: speakers do know what other speakers mean (Publicness), on the basis of observation (Empiricism), and whatever such knowledge they can have, on the basis of the ordinary (non-scientific) observations they do make, can be had (and probably is had) on the basis of observation of behavior in observable circumstances (Behavior Relevance).

[18]   Edward Craig has made a similar point about the reliability conception of knowlege in connection with publicness of meaning, in "Meaning, use and privacy", *Mind* 91, pp. 554–555, note 2, discussing Dummett's views, and in "Privacy and rule-following", in Jeremy Butterfield (red), *Language, Mind and Logic*, Cambridge University Press, Cambridge 1986, pp. 174–76, discussing rule following and the private language argument.

[19]   You can still ask whether the situation I imagine is *plausible*. You may think that it is quite unreasonable to suppose that two speakers may have the same speech dispositions, and yet mean different things by their sentences. In fact, I sympathize with this view, but only provided "speech dispositions" is taken in a sufficiently wide and unspecified sense. As soon as the notion of speech disposition is delimited in any reasonably sharp and reasonably narrow way, there are, I strongly suspect, factors which do affect speakers' interpretation of other speakers but which aren't included under that specified notion. What is really implausible, I would say, is the idea of a specifiable totality of behavioral facts which are relevant to meaning.

[20]   *Pursuit of Truth*, p. 43.

[21] *Pursuit of Truth,* pp. 46–47.
[22] *Word and Object,* pp. 32–33.
[23] *Word and Object,* pp. 42, 46 and 55, respectively.
[24] *Word and Object,* p. 57.
[25] *Word and Object,* p. 68.
[26] The gavagai-example in *Word and Object,* section 12, and the introduction of "proxy"-functions in "Ontological reduction and the world of numbers", *The Ways of Paradox,* Harvard University Press, Cambridge, Mass. 1979, pp. 217–220, and "Ontological relativity", in *Ontological Relativity and Other Essays,* pp. 55–58.
[27] Gerald Massey, "The indeterminacy of translation: a study in philosophical exegesis", *Philosophical Topics,* Vol 20, 1992. It may be noted that Massey's alternative conditions are no less behavioristic than Quine's.
[28] In his book Robert Kirk tries to render the indeterminacy thesis more empirical by letting it concern what real field linguists might do, but he is forced to add clauses about freedom from prejudice etc., since otherwise the thesis would lose interest. And conditions of correctness reenter with such qualifications.
[29] *Word and Object,* p. 74.
[30] See e.g. "Epistemology naturalized", in *Ontological Relativity and Other Essays,* pp. 82–83, *Pursuit of Truth,* pp. 19–22, 28.

REFERENCES

Bar-On, Dorit, "Semantic Verificationism, linguistic behaviorism, and translation", *Philosophical Studies* 66, 1992, pp. 235–59

Craig, Edward, "Meaning, use and privacy", *Mind* 91, 1982, pp. 541–564

Craig, Edward, "Privacy and rule-following", in Jeremy Butterfield (ed), *Language, Mind and Logic,* Cambridge University Press, Cambridge 1986

Davidson and Hintikka (eds), *Words and Objections,* Reidel, Dordrecht 1969

Føllesdal, Dagfinn, "Indeterminacy and mental states", in Robert Barrett och Roger Gibson (eds), *Perspectives on Quine,* Basil Blackwell, Oxford 1990

Føllesdal, Dagfinn, "In what sense i language public", in Paolo Leonardi och Marco Santambrogio (eds), *On Quine,* Cambridge University Press, Cambridge 1995

Kirk, Robert, *Translation Determined,* Clarendon Press, Oxford, 1986

Massey, Gerald, "The indeterminacy of translation: a study in philosophical exegesis", *Philosophical Topics,* Vol 20, 1992.

Quine, W.V.O., *Word and Object,* MIT Press, Cambridge, Mass. 1960

Quine, W.V.O., *Ontological Relativity and Other Essays,* Columbia University Press, New York 1969

Quine, W.V.O., "Ontological Relativity", in Quine, *Ontological Relativity and Other Essays*

Quine, W.V.O., "Epistemology naturalized", in *Ontological Relativity and Other Essays*

Quine, W.V.O., "Reply to Harman", in Davidson and Hintikka (eds), *Words and Objections,*

Quine, W.V.O., *The Roots of Reference,* Open Court Publishing, La Salle, Ill.1973

Quine, W.V.O., *The Ways of Paradox and Other Essays,* Harvard University Press, Cambridge, Mass. 1976.

Quine, W.V.O., "Linguistics and philosophy", in *The Ways of Paradox and Other Essays*

Quine, W.V.O., "Ontological reduction and the world of numbers", in *The Ways of Paradox and Other Essays*

Quine, W.V.O., "Indeterminacy of translation again", *Journal of Philosophy* 84, 1987

Quine, W.V.O., *Pursuit of Truth,* 2nd ed., Harvard University Press, Cambridge, Mass. 1992

Stjernberg, Fredrik, *The Public Nature of Meaning,* Almqvist & Wiksell, Stockholm 1991.

Zabludowski, Andrzej, "On Quine's indeterminacy thesis", *The Philosophical Review,* Vol. XCVIII, 1989, pp. 35–63

FREDERICK STOUTLAND

INDIVIDUAL AND SOCIAL IN QUINE'S PHILOSOPHY OF LANGUAGE

INTRODUCTION

Anyone who works through Quine's philosophy of language carefully and without malice cannot fail to be impressed by his achievement. Most of his predecessors appealed to meanings as entities of some kind, abstract or mental, and attempted to explain linguistic behavior on that basis. Quine, on the contrary, made our *use* of expressions basic and let meanings fall out as they may, showing, whatever one's objections, how it might be done. He consistently argued that meanings cannot explain our use of expressions because whatever there is to meaning can be explained only in terms of use.

Putting it this way presumes parallels between Quine and Wittgenstein. I don't know how much Quine learned from Wittgenstein; while he must have learned something, he surely worked out his main ideas independently, and his formulations and arguments are often quite different. Yet the parallels are striking – both reject Platonist and mentalist conceptions of meaning and identify similar issues as crucial – and much can be learned by using each to interpret the other.

One major difference is that, unlike Wittgenstein, Quine gave us a systematic and comprehensive account of language. He identified a core use of expressions – explicated in terms of the notion of shared dispositions to assent to or dissent from queries – and used that as a basis for giving an account of the rest of language. Defenders of Wittgenstein often criticize Quine just because his account is systematic and comprehensive. I think that is off the mark: Wittgenstein's insistence that one think for oneself requires that his insights be developed beyond the aphoristic form they take in his own writings, and Quine shows one way to do that, which merits close attention.[1]

In this paper I focus on Quine's treatment of the individual and social dimensions of language and their relation to normativity. Here the parallel between Quine and Wittgenstein is particularly illuminating, helping to interpret both and to focus questions of increasing significance for philosophers. Any plausible account of language will admit that language is both individual – only individuals, for instance, learn and speak a language – and social – people cannot learn or speak a language all on their own – but accounts differ in the way they construe these dimensions of language and how they relate to normativity.

Some accounts are *individualistic*: the social dimensions of language depend on and are explicable in terms of its individual dimensions, but its individual dimensions neither depend on nor are explicable in terms of its social dimensions. It is

Alex Orenstein and Petr Kotatko (eds.), Knowledge, Language and Logic, 181–194.

competencies which are either innate in individuals or which they could acquire independently of interaction with other speakers which make language as a social phenomenon possible. The social is the inter-subjective or the inter-individual: it brings together individuals whose linguistic competencies are already formed but which need to be re-formed so as to coincide with the competencies of other individuals. Even if some dimensions of language – its normativity, for example – are necessarily social, those dimensions depend entirely on competencies individuals could have by themselves. The view is, as Dummett puts it, that "a language can only be explained as the common overlap of many idiolects."[2]

Other accounts are (for lack of a better term) *non-individualistic*: some of the individual dimensions of language depend on and are explicable only in terms of its social dimensions. It may be argued, for example, that the notion of a solitary individual having linguistic competence is incoherent because the normativity such competence presumes depends on interaction with other speakers. Or it may be argued that the linguistic competence of individuals depends on their having learned to participate in various practices and institutions which are necessarily social. The view is, to quote Dummett again, that "one cannot so much as explain what an idiolect is without invoking the notion of a language considered as a social phenomenon."[3]

In what follows, I begin with a rather free-wheeling exposition of the role of the social and the normative in Quine's philosophy of language, focussing on observation sentences and arguing that his view of the social is individualistic. In the next section I state some objections to his view. In the final section I sketch a couple of anti-individualist alternatives, not to establish them but to help undermine the received view that individualism *must* be right.

1. QUINE ON THE SOCIAL DIMENSIONS OF LANGUAGE

First, some preliminary comments. Conceptions of language which make use primary often fail to give an adequate account of the *kind* (or aspect) of use persons who are linguistically competent have mastered. Some fail by giving an account of use which is much too broad. Expressions can be used in all sorts of ways, many of which are only loosely connected with their meaning and hence with a speaker's competence in the language. An adequate account must, for example, make a distinction between knowing how to use an expression to frighten the cat or express one's feelings and knowing how to use it to say something. Behaviorist approaches to language which characterize use in terms of conditions necessary and sufficient to utter or respond to a sentence fail in this way because (language being stimulus-independent) such conditions are too broad to mark out language from other forms of behavior, to define the relevant sense of meaning or to specify linguistic competence.

Other accounts fail by giving an account of use which is too narrow in that they try to characterize a notion of purely linguistic use, that is, a kind (or aspect) of use which depends *only* on knowing what expressions mean in a language.[4] Such a notion of use presumes a distinction in principle between use due to what speakers take expressions to mean and use due to their beliefs about the world, a distinction undermined by Quine's critique of the analytic-synthetic distinction. While there may be a point to that distinction for particular purposes within a language, it cannot be used in an overall account of the meanings of expressions or of what it is to be a competent speaker of a language (and hence it cannot be used in radical translation, which is the context Quine sets up in order to give such an overall account). The way expressions are used depends both on what they mean and on what the speakers who use them believe, and there are no principles which enable us to make a general distinction between aspects of use which depend on the former and those which depend on the latter.

While Quine uses behaviorist concepts to give an account of the relevant kind of use, he does so in an ingenious way, which avoids an account either too broad or too narrow. Instead of specifying necessary and sufficient conditions for uttering or responding to a sentence, he appeals to dispositions to assent or dissent to a sentence if queried about the sentence under diverse stimulus conditions, thereby characterizing a kind of use whose mastery is necessarily connected with linguistic competence and meaning. At the same time, this avoids a too narrow approach, for it presumes no principled distinction between assent (or dissent) which depends on what the speaker takes the sentence to mean and assent (or dissent) which depends on what she believes – on "collateral information". We don't need a general distinction between response due to meaning and response due to collateral information in order to characterize a speaker's behavior as manifesting understanding of what expressions mean in a language. A speaker is competent is a language just in case she responds to queries about a sentence as a competent speaker should; indeed, her competence *consists* in her having dispositions to respond as a competent speaker should.

I just wrote of responding "as a competent speaker should", and I take that "should" ("ought" would do as well) to be normative. The notion of understanding a language is, in Quine's view, normative, as is the notion of meaning, and that strikes me as a merit of his point of view.[5] What expressions mean in a language is a matter of how they should be used in that language, and to understand a language is to know how to use expressions as they should be used. In this respect, language is like a game: that tokens are chess pieces is a matter of how they should be ("must be", in the normative sense) used in a chess game, and to know how to play chess is to know how to move certain pieces as such pieces should be moved. Even if language is unlike chess in that there are no game-like rules which determine meaning or linguistic competence, it does not follow that language is not normative, for normativity does not require rules (or conventions). It requires a distinction between doing something correctly – as it should be done – and doing it incorrectly, but that is not equivalent to a distinction between following and failing to follow a rule.

Quine's views about the normativity of language are best seen in his conception of observation sentences, which are at the periphery of the web of sentences which constitute a language, tying sentences to the empirical world and giving them the empirical meaning every meaningful sentence must have to one degree or another.

They play their roles at the edge of language, where language begins – whether for the species, the individual speaker, the translator, or the seeker after knowledge.[6]

While observation sentences are at the edge of language, Quine thinks there must be something language-like even further out. This is stimulus meaning, which is analogous to Grice's speaker meaning in that it is what an individual speaker means by a sentence; in Quinean terms, it is the dispositions to assent or dissent to a sentence an individual speaker happens to have at a given time.[7] It contrasts with what the sentence means in a language, that is, the way it *should* be used by a speaker of that language, so-called linguistic meaning. Like Grice, Quine wants to build *linguistic* meaning out of a kind of proto-linguistic speaker meaning. The crucial difference is that Quine rejects Grice's mentalism: he wants a behaviorist account both of the *proto-languages* of single individuals and of the *real* languages spoken by persons who have learned a language which enables them to communicate with others.

Quine defines stimulus meaning in terms of assent-dissent dispositions relativized to an individual and a time. The stimulus meaning of a sentence for an individual is (roughly) the class of stimuli which would prompt her to assent to, and the class of stimuli which would prompt her to dissent from, a query about the sentence at a given time. This is not *linguistic* meaning for a number of reasons, of which the one that interests me here is that there is nothing normative about it: the stimulus meaning of a sentence is constituted by *whatever* stimuli would prompt a speaker to assent or dissent at a time, so that any disposition to respond to stimuli is as correct as any other, which means the notions of correct or incorrect have no application.[8] Because the notion of linguistic meaning is normative, however, notions of correct or incorrect must apply: there must be a distinction between using a sentence as it should be used and not so using it, between having the dispositions one should have and the dispositions one should not have.

Quine calls stimulus meaning "meaning" (I presume) because he thinks linguistic meaning can be built out of it. His view is that when the stimulus meanings of a sentence coincide (approximately) for all (or most) speakers of a language, then those stimulus meanings determine how it *should* be used and hence constitute its linguistic meaning. Stimulus meanings which are intersubjective in this sense constitute linguistic meaning, for then there is such a thing as using a sentence correctly (or incorrectly), namely, in accordance with (or contrary to) the ways others use it. Proto-languages which coincide for speakers thereby constitute a real language with norms about how sentences should be used. Now we can speak of *understanding* a sentence, namely, having the dispositions a speaker *should* have with respect to the sentence – that is, dispositions to assent or dissent under the same (or similar) stimulus conditions as other speakers. Now we can speak of what the sentence (rather than the speaker) means, namely the set of stimulus conditions which *should* prompt a speaker of the language to assent or dissent, that is, those which do prompt competent speakers generally.

On this view, the normativity necessary for linguistic meaning and competence derives entirely from the fact that speakers of a language have come to respond uniformly to certain sentences, so that those who conform use the sentences correctly, those who deviate use them incorrectly. These sentences are such that speakers become competent in the language by acquiring the same dispositions to assent or dissent to them, so that a speaker who does not share those dispositions does not understand what the sentences mean. What a speaker *should* do is determined by what most speakers *would* do, and the dispositions a speaker *should* have are determined by the dispositions most speakers *do* have.

Language is, therefore, normative only insofar as there are enough speakers whose verbal dispositions have come to coincide sufficiently to establish a social consensus which determines what is normal. This is a conception of language as necessarily social although in an individualistic sense of the social: the language of a linguistic community consists of the proto-languages of its individual members – of the dispositions of individual speakers which have come to coincide. It is in this respect like Frege's conception, as characterized by Dummett: "The sense which an expression has in some language [can only be the sense] which all, or most, individual speakers of the language attach to it.... The basic notion is really that of an idiolect, and a language can only be explained as the common overlap of many idiolects."[9]

While all sentences have stimulus meanings, the linguistic meanings of most sentences cannot be built out of their stimulus meanings.[10] The basic reason is that the stimulus meanings of most sentences will not coincide for speakers of a language, so that the normativity of linguistic meaning cannot be constituted by the kind of consensus just discussed. The latter is possible only for *occasion* sentences and only for those whose stimulus meanings "vary none under the influence of collateral information" [WO, p. 42], the reason being that since collateral information varies enormously from speaker to speaker, stimulus meanings which depend on it cannot coincide. The only sentences whose stimulus meanings can coincide so as to yield linguistic meaning are Quine's observation sentences, one of whose definitions I quoted above: an observation sentence is an occasion sentence whose stimulus meanings vary none under the influence of collateral information.

Here two lines of argument coincide. On the one hand, an observation sentence is an occasion sentence whose use is supposed to depend only on the way our senses are affected, hence one connected with dispositions to assent or dissent which do not vary with collateral information. The only possible way to determine whether a sentence meets that condition, however, is to observe whether different speakers respond in the same way to the same external stimuli. On the other hand, stimulus meanings can constitute a sentence's linguistic meaning (making possible incorrect, hence correct, ways of using the sentence), only if they coincide for all (or most) speakers of the language, and that means they must coincide regardless of individual differences among the beliefs speakers have (or other sentences they understand). Hence the only kind of sentences whose linguistic meanings can be built out of stimulus meanings are observation sentences.

Being wary of talk of meanings, Quine follows this with a *behavioral* definition of an observation sentence: "An occasion sentence may be said to be the more observational the more nearly its stimulus meanings for different speakers tend to coincide." [WO, 43] This drops the reference to collateral information but insofar as that varies from speaker to speaker, it is taken care of by requiring that stimulus meanings coincide. That does not take care of collateral information *most* speakers have, but it should not, for it is not possible to sort out the role of stimulus meanings from the role of widely shared beliefs without appealing to the analytic-synthetic distinction in a form Quine firmly rejects. "I suspect that no systematic experimental sense is to be made of a distinction between usage due to meaning and usage due to generally shared collateral information". [WO, 43]

It is only observation sentences, therefore, whose normative status is directly constituted by the social coincidence of stimulus meanings. They are necessarily social in that their very definition requires such coincidence, and hence no expression uttered by an always solitary creature could be observation sentences.[11] But they are social in the individualistic sense that their meaning in the language is just their meaning for each individual, provided the latter is the same for every individual. As in utilitarianism, the common good is the sum of individual goods, so the language of a community is the sum of individual proto-languages.

In *Pursuit of Truth*,[12] Quine traces his shifting views about observation sentences and suggests they need not be social at all since they can be defined for a single speaker. An observation sentence is an occasion sentence which meets the following condition: "If querying the sentence elicits assent from the given speaker on one occasion, it will elicit assent likewise on any other occasion when the same total set of receptors is triggered; and similarly for dissent." An observation sentence *for a community* is defined as one which is observational for each member, provided "each would agree in assenting to it, or dissenting, on witnessing the occasion of utterance". [p. 43]

This is not, however, a substantive change about what it is for a sentence to belong to a real language. Nothing could count as correct or incorrect about the meaning of a sentence which is observational only for a single speaker. If a speaker S assented to a query about P when one total set of receptors is triggered and then dissented to another query when the same total set of receptors is triggered, that would show that P does not meet the criterion for being an observation sentence for S, not that S had responded wrongly. There is still no room, on Quine's view, for a single speaker responding wrongly or rightly; that distinction requires several speakers.[13] Sentences which are observational for a single speaker belong to the same proto-language as sentences whose stimulus meanings are not intersubjective. To belong to a real language, they have to be social. That the shift in Quine's position is not substantive but verbal is seen when he reaffirms his earlier claim that an observation sentence "is an occasion sentence on which speakers of the language can agree outright on witnessing the occasion" [p. 3] and goes on to say that "language is where intersubjectivity sets in". [p. 44]

Only observation sentences have their normative status directly constituted by the social coincidence of stimulus meanings. But Quine applies this concept of the normative as constituted by socially shared dispositions to every expression with a determinate meaning. It does not apply to expressions whose meaning is not determinate since they do not have normative status in the required sense. Expressions whose meaning is indeterminate cannot (by definition) be determinately translated, which means there are numerous (Quine says "countless") incompatible ways of translating them, none of which is any more correct (or incorrect) than any of the others, which just means that the concept of correctness has no application to their translation. Although tradition or other pragmatic values may favor one translation over another, such values are not relevant to the normative question of which translation is correct. Given that meaning must (at least) be what is preserved in translation, the notion of meaning which is not determinate is, therefore, a kind of oxymoron since Quine holds that not even truth values have to be preserved in translating sentences whose meaning is not determinate. To put the point in other words, there can be no linguistic meaning where there are no norms distinguishing correct from incorrect translation.

One class of expressions which are not observation sentences but whose meaning is determinate and hence constituted by norms established by socially shared dispositions is the truth functional connectives. They admit of determinate translation partly because they have "semantic criteria", specifiable in terms of assent-dissent dispositions, first for an individual and then, by summation, for a community: "The semantic criterion of negation is that it turns any short sentence to which one will assent into a sentence from which one will dissent, and vice versa" [WO, 57], and so on for the other connectives.

If it turns out that the natives seem not to conform to these criteria and assent to a sentence we translate as a self-contradiction, we should not conclude that they are wrong about the meaning of the logical connectives because they do not accept our semantic criteria (that would be an illegitimate appeal to meaning) or that they are illogical. We should conclude, by an appeal to the principle of charity, that we have mistranslated what we take to be their sentential connectives. "The maxim of translation underlying all this", Quine writes, "is that assertions startlingly false on the face of them are likely to turn on hidden differences of language." He goes on, "The common sense behind the maxim is that one's interlocutor's silliness, beyond a certain point, is less likely than bad translation." [WO, 59]

The appeal to semantic criteria is social, for the sentential connectives are part of a language only when the criteria are generally accepted by speakers of that language. So is the principle of charity, for its application to a single speaker is, on Quine's ground, highly dubious. Although Quine's use of it is much narrower than Davidson's, he does not limit it to truth-functional logic. If we translate the natives so as to make them assent to such silliness as that no one has any arms or that there are no insects, we also violate the principle of charity. But a single native might assent to such things for some perverse reason; what makes for real silliness is that we should think that a whole community might assent to them. The principle of charity is a social principle, construed by Quine as summing up the dispositions of speakers in a community to assent or dissent to various sentences.

Quine holds that beyond observation sentences and truth functional connectives, translation turns on such pragmatic matters as how expressions have traditionally been translated. But he also holds that a traditional translation practice is not to be confused with constitutes a correct translation. Dummett has argued that "the existence of a socially agreed convention for translating from one language into another is itself a linguistic fact, to which, ultimately, speakers of either language are also responsible."[14] That is, a practice of translating between languages may be a matter of social consensus – analogous to that required for observation sentences – and when that is so, there will be an analogous sense of the socially normative about such a practice: there will be a fact of the matter about the correctness of the traditional translation of any sentence.

The analogy fails, however, for there are no socially inculcated *dispositions* to translate in one way rather than another. Quine's view about social consensus for observation sentences is that what makes for the correct-incorrect distinction required for real meaning or use is not *theoretical* agreement but agreement in practice, that is, in speech dispositions. Language is shared *practical know-how*; translation is a matter of a *theoretical* grasp of such know-how. Agreement must be, as Wittgenstein said, in action or practice, not in theory or thought.

Bi-linguals are not an exception, for they have dispositions to speech behavior in *each* of two languages, but need have no *dispositions* to *translate* from one to another. Even when bi-linguals do have such dispositions to translate, there would be nothing about their dispositions which makes them more correct or incorrect than rivals. Observation sentences involve dispositions to respond differentially to non-verbal stimuli; if beliefs and other sentences play a role, they are shared by everyone so that what accounts for a speaker's response on a particular occasion is the sensory stimuli he receives on that occasion. Translation dispositions would be dispositions to respond to *verbal* stimuli, and there can be alternative sets of them without any of them being more or less correct than any other.

2. DIFFICULTIES IN QUINE'S VIEW OF THE SOCIAL ROOTS OF THE NORMATIVE

The previous section was implicitly critical of some of Quine's views. I want now to make some of those criticisms explicit, centering on the claim that, although Quine rightly makes language an inescapably social phenomenon, his individualistic way of construing the social roots of the normative is inadequate.

Stimulus meanings per se cannot account for *linguistic* meaning or competence because (among other things) they consist of the present dispositions of individual speakers, about which there is nothing normative. No one can respond incorrectly in the sense of violating her own dispositions, and hence no one can respond correctly either. If whatever response a speaker has a disposition to make is *thereby* correct, then the distinction between correct and incorrect has no application, and the speaker's behavior is not subject to the norms required for behavior to manifest understanding of a language.

On my interpretation of Quine, the normativity required for language comes on the scene only when the dispositions of many speakers coincide, thus making possible observation sentences, "whose stimulus meanings for different speakers tend to coincide." [WO, 43] A response to a query about a sentence is incorrect when it deviates from other speakers' dispositions to respond, correct otherwise. As for non-observation sentences, to the extent that their meaning is not determinate, there are no norms governing correct or incorrect use. To the extent that their meaning is determinate, the norms governing their use derive from their connection to observation sentences or depend in some other way on dispositions shared generally by speakers of the language. In any case the norms required for linguistic meaning and competence are constituted by the coincidence of the dispositions of individual speakers: the normativity of language is rooted in social consensus among individual speakers.

It is an immediate consequence of this view that social consensus itself is not subject to linguistic norms.[15] The dispositions of individual speakers are correct or incorrect depending on whether they conform to or deviate from the shared dispositions which constitute correctness, but there is nothing from which the shared dispositions themselves can deviate and hence nothing to which they can conform. What a linguistic community takes to be the correct way to use an expression is *thereby* the correct way to use it. But if any way of using an expression is correct, then the notion of correctness does not apply, and hence the shared dispositions which constitute how expressions should be used by each competent speaker of a language are neither correct nor incorrect. While the source of the norms which constitute linguistic meaning and competence, they are not themselves governed by linguistic norms.[16]

The fact that the source of normativity is not itself subject to norms is not objectionable. Indeed, requiring that every source of norms be subject to norms, would generate a vicious regress. The difficulty is that the kind of social consensus Quine appeals to as the source of normativity *ought* to be subject to normative assessment, (which shows it is not the ultimate source of linguistic normativity). There are at least two ways in which linguistic norms figure at the level of social consensus.

The first is that to identify the dispositions each speaker ought to have with the dispositions speakers in general do have leaves open the question of what is meant by "speakers in general". It cannot literally mean *all* speakers, for then deviance would be ruled out by definition; the shared dispositions which constitute norms would fail to do so if any speaker did not share them. If it is taken to mean all *competent* speakers, two problems arise. The first is an evident circularity: a speaker is competent just in case her dispositions coincide with those of others, but the others to which her dispositions must coincide are just the competent speakers. The second arises if we try to avoid the circularity by offering an independent account of what it is to be a competent speaker, for example along the line of Putnam's notion of "division-of-labor" about meaning, which appeals to experts: physicists are in the best position to explain the meaning of "gold", biologists to explain the meaning of "beech", and so on. The problem is that norms are needed to determine who should have the status of experts to determine for the rest of us what is linguistically correct or incorrect, but these norms obviously cannot rest on social consensus.

Another possibility would be to take "speakers in general" to refer to *most* speakers. But how many count as "most" and how are they determined? What about a small group whose dispositions deviate from that of most other speakers? Are they speaking the language but doing so incorrectly, are they speaking another language, or are they speaking the same language but as members of an obscure cult with deviant beliefs? What kind of relationships must be present in a community so that dispositions individual speakers have come to share constitute expressions as meaningful? These are normative questions, which are intelligible and which bear on what it is to use an expression correctly or incorrectly; but they cannot be answered unless sources of normativity other than socially shared dispositions are smuggled in.

The second way in which linguistic norms figure at the level of social consensus is that they make it possible to raise questions about whether a whole linguistic community may be mistaken about how certain expressions should be used, about whether even socially shared dispositions are mistaken. If what a linguistic community takes to be the correct way to use an expression is *thereby* the correct way to use it, then such questions cannot even be raised. I contend, however, that they are perfectly intelligible questions and that we can conceive of situations where linguistic communities as a whole might use expressions incorrectly.

My point is not merely that it should be possible for a community to have a wrong *theoretical* understanding of some expressions; most members of a community probably have no theoretical account at all to offer, and those who do surely often offer wrong ones. I have in mind the possibility of a wrong *practical* understanding of expressions, the possibility that most members of the community *use* them wrongly, even those who are thought to be experts. This cannot happen to *many* expressions because meaning is holistic, and widespread error would undermine any meaningfulness whatever. There is no room for wholesale skepticism of the Cartesian or solipsistic versions; but there should at least be the possibility of community error on the meaning of some expressions. The fact that there is not even conceptual space for such a possibility on the social consensus view of normativity is a reason to reject that view.

The chief argument that it must be possible for a whole community to misuse an expression appeals to Quine's central claim that there is no general distinction in principle to be made between use due to what speakers take an expression to mean and use due to what they believe about the world. If a whole community may be wrong on some beliefs, which can surely not be denied, then its use of expressions involving those beliefs may also be wrong. Or to put it in other terms, it may have some concepts so defective that the use members of the community make of expressions which involve those concept is just wrong. A whole community of speakers may for such reasons be so deeply confused in how they use certain expressions, that the best thing to say would be that there is collective misuse.

It cannot be responded that what accounts for the expressions being used incorrectly by speakers generally is not that speakers are mistaken about what the expressions mean but that they are mistaken in their beliefs. This is exactly the kind of situation where such a response cannot be made because it would require general principles about the meaning-belief distinction which we do not have. Since the distinction cannot be made, there is no reason to ascribe such community-wide misuse to mistakes about what to believe rather than to mistakes about what expressions mean and how they should be used. Community-wide misuse is no doubt uncommon and no doubt in general unrecognizable to speakers who are members of the community (though not necessarily to other speakers), but that it is not incoherent shows that social consensus cannot be the source of linguistic norms.[17]

These difficulties in the social consensus conception of linguistic normativity suggest a more fundamental objection to that conception, which is that consensus, no matter how strong, never establishes norms of any kind except in a context which is already norm-laden. Take, for example, a game; players might by consensus establish a rule that a move must not take more than one minute. This is an arbitrary rule, established simply by players agreeing to it, which appears to be the very paradigm of a norm constituted simply by consensus. But there is more to it than that; the consensus would not establish anything if there not a game in progress, if certain persons were not identified as players, if a norm for what should count as a move were not in force, and so on. Those factors indicate that there is a norm-laden context within which the consensus was reached, and it is only in such a context that consensus could establish any kind of rule.

This is the general case: consensus does not establish norms of any kind except in a context which is already normative, and which, therefore, makes it possible (among other things) to deal intelligibly with (even if not to answer definitively) questions such as who should participate in a consensus, what should be settled by consensus, and so on. Sheer consensus never constitutes any norms, and if it appears to do so, it is only because the normative setting for the consensus is overlooked.

Quine's stimulus meanings are by design norm-free. They are meant to be accessible to any observer who possesses concepts which apply to complex behavior of any kind, even that of machines or chemical reactions. They are as far from the normative as the neural mechanisms in terms of which Quine defined them in *Word and Object*, and neither their recognition, their acquisition, nor their modification require a normative context. Any consensus that involves simply the coincidence of stimulus meanings can have no normative status or generate any norms about what should be done, because there is no norm-laden context for the development of the consensus. The origin of linguistic norms cannot be in the coincidence of stimulus meanings but elsewhere, in places not made explicit in Quine's philosophy of language.

This does not mean that dispositions which are broadly like Quine's stimulus meanings are irrelevant to linguistic competence. The problem is that they are misplaced in Quine's scheme. Even dispositions which coincide across a community cannot ground the normativity required for language because they yield only social regularities, which are distinct from norms. Such matter of fact regularities are *presupposed* by normative linguistic competence, but they are not *constituitive* of it. Without the sort of social regularities Quine characterizes and which are largely produced by behavioral conditioning (which Wittgenstein called "training"), there would be no language; but these regularities do not constitute linguistic meaning or competence. They belong to a level of discourse which might be spelled out in terms of how atoms in motion impact on our skins to produce neural processes, but that is not linguistic normativity even of the thinnest kind. Linguistic meaning and competence belong to a different level of discourse and require different concepts, not only to characterize language but also the kind of world necessary to learn and speak a language.[18]

3. ALTERNATIVES TO QUINE'S VIEW OF THE NORMATIVE

In this final section I will make some brief remarks about three alternatives to Quine's way of construing the social character of linguistic normativity. The first is a different kind of individualism: since you cannot get normativity by summation of non-normative individual dispositions, there must be normativity intrinsic to the pre-linguistic mental capacities of individuals. This is the view of mentalists of all stripes, whether dualists or physicalists. It is held by those who speak of mental states as having intrinsic intentionality or meaning, hence as being intrinsically subject to normative assessment. It is the view taken by functionalists who think that mental states with content or meaning are states which function properly in individual speakers. It is maintained by those who think the right causal relations between objects in the world and symbols in an individual's language of thought will endow those symbols with content or meaning. Quine is rightly untempted by any of these views, which are far more individualistic than his.

The second alternative has been formulated by (among others) Davidson, who attacks Quine's notion of stimulus meaning, arguing that the primary level of meaning should not be construed in terms of neural uptake but of readily observable physical objects or events. But the latter require, he argues, a second speaker who is capable of interpreting the first, for otherwise we could not identify the objects or events which, because of the way they cause the speaker to hold true a sentence (Davidson's equivalent for Quine's assent-dissent responses), are what the sentence is about. Quine separates the cause of assent or dissent to a sentence from what the sentence is about: the cause consists of neural uptake, but the sentence is about "ordinary things". Davidson rejects this separation, arguing that what causes assent or dissent (which are evidence for the sentence a speaker holds true) must also be, in general, what the sentence is about.

I won't discuss Davidson's arguments for this view. Although it is a significant step toward a viable non-individualistic view of the social dimensions of language, it does not, in my view, go far enough. The kind of interaction Davidson thinks required for language is itself proto-linguistic and therefore assumes speakers already capable of a proto-language. As Davidson put it in an address to the 1988 World Congress of Philosophy: "It's a matter of two private perspectives converging to mark a position in intersubjective space." To have a private perspective in this sense is already to have proto-linguistic capacities which, on this view, must exist prior to social interaction. The social is, again, constituted from the capacities and activities of individuals.[19]

The third alternative has been held by philosophers such as Dewey, Heidegger, or Wittgenstein. It holds that the linguistic capacities of individuals are grounded on and impossible without the social. Individual capacities and activities cannot constitute the social, for the capacities and activities which could do that are already social. An individual can learn a language only because there already exists a language to be learned. A language exists, of course, because there are speakers who speak it but, just as importantly, it exists because there is a social world which consists of objects and events not intelligible apart from language or language-like norms: modes of transportation, types of housing, trade and production, tools and toys, ways of dressing and eating, furniture, art, and so on. In learning a language, we learn to respond to these things, just as learning to respond to these things is to learn a language. None of this makes sense apart from a community, and to characterize it in terms of what individuals do is an abstraction – albeit a useful one for many purposes.

John Dewey articulated this view as explicitly and persuasively as anyone, so let me conclude with some quotations from the book he called *Logic*, a brilliant book, even if its subject would be unrecognizable to any contemporary logician.[20]

In every interaction [with physical surroundings] that involves intelligent direction, the physical environment is part of a more inclusive social or cultural environment.... The occasions in which a human being responds to things as merely physical in purely physical ways are comparatively rare.

Man is social in another sense than the bee and ant, since his activities are encompassed in an environment that is culturally transmitted, so that what man does and how he acts, is determined not by organic structure and physical heredity alone but by the influence of cultural heredity, embedded in traditions, institutions, customs and the purposes and beliefs they both carry and inspire. Even the neuromuscular structures of individuals are modified through the influence of the cultural environment upon the activities performed.... To speak, to read, to exercise any art, industrial, fine, or political, are instances of modifications wrought *within* the biological organism by the cultural environment.

Language is a distinctive cultural institution in that 1) it is the agency by which other institutions and acquired habits are transmitted, and 2) it permeates both the forms and the contents of all other cultural activities.

St. Olaf College, University of Helsinki

NOTES

[1] The least that is required, it seems to me, is a theoretical account of language. "Theoretical", however, is ambiguous. It may refer to a set of systematically interrelated claims which imply (empirical) consequences by which alone they are tested. Or it may refer to the opposite of "practical", in which case a theoretical account of language is one which attempts to *make explicit* in articulate discourse what people do in practice, though not necessarily in the guise of a theory , that is, not necessarily "theoretical" in the first sense. To make use of Wittgenstein, we *must* be theoretical in the second sense, but we must be cautious about being theoretical in the first sense.

[2] Michael Dummett, 'The Social Character of Meaning' in *Truth and Other Enigmas* (Harvard, 1978), p. 424.

[3] *Ibid.*

[4] Dummett is a prominent example with his insistence on a sharp distinction between force (as use dependent on meaning) and point (as use dependent on intention).

[5] This raises a host of issues, currently under intense discussion by philosophers, many of whom deny that language is normative. I agree with many of the critiques of the way concepts of the normative are applied to language, and this paper develops one line of criticism. I do not agree, however, that language can be understood without normative concepts, though defending that claim against its critics is beyond the scope of this paper.

[6] In 'In Praise of Observation Sentences' [*Journal of Philosophy*, March 1993, pp. 110f.], Quine lists seven vital roles they play. First, "they were probably the origin of language.

Second, they are the infant's entry to language. For much the same reason, they are the radical translator's way into the jungle language. [Fourth, they are] the vehicles of evidence for our knowledge of the external world.... [Fifth, they enable diverse theories to be commensurable. Sixth,] they are the primitive source of the idioms of belief and other propositional attitudes. [Seventh] their holophrastic role bears significantly on the epistemology of ontology." That is, since they play their role as whole sentences, we are free to reconstrue the reference of terms in many ways.

7   "The stimulus meaning of an occasion sentence is by definition the native's total battery of present dispositions to be prompted to assent or to dissent from the sentence." *Word and Object* (MIT Press, 1960), p. 39 (henceforth 'WO').

8   This is a point stressed by Wittgenstein in, for example, his discussion of rule-following and the possibility of a private language. The point is not that we cannot tell what is correct or incorrect but that there is no distinction to be made. Cf. *Philosophical Investigations* #258: "One would like to say: whatever is going to seem right to me is right. and that only means that here we can't talk about 'right'."

9   "The Social Character of Meaning", p. 424.

10   For Quine's discussion of this cf. WO, p. 36f.

11   "The notion of observationality is social. ... [It] turns on similarities of stimulus meanings over the community. What makes an occasion sentence low on observationality is, by definition, wide intersubjective variability of stimulus meaning." [WO, 45]

12   Harvard University Press, 1990, pp. 40ff.

13   Or it requires that the sentence to which S responds *already* has a linguistic meaning, which presumably requires that its stimulus meanings coincide for several speakers.

14   'The Significance of Quine's Indeterminacy Thesis' in *Truth and Other Enigmas*, p. 403.

15   By 'linguistic norms' I mean whatever norms are necessarily connected with being a competent speaker of a language or with concepts without which we cannot make sense of language. They need not be a matter of rules or conventions, and they do not presume that there is such a thing as distinctively linguistic use.

16   This does not mean shared dispositions cannot be *evaluated* in some sense or other, for the suitability of their sound for poetry or opera, for instance, or for being prolix or unclear. But these are distinct from the normative considerations I am discussing, which concern what counts as using an expression correctly or incorrectly and how far that distinction can extend.

17   This is a more complex question than I can adequately deal with here, among other reasons because of its connection to Quine's *neural* understanding of stimulus meaning, which he developed in *Word and Object* (which I have not discussed in this paper) and which, as he saw it, sustained "the philosophical doctrine of infallibility of observation sentences", [p. 44] which means that Quine grounds meaning on sentences which permit mistakes neither about meaning nor about belief. However, I do not find the neural conception of stimulus meaning at all plausible, and the doctrine of the infallibility of observation sentences is much attenuated by Quine's characterizing is as holding that there is no scope for error "insofar as verdicts to a sentence are directly keyed to present stimulation", which only raises the question of how we can tell which verdicts are thus keyed and which depend essentially on collateral information which is widely shared. It will not do to say that we cannot be in error about observation sentences, though it is always possible that we are in error about which sentences are in fact observational.

18   This oversimplifies a complex situation; for some of the complexities, cf., for example, John McDowell, 'Wittgenstein on Following a Rule', *Synthese* 58 (1984).

19   A version of this alternative which is not subject to this particular criticism has been worked out by Robert Brandom in *Making It Explicit* (Harvard University Press, 1994). He calls it an "I-thou" view of the social dimension of language, contrasting it with an "I-we" view, which, as he characterizes it, is pretty much Quine's.

20   *Logic: the Theory of Inquiry* (Henry Holt and Company, 1938), pp. 20 and 42f.

ALEX ORENSTEIN

PLATO'S BEARD, QUINE'S STUBBLE AND OCKHAM'S RAZOR

Once upon a time there was a thesis, an antithesis, and a synthesis. The dialectic of this particular Hegelian story is the relation of natural language to logical theory. The thesis was provided by that early stage in analytic philosophy wherein when logical theory clashed with natural language, it was natural language that suffered. An epidemic of charges of meaninglessness occured. Among those charged as linguistic deviants were such purported perversions of use as singular existentials, strings with vacuous singular terms, and the improper mating of objects or expressions of the wrong type. The title of Ryle's famous essay "Systematically Misleading Expressions" captures the ethos of that period. That essay documented purported cases of natural language, usage which were perceived to be at odds with certain logical forms provided at the time, and predictably for that period, the fault was located in natural language, not in the logical forms suggested by Principia Mathematica. The antithesis in this dialectic was supplied by ordinary-language philosophy, where such clashes lead to downplaying the role of logical theory and upgrading natural-language considerations. A favored practice of the period consisted of dissolving philosophical problems by illustrating that they had their roots in the misuse of ordinary language. The problem would disappear upon abandoning some theoretical infringement on natural language and by carefully sticking to ordinary language. The synthesis (the hero in Hegelian fictions) is the present period, and especially the position taken by the author of the fiction. Here natural language considerations and those of logic go hand in hand. This is due to a number of factors: a growth in logical theory, a more flexible attitude towards logical forms (competing theories of logical form are tolerated) and the growth of linguistics as a theoretical and somewhat formal theory of natural languages.

PROCRUSTEAN LOGICAL FORMS

The goal of this paper is to enforce with respect to certain issues of logical form the maxim of minimal mutilation. This maxim operates in tandem with other requirements for deciding between theories, such as that one explanation is superior to another if it has greater explanatory power. The maxim says that mutilations should not be made beyond necessity. Explanations should not rule out more than is necessary of previously accepted beliefs. Of two theories other things being equal the one that clashes least with background beliefs is to be preferred. Ruling out cases of the expressive force of natural language and/or intuitively acceptable-plausible

Alex Orenstein and Petr Kotatko (eds.), Knowledge, Language and Logic, 195–212.
© *2000 Kluwer Academic Publishers. Printed in Great Britain.*

logical inferences is inflicting mutilations. It can consist of imposing Procrustean logical forms ("Procrustes" The Oxford Universal Dictionary: "a fabulous robber of Attica who made his victims conform to the length of his bed by stretching or mutilation". Hence "Procrustean" The Random House College Dictionary: "pertaining to Procrustes, tending to produce conformity by violent or arbitrary means".)[1]

My concern is with a family of cases where natural language is unduly mutilated in the cause of assigning certain logical forms. These cases involve singular sentences, especially those with vacuous-empty subjects, predication, negation, and generalizations on them. By "logical form" I intend a minimal conception, that of providing a framework in a logical theory for intuitively acceptable inferences in natural language: explaining our intuitions as to which are valid and which aren't. This I take it is what has taken place historically, e.g., Aristotle on categorical sentences, Frege and Peirce on multiple quantification and relational notions, work on modal logic, proposals such as Davidson's for action sentences, ongoing work on the logical form of belief sentences, etc. So construed, logical form is in its essence a theory of the valid and invalid inferences sentences enter into. That is the sense in which this paper is concerned with it. Others add on that it provides an account of ontological commitment in terms of quantification. I will argue against this view and will propose an alternative. Yet others say a theory of logical form should provide an account of how we acquire or process language. I will forego this question entirely. It is not clear to me that logical form as a theory of inference is the same as a theory of how we acquire or process language. (Segal, p. 128)

My methodological assumption is: the fewer natural language sentences ruled deviant the better, and the more intuitively valid inferences recognized and accounted for the better. So "Pegasus is a flying horse", "He doesn't exist", "Vulcan is a planet", "Vulcan exists" or "Vulcan doesn't exist", "Deno forced me to commit the murder",[1] "Deno lives in the Bronx", "Deno [Ossian] exists", "Nessie doesn't exist" are all intuitively meaningful.[2] They have "street-cred". Dismissal of arguments containing such sentences, e.g., "Pegasus doesn't exist so something doesn't exist", is an unacceptable Procrustean solution. In these cases I am not concerned with such terms as they bear on problems pertaining to fiction as a literary genre.

TWO ACCOUNTS OF PREDICATION

There are several different accounts of predication. The two I wish to contrast cut across other ways of discussing this topic. My use of "predication" focuses on sentences of the form "Fa", where "F" is a position that can be filled by predicates of the type that occur in base clauses of truth or satisfaction conditions, e.g., "is a human", "is white", "runs". For ease of exposition and since it is not relevant to the arguments of the paper I confine myself to one-place predicates, but the points apply as well to many place predicates – relational expressions such as "is taller than" and "is between and". The "a" position can be held by singular terms, i.e., names or variables. I use the notion of predicate in the Fregean-Rheme sense, as roughly speaking, everything in a true/false atomic sentence other than the singular terms. The issues I am interested in can, to a certain extent, be discussed equally well as bearing on truth conditions for atomic sentences and their negations.

The first account is tailored to fit a Tarskian account of satisfaction. It has the consequence that negations of atomic sentences have existential import/ontological commitment.[3] The second account is found in Terminist logicians such as Ockham, Buridan, and the Psuedo Scotus; and in the Lesniewskian tradition. This second view does not have this existential consequence. If one wished to be historical, there is a sense in which the first view might be dubbed *Platonic* and the second *Terminist*. It was Plato who insisted that negations, even those denying that non-being is, have existential import. As Ruth Marcus puts it:

"Plato, in arguing against the Sophists' claim that erroneous beliefs are not about anything, says, "Whenever there is a statement it must be about something", and that, he claimed, holds for false statements as well as true ones... . Statements true or false, speak of objects". (Marcus, p. 112).

The Tarskian inspired view dovetails with the Tarskian theory of truth. That theory employs the semantic relation of satisfaction as found in satisfaction conditions used to define truth. What concerns us here are the relevant accounts of satisfaction for base (atomic) clauses, e.g., "x is human", and for their negations, e.g., "¬ x is human". From these, the tailored-to-fit predications will follow. Both satisfaction and predication are semantic relations between words and objects. The expressions involved in the satisfaction relation are open sentences, a predicate and a free variable, e.g., "x is human", "x is white", "x runs", and their negations, conjunctions, etc. The units satisfying an open sentence are sequences of objects or, allowing further simplification for the purposes of this paper, the objects in the sequences themselves. So, for example, an open sentence "x is human" is satisfied by the author (or an appropriately ordered sequence containing the author and his shirt). Satisfaction of this atomic open sentence involves the existence of an object. The satisfaction clause for a negation (I restrict myself to negations of atomic open sentences, though, of course, the negation clause is of much broader application) such as "¬ x is human" is satisfied by some non-human object such as the author's shirt (or an appropriately ordered sequence of objects made up of the shirt, the author, etc.). Once again the negation of an open sentence, such as "¬ x is human", has existential import, involving the existence of some non-human object.

A predicate is part of an open sentence, and to speak of its semantic relation to objects one speaks of the predicate "is human" or "runs" as applying to objects, or being true of objects, or multiply denoting objects. A predication based on satisfaction would amount to saying that the closed sentence obtained by replacing the variable "x" in the open sentence "x is human" by a singular term such as "Alex" is true, i.e., that the predicate "is human" applies to the object which the term "Alex" singly denotes. Such predication inherits the ontological commitment which attends the satisfaction relation. This treatment of atomic sentences might have been the one Quine intended in his account of predication. "Predication joins a general term and a singular term to form a sentence that is true or false according as the general term is true or false of the object, if any, to which the singular term refer" (Quine, 1960, p. 96).

Such predications embedded in negation also have existential import. Thus the negation of an atomic sentence, e.g., "¬ this shirt is human" is true because the open sentence "¬ x is human" is satisfied by some non-human object, e.g., this shirt. As Quine, speaking of satisfaction puts it: "The relevant logical trait of negation is not just that negation makes true closed sentences out of false ones and *vice versa*. We must add that the negation of an open sentence with one variable is satisfied by just the things that sentence was not satisfied by; ..." (Quine, 1986, p. 36)

The second view, the Terminist-Lesniewskian one, of predication and of truth conditions for atomic sentences and their negations accords existential commitment to atomic sentences but not to their negations. Consider how an atomic sentence could be said to be true. One might say somewhat figuratively that it corresponds to reality or that things are as they are described in the sentence or (a sixties version) "telling it like it is". Such remarks can with generosity be construed as metaphorical versions of somewhat more careful accounts. The sentences "Socrates is human" and "This shirt is made of cotton" are true when the subject and the predicate stand for the same thing, i.e., the subject's referent, its singular denotation, is one of the things the predicate applies to. A set-theoretic variant says that the sentence is true when the semantic value of the subject "Socrates" is a member of the set that is the semantic value of the predicate "is human". Another variant is that the individual the subject denotes has the property the predicate expresses, i.e., Socrates has the property of being human. These three approaches differ in their ontological assumptions but for the purposes of this paper any one of them will do. As on the Tarskian theory, an atomic sentence "Fa", "Socrates is human" requires for its truth that a (Socrates) exists and that F's (humans) exist.

But now consider four ways in which such sentences, atomic ones, can be false and so their negation "¬Fa" be true:

1.   Socrates ate pepperoni pizza.
2.   Socrates is a flying horse.
3.   Pegasus ate pepperoni pizza.
4.   Pegasus is a flying horse.

1 is false because, though Socrates exists tenselessly and pepperoni pizza consumers do too, Socrates is/was not one of them. So the negation of 1, ¬ (Socrates ate pepperoni pizza), is true.

2 is false because, there are no and never have been any flying horses so Socrates is not one of them. Thus, the negation of 2, ¬ (Socrates is a flying horse), is true.

3 is false because, though pepperoni pizza consumers exist, Pegasus does not; so he cannot be one of them. The negation of 3, ¬ (Pegasus ate pepperoni pizza), is true.

4 is false for either of the reasons involved in 2 and 3. The negation of 4, ¬ (Pegasus is a flying horse), is true.

A crucial difference has emerged between this account and the Tarskian inspired one. On this second view a sufficient condition for the falsity of an atomic sentence is the vacuity of its subject or its predicate. This also suffices for the truth of its negation. Since such negations can be true because of the vacuity of their parts, they have no existential import. They involve no ontological commitment and the same holds for sentences that are logical consequences of them. This is the basis for the maxim of Aristotelian logic that quality (the affirmative/negative distinction) determines existential import and not quantity (the universal/particular distinction).[4]

Buridan put this point well (perhaps so elegantly that some might think it applies only to negative existentials). It is remarked on as being a sufficient condition for the truth of the negation of a singular affirmative sentence (or any sentence grounded on such a negation), which in predicate logic notation would take the form of a negation of a singular-atomic sentence with a base-clause predicate. "I agree that Aristotle's horse [a vacuous term] does not exist.... It is true because the subject stands for nothing, so the subject and predicate do not stand for the same thing, which suffices for the truth of a negative" (Buridan, 1966, pp. 94–6).

This non-Tarskian approach bears investigation. It fits in with our intuitions that singular sentences with vacuous subjects are false (there is no reality for them to correspond to) and that such false singular sentences and their corresponding negations do not ontologically commit us to anything. These intuitions suggest that it is the less Procrustean and hence the more plausible of the two accounts of predication.

The thesis that quality and not quantity determines the existential import (ontological commitment) of a sentence contains a positive and a negative claim. The positive one is that quality (the affirmative/negative distinction as per atomic sentences, i.e., atomic sentences and their negations, and, consequently, the sentences depending on them) determines existential import. The negative claim is that quantity (whether a generalization is a universal quantification or a particular quantification – so called "Existential" quantification (though only called so since Frege), does not determine existential import.

THE POSITIVE CLAIM: WHAT EXISTENTIAL IMPORT IS

The positive thesis asserts that a given sentence has existential import if and only if the truth of an atomic sentence is a necessary condition for that given sentence's truth. Existential import is a matter of predications in affirmative atomic sentences. The case for this thesis proceeds in two stages. The first is to argue that atomic sentences and those requiring them have existential import. The second stage argues that only such sentences have existential import. It is evident from the different versions of the truth conditions for atomic sentences that both the singular subject and the predicate cannot be vacuous. This non-vacuity is the source of existential import. Any sentence that requires for its truth the truth of such an atomic sentence inherits the ontological commitment/existential import of that atomic sentence.

To establish that only the sentences requiring for their truth atomic sentences have existential import, we make use of the assumption that every sentence "depends" on either an atomic sentence or its negation. Those sentences that "depend" on negations of atomic sentences in the sense that the truth of such a negation suffices for the truth of the sentence in question do not have existential import. Since the negation of an atomic sentence does not have existential import, any sentence following from such a negation can be true without requiring the existence of anything. We are left with the conclusion that only sentences that don't "depend" on such negations have existential import. These are the sentences whose truth doesn't follow from such negations. They "depend" on something else for their truth, they require the truth of an atomic sentence. So only sentences requiring the truth of atomic sentences have existential import. Quality determines existential import.

A sentence, S, has existential import iff there is an atomic sentence, A, required for S's truth (the truth of an atomic sentence is a necessary condition for S's truth):

S has existential import $\leftrightarrow$ (S $\rightarrow$ A), where A is an atomic sentence

A sentence, S, doesn't have existential import if a negation of an atomic sentence, $\neg$ A, suffices for S's truth (if the negation of an atomic sentence implies A):

$\neg$ (S has existential import) $\leftrightarrow$ (– A $\rightarrow$ S).

THE NEGATIVE CLAIM: WHAT EXISTENTIAL IMPORT IS NOT

The negative claim denies the entrenched view of ontological commitment that existence is what so called existential quantification expresses. The prefix expression "existential" in "existential quantifier" is question begging. On this view all sentences governed by the some/particular quantifier (or implying such sentences) have existential import. One can construct a family of arguments to the contrary. Take as the premise a false vacuous singular sentence and assuming bivalence derive the truth of the vacuous singular sentence's negation. Derive a particular generalization from this negation by the rule of particular generalization (so-called "existential" generalization). When the conclusion is construed as having existential import, there is a paradox. How can a conclusion asserting existence follow from a premise whose truth doesn't require existence? (see Orenstein, 1995).

This type of argument takes a particularly well focused form when the premise is a singular denial of existence. To start with take a singular existential. Contrary to the tradition of mutilation which excludes such sentences and their denials we allow them as premises. Singular existentials and their denials are meaningful, e.g., Pegasus [Vulcan, Nessie, Deno] does not exist. Using any one of these denials of singular existentials one can generalize to the conclusion that something doesn't exist. The premise and the conclusion are not deviant. They express contingent truths. Moreover the conclusion follows from the premise. We don't want to mutilate these natural-language and common-sense observations and we want to be able to make valid inferences involving them as they stand. What I mean by "as they stand" is that the logical form which stays closest to the surface grammar of the argument is the preferable one. The argument wears its logical form on its sleeve. An account of logical form which requires equating particular generalizations with existential claims, renders the conclusion, "Something does not exist", problematic. On the existential reading it is interpreted as saying that there exist things that don't exist. This is "a contradiction in terms" (Quine, 1940, p. 50) and as such false though apparently following by truth-preserving rules from a true premise. Quantification, when construed as expressing existence, as having existential import, distorts – mutilates – natural language, forcing the existentials and generalizations of natural language onto a Procrustean bed.

If one were writing from the perspective of our antithesis, the period of ordinary-language philosophy, one might claim to have dissolved the problems associated with such arguments. The ordinary-language argument dissolves the puzzle and does so by conforming to intuitions based on natural language, e.g., "Pegasus doesn't exist so something doesn't exist" is intuitively sound.

Unlike the ordinary-language stage of twentieth-century philosophy, we should provide sufficient logical theory to support these natural-language intuitions.

The argument so far is that natural-language reasoning from "Pegasus doesn't exist" to "Something doesn't exist" is valid, indeed sound, and that a non-Procrustean account should capture its logical form as it stands, i.e., its surface grammar should dictate its logical form. We should be on guard against three ways in which our intuitions can be mutilated.

1.   The premise is a contingent truth.
2.   The conclusion is a contingent truth.
3.   The argument is valid as it stands.

A Terminist treatment can preserve all of these intuitions. It provides truth conditions whereby the premise (involving a negation of an atomic sentence) is true and the conclusion (a particular generalization), also true, follows from the premise. The truth condition for an atomic sentence in the true case can be the model theoretic one. However, unlike the usual model-theoretic account, when either the singular term or the predicate is vacuous the sentence is false. The latter suffices for the truth of the negation of such an atomic sentence. All other aspects of truth-functional connectives remain the same. The vacuity of "Pegasus" in the premise guarantees its truth whether we assume "exists" is a predicate *ab inito* or "a exists" is defined. If it is defined in terms of atomic predicates, then vacuity in those atomic contexts guarantees the premise's truth. As examples consider defining "exists" in terms of identity, a Lesniewskian copula, or base-clause predication. In each case "a exists" is true iff some atomic predication is true, e.g., "a = b" or "a is a b" [Lesniewskian "is a"] or "Fa" [base-clause "F"]. Where a term is vacuous the atomic sentence is false and the object is said not to exist as is required for the truth of the premise.

The truth conditions given for generalizations (and so for the conclusion) are a hybrid combining substitutional and non-substitutional approaches. We want to capture the advantages of each and avoid each's drawbacks. Among the advantages of the substitutional interpretation of quantification (as it is typically construed) is its ability to readily allow substituends that have no denotations. These include vacuous names which, while of the same category of names, have no denotations, and substituends such as predicates, sentences, and connectives which are not of the category of denoting expressions. The disadvantage of substitutional quantification is the problem of not being guaranteed enough names for all the objects in a quantifier's domain. An advantage of a non-substitutional approach (as it is typically construed) is that it accounts for all of the objects in a domain without requiring that each have its own name. The disadvantage may be in the treatment of non-denoting variables, i.e., vacuous names, non-names, e.g., variables for predicate positions.

The solution is to say that a universal generalization is true iff its substitutional interpretation conditions are met and its non-substitutional conditions are met. A particular generalization is true iff either the substitutional or the non-substitutional conditions are met. The non-substitutional account we employ is the method of beta-variants found in Mates (I assume here that it is non-substitutional). A universal/particular generalization is true in a language with at least one individual constant iff the instance of that generalization is true in every/at least one beta-variant. A beta-variant is a new interpretation where the given individual constant is assigned a new object from the domain. We go on reinterpreting the individual constant, assigning different objects to it in different interpretations and the universal/particular generalization is true as per being true in every/at least one such interpretation.

On our Terminist account, a universal generalization is true iff every substitution instance and every beta-variant is true. A particular generalization is true iff at least one instance is or at least one beta-variant is. The conclusion "Something does not exist" is true since one substitution instance is true, viz., the premise: "Pegasus doesn't exist".

Quine, like many others, has considered this *Plato's Beard* argument. It is discussed in *Word and Object* as a puzzle/problem on the way to quite a different conclusion than that just given (Quine, 1960, pp. 176–186). Let us organize the puzzle into the following steps:

True 1. "Pegasus exists" is false. Since there is no such thing as Pegasus, there is nothing for the false sentence to correspond to.

True 2. $\neg$ Pegasus exists.   From 1

True 3. $\neg$ Pegasus exists $\vee \neg$ a exists $\vee \neg$ b exists $\vee$ etc.,
From 2 disjunctive add.

True 4. $(\exists x)(\neg$ x exists$)$ From 2 and an analogue of 3

False 5. There exists an object x such that x does not exist.
5 is an existential reading of 4.

      6. Something doesn't exist.   6 is a natural-language
reading without existential force of 4.

The puzzle is that 1 is true because nothing exists to correspond to the false sentence, but 5 which seems to follow from 1 on the existential reading of the quantifier says that something exists. It is like getting something from nothing. Moreover 5 is a contradiction in terms while 6 is a contingent truth.

A number of different approaches have been offered for dealing with this puzzle:

I. Object-dependent truth vehicles, e.g., Evans, Strawson, Van Fraassen, Salmon. Singular sentences with vacuous subjects are meaningless or fail to yield truth vehicles;
II. Singular existentials are meaningless, e.g. Frege, Russell;
III. Non-existential reading of the particular quantifier, *via* Meinongian quantification;
IV. Free Logic, e.g., Lambert, Bencivenga;
V. Russell-Quine ("Quinize the name and Russell away the description");
VI. Terminist: A non-existential reading of the particular quantifier.

I will understand "object dependence", I, in a rather comprehensive way, as the view that maintains that if the object the subject term is used to refer to doesn't exist, then there is no bivalent truth vehicle. This construal covers Evans' object dependence as such a no-object-so-no-proposition view, where the object is not part of the proposition; Russellian singular propositions as in Salmon where the object constitutes part of the proposition/truth vehicle (so, no object, then no proposition); Strawson's view interpreted as saying no object, so no statement/truth vehicle; as well as Strawson construed by Van Fraassen as saying that if there is no object, then bivalence is suspended and there is a non true/false vehicle. All of the above violate to different degrees the intuition that such sentences express contingent truths. In addition, a multitude of cases exist where such sentences are embedded in negations, conditionals, propositional attitudes. There is also the role of such sentences in arguments for existence or non-existence, e.g., God exists, and arguments in law courts (Does Deno exist?), science (Vulcan does not exist), etc. Object-dependent views do not so much solve the problem of vacuous terms as sweep it under the rug by declaring that a string purporting to be a premise fails to be one, claiming that it is meaningless, not a truth vehicle, or not a bivalent one. Without the premise there is no plausible argument which conflicts with our intuitions that there are arguments containing vacuous singular terms. Another version, II, of the meaningless-string approach stems from Frege's and Russell's view that singular existentials are deviant. This view is equally counterintuitive, e.g., I exist, God does (or does not). (Orenstein, 1995, Sinisi-Wolinkski).

Though Meinongian approaches, III, vary, I will assume that what they have in common are quantifiers which range over non-existent objects. This allows for a solution in accord with our intuition that the conclusion "Something doesn't exist" is a contingent truth. On my reading Meinong is tough-minded (robustly realistic) about the use of the word "exists". He would agree with Russellians, Quinians, Lesniewskians, Terminist logicians etc. (putting aside considerations about time and about abstract objects) that Pegasus, Vulcan, Ossian, Nessie (probably), Deno, etc. do not exist. The difference between Meinong and the others is in his notion of an object. Whereas the others make the expressions "exists" and "is an object" coextensive, he makes the notion of an object a more inclusive one. Pegasus, Nessie, etc. do not exist but they are objects and the conclusion that some things (some objects) don't exist is as such true and justifiable within the standpoint of this theory. Two considerations lead me to not being Meinongian on such puzzles. The first is parsimony. It is worth seeing whether one can provide a non-Procrustean treatment without positing objects over and beyond existing ones. A word on terminology. When Russell employed the expression "robust sense of reality" to contrast with Meinongian views it may partly have been in the spirit of favoring a more parsimonious view. His other criticisms have not fared as well (see Parsons, Lambert and Routley/Sylvan for rejoinders to Russell). A confusion of terminology arises with the expression "realism" as per realism versus anti-realism controversies. So Russell's "robust realism" sound strange at present since Meinong is now dubbed the realist on this issue and Russell the anti-realist.

In addition to the parsimony point we can construct a dilemma. It consists of inquiring whether vacuous singular terms in a special extended sense are allowed or not. If vacuous singular terms are allowed and by these I mean singular terms which not only don't have an existent that they denote but on Meinong's extended notion of an object, these vacuous terms don't even have objects that they denote. Let us use the expression "vacuous as to existence" for ordinary vacuity and "vacuous as to objecthood" for the vacuity I have in mind as posing a problem for Meinongians. As seen earlier, the Meinongian solution to the Plato's-beard-type puzzles consists of two moves: 1) of having no vacuous-as-to objecthood singular terms, and 2) of construing quantification in a Meinongian spirit, i.e., "Some objects don't exist", and thus not as producing the contradiction in terms, "Some existing things don't exist". A parallel puzzle arises when the Meinongian allows for vacuous-as-to-objecthood singular terms. Let us assume that "Pppegasus" is such a term (pronounced by sounding out the separate "P"s). Perhaps it is a term with a history which parallels to non-objecthood in a Meinongian universe the history of "Pegasus" in a non-Meinongian universe. Assume there is a Pppegasus story and "Pppegasus" is a bona-fide meaningful singular term). We now have a true puzzle premise:

Pppegasus is not an object.

By the second move of the Meinongian strategy we construe the quantifiers as ontologically committing us to Meinongian objects. We get a Meinongian contradiction in terms as a conclusion:

Some objects are not objects

The puzzle remains or recurs, if Meinongians allow for vacuous-as-to-objecthood singular terms.

On the second horn of the dilemma, the Meinongian will deny that any syntactically meaningful terms can be vacuous *simpliciter* (with respect to objects as well as existents). This is to subscribe to the policy that there are no vacuous singular terms (or perhaps more strongly that there couldn't be any). Now the premise: "Pppegasus is not an object" is false and the false conclusion (contradiction in terms) seems less of a problem. But a question remains as to why a singular term, if part of a language – a linguistic entity (or even a representation of some sort) –

should always have a denoted item. Singular terms, if construed as linguistic items, are customarily distinguished from the non-linguistic items the singular terms denote. What argument is there that being a member of the set of such linguistic items guarantees that there is a member of the set of denoted (or possibly denoted) non-linguistic items for each term to denote? In today's realist/anti-realist parlance, Meinongians are construed as holding a realist thesis that certain items really are in some independent sense the subject matter of true-false attributions. The more one poses this realist construal of Meinong (the more realistic in the current sense – the less robustly realist in the Russellian sense), the greater the difficulty in framing an argument for there being Meinongian denotations for all singular terms.

This argument could and perhaps should have been pursued earlier in the discussion of Meinong. What guarantees that all names have a denotation, even granting that there are objects that don't exist? What argument is there on realist grounds that all names have objects that they denote? If not all names have objects, then the puzzle reappears and if they all do, then we are owed an explanation of this. The problem of vacuous names remains unsolved.

To understand, IV, the free logicians attempt to resolve the puzzle, a word must be said about their stipulative definition of "free logic". The leading spokesmen for free logicians, Lambert and Bencivenga, stipulate that the phrase applies exclusively to logics that allow vacuous names and that read the particular quantifier, "$(\exists x)$", existentially (Lambert, Bencivenga). Before this restricted nomenclature of "free logic" was adopted, the term had the broader connotation of a logic free of existence assumptions. On this earlier connotation Lesniewskian views such as those in this paper would be regarded as free logics. After all, what could count more as a logic free of existence assumptions than a Lesniewskian style account which frees the particular quantifier of existential significance and which also allows for vacuous names. The usual direction free logic in the narrow stipulated sense takes is to deny the ordinary and intuitive particular/"existential" generalization rule and the universal instantiation rule. Instead of the familiar "Φa", therefore, "$(\exists x)\Phi x$", the free-logic rule tends to be a variant of "Φa" and "a exists", therefore, "$(\exists x)\Phi x$". Here the inference from "a is an Φ" to "Something is an Φ" is invalid. According to the free logician's solution, the argument from "Pegasus does not exist" to "Something does not exist" is invalid. Not only does this violate our intuition that the argument is valid. It also violates our intuition that the familiar rule is valid. This way of forcing the existential reading on the particular quantifier and allowing for vacuous names leads to denying the otherwise solid intuition that is embodied in the standard rule. In addition, it violates the analogies that should hold between conjunction/disjunction and universal/particular generalizations. Particular generalization is analogous to disjunctive addition and this analogy requires that just as "Φa" implies "$\Phi a \vee \Phi b \vee \Phi c \vee$ etc.", so "Φa" implies "$(\exists x)\Phi x$". Note that these implications hold without additional premises. It is as though the free logicians' version of sentence logic would maintain that standard disjunctive addition is invalid and requires an additional premise. The free-logic version of particular generalization destroys this analogy. So the argument as it stands is invalid for the free logician. When we supply the purportedly missing premise to turn the invalid argument into a valid instance of free particular generalization we get

Pegasus doesn't exist.
Pegasus exists.
Therefore, something doesn't exist.

But this argument should strike one as valid because of its grosser sentence logic form, viz., *ex falsum quodlibet* (from a contradiction everything follows). This hardly helps to save our original intuition that the argument was valid as it stood. Not to mention that it requires supplying a false premise, thus making this version mutilate even further our original intuition that the original argument was sound. Last of all, the conclusion, in the invalid and the valid forms, is, as Quine put it, a contradiction in terms, given the existential reading of the quantifier.

Let us turn to Quine's solution to the puzzle, V. Quine has been an important corrective force as to charges of meaninglessness. He has not wanted to extend the concept of meaninglessness beyond strict violations of syntax. Charges of meaninglessness have been challenged by Quine in at least two ways. The first is the avoidance of type theory and its philosophical spinoffs. Russell's solution to his own paradox involved multiplying cases of meaninglessness for type violations. This furnished a precedent in the thesis period and beyond for talk of category errors and meaninglessness. Quine has argued that type violations/category errors can be regarded quite simply as obvious falsehoods. Secondly, one can interpret Quine's criticisms of verifiability tests of meaningfulness as the more modest claim that assertions about God, the Absolute, etc. do no work in empirical inquiries. These are more modest charges than claiming meaninglessness, but they should be sufficiently damning. Non-syntactical meaninglessness in its several forms is Procrustean overkill.

So Quine's views on vacuous singular terms have not involved holding that the sentences containing them are meaningless. If there are Procrustean elements in his solution to such puzzles it is not on this score.[5] His most distinctive way of dealing with names, vacuous or not, is to accord them no status in his canonic notation. They are supposed to be in some serious sense defined away in favor of predicates and variables of the category of singular terms. Wherever there is a name in natural language we form instead a predicate which applies to the object (if any) that the name applies to and then use Russell's theory of definite descriptions in connection with the predicate version of the name. As David Kaplan has aptly quipped "we Quinize the name and Russell away the description". (Kaplan) The starting point of the puzzle "Pegasus exists" is in turn treated as "There is one and only one object which pegasizes" and in canonic notation appears as "$(\exists x) (Px \,\&(\forall y)(Py \rightarrow y = x))$". It is false and so we can take its denial as Quine's version of the second line of our puzzle: $\neg (\exists x) (Px \,\& (\forall y) (Py \rightarrow y = x))$. The only singular terms present are variables (and bound ones at that). There is no troublesome vacuous name to "existentially" generalize on and yield the puzzle. But then there are no names for the natural deduction rules of generalization to apply to. This violates our intuition that inferences with names be allowed as they stand and in particular that the puzzle inference is sound as it stands. If there is a problem about vacuous names, a less Procrustean way of dealing with it would be preferable.

There does appear to be a semblance of an inconsistency in Quine's treatment of the puzzle. In *Word and Object* (p. 176), he uses intuitive inferences with names to create the puzzle, thereby implicitly acknowledging their intuitive appeal, and then offers a solution which would deny that one can make such inferences. Names even appear in a somewhat more systematic fashion on the way to a final solution which would dispense with them. Schematically "Fa", where "a" is a name or schema for a name, becomes "$(\exists x)(x = a \ \& \ Fx)$"; then "= a" is treated as though it were a simple predicate without any internal structure. Let us try to do justice to Quine and attempt to make a case for his using natural language inferences as a ladder to be abandoned once he has reached a higher goal. The issues involved parallel an argument of Russell's: common sense leads to physics, and if physics is right then common sense is wrong, so if common sense is right, then it is wrong; therefore it is wrong. Should we consider Quine as replying to common-sense intuitions as to reasoning with names as follows: natural language leads to logical theory and logical theory (at least Quine's version) if right shows that natural language is wrong, so if natural language is right then it is wrong; therefore it is wrong. Take the puzzle and the problems concerning vacuous names as a case in point for redeploying Russell's common-sense-to-physics argument. So used, it purports to justify a Procrustean point of view: Sacrifice our common sense intuitions for logical theory. But there are differences between the folk physics and the folk logical theory cases. The greater explanatory power gained by mutilating folk physics compensates for that mutilation. It is not at all obvious that the mutilations or sacrifices made on behalf of Quine's solution to the puzzle of vacuous names are compensated for by an increasing gain in logical theory. *Pursuit of Truth* has two opening quotations which favor empiricism. The first is Plato's "save the phenomena/appearances". The second is a pun on a paint company's advertisement: "Save the surface and you save all". Quine requisitions it as an advertisement for empiricism. Lets add "Save the surface grammar".

I interpret Quine as holding the view that Quinizing names (replacing names with predicates and variables) and Russelling away the associated descriptions is in some sense the best theory for dealing with the problem of vacuous names. How should we understand the claim that he has dispensed with names?

One point to note about Quinizing names and Russelling away the descriptions is that it is a species of no-empty singular-terms solutions. On Quine's approach there are no names in the ordinary sense, hence no empty names. The only singular terms are variables and these are never vacuous. Variables are construed in a Tarskian spirit. Variables as per the open sentences in which they occur are satisfied by objects. There are no vacuous variables. Thus no singular terms, i.e., variables or ordinary names, are vacuous. It may prove of interest to compare this feature of Quine with other accounts, such as substitutional ones, which allow for variables having substituends which are vacuous.

A different point worth stressing is that there is only a cosmetic terminological difference here between names, in the ordinary sense and as used so far in this paper, and variables. This is a matter which has not gone unnoticed. Dummett remarks: "In regard to any open sentence, such an assignment confers upon the free variables occurring in it the effective status of individual constants or proper names" (p. 16). Shaughn Lavine says something to the same effect: "In Quine's case, generality is assured because any object can be assigned to x.... . Thus, even Quine in effect makes use of the notion of a tag [name of a sort], under the guise of an assignment" (Lavine, p. 271).

My own recognition of this point arose independently of the above authors in connection with comparing Tarski on satisfaction and the method of beta-variants used for giving truth conditions by Mates in his *Elementary Logic*. I take it that an individual constant is to an artificial language what names in some paradigmatic sense are to natural languages. The method of beta-variants gives truth conditions for atomic sentences in the usual way, i.e., "Fa" is true if and only if the semantic value of "a" is a member of the semantic value of "F".[6] The method of beta-variants distinctiveness lies in its truth conditions for generalizations. A universal/particular generalization is true if all/some of its beta-variants are. To repeat, the idea is that a generalization such as "(x)(x is in space)" is true if and only if we form an instance of the generalization and we keep reinterpreting the individual constant so that on each interpretation (beta-variant) it is assigned a different object. A universal generalization is true if, given an instance of it, that instance remains true under every new interpretation of (assignment to) the individual constant in question. A particular generalization is true, if given an instance of it, that instance remains true under at least one interpretation of the individual constant in question. These conditions for generalizations involve quantifying over interpretations. Tom Baldwin argued for the naturalness of this approach by explicating the truth of generalizations, e.g., "Everything is in space", by appeal to the truth, e.g., of instances containing demonstratives "That is in space", "That is in space", etc. where the demonstrative "That" is assigned different objects. One does not appeal to open sentences; they can be treated as ill-formed strings. (This is also a plus, in keeping with a worthy tradition in logical theory which does not sanction open sentences).

On Tarskian semantics, satisfaction is a relation between open sentences and objects (sequences of objects, to be more exact). A generalization is satisfied if all/some objects (sequences) satisfy an open sentence. On the Taskian account we quantify, so to speak, within one interpretation over sequences which involve different assignments to the same variable. The universal generalization "(x)(x is in space)" turns out true when the open sentence "x is in space" is satisfied by every object (sequence of objects to be more exact). The expression "x" in the open sentence is assigned different objects.

What is the difference between the two methods and what does it indicate about first-order variables and names? The Tarskian account quantifies over objects or sequences of them within, so to speak, one interpretation. The beta-variant approach quantifies over interpretations. Both seem to involve assigning different objects to a singular term, in Tarski to a variable and in Mates to an individual constant. In Mates the connection of variables to constants/substituends is brought to prominence. In Quine-Tarski the relation is one of variables to objects without the intermediary substituend-constant. Many of those familiar with Taski, on first hearing of the beta-variant view, strenuously maintain that it is the same as Tarski's. Perhaps what provokes this reaction is in part the recognition of there not being much difference between variables assigned different objects in Tarski and the individual constants reinterpreted in Mates. Tarskian style variables are misleadingly categorized when they are thought of as somehow seriously different from names. Perhaps the fault is due to a way of considering open sentences. When they are considered without assigning an object to "x" in "x is in space", the open sentence is in this respect disinterpreted and seems to suggest that the expression "x" is not namelike. But let us take a lesson from the Quine of "Truth by Convention" about not being misled as to the philosophical-semantical significance of disinterpretation. A disinterpreted string tells us little about the semantic status of its constituents. The question then would appear to be one of the difference, if any, between the variable "x" in "x is human" under an assignment and individual constants/names. In the Mates account individual constants, the artificial-language correlate of names, are explicitly present, syntactically and semantically. The truth conditions for quantifications and, so to speak, variables essentially involve names. On Quine's Tarskian account, syntactically there are no individual constants. But the individual variables semantically seem to be quite namelike. It is hard to see more than a cosmetic terminological difference between a variable under an assignment and a name. What is the difference between how "x" functions when "x" is assigned Alex (or the sequence of which Alex is the significant element) in the open sentence "x is in space" and the individual constant/name "Alex" in "Alex is in space"? To label "x" under an assignment a variable and so argue that it is not a name, has much in common with arguing that the glass of water is half empty and so is not half filled. It seems merely terminological whether to classify variables as being opposed to names or as being a variety of names.[7] The remark that Quine "dispenses with names" or "that names are defined away (where definition is elimination)" must be taken with some qualification.

By way of summary, we note the undesirable Procrustean elements in Quine's account. Our intuitions are that the ordinary-language argument is sound as it stands and should be regarded as wearing its logical form on its sleeve. Save the surface grammar and you save the whole inference. On Quine's view however, premise, conclusion and inference are mutilated. 1) While the first premise is treated as a contingent truth, the logical form accorded it is not as close to the original as one would desire. There are no names (in the somewhat superficial cosmetic names-versus-variables sense) and hence no valid inferences with them. 2) Given the existential reading of particular generalizations, the intuitively contingently true conclusion "Some things do not exist" is mutilated by being construed as a contradiction in terms which seems to violate the principle of charity in our own community. 3) The original argument was valid, in fact sound, but

Quine's version makes it invalid. The canonical version of the English sentence, i.e., "It is not the case that exactly one thing pegasizes" is false. Since it contains no singular terms fit for "existentially" generalizing on, the conclusion does not validly follow. By contrast, the terminist inspired account preserves our beliefs that the argument is sound and the intuition that there is no puzzle, that the contingently true premise containing a name, a vacuous one at that, formally implies the contingently true conclusion.

PLATO'S BEARD AND QUINE'S STUBBLE

Early in "On What There Is" we are told: "This is the old Platonic riddle of nonbeing. Nonbeing must in some sense be, otherwise what is it that there is not? This tangled doctrine might be nicknamed Plato's beard ..." (Quine, pp. 1–2). Ruth Marcus puts such issues in a somewhat broader context about erroneous beliefs (which I construe as bearing on negations, and in particular, denials of nonbeing). "Plato, in arguing against the Sophists' claim that erroneous beliefs are not about anything, says, "Whenever there is a statement it must be about something", and that, he claimed, holds for false statements as well as true ones... . Statements true or false, speak of objects (Marcus, p. 112)".

The Terminist view is that a denial of a singular sentence should imply nothing substantive. Denying that Fa, does not imply anything substantive. To illustrate this point, consider a Carnapian state description account or a modest modal view where the falsity of an atomic sentence does not imply anything substantive, i.e., some other atomic sentence, its denial, or anything non-tautological that follows from these. To deny that a exists does not intuitively imply anything other than itself or tautlologies, even that anything at all exists or that something other than a exists and is a non-F. From the denial of Pegasus is a flying horse it does not follow that there is something other than Pegasus and that it is a non-flying horse. But on Quine's view, Russelling away a name and doing it in a Quine/Tarski manner, has a substantial consequence. Accept a Quinian denial of nonbeing and you counter-intuitively acquire a substantive assertion of being.

Quine's solution to the riddle of Plato's beard is not as deflationary as one might have been lead to believe. Quine, in his Tarski-based use of Russell, provides a special case of a Platonic approach. Ruth Marcus' version of Plato's position that both false as well as true statements must be about some object suggests the following Platonic alternatives: 1) in denying nonbeing we are implying nonbeing; 2) in denying nonbeing we are implying being. Quine sees himself as being anti-Platonic. But he is so only in the first sense. He is Platonic in the second sense. The Terminist-Lesniewskian inspired account of this paper denies both versions of the Platonic view. If the former Platonic view is dubbed Plato's beard, the latter might be nicknamed Quine's stubble. It is the counter intuitive doctrine that denials of nonbeing involve being.[8]

Queens College and the Graduate Center, City University of New York

ACKNOWLEDGEMENTS

I should like to thank my friend and colleague, Jack Lange, for his careful reading of this paper and for his many helpful comments. I am also indebted to Joan Meyler for her keen eye in spotting errors and infelicities. Thanks too to Queens College for a presidential research grant.

NOTES

[1]   The image of Procrustes was shoplifted from Quine (*Quiddities* p. 158); "Despite such exclusions [modalities, contrary to fact conditionals, etc.], all of austere science submits pliantly to the Procrustean bed of predicate logic. Regimentation to fit it thus serves not only to facilitate logical inference, but to attest to conceptual clarity. What does not fit retains a tentative and more provisional character".

[2]   Deno is a purported-to-exist drug pusher who purportedly forced someone to commit a murder. This was the defense claimed by the accused and reported on in several N.Y. newspaper reports.

[3]   I equate the traditional talk of existential import with talk of ontological commitment.

[4]   I am obviously indebted to Manley Thompson's work. However his use of this maxim differs from that in this paper. Here quality, the affirmative/negative distinction, is tied to atomic sentences and their negations, whereas Thompson ties it to the four categorical sentences.

[5]   In *Mathematical Logic* he did offer an argument from the syntactical form of quantification claims and the view that existence is what existential quantification expresses that singular existentials construed as a quantifier concatenated with a singular term, e.g., "$(\exists x)$ a" is not a well formed formula. However this should be distinguished from the type-theoretical arguments offered by Fregeans and Russell. I think that Quine now agrees with more current views that assign a different and *bona fide* logical form to singular existentials, e.g., "$(\exists x)(x = a)$". In order to deal with the *Word and Object* form of the puzzle special provisos would be made, in effect making the singular term inaccessible to quantification. This may well be a Procrustean move in not countenancing ordinary inferences involving names. Moreover, as indicated earlier, the intuitively meaningful and contingently true conclusion is transformed via the existential reading of the quantifier into the contradiction in terms "There exist things which don't exist".

[6]   Mates' use of this condition differs from my use in this paper. He assumes that all the individual constants are going to be interpreted so as to be non-vacuous. Let us put aside the topic of vacuity for just a moment.

[7]   A more significant account of differences between variables and names may arise on some substitutional accounts of truth conditions.

[8]   We are also left with some anachronistic and unanswerable questions of scholarship. Would Russell have agreed to embedding his theory of definite descriptions in Quine and Tarski's accounts of satisfaction-predication? Would he have acquiesced in the view that Pegasus' not existing implied the existence of anything else? Was Russell Platonic or Terminist-Lesniewskian?

REFERENCES

Baldwin, T. "Interpretations of Quantifiers", *Mind*, (350), 1979.
Biencivenga, E. (1986) "Free Logic" in Gabbay and Guenther, Eds. *Handbook of Philosophical Logic* Vol. 3, Dordrecht, Reidel.
Buridan, J. (1966) *Sophisms on Meaning and Truth*, New York, Appleton Century Crofts.
Dummett, M. (1973) *Frege Philosophy of Language*, London, Duckworth.
Kaplan, D. (1970) "What is Russell's Theory of Descriptions?" in Davidson and Harman, Eds. *The Logic of Grammar*. Encino, Dickinson Publishing Co.
Lambert, K. (1983) *Meinong and the Principle of Independence*, Cambridge, Cambridge University Press.

Lavine, S. "Review of Ruth Marcus' *Modalities*", *British Journal for the Philosophy of Science*, 46, 1995.

Mates, B. (1972) *Elementary Logic*, New York, Oxford University Press.

Marcus, R. (1993) *Modalities*, Oxford, Oxford Univ. Press.

Orenstein, A. (1995) *Proceedings of the Aristotelian Society* Vol. XCV, Oxford, Blackwells.

Orenstein, A. (1995) "Existence Sentences" in Sinisi and Wolinski, Eds., *The Heritage of Kazimierz Ajdukiewicz*, Atlanta, Rodopi.

Parsons, T. (1980) *Nonexistent Objects*, New Haven, Yale University Press.

Quine, W.V.O. (1960) *Word and Object*, New York, John Wiley and Sons.

Quine, W.V.O. (1936) "Truth by Convention" in *The Ways of Paradox and Other Essays*, New York, Random House.

Quine, W.V.O. (1940) *Mathematical Logic*, Cambridge, Harvard University Press.

Quine, W.V.O. (1948) "On What There Is" in *From a Logical Point of View*", Cambridge, Harvard University Press.

Quine, W.V.O. (1961) *From a Logical Point of View*, Cambridge, Harvard University Press. (Includes papers Quine 1937, Quine 1948, Quine 1951).

Quine, W.V.O. (1966) *The Ways of Paradox and Other Essays*, New York, Random House. (Includes papers Quine 1936 and Quine 1954).

Quine, W.V.O. (1986) *Philosophy of Logic*. Cambridge, Harvard University Press.

Quine, W.V.O. (1987) *Quiddities*, Cambridge, Harvard University Press.

Quine, W.V.O. (1992) *Pursuit of Truth*, Cambridge, Harvard

Routley, R. (1980) *Beyond Meinong's Jungle*, Canberra, Australian National University.

Segal, G. (1994) "Priorities in the Philosophy of Thought" *Proceedings of the Aristotelian Society*. Supplementary Vol. LXVIII, Oxford, Blackwells.

Thompson, M. (1953) "On Aristotle's Square of Opposition", *The Philosophical Review*.

TERENCE PARSONS

INDETERMINACY OF IDENTITY OF OBJECTS: AN EXERCISE IN METAPHYSICAL AESTHETICS

"A good scientific theory is under tension from two opposing forces: the drive for evidence and the drive for system."

"My inclination is to adhere to [bivalence] for the simplicity of theory that it offers."

"One might ... despair of bivalence and proceed disconsolately to survey its fuzzy and plurivalent alternatives in hopes of finding something viable, however unlovely."

... W.V. Quine[1]

This essay and the next are meant to form two parts of a study of the indeterminacy of identity of object and sets. This essay is concerned with indeterminacy of identity of objects; the next focuses on sets.[2] The purpose of these essays is to explore the idea that the world might be indeterminate, indeed, so indeterminate that even identity may be indeterminate. That is, that certain questions of identity have no answer, not because of an inadequacy in the language in which they are framed, but because of genuine indeterminacy in the world. According to Professor Quine, we thus choose empirical adequacy over simplicity of theory. But the theory is not all that complex after all, and it is certainly not complex when compared with the bivalent alternatives. The picture we propose is neither fuzzy nor (literally) plurivalent; we also find it lovely; thus our subtitle.

THE ISSUE

A curious thing about the identity problem is its simplicity. In can be put in three Anglo-Saxon monosyllables: "What is it?" It can be answered, moreover, in a word – "Itself" – and everyone will accept this answer as true. However, this is merely to say that a thing is what it is. There remains room for disagreement over cases; and so the issue has stayed alive down the centuries.[3]

The hypothesis of indeterminacy of identity arises naturally as a plausible solution to cases in which there is persisting disagreement regarding identity questions. Our hypothesis is that the disagreements persist because there are indeed no answers. Examples in the literature fall into several classes. First, there is a question of identity over time when there is a simple disruption of some kind. For example, a person receives a new brain having the old memories, or else a new set of memories is inserted into an existing brain, or ... There seems to be a single person under discussion before the disruption, and a single person under discussion after the disruption, and the question arises as to whether we are dealing with one person or two. Put in terms of identity, is person a (the person before the disruption) identical

213

Alex Orenstein and Petr Kotatko (eds.), Knowledge, Language and Logic, 213–224.
© 2000 Kluwer Academic Publishers. Printed in Great Britain.

with person *b* (the person after the disruption)? Cleverly designed cases will undermine any definite answer to the question.

We will focus on another sort of case. Recall this famous quote:

Wie Schiffer sind wir, die ihr Schiff auf offener See umbauen müssen.[4]

Suppose then that we steal such a ship, and we repair it at sea by replacing its parts one at a time. Further, we are followed by an entrepreneurial salvage company that retrieves the replaced parts and assembles them, and then takes the assembled ship to the original owners. Did the owners get their ship back, or are we still in possession of stolen goods? We suppose that there is no answer to this, and that is because of how the world is.

We propose to take a Peircean approach, and begin not with fake Cartesian doubt, but rather with real belief and real doubt. What we really believe is that there is one ship at place A at the beginning of the trip, the original ship. There is one ship at place B at the end, the repaired ship. And there is one ship at place C at the end, the assembled ship. The repaired ship and the assembled ship are located at distinct places at the same time, and are thus distinct. It is problematic, and subject to doubt, whether the original ship is the repaired ship, or is the assembled ship, or neither. Perhaps there *is no answer*, and these identities are indeterminate.

Our strategy will be to assume that there is indeterminacy in the world, and try to describe how one can theorize about it in a way that is both intellectually and aesthetically pleasing. That is our overall task. But we begin with a side issue. If there is genuine indeterminacy in our statements about the world, many people are inclined to take for granted that this is because of some deficiency in our conceptual apparatus. It is natural to assume that the world is fully determinate, and our concepts are simply not sufficiently refined to make full contact with it; thus the indeterminacy in our judgments. With sufficiently refined concepts, all indeterminacy would vanish. We begin by describing this "everything is determinate" view; we will argue that it is in fact neither intellectually nor aesthetically pleasing. We will then turn to describing our own view.

CLARIFYING CONCEPTS

The idea is that apparent indeterminacy of identity is due to unclarity in our concepts, and indeterminacies vanish if we *refine* those concepts.[5] When there is a genuine puzzle about identity, this is because the key concept(s) appealed to in the puzzle can be clarified in different ways. For example, the puzzle about ships arises because our concept of a ship is unclear. It may be made more clear. In fact, it may be clarified in one way, in which we identify the original ship with the reassembled ship, and it may be clarified in another way, which identifies the original ship with the repaired ship, or in a third way in which it is identified with neither. This is not quite accurate, since in terms of unclarified concepts the phrase 'the original ship' does not pick out a ship at all. Instead, we can refer to the original $ship_1$, the original $ship_2$, the original $ship_3$, and so on.

We think that this view requires a problematic ontology. Recall, we began with the assumption that the total number of ships is at most three. (It is not clear whether there are two or three.) The refined concepts view requires *more* objects. The notion of *ship* can be refined so that the same $ship_1$ is in both places A and B and a different $ship_1$ is in place C. It can also be refined so that the same $ship_2$ is in places A and C, and a different $ship_2$ is in place B. And it can also be refined so that there are distinct $ship_3$'s in all three places. Further, we don't have to choose between these refinements; in fact, we must not. All are actual, and the result is that there are at least three "ships" at place A, at least two at place B, and at least two at place C, which, when you factor in the possible cross-identities, yields at least five "ships" (using the unrefined notion of *ship*). This leads to a bloated ontology. We asked for relief from slooplessness; we got a whole fleet! We might have settled for yachts that are longer than they are; but these yachts are greater in number than they are!

The concept-refiner's overpopulated universe is in many ways unlovely. It offends the aesthetic sense of us who have a taste for open seas.[6]

The issue is not the number of entities involved. We accept the fact that every filled spatio-temporal region contains a distinct physical entity. So there are at least five ocean-floating physical entities in this situation. The problem is how we are to classify these entities into ships – into refined ships, that is. For in our own epistemological dealings with the world we do not do much with the concept of filled spatio-temporal region, but we do deal with the concept of a ship. And we need to be clear about how many ships there are that we propose to go sailing on. It is disconcerting to find five of them here.

But perhaps that is just our naïvete speaking. After all, if you stand at the original place you are not on several $ship_1$'s, you are only on one. And you are only on one $ship_2$. And so on. But that's like being on one yacht, *and* being on one sloop, *and* being on one boat, while being told that they are all *distinct* from one another. When your friend shows you his fleet of ocean-going vessels, you apparently see only one. It takes some metaphysical sophistication to see them all, while realizing that you are seeing several.

This multiplicity of refined ships is plausible only because people don't know much about ships and care less. If we consider something we do know about and care about, this kind of burgeoning ontology is more troublesome. So let us instead refine the notion of *person*. Let us ask first what constraints are there on the conditions for refinement of the notion of *person*. Some natural answers that come to mind are that people are things that think, they are things that pay taxes, and they are things that we marry. If these sorts of paradigms do not hold then it is not plausible to claim that we are *refining* the notion of person, we are simply changing the subject. Now refinements in the notion of *person* will not end up with there being multiple *persons* in one place at a time, since our concept of person stands in need of refinement, and should be dispensed with. But (like the ship case) the refinements will require there to be in a single place both a person₁, and a person₂, ... If these "refined people" have the characteristics that are essential to refinements of persons, then in a given place will be located two or more things that think and feel and desire, two or more things that pay taxes (do they both pay the same taxes, or have there been duplicate payments, or...?), and so on. The case I worry most about is whether there are two or more refined persons to whom I am married. This is disconcerting to say the least.

There are two problems here. The immediate problem is that the view under consideration is not a developed view at all; we don't really have an option to consider, but many options, depending on what refined things are supposed to be like. The options need to be spelled out. We need to know, for example, whether we get two person-like things that share a body, or two bodies as well. This is a proposal to reform our ontology, without instructions to how to do it. The instructions say "make it clear," but it's not clear *which* of the clear options we are to choose. Recall, the issue is *not* whether there are filled regions of space-time; we are already clear about that. The refiner's proposal is not exhausted by this assumption. When we clarify our concepts, we also need to decide, for example, *which* such regions are occupied by refined people, and how many? For example, which such region contains my wife? Of course, I speak here of my unrefined wife, and I am not supposed to do that. The instructions are to abandon the belief that I have a wife, and acquire certain other beliefs. One will be to believe that I have several wives-sub-14, all of whom I married and all of whom ceased to exist before 8 AM this morning. I also have several wives-sub-72 that I married and that will not cease to exist until just before dinner. I may also have some wives-sub-62 whom I never married because they came into existence in the same place as the ones I did marry, just after the ceremony. Or does that mean I am not married to them (but I am sleeping with them)?

Imagine for a moment that my unrefined wife is now standing in that doorway. Then:

Take, for instance, the actual woman-sub-14 in that doorway; and again, the actual woman-sub-18 in that doorway. Are they the same actual entities, or two actual entities? How are we to decide? How many actual woman-like entities are there in that doorway to whom I am married? Are there more to whom I am not married than to whom I am married? How many of them are married to me? Or would being married to me make them one? .... These elements are well-nigh incorrigible.[7]

Here is what is really wrong with this idea: It ignores our epistemological position in the world. You get the impression that if we were just to abandon our current notions of ship, person, wife, and so on, and replace them with clearer ones, we would be better off. But that ignores the wisdom with which this quote was selected by Professor Quine:

Wie Schiffer sind wir, die ihr Schiff auf offener See umbauen müssen.

We don't get to leave our foundering ship like rats, and swim over to a better one, even if the brass is cleaner on the other side. We would drown on the way.

We deal with a world peopled with unrefined ships, unrefined oceans, and unrefined people. We can't even locate a refined ship except parasitically upon the procedures that we presently use to located unrefined ships. I can't locate a refined wife except parasitically upon the procedures that I now use to locate my unrefined wife. The proposal to "refine your concepts and the indeterminacies will vanish" is a proposal to leap overboard and climb onto a new ship that hasn't yet been built. We see how it might go, of course, and we see the kinds of decisions that would need to be made along the way. But that's like jumping overboard thinking you're going to be kept afloat by a blueprint of a ship. Worse, you're climbing onto a bale of blueprints of ships of different kinds, incompatible with one another, none of which have yet been built, and all of which need to be built in the same place.

Unlovely!!!

INDETERMINACY IN THE WORLD

Our view is this:

The world consists of some objects, and some properties and relations, with the objects possessing (or not possessing) properties and standing in (or not standing in) relations. Call these possessings and standings-in "states of affairs." Then some but not all such states of affairs are determined by how the world is.

We accept the traditional Leibnizian definition of identity between a and b as indistinguishability of a from b in terms of the actual properties and relations that they have or stand in. Whether this holds in a particular case may be undetermined by the facts. If so, a and b will actually be neither determinately identical nor determinately distinct. This will be a case of indeterminacy of identity. The framework of properties and relations sketched above is neutral about whether this ever actually happens. It will happen if a definitely has every property that b definitely has (and vice versa), and a definitely lacks every property that b definitely lacks (and vice versa), but there is some property that a definitely has or definitely lacks when it is indeterminate whether b has it (or vice versa).

But what sense can be found in talking of entities which cannot meaningfully be said to be identical with themselves and distinct from one another? These elements are well-nigh incorrigible.[8]

Our view is that it *is* meaningful to say of any x and any y that x=y.
And each entity is identical with itself: $\forall x(x=x)$.
And each entity is distinct from every other: $\forall x\forall y(y \neq x \rightarrow x$ and y are distinct).

These desiderata are all consistent with the view that identity is sometimes indeterminate.

CANONICAL NOTATION

Professor Quine has taught us to be careful of how we design the canonical notation within which we philosophize. Here, for present purposes, is ours:

1. Each atomic predicate is true of certain objects, and is false of certain other objects, and is thus neither true nor false of the remainder, if any. An atomic predicate may or may not express a property. *If* it expresses a property, then it must be true of exactly those objects that have the property and false of exactly those objects that lack the property. (Similarly for relations.)

2. Satisfaction:
    (i) An object o satisfies an atomic formula "Px" if "P" is true of o; o dissatisfies "Px" if "P" is false of o, and otherwise o neither satisfies nor dissatisfies "Px". (And similarly for relational predicates.)
    (ii) "$x=y$" is satisfied by a pair of objects if they both have and lack the same properties, dissatisfied if one of them has a property that the other lacks; otherwise it is neither satisfied nor dissatisfied by that pair of objects.

3. Connectives and quantifiers:
    $\neg A$ is true if A is false, false if A is true, and truth valueless if A lacks truth value.
    $A\&B$ is true if A and B are both true, false if either A or B is false, and otherwise lacks truth value.
    $(\exists x)A$ is true if A is satisfied by at least one object, false if A is dissatisfied by every object, and otherwise neither true nor false.

4. Indeterminacy:
    ∇A is true if A lacks truth value, and is otherwise false.

The logic that we take for granted is not mysterious at all; it is exactly what you expect if some sentences can lack truth value. For example, from the conjunction "A&B" you may infer "A", because if a conjunction is true, so is either part. But certain other principles fail. For example, if an argument has a valid form, you can't take for granted that its contrapositive form is also valid. For example, this is valid:

$S /\therefore \neg\nabla S$

but its contrapositive is not even remotely plausible:

$\nabla S /\therefore \neg S.$

What we propose is not fuzzy. But is it plurivalent? Well, yes and no. We do not propose that there are multiple truth values: you have true and false, and that's it. But there are three *options*: a sentence may be true, false, or neither. So we plead guilty to trivalency of options, but that is not so bad. Our thesis can now be put in one simple assertion: $\exists x\exists y\nabla x=y$.

CONCEPTUAL TOOLS FOR DESCRIBING THE WORLD

What about simplicity? That is partly a question of what the theory says, and partly a question of the conceptual tools that are available for using it. So we want to talk briefly about Venn diagrams. These are a conceptual tool for dealing with the simple non-relational logic of determinacy. They can be easily modified to deal with indeterminacy. We picture mundane indeterminacy using diagrams similar to Euler or Venn diagrams, in which objects are represented not by points but by small filled regions. We use circles to represent the extensions of properties, exactly as in Venn diagrams. Then an object is pictured as having a property if its image is wholly inside the property's extension, and it is pictured as lacking a property if its image is wholly outside a property's extension. When its image overlaps the boundary of the property's extension, that means that it is indeterminate whether the object pictured has the property:

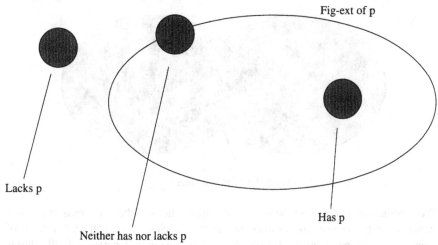

It is easy to extend this kind of picturing to identity of objects; objects *a* and *b* are pictured as identical if they are represented by the same image, and they are pictured as distinct if they are represented by disjoint images. If their images properly overlap, this pictures (the state of affairs) that it is indeterminate whether *a* and *b* are the same:

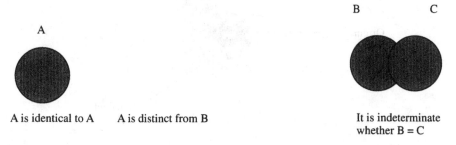

With this convention, Leibniz's definition of identity is nicely captured. It is easy to see that if the images picture two objects as distinct, then the images do not overlap, and a circle can be drawn to represent the extension of a property that one of them wholly has and that the other lacks. So objects represented as distinct will differ in their properties. If the images coincide then no circle representing the extension of a property can distinguish the object(s) in any way. So objects represented as identical must agree completely in their properties. Suppose that the objects are pictured by means of overlapping images. Then no property can wholly distinguish one from the other (since any circle that encloses one image must enclose at least part of the other), but there are properties that one has or lacks and it is indeterminate whether the other has or lacks it (just draw a circle that totally encloses one image but not all of the other image).

Nothing about these diagrams excludes the possibility that one object image might fall totally inside another, or inside a group of others. By fiat, we prohibit this from happening for the applications discussed in this paper:

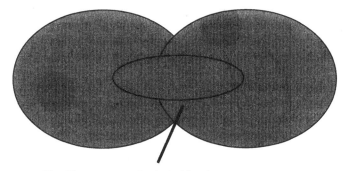

No objects are completely inside others

We prohibit these cases because in all of the applications of the hypothesis of indeterminate identity to solve puzzles from the literature such (onto)logical containments of objects in others never arises as a possibility. Such overlaps generate radically different conceptual schemes; these are worth exploring, but we will not do so here.

With diagrams of this sort we picture the ships as follows:

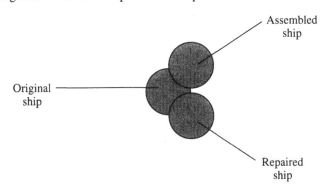

Assembled
ship

Original
ship

Repaired
ship

A MORE RIGOROUS FORMULATION

Venn diagrams are two-dimensional pictures. But what is being pictured is not two-dimensional. If you abstract away from the two-dimensionality of the above diagrams, you have a general way to model situations involving indeterminacy with determinate models. Here is a way to generalize the picturings.

Suppose that the image space of the picture consists of a set D of points, called *ontons*. The set of ontons is any set, with no geometric notion of dimensionality assumed. The term 'onton' is purely suggestive; the points form the *onto*logical basis of the space. Ontons are purely artificial, much as possible worlds are artificial in modal model theory.

In any total picturing of reality there are a number of images of objects; each image is a *set* of ontons. The only constraint we impose on such a modeling is that each object image contain some ontons (at least two) that are not in any other object.

Object images represent identical objects if they consist of the same ontons, they represent distinct objects if they share no ontons, and they represent indeterminately identical objects if they share some ontons but do not share some others.

Each property on D has a figurative extension, which is a set of ontons. An object image i pictures its object as *having* a property p iff every onton in i is in the figurative extension of p. An object image i pictures its object as *lacking* a property p iff no onton in i is in the figurative extension of p. And an object image i pictures its object as neither having nor lacking a property p iff some of its ontons are in the figurative extension of p and some are not. (Similar conditions apply to pairs of objects and relations.)

Finally, we suppose that there is such a picture that accurately pictures reality. That is, we assume that there are indeed objects and properties, and that any object image pictures a unique object, and that every property has a figurative extension. An object has (or lacks) a property iff it is pictured as having (or lacking) it.

These picturings validate a set of logical rules, which follow. (These are due to Peter Woodruff.) A set of sentences is provable by those rules iff they are all true in every picturing of the sort we described above. These rules take as primitive the non-falsity operator; i.e. "$\lhd \Phi$" is true when "Φ" is not false.

Rules for Conjunction

$$
\begin{aligned}
&\&\text{intro:} && \phi, \Psi \vdash \phi \& \Psi \\
&\&\text{elim:} && \phi \& \Psi \vdash \phi \\
& && \phi \& \Psi \vdash \Psi \\
&\lhd\&\text{intro:} && \lhd\phi, \lhd\Psi \vdash \lhd(\phi\&\Psi) \\
&\lhd\&\text{elim:} && \lhd(\phi\&\Psi) \vdash \lhd\phi \\
& && \lhd(\phi\&\Psi) \vdash \lhd\phi
\end{aligned}
$$

Rules for Negation

$$
\frac{\Sigma, \lhd\neg\phi \vdash \Psi \qquad \Sigma, \lhd\neg\phi \vdash \lhd\neg\Psi}{\Sigma \vdash \phi}
$$

$$
\frac{\Sigma, \lhd\neg\phi \vdash \lhd\Psi \qquad \Sigma, \lhd\neg\phi \vdash \neg\Psi}{\Sigma \vdash \phi}
$$

$$\frac{\Sigma, \neg\phi \vdash \Psi \quad \Sigma, \neg\phi \vdash \lhd\neg\Psi}{\Sigma \vdash \lhd\phi}$$

$$\frac{\Sigma, \neg\phi \vdash \neg\Psi \quad \Sigma, \neg\phi \vdash \lhd\Psi}{\Sigma \vdash \lhd\phi}$$

Rules for Non-Falsity

| $\lhd$ intro: | $\phi \vdash \lhd\phi$ |
| $\lhd$ elim: | $\lhd\lhd\phi \vdash \lhd\phi$ |

Rules for Quantification

∃intro: $\phi(t/x) \vdash \exists x\phi$

∃elim: $\dfrac{\Sigma, \phi(y/x) \vdash \Psi}{\Sigma, \exists x\phi \vdash \Psi}$ (provided y is not free in Σ, ϕ or Ψ).

Rules for Quantification with non-falsity:

$\lhd \exists x\phi \vdash \exists x\lhd\phi$
$\exists x\lhd\phi \vdash \lhd\exists x\phi$

Reflexivity: $\vdash t{=}t$

Leibniz' Law: $\phi(s/x), s{=}t \vdash \phi(t/x)$

Weak Symmetry: $\lhd(s{=}t) \vdash \lhd(t{=}s)$.

The claim that there is indeterminacy of identity (the thesis: "$\exists x\exists y\nabla x{=}y$") is consistent within this set of rules.

It is essential that Leibniz's Law hold – so long as the language does not contain non-extensional contexts. As many people have pointed out, if Leibniz's Law does not hold, then it is doubtful whether we are talking about identity and saying something radical about it, or whether we have simply changed the subject and are talking about some other weaker kind of equivalence relation. We are happy to say that the picturings under discussion validate Leibniz's Law in its full generality:

a=b
$$\frac{\Phi(a)}{\therefore \ \Phi(b)}$$

But the contrapositive of Leibniz's Law (which we call the "principle of definite difference") is another matter. This does *not* hold in general:

DDiff:

$$\Phi(a)$$
$$\underline{\neg\Phi(b)}$$
$$\therefore a{\neq}b$$

Why would someone think that this would hold? Well, what about Leibniz's definition of identity in terms of possessing all the same properties and relations? If a and b disagree in the properties they possess, then they should be different, shouldn't they? Yes indeed, they should. And in some cases that is what DDiff will be saying. These will be the cases in which Φ expresses a property. When Φ expresses a property, DDiff will hold, and when Φ does not express a property then it need not.

When will Φ not express a property? It is easy to make up examples using the indeterminacy operator. Here is one; suppose that $\Phi(a)$ is:

$$\nabla a{=}b.$$

This says that a certain identity statement fails to express a truth value. Can such a claim be reformulated in terms of an object possessing a property? Not if properties are the sort of thing possession of which is constitutive of identity. In particular, what about this formula:

$$\nabla(x{=}a).$$

Does this express a property of x? No it does not. It is a meaningful formula, but language cannot create properties any more than a comprehension axiom can create sets. In fact, ignoring intensionality, these are the same points. Tricks with language cannot put things into our ontology.

In general, a formula will be capable of expressing a property when it (determinately) satisfies this condition:

$$\neg\exists x\exists y[\nabla x = y \ \& \ \Phi(x) \ \& \ \neg\Phi(y)]$$

(One might consider making this stronger: one might consider saying that a formula expresses a property when that condition is necessary, but in honor of Professor Quine we will refrain from introducing modalities.)

ORIGINS OF INDETERMINACY OF IDENTITY

Indeterminacy of identity arises naturally as a result of the independence of the sciences from one another and from our everyday views. Consider physics, as an investigation of those systems that obey physical laws. And consider botany as the study of plants. Both sciences include a drive for systematicity, and one might argue that the best view of each is that it does not countenance indeterminacy of identity. But their overlap is another matter. Consider an orchid in a pot, and suppose that ordinary principles of botany require that there be exactly one such orchid. Now consider instead the many different various physical systems that co-incide as closely as we can determine to the orchid. They differ from one another in that one includes a molecule that is dangling off the edge of a brown spot on the tip of a dead leaf, and another does not. And so on. It will be indeterminate which of these physical systems is identical with the orchid. Simplicity of theory is one thing; simplicity among theories is another. Each science may eliminate indeterminacy of identity within its purview, but only metaphysics can eliminate indeterminacy of identity where they overlap. Perhaps metaphysics is continuous with the sciences, but the facts that it addresses are different because it deals with their relations to one another. Simplicity might require indeterminacy of identity *between* the ontologies of the various sciences even if it rejects it *within* each.

INDETERMINATE SETS?

I have been talking freely of properties, and this brings us dangerously close to what a certain famous person has labeled "dabbling with intensions". No dabblers we! Everything we say is consistent with the assumption that properties are identical when they are possessed and dis-possessed by the same objects, distinct when they are possessed by some objects and dis-possessed by others, and indeterminately identical otherwise. In short, everything we are saying is consistent with the view that properties are just sets – but sets whose own identities may be indeterminate.

Indeterminacy for sets too? Is this a pandora's box? No, for the indeterminate identity of sets is also a thing of beauty, as Peter Woodruff explains in the next essay in this volume.

University of California, Irvine

NOTES

[1] W.V. Quine, "What Price Bivalence?" *The Journal of Philosophy 78*, 1981, 90–95.
[2] In this essay, Parsons speaks mostly on behalf of both himself and Woodruff; in the next, Woodruff speaks mostly on behalf of himself and Parsons. The essays are based on joint work but are mostly independently authored. We have also written on these topics in two other papers: "Worldly Indeterminacy of Identity," *Proceedings of the Aristotelian Society*, Winter, 1995, and "Indeterminacy of Identity of Objects and Sets," forthcoming in *Philosophical Perspectives*, 1996.
[3] A distortion of the opening paragraph of W.V. Quine, "On What There Is," in *From a Logical Point of View*, Harper & Row, New York, p. 1.
[4] "We are like sailors who must repair our ships upon the open sea." Quoted from Otto Neurath on page vii of W.V. Quine, *Word and Object*, Wiley, New York, 1960.
[5] For a view like this see R. Stalnaker, "Vague Identity," in Austin, D.F. *Philosophical Analysis*, Kluwer, 1988, 349–60.
[6] A distortion of a passage in W.V. Quine, "On What There Is," *From a Logical Point of View*, *op. cit.*, page 4.
[7] A distortion of a passage in W.V. Quine, "On What There Is," *From a Logical Point of View*, *op. cit.*, page 4.
[8] A distortion of a passage in W.V. Quine, "On What There Is," *From a Logical Point of View*, *op. cit.*, page 4.

PETER WOODRUFF

INDEFINITE OBJECTS OF HIGHER ORDER

The previous essay outlines an account of indeterminate identity, and offers some preliminary illustrations of its application to puzzles about identity through time. We wish now to consider an application to a particular, rigorously formulated theory. The theory of sets is especially apt for this role, since it brings into prominence the Quinean theme of identity conditions: "No entity without identity." At the same time, our discussion will confront what I will call the Basic Objection to many-valued theories, also due to Quine: that they are simply too messy.[1]

1 CHOOSING PRINCIPLES: A PICTURE AND A THEORY

Let us begin by recalling the basic principles of classical set theory: comprehension and extensionality.

Classical Comprehension For every well-defined condition $\phi(x)$,

$$\forall t_1 \dots \forall t_k \, \exists B \, \forall x[x \in B \leftrightarrow \phi(x)]$$

Classical Extensionality

$$\forall A \, \forall B \, [\forall x \, (x \in A \leftrightarrow x \in B) \to A = B]$$

The qualification "well-defined condition" is of course intended to provide protection against the paradoxes of set theory. For present purposes we will confine ourselves to the study of sets of objects (non-sets), so that we need not concern ourselves with spelling out the qualification beyond a consistent use of capitals on the right side of "$\in$", as in the principles just given.[2]

The problem confronting us now is to specify corresponding principles for set theory with indeterminate identity (and therefore, by Quine's dictum, with indeterminate objects). More precisely, we must choose connectives of three-valued logic to play the role of the conditional and biconditional in the above axioms. There are two ways to proceed. First, we might suppose that there is an intuitive notion of indeterminate set that we are aiming to capture; then there are objectively right and wrong choices. On the other hand, we might simply say that different choices determine different concepts of "set". We will begin with the first perspective, but add some remarks from the other point of view in the last section of the paper.

Whence, then, our intuitive notion? The answer is ready to hand if we suppose that sets are extensions of properties. For we had a picture of such extensions in the previous essay. There a property is represented by a region of the plane, its "figurative extension". Individuals which are candidates for belonging to the extension

225

Alex Orenstein and Petr Kotatko (eds.), Knowledge, Language and Logic, 225–235.
© *2000 Kluwer Academic Publishers. Printed in Great Britain.*

are also represented by regions. An individual was said to have the property if it is included in the figurative extension of the property, to lack the property if it is wholly excluded, and otherwise (the overlap case) to neither have nor lack the property. It is now just a short step to conceive the figurative extension as representing the extension of the property; if the property is non-bivalent, then its extension will be a candidate for a "non-bivalent" set. The problem of specifying comprehension and extensionality principles now becomes the problem of saying when an expression picks out a property, and when two properties determine the same extension.[3]

It should be noted immediately that these problems are more complicated than in the bivalent case, in the following sense: it is not in general enough to say when it is (definitely) true that an expression picks out a property, or that two properties have the same extension; we must also independently specify when these things are indeterminate, or equivalently, when they are definitely false. This is an example of the sort of "messiness" abhorred by Quine, but also suggests a response to the Basic Objection: we can make finer distinctions. For two theories of sets may agree in their "definitely true" part but disagree elsewhere, and this distinction could not be made in a bivalent theory.

Let us begin then with the following: given the intuitive model, when does an expression definitely pick out a property? The first thing to note is a certain interaction between our modelling of extensions of properties and our modelling of individuals. For suppose that a has the property P, and that b is indefinitely identical with a, as in the following diagram. Then clearly the representation of b must also overlap the figurative extension of P, and hence b cannot lack P. Thus if $\phi(x)$, say, picks out a property, then whenever $\phi(a)$ is definitely true and $a = b$ is not definitely false, $\phi(b)$ cannot be definitely false. Let us say that if $\phi(x)$ satisfies this condition it is transparent.[4] Not all expressions $\phi(x)$ meet the test of transparency: for instance, $\triangleright x = a$ does not, for if b is only indefinitely identical with a, we will

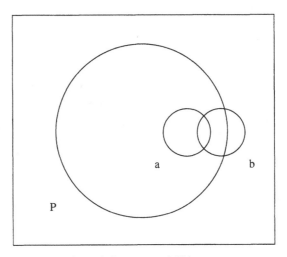

Figure 1: Property and Objects

$\Leftrightarrow$	t	u	f		$\cong$	t	u	f
t	t	u	f		t	t	f	f
u	u	t	u		u	f	t	f
f	f	u	t		f	f	f	t

Figure 2: Two three-valued equivalence operators

have $\triangleright a = a$ but also $\neg\triangleright b = a$. Let us suppose that this is the only test we impose on ϕ over and above those we required for the avoidance of paradox. Then we will want to have a set as extension of $\phi(x)$.

Now we must ask how to state the instance of the comprehension axiom which postulates this set. Our basic intuition suggests that an individual should definitely belong to the extension of a property if it has the property, definitely not belong if it lacks the property, and otherwise belong indefinitely. There are a number of three-valued equivalence connectives which will serve to express these conditions; here are truth-tables for two of them:

The first of these connectives is Lukasiewicz' equivalence, the second may be called strict equivalence and essentially asserts the definite truth of the first. Since for us assertion is always assertion as true, there will be no difference in asserting equivalences of the two sorts, but differences emerge in other contexts; for instance, when they denied rather than asserted.[5] For present purposes we choose L-equivalence, and so are led to the following version of comprehension:

Comprehension   For every well-defined and transparent condition $\phi(x)$,

$$\forall t_1 \ldots \forall t_k \exists B \, \forall x[x \in B \Leftrightarrow \phi(x)]$$

Comprehension gives a sufficient condition for definite set existence; is that condition also necessary? That is, if $\phi(x)$ definitely picks out a set, must it be transparent? Nothing we have said so far guarantees that this will be the case unless we know that the condition $x \in S$ is transparent. Our intuitive model does in fact satisfy this requirement. We therefore add the following postulate:[6]

Transparency

$$\forall A \forall x \forall y (x \in A \ \& \ \triangleleft x = y \rightarrow \triangleleft y \in A)$$

Next let us consider identity conditions for sets: when are the extensions of two properties definitely the same, and when are they definitely different? Again our intuitive model suggests answers. For it was part of the intended modelling that any two pictures which exhibit the same pattern of inclusions, overlaps and exclusions represent exactly the same figurative extension. Thus if A and B agree in definite, indefinite and definite non-members, they should be definitely the same. But such agreement can be represented by (the truth of) L-equivalence; thus we should have that $A = B$ is definitely true if (and, by the simple logic of identity, only if) $\forall x(x \in A \Leftrightarrow x \in B)$ is definitely true.

The question of when A and B are definitely distinct (i.e., $A = B$ is definitely false) is more delicate. We cannot appeal to simple features of the intuitive model, since we have no way of representing identity of extensions directly.[7] However, on the assumption that sets are "nothing but" a collection of members, it is plausible to suppose that definite difference of sets must be explained in terms of definite difference of membership, so that S and T are definitely different if and only if some definite member of the one is definitely not a member of the other. The same conclusion can be arrived at by a different route: if we think that identity between sets can be reduced to equivalence of membership, then the left and right sides of the extensionality axiom should express the same proposition, and so a fortiori have the same truth conditions.[8] Thus we shall stipulate

Extensionality

$$\forall A \forall B [\forall x (x \in A \Leftrightarrow x \in B) \Leftrightarrow A = B]$$

It is a corollary of this postulate that $\in$ will also be transparent "on the right"; i.e., that if $a \in A$ and $\lhd A = B$, then $\lhd a \in B$. For if not, then we have $a \in A$ definitely true and $a \in B$ definitely false, so $a \in A \Leftrightarrow a \in B$ definitely false, and hence (by Extensionality) $A = B$ definitely false, contrary to hypothesis.

This completes our preliminary exposition of the theory; we turn now to some applications.

2 FINITE SETS OF INDIVIDUALS

The most elementary sets are collections of individuals specified by indication. These are usually indicated by expressions of the form $\{a_1, \ldots, a_n\}$, with the a_i being terms designating individuals. We introduce such sets by (closures of) definitions of the following sort (where x and all y_i are distinct variables)[9]:

$$\{y_1, \ldots, y_n\} = A \Leftrightarrow \forall x (x \in A \Leftrightarrow x = y_1 \vee \ldots \vee x = y_n)$$

That these definitions are proper follows readily from our axioms.[10] From them we can easily derive the following identity and membership conditions for small collections:

Unit Sets: $x \in \{a\} \Leftrightarrow x = a$
                  $\{a\} = \{b\} \Leftrightarrow a = b$
Pair Sets: $x \in \{a, b\} \Leftrightarrow x = a \vee x = b$
                  $\{a, b\} = \{c, d\} \Leftrightarrow (a = c \ \& \ b = d) \vee (a = d \ \& \ b = c)$

In particular, $\{a, b\} = \{a, a\} = \{a\} \Leftrightarrow a = b$.

It is possible (and suggestive) to picture sets such as this by identifying the region of the plane corresponding to the set with the union of the regions correspondig to the members.[11] If we recall from the previous essay that an individual can be contained in a union of others only if it is identical with one of them, we obtain precisely the above membership conditions. On this understanding Figure 3

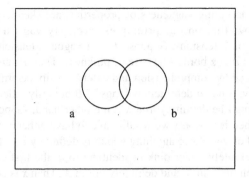

Figure 3: Sets of individuals

represents a situation in which we have two objects a and b, with a definitely belonging to both $\{a\}$ and $\{a, b\}$ but only indefinitely to $\{b\}$; it is of course precisely a picture of the indeterminacy of $a = b$, as the last equivalence above requires.

3 ORDERED PAIRS AND THE WIENER/KURATOWSKI REDUCTION

Ordered pairs are an interesting testing ground for intuitions about indeterminate identity. It is clear that we want two ordered pairs to be definitely identical iff their respective left and right members are definitely identical, but it is possible to take various attitudes towards their definite difference (or, equivalently, their weak or "non false" identity). Perhaps the most natural view declares $< a, b >$ definitely distinct from $< c, d >$ iff either a is definitely different from c or b from d. In any event, it would seem that whatever conditions we choose should be symmetrical with respect to the left and right sides. Curiously, the well known reduction of ordered pairs to sets due to Wiener and Kuratowski, which identifies $< a, b >$ with $\{\{a\}, \{a, b\}\}$, fails this test. For given this reduction, we can calculate from the definitions of the previous section that if a is definitely different from c then the same is true of $< a, b >$ and $< c, d >$, but that if b is definitely different from d the same may not be true of the two pairs. Indeed, consider the special case where a and c are definitely identical. Then it is consistent with the definite difference of b and d that neither be definitely different from a; but then $\{a\}$ will not be definitely different either from $\{a, b\}$ or from $\{a, d\}$, hence the two ordered pairs $< a, b >$ and $< a, d >$ will be weakly identical despite the definite difference of their second members.

The argument just given illustrates one compensation for the added complexity of three-valued set theory: it provides a richer testing ground for proposed reductive definitions.

4 TWO ARGUMENTS

In this section we apply the theory to the analysis of two arguments, one in favor of indefinite identity, and one against. The first argument (our own) is directed at those who, while accepting vagueness of properties, are skeptical about indefinite identity.[12] It shows that on our principles, property vagueness will produce indefinite identity of extensions. Suppose P is a vague extensional predicate (e.g. "is pink"), and that a is a borderline case of pinkness. Then Px and $Px \vee x = a$ will be transparent, hence by comprehension and extensionality determine unique sets A and B, respectively. Since a definitely belongs to B but only indefinitely to A, $\forall x(x \in A \Leftrightarrow x \in B)$ cannot be definitely true. On the other hand, suppose the latter were definitely false. Then for some x we would have to have either x definitely in A and definitely out of B or v.v. Since anything which is definitely in A must be definitely pink and hence definitely either pink or identical to a, the first case cannot occur. So suppose x is definitely in B, and definitely not in A. Then x is definitely not pink, but also definitely either pink or identical to a, and so must be definitely identical to a. Since *ex hypothesi* a is a borderline case of pinkness, this implies that x is so as well, and thus not definitely not pink. Thus the second case cannot occur either, and so $\forall x(x \in A \Leftrightarrow x \in B)$ cannot be false, either. Therefore it is indeterminate, and so (by Extensionality) the same must be true of $A = B$. Is this a knock-down argument for worldly indeterminacy of identity? We are not sure; consider the following dialogue.

SKEPTIC: "The conclusion of your argument is supposed to be that there is real indeterminate identity in the world. But all you have shown is that $A = B$ is indeterminate, and it is not clear to me that A and B are in the relevant sense "in the world". After all, they are abstract entities."

US: "Well yes, but we are only talking about collections of individuals, and surely they are just about as much in the world as their members. Certainly they are not linguistic, nor are they properties."

SKEPTIC: "I think I made the wrong point. What I really want to maintain is that indeterminacy of identity always has to do with the way in which the terms of the identity pick out something, rather than with the things picked out. Now you can claim that COMP provides us with a definite set picked out by any transparent condition. But it seems to me that I could legitimately ask you which set is picked out, where the answer was to be given in terms of pointing out the objects belonging to the set. This doesn't mean *per se* that a set couldn't have borderline members; it only means that such members would have to be individuals which were indefinitely identical with something pointed out. And since I am already skeptical about that kind of identity, I have no more reason to believe in the indefinite identity of sets."

US: "But why is it not legitimate to point out not only the things definitely belonging to the set but also those things which are borderline members, as a "direct" way of specifying the set?"

We leave the dialogue at this point, but note that it would be possible to have a theory of sets which gives the skeptic what he wants. It would be a theory which allows only sets $\{a, b, \dots\}$ as above, and combines this with a "direct reference" theory of names according to which the latter acquire their reference without any conceptual intervention.[13] The picturing mentioned in the previous section makes this approach vivid.

The second argument we consider is Nathan Salmon's (in Salmon [3], pp. 243–244) and was intended as a refutation of indeterminate identity (in his case, indeterminate transworld identity). The argument originally involves ordered pairs, but we adapt it here to unordered pairs. "Suppose that there is a pair of entities x and y such that it is vague (neither true nor false, indeterminate, there is no objective fact of the matter) whether they are one and the very same thing. Then this pair x, y is quite definitely not the same pair as x, x, since it is determinately true that x is one and the very same thing as itself. It follows that x and y must be distinct. But then it is not vague whether they are identical or distinct." The argument may be formalized in our framework as follows.

1. Suppose $\nabla x = y$
2. But $\rhd x = x$
3. So $\neg \nabla x = x$
4. So $\{x, y\} \neq \{x, x\}$
5. So $x \neq y$
6. So $\neg \nabla x = y$

There is an evident gap in this argument at step 4. It is not clear from the text above how Salmon means the gap to be filled. What we *can* deduce in our theory from 1, and the principles previously discussed is

$$\nabla \{x, y\} = \{x, x\}$$

This entails

$$\neg \rhd \{x, y\} = \{x, x\}$$

but not the stronger 4, from which we could infer

$$\rhd \neg \{x, y\} = \{x, x\}$$

And in fact, the diagram of the previous section sketches a countermodel to the argument.

5 HEREDITARILY FINITE SETS, PURITY AND DETERMINACY

To this point we have been restricting ourselves to sets of individuals. In this section we will relax that restriction very slightly, to consider on the one hand the empty set and on the other sets built up from individuals and the empty set by repeated but finite application of collection. Our interest will be to trace the sources of indeterminacy in such sets. We now allow as "well-defined conditions" all and only conditions of the form

$$v = t_i \vee \dots \vee v = t_n$$

where t_i may be any individual or set variable not identical to v, and where we understand the condition to be $v \neq v$ if n is 0. Furthermore, we make explicit that lowercase variables may be instantiated to set variables. We note that all of these conditions are transparent, so that comprehension will deliver sets for them. In particular this will be true for the null condition $x \neq x$, which delivers the null set:

$$\emptyset = A \Leftrightarrow \forall x(x \in A \Leftrightarrow x \neq x)$$

A pure set is one which can be built up from the null set alone by repeated collection; thus , $\emptyset$, $\{\emptyset\}$, $\{\emptyset,\{\emptyset\}\}$ and so on are pure sets. While we have not allowed ourselves sufficient resources to define the set of pure sets within our theory, we can talk (in our classical metalanguage) about the set of all pure set terms, defined recursively in the obvious way (fill in the "and so on" above). Each such term will have a "level" determined by the maximum nesting of braces within it; only the null set is of level zero, and only $\{\emptyset\}$ is of level one, but both $\{\emptyset, \{\emptyset\}\}$ and $\{\{\emptyset\}\}$ are of level two. The terminology can be extended to impure set terms (containing names of individuals), with names counted as of level 0. It will also be clear that for any set term we can determine the set of individual names which it contains; we will call this the basis of the term. Pure sets are purely abstract; they contain no admixture of the mundane. It is plausible to suppose that all indeterminacy is mundane, arising out of the messiness of spatio-temporal identity conditions. This thesis could be taken to have two implications; first, that any identity between pure set terms should be determinate and secondly that any identity between pure and impure sets should be determinately false. A stronger thesis, which implies the previous two, is that any indeterminacy of identity between sets should be traceable to indeterminacy of identity between some individuals out of which they are constructed.[14] Strikingly, the theory as we have elaborated it up to now fails to substantiate this position right at the beginning; nothing we have said rules out the possibility of indeterminate or, for that matter determinate identity between an individual and any set. However, if we fill that gap by the following postulate:

Concreteness $t \neq a$, for every set term t and individual name a

then it does indeed hold. We now sketch a proof of this result.

Proposition 1   *For any two set terms s and t, $\vdash \nabla s = t \rightarrow \bigvee_{ij} \nabla s_i = t_j$, where the s_i and t_j are the names in the basis of s, t, respectively, or $\vdash \neg \nabla s = t$ if either basis is empty.*

Proof: By induction on the maximum of the levels of s, t. Suppose the proposition holds for terms of lower maximum level, and suppose s, say, is of maximum level. If that level is 0, then if both bases are non-empty s and t are the only names in their bases, so the right hand side (rhs) is the same as the left. If both s, t are $\emptyset$, then we have $\vdash \neg\nabla\emptyset = \emptyset$. And if one is $\emptyset$ and the other a name, we use Concreteness. If the level of s is greater than 0 let s be $\{u_1, \ldots , u_k\}$. Again, if t is a name we use Concreteness, while if t is $\emptyset$ we use the fact that $\vdash \triangleright u_1 \in s$ and $\vdash \triangleright \neg u_1 \in \emptyset$ together with Extensionality, to establish $\triangleright s \neq t$ and hence $\neg \nabla s = t$. Finally, suppose t is $\{w_1, \ldots , w_m\}$. Then $\vdash s = t \rightarrow \bigvee_f \bigwedge_i u_i = w_{f(i)}$, where f ranges over all permutations of $\{1, \ldots , m\}$. Hence if $s = t$ is indeterminate, none of the conjunctions can be true, while at least one of them must be not definitely false; that one can contain no definitely false conjunct, but must have at least one conjunct which is not true; say this conjunct is $u = w$. Then the maximum level of u and w is at least one less than that of s and t, so by inductive hypothesis we must have an indeterminate identity between some elements of their bases, but those elements will also be elements of the bases of s and t, respectively.

As indicated in our earlier discussion, the situation changes sharply if we allow sets to be defined from "empirical" predicates, since then we could even have $\nabla\emptyset = \{x : Px\}$, if P were a predicate which was indefinitely true of something but definitely true of nothing.[15]

6 COLLECTIONS vs. CONCEPTS: AN ALTERNATE THEORY

It is a familiar idea that sets may be thought of in two different ways: combinatorially, as collections of objects determined iteratively in a series of stages, or logically, determined as extensions of properties. If we take Gottlob Frege[16] as representative of this latter interpretation, we are led to regard sets as courses of values of concepts, the latter being functions which map objects into truth-values. In classical logic this way of thinking of things leads to a theory isomorphic to the theory of collections (at least, in the simple cases we are considering). But in the present case there is a plausible argument that the theories are distinct. For suppose that two concepts are false of exactly the same things, but that one is true of something the other is indeterminate about, and perhaps vice versa. Then the collections they determine will be counted indefinitely identical by the theory of the preceding sections, but it seems the same should not be so for the concepts; since they produce different values for the same object, they definitely cannot be the same function. In any case, we can readily provide a theory which reflects this intuition: we keep all of our postulates except for extensionality, which we replace with

Strict Extensionality

$$\forall x(x \in A \cong x \in B) \Leftrightarrow A = B$$

and add

Determinacy of Concepts

$$A = B \vee \neg A = B$$

As we observed in note 7, on this theory it will follow from the indefinite identity of a and b that they have definitely distinct unit sets, and also that their pair set is definitely distinct from the unit set of either. Nevertheless, Salmon's argument still fails in the new theory: the inference to step 4 is now valid, but the inference from 4 to 5 fails (though its contrapositive is valid).

University of California, Irvine

NOTES

[1]   See Quine [2], pp. 85–86.
[2]   It is important to note tht we are *not* invoking three-valued logic as a way of avoiding the paradoxes, as some have thought to do (e.g. Skolem [4]).
[3]   There is actually an option here: we could instead say that any expression picks out a property and then ask when a property has an extension. But the practice begun in the previous essay is to count all properties as having extensions; our form of the question is thus consistent with this practice.
[4]   The term is intended to remind the reader of Quine's notion of referential transparency. A position is referentially transparent, for Quine, when substitutivity of identity salva veritate holds there; for us, when substitutivity of weak (non-false) identity never takes us fro truth all the way to falsity.
[5]   Thus if p is true and q indefinite, $p \Leftrightarrow q$ will be indefinite and $p \cong q$ definitely false, so it will be correct to deny the latter but not the former, if denial is assertion of the negation.
[6]   The operators $\triangleright$, $\triangleleft$ and $\rightarrow$ are defined from ∇ as follows

Definite Truth: $\triangleright\phi =_{df} \phi \& \neg\nabla\phi$
Non-falsity: $\triangleleft\phi =_{df} \neg\triangleright\neg\phi$
If true: $\phi \rightarrow \psi =_{df} \triangleright\phi \supset \psi$

We choose $\rightarrow$ among many possible three-valued implication connectives because it satisfies the deduction theorem. A logic adequate for our language was given in the preceding essay [1], and we shall use logically valid inferences without comment in the ensuing arguments.
[7]   One should not assume that because extensions are represented by regions of the plane, as are individuals, that extensions qua unities can be treated like individuals, with overlap indicating indefinite identity and so on. The similarity is a mere artifact of the model.
[8]   This is not quite as strong an argument as it seems. For the decision to represent the left side of this axiom by $\forall x(x \in A \Leftrightarrow x \in B)$ was not forced; the considerations given would have allowed us also to use strict equivalence: $\forall x(x \in A \cong x \in B)$. Had we made this choice, the reduction argument would have different consequences. For instance, we would then be able to infer that if a is not definitely identical to b, their unit sets are definitely distinct.
[9]   We adapt Suppes' account of definitions [5] to the present system as follows: a definition of a singular term is proper (in a given theory) iff it is a closure of a formula $f(v_1 \ldots v_n) = v \Leftrightarrow \phi$, where every variable free in ϕ is either v or some v_i, f does not occur in ϕ, and we have as theorems of the theory (without the definition) the closures of the existence condition

$$\exists v\phi$$

uniqueness conditions

$$\phi(v) \& \phi(w) \rightarrow v = w$$

and transparency condition

$$\phi(v) \& \neg\phi(w) \rightarrow v \neq w$$

When these are satisfied, the definition will meet the criteria of eliminability and non-creativity. In the present application we treat $\{v_1, \ldots, v_n\}$ as an abbreviation of $\{\ \}$ $(v_1 \ldots v_n)$.
[10]   The condition $x = y_1 \vee \ldots \vee x = y_n$ is easily seen to be transparent and (by what we said earlier) well defined, so comprehension provides us with an A satisfying the right hand side of the definition, while extensionality assures us that the r.h.s will satisfy uniqueness and transparency.
[11]   This means that an individual and its unit set will have the same representation, which accords well with another familiar Quinean theme: the identification of individuals with their unit sets.
[12]   Note that we say: vagueness of properties, not: vagueness of predicates. It is an old theme that all vagueness is linguistic vagueness; underdetermination of truth-value is then explained as insufficiency in the linguistic conventions for determining true-value. We are not concerned here to argue against that position.
[13]   Remarks of Peter Van Inwagen [6], pp. 217–218, suggest that this is the theory he would espouse. A similar theory has been proposed in the context of modal logic by Ruth Marcus.
[14]   This is essentially the view mentioned just after the dialogue of the previous section.

[15] In discussion, Professor Quine offered the following nice example: we discover a strange breed of antelope in the remote regions of the Sahara with a single horn; now $\nabla\emptyset = \{x : x \text{ is a unicorn}\}$.

[16] See e.g. "On Sense and Reference" and "Concept and Object".

REFERENCES

[1] Terence Parsons. "Indefinite Objects of First Order", this volume, pp.
[2] W.V.O. Quine. *Philosophy of Logic*, 2d ed., Harvard University Press, Cambridge 1986.
[3] Nathan Salmon. *Reference and Essence*. Basil Blackwell, Oxford 1988.
[4] Th. Skolem. "A set theory based on a certain 3-valued logic", *Zeitschrift für Mathematische Logik und Grundlagen der Mathematik* **6** (1960).
[5] P. Suppes *Introduction to Logic*. van Nostrand, New York 1957.
[6] Peter van Inwagen. "How to Reason about Vague Objects", *Philosophical Topics* **16** (1988); 255–84.

STEPHEN NEALE

ON A MILESTONE OF EMPIRICISM

Thus the theory of description matters most.
It is the theory of the word for those
For whom the word is the making of the world,
The buzzing world and lisping firmament.

– WALLACE STEVENS

Quine has presented powerful arguments against the intelligibility of any statement involving quantification into the scope of a non-extensional sentence connective purporting to express logical necessity, logical possibility, or strict implication. He has also presented a fully *general* argument against the possibility of satisfactory non-extensional connectives. It is easy to be lured into thinking that if definite descriptions (and also, perhaps, proper names) are analysed in accordance with Russell's Theory of Descriptions – an idea to which Quine is highly sympathetic – then facts about substitution and deduction in primitive notation yield all that is needed to circumvent Quine's formal arguments. I have claimed as much myself, and so have Carnap, Church, Fitch, Føllesdal, Kripke, Marcus, Myhill, Prior, Sharvy, Smullyan, and others. But as far as logical modality is concerned this is a mistake. Only Quine's *general* argument is thus undermined, and seeing why leads to an interesting logico-semantic hypothesis, suggested by Quine and Gödel, and, as far as I know, hitherto unproved. Furthermore, it would seem that the only way to answer Quine's original challenge to quantified modal logic is to make the controversial assumption that names of the same object are *synonymous* and hence intersubstitutable in *all* contexts. But even this does not provide a full answer; and in fact what has happened is that ideas in the work of Føllesdal, Hintikka, Kanger, Kripke, and Quine, have resulted in a metamorphosis of subject matter.

My essay comes in two parts. In the first, I set out a number of important logical points about descriptions, sentence connectives, and extensionality; and then I draw some morals for those would posit non-extensional connectives by producing two deductive proofs of constraints any such connective must satisfy, proofs that have their origins in work by Quine, Church, and Gödel in the early 1940s. In Part II, I scrutinise various aspects of the history of modal logic – paying particular attention to the period from 1941 to 1947 (by which time all of Quine's main points had been made) and the period from 1957 to 1972 (by which time a notion of *metaphysical* necessity is firmly in place) – and the relationship between slingshot arguments and Quine's attack on logical modality.

237

Alex Orenstein and Petr Kotatko (eds.), Knowledge, Language and Logic, 237–346.
© *2000 Kluwer Academic Publishers. Printed in Great Britain.*

§1. Paraphrasis and contextual definition

The idea of syncategorematic words, defined in linguistic context rather than in isolation, can be found in nominalistic strands of mediaeval thought. A later nominalist, Jeremy Bentham, used what he called "definition by paraphrasis" in connection with seemingly categorematic expressions on the grounds that many of them refer to "fictions", amongst which he included qualities, relations, classes and indeed all of the entities of mathematics. Bentham's strategy was to provide a systematic method for converting any sentence containing a dubious term α into a certain type of sentence that is α-free, thereby revealing α to be syncategorematic (despite grammatical appearances). Embarrassing ontological commitments could then be avoided while still getting some basic grammatical work out of terms of dubious denotation.

It was the shift of semantic focus from terms to sentences, says Quine (1981), that made this possible, a shift he regards as one of five "milestones of empiricism." The method of paraphrasis, now commonly called the method of *contextual definition,* had its impact on philosophy largely through the work of Frege and then Russell, most famously through the latter's Theory of Descriptions. Quine is very clear on this impact:

> Contextual definition precipitated a revolution in semantics: less sudden perhaps than the Copernican revolution in astronomy, but like it in being a shift of center. The primary vehicle of meaning is seen no longer as the word, but as the sentence. Terms, like grammatical particles, mean by contributing to the meaning of the sentences that contain them. The heliocentrism propounded by Copernicus was not obvious and neither is this....the meanings of words are abstractions from the truth conditions of sentences that contain them.
>
> It was the recognition of this semantic primacy of sentences that gave us contextual definition, and vice versa. I attributed this to Bentham. Generations later we find Frege celebrating the primacy of sentences, and Russell giving contextual definition its fullest exploitation in technical logic. (1981, p. 69)

On Quine's account, ordinary singular terms – put variables aside for just a moment – are ultimately redundant: any sentence containing an ordinary singular term can be replaced without loss – and in a manner suggested by Russell's Theory of Descriptions – by a sentence that is term-free. On such an account, ordinary singular terms, e.g., names, are no more than "frills" (1970, p. 25) or convenient "conventions of abbreviation" (1941a, p. 41); and contextual definition is our "reward", as Quine sees it, for "for recognising that the unit of communication is the sentence not the word." (1981, p. 75) This is a rich idea, I believe, and goes well beyond anything Frege or Russell envisaged – indeed, on occasion it brings him into conflict with both).

It is sometimes suggested that Quine is trying to have his terms and eat them, that he plays fast and loose with Russell's Theory of Descriptions (endorsing it for one purpose, rejecting it for another), that he cannot champion the elimination of singular terms by paraphrasis and at the same time attempt to draw philosophical and formal conclusions from arguments involving the substitution of purportedly co-referential terms within the scopes of purportedly non-extensional sentence connectives. This is incorrect; I shall argue the following:

(A) The suggestion that Quine is trying to have his terms and eat them is without foundation; it misses the fact that individual variables occupy term positions and require interpretation, as do matrices that contain them.

It has been quite common since the publication of work by Church (1942), Carnap (1947), Smullyan (1947, 1948), Marcus (1948), Fitch (1949), and others to claim that Quine does not always fully appreciate certain facets of Russell's Theory of Descriptions and so gets himself into serious logical difficulties, particularly when he is arguing that modal contexts are "referentially opaque" and when responding to Smullyan in 1953, 1961, 1969, and 1980. I shall argue the following:

(B) Quine does sometimes err in connection with Russell's theory, but only once does this have important ramifications, and these do not pertain to his critique of quantified modal statements *per se* or to the real nature of his response to Smullyan. Critics of Quine's critique, by contrast, make fatal mistakes when invoking Russell's theory.

We need to distinguish two forms of argument Quine has used in connection with non-extensional sentence connectives.
(i) Using one form of argument, Quine (1941a, 1943, 1947, 1953a, 1953c, 1960) draws conclusions about the *particular* modal connectives $\square$ and $\lozenge$ – understood as expressing logical necessity and logical possibility – including the conclusion that they induce referentially opacity and do not permit variables inside their scopes to be bound by quantifiers outside.[1]
(ii) Using a different, but historically related form of argument, a so-called "slingshot", Quine (1953a, 1953c, 1960) draws a powerful conclusion about non-extensional connectives quite *generally*: there aren't any; plausible connectives must be extensional.[2]
It is widely held that both the particular and the general arguments make crucial use of the idea that definite descriptions (or class abstracts) are singular terms and that this is their downfall. I will argue the following:

(C) The particular argument concerning $\square$ and $\lozenge$ need not (and sometimes *does not*) use descriptions (or abstracts); so it is impossible to avoid Quine's conclusions – that modal contexts are referentially opaque and that quantifying into such contexts is unintelligible – by brute appeals to facts about the semantics of definite descriptions or to the fact that Quine (1940, 1941a) endorses Russell's theory. Church (1942, 1987), Carnap (1947), Smullyan (1947, 1948), Marcus (1948, 1962, 1968, 1990, 1993), Fitch (1949, 1950), Myhill (1958, 1963), Føllesdal (1965, 1969), Prior (1963), Kripke (1971), and – rather less importantly for you than me – Neale (1990, 1993, 1995) are all wrong about *that*. Matters are considerably more complex.

In my own case, my entire 1990 discussion was based on a misconception of Quine's (1943, 1947, 1953a) main philosophical complaint about quantified modal logic, a misconception engendered by what I now view as an anachronism: the false belief that by the late 1940s and early 1950s something close to what we are today inclined to call *metaphysical* necessity – distinct from *strict* or *analytic necessity* – was already in the air; and that in line with the view (expressed by, e.g. Russell) that logic is ultimately about the structure of reality, some of those who were interested in logical modality were peeling away its verbal garb with a view to uncovering something deeper than analytic or linguistic truth. I redress my own failings in this area in sections 10 and 11 in the context of a history leading up to Quine's *general* argument against non-extensional connectives.[3]

This leads to my next claim:

(D) Quine's general (slingshot) argument *does* require the use of a description (or abstract). The argument fails in so far as it is meant to demonstrate the impossibility of non-extensional connectives. However, the argument can be tidied up so as to yield an interesting conclusion, albeit weaker than the (eliminative) one Quine was seeking.

Substitution (interchange) arguments alone are inherently incapable of undermining non-extensional logics in general, but they can and do impose interesting constraints on the structures of such systems and on the interpretation of particular non-extensional connectives. This leads to another claim:

(E) A slingshot argument due to Gödel (1944) can be put in Quinean "connective format" and shown to place a more serious constraint on non-extensional connectives than Quine's own slingshot.[4]

We need to begin by examining in detail precisely what is involved in Russell's contextual definition of definite descriptions and the use to which Quine puts Russell's theory. We can then turn to issues of extensionality, scope, substitutivity, and so on, to the general arguments of Quine and Gödel, and finally to Quine's attack on quantified modal logic.

§2. Abbreviation and description

Consider a language L containing a handful of one-place predicates, the usual supply of variables $x_1 \ldots x_n$, devices of punctuation "(" and ")", and the following logical vocabulary:

(1)   "∃", "∨", "~", "=".

Assume a rudimentary Tarskian truth definition for L (where "s" ranges over infinite sequences of objects in the domain, "k" ranges over the natural numbers, s_k is the object in the kth position in s, and "ϕ" and "ψ" range over formulae of L):

(2) $\forall s \forall k$ (with respect to s, the referent of "x_k" $= s_k$);

(3) $\forall s \forall k \forall \phi$ (s satisfies "$\exists x_k \phi$" iff ϕ is satisfied by at least one sequence differing from s at most in the kth position);

(4) $\forall s \forall \phi$ (s satisfies "$\sim\phi$" iff ϕ is not satisfied by s);

(5) $\forall s \forall \phi \forall \psi$ (s satisfies "($\phi \vee \psi$)" iff s satisfies ϕ or s satisfies ψ).

(I will not be using corner quotes. Where necessary I use ordinary quotes instead, sometimes redundantly. Sometimes I leave quotes out altogether, as context or mood affects me. No confusion should arise.) Now suppose we were to find ourselves using a great number of formulae involving negation and disjunction (because we wanted to express conjunction) and wanted to save some ink. Within the broad range of things we could do, we must distinguish (i) introducing some metalinguistic shorthand and (ii) adding new symbols to L.

(i) An attractive way of introducing metalinguistic shorthand would be to use "quasi-formulae" to represent genuine formulae of L. For example, we could use "($\phi \cdot \psi$)" as a convenient shorthand for

(6) $\sim(\sim\phi \vee \sim\psi)$

without actually adding "$\cdot$" to L. The method of paraphrasis is at work, albeit outside the realm of terms. As Quine (1941) puts it when discussing such contextual definitions in *Principia Mathematica*, they are "conventions of abbreviation... "$p \cdot q$" is mere shorthand for "$\sim(\sim p \vee \sim q)$"." (p. 141)

(ii) a. One obvious way of adding to L itself would be to introduce a two-place connective "$\cdot$" by way of an appropriate syntactical rule and a new semantical axiom involving the connectives "$\vee$" and "$\sim$":

(7) $\forall s \forall \phi \forall \psi$ (s satisfies "($\phi \cdot \psi$)" iff s satisfies "$\sim(\sim\phi \vee \sim\psi)$").

(ii) b. Alternatively, we might prefer to add "$\cdot$" to L more directly, i.e. by way of an appropriate syntactical rule and a new semantical axiom that does not involve any other expressions of L:

(8) $\forall s \forall \phi \forall \psi$ (s satisfies "($\phi \cdot \psi$)" iff s satisfies ϕ and s satisfies ψ).

Similarly, any of these methods could be used in connection with other symbols we might consider, for example $\supset$, $\equiv$, $\neq$ and $\forall$.

Consider now a language L', just like L but containing the following logical vocabulary:

(9) $\exists, \forall, \sim, \cdot, \vee, \supset, \equiv, =, \neq$.

There is nothing that can be said in L' that cannot be said in L, but obviously there are many things that can be said with fewer symbols in L' and in ways that are more readily understandable.

Within a middle range, there is a trade-off between economy of symbols and ease of interpretation. Obviously economy of symbols is not the only criterion we invoke in designing formal languages with which we work. If we are doing the metatheory of first-order logic, we are naturally drawn to the economy afforded by a language with fewer symbols; when we come to *use* (rather than talk about) first-order logical formulae, we are naturally drawn to the economy of length and simplicity afforded by a language with a greater number of symbols. (Imagine how difficult it would be to train ourselves readily to understand sentences whose logical vocabulary contains just $\exists$, =, and the Scheffer stroke. It is worth noting that in the introduction to the second edition of *Principia*, Whitehead and Russell are impressed by the fact that negation is more primitive psychologically than the stroke.)

Now consider a language L'', just like L' but containing three individual constants "a", "b", "c", understood as primitive singular terms (names if you like). Suppose we were to find ourselves wanting to talk about things in the domain not named by "a", "b", or "c", or about things satisfying certain descriptive conditions and not known by us to be named by "a", "b", or "c". We could render the statement that the unique thing satisfying the predicate "F" also satisfies the predicate "G" (i.e. the statement that the F is G) as follows (to ease the eye, in place of x_1, x_2, etc., henceforth I shall use x, y, etc. for variables wherever possible):[5]

(10) $\exists x(\forall y(Fy \equiv y=x) \cdot Gx)$.

Again, there are things we could do to save ink and present ourselves with more readily interpretable formulae:

We could introduce some shorthand. For example, we could follow Russell, who viewed (10) as rendering the logical form of the English sentence

(11) The F is G.

Adapting Peano's *iota* notation, Russell represents a definite description "the F" by an expression of the form

(12) $\iota x Fx$

which can be read as "the unique x such that Fx". The *iota*-operator *looks like* a variable-binding operator for creating a term from a formula ϕ: a simple one-place predicate symbol G may be prefixed to a description "$\iota x\phi$" to form a quasi-formula of the form of (13), which is shorthand for a genuine formula of the form of (14) (where $\phi(y)$ is the result of replacing all occurrences of x in ϕ by y):

(13) $G\iota x\phi$
(14) $\exists x(\forall y(\phi(y) \equiv y=x) \cdot Gx)$.

It is important to see that for Russell, a phrase of the form "$\iota x\phi$" *is not a genuine singular term*; it is an *abbreviatory device* that permits (provably legitimate) short-cuts in the course of proofs, and the use of quasi-formulae that are usually easier to grasp than the genuine formulae for which they go proxy. A quasi-formula of the form of (13) is, to use Quine's (1941) wording again, "mere shorthand" for a genuine formula of the form of (14). The important point for present concerns is that *the iota operator has not been added to L'' itself*. (This fact is going to be extremely important when we come to discuss quantified modal statements.)

At this point, then, the following contextual definition of definite descriptions might be considered, where $\Sigma(\iota x\phi)$ is a sentence containing $\iota x\phi$ and $\Sigma(x)$ the result of replacing an occurrence of $\iota x\phi$ in $\Sigma(\iota x\phi)$ by x:

(15) $\Sigma(\iota x\phi) =_{df} \exists x(\forall y(\phi(y) \equiv y=x) \cdot \Sigma(x))$.

But this will not suffice once more complex examples are examined. Whitehead and Russell bring this out with the quasi-formula (16):

(16) $G\iota x\phi \supset \psi$.

Depending upon how (15) is applied, (16) could be viewed as shorthand for either (17) or (18), which are not equivalent in respect of truth conditions:

(17) $(\exists x(\forall y(\phi(y) \equiv y=x) \cdot Gx) \supset \psi)$
(18) $\exists x(\forall y(\phi(y) \equiv y=x) \cdot (Gx \supset \psi))$.

So unlike any genuine formula of L'' (or *Principia*), a quasi-formula of the form of (16) is, at the moment, ambiguous. Consequently, some sort of modification or supplementation to the system of abbreviation is required if it is to be of service.

The point is worth stressing in connection with another sort of example that is of more interest to the concerns of the present essay. Depending upon how (15) is applied, the quasi-formula $\sim G\iota x\phi$ could be viewed as shorthand for either (19) or (20), which are not equivalent in respect of truth conditions:

(19) $\sim\exists x(\forall y(\phi(y) \equiv y=x) \cdot Gx)$
(20) $\exists x(\forall y(\phi(y) \equiv y=x) \cdot \sim Gx)$.

(The former, unlike the latter, can be true if nothing uniquely satisfies ϕ.) Whitehead and Russell adopt a rather cumbersome supplementation of their shorthand in order to eradicate ambiguity in their quasi-formulae: they place a copy of the description inside square brackets at the front of the formula that constitutes the scope of the relevant existential quantifier. Thus (19) and (20) are abbreviated as (19') and (20') respectively:

(19') $\sim[\iota x\phi]\, G\iota x\phi$
(20') $[\iota x\phi]\, \sim G\iota x\phi$.

Since they are concerned mostly with formulae in which descriptions have minimal scope, they allow themselves to omit the square-bracketed copy of the description if the scope of the description is understood to be minimal. Thus (19'), but not (20'), can be simplified to $\sim G\iota x\phi$, now understood as unambiguous. (This is, after all, just an abbreviatory notation: the primitive language is unaffected by these conventions.)

The Theory of Descriptions can be summarised in two succinct propositions. The central proposition is the following contextual definition:

*14.01 $[\iota x\phi]\Sigma(\iota x\phi) =_{df} \exists x(\forall y(\phi(y) \equiv y=x) \cdot \Sigma(x))$.[6]

The second proposition is relevant only to statements that seem to involve talk of existence. On Russell's account, there is no possibility of a genuine singular term failing to refer, so no predicate letter in the language of *Principia* stands for "exists". But a statement of the form "the F exists" is meaningful and Russell introduces an abbreviatory symbol "E!" that may be combined with a description "$\iota x\phi$" to create a second type of quasi-formula "E!$\iota x\phi$", which is also to be understood in terms of a contextual definition:

*14.02 E!$\iota x\phi =_{df} \exists x\forall y(\phi(y) \equiv y=x)$.

The Theory of Descriptions claims that any well-formed formula containing a definite description (regardless of the complexity of $\Sigma(\iota x\phi)$ in *14.01) can be replaced by an equivalent formula that is description-free. It is clear that using Russell's abbreviatory convention does not add to the expressive power of L''.

Once certain derived rules of inference have been proved for truth-functional contexts, most importantly:

*14.15 $(\iota x\phi=\alpha) \supset \{\Sigma(\iota x\phi) \equiv \Sigma(\alpha)\}$
*14.16 $(\iota x\phi=\iota x\psi) \supset \{\Sigma(\iota x\phi) \equiv \Sigma(\iota x\psi)\}$

it is extremely useful for purposes of proof to be able to regard the definite descriptions "$s(x)$", "$\sqrt{x}$", "log x", "sin x", "$x + y$" and so on as functioning like singular terms (this matter is discussed in detail later).[7]

Another useful theorem for truth-functional contexts is *14.18 (which will crop up later):

*14.18 E!$\iota x\phi \supset \{\forall x\Sigma(x) \supset \Sigma(\iota x\phi)\}$.

This says that if there exists exactly one thing satisfying ϕ, then that thing "has any property which belongs to everything." (p. 174) The description $\iota x\phi$ "has (speaking formally) all the logical properties of symbols which directly represent objects...the fact that it is an incomplete symbol becomes irrelevant to the truth-values of logical propositions in which it occurs." (p. 180).

The fact that Whitehead and Russell prove some interesting theorems about descriptions occurring in truth-functional contexts should not obscure the quantificational character of the Theory of Descriptions, which comes through clearly not only in the contextual definitions *14.01 and *14.02 but also in Russell's talk of *general* propositions and *general* facts. By the time of *Principia*, having more or less given up on the notion of propositions as non-linguistic entities of theoretical utility, Russell takes a true sentence to stand for a fact. But in order to avoid unnecessary distractions that emerge in connection with false sentences, it will be convenient, for a moment, to go back a few years, to when Russell was first presenting the Theory of Descriptions, to a period when he took propositions more seriously as non-linguistic entities.

For Russell, a singular term α may be combined with a one-place predicate phrase "( ) is G" to express a proposition that simply could not be entertained or expressed if the entity referred to by α did not exist. Russell often puts this by saying that the referent of α is a *constituent* of such a proposition, a so-called *singular* proposition whose existence is contingent upon the existence of the referent of α.

A sentence of the form "the F is G" does *not* express a singular proposition; it expresses a *general* proposition, a proposition that is not *about* a specific entity (described by "the F"), and whose existence is not contingent upon the existence of the entity which in fact satisfies the predicate F (if anything does).[8] I belabour this point because many people who appeal to (or at least claim to endorse) the Theory of Descriptions seem not to appreciate it. If one does not see that on Russell's account "the F is G" expresses a *general* proposition, that "the F" *never* refers, one simply does not understand the theory.

To say that the proposition expressed by a sentence S is singular is really just to say that the grammatical subject of S *stands for* an object and *contributes* that object to the proposition expressed by an utterance of S (or, if you prefer, contributes that object to the truth conditions of an utterance of S). To say that a sentence S expresses a *general* proposition is just to say that the grammatical subject of S is not the sort of expression that stands for an object or contributes an object to the proposition expressed by (or the truth conditions of) an utterance of S. This is the heart of Russell's position that English phrases of the form "the F", as well as those of the form "every F", "some F", and so on – i.e. all "denoting phrases" – are *incomplete* symbols: they are incomplete because they do not "stand for" or "directly represent" objects.[9]

§3. Scope (I)

A truth-function, according to Whitehead and Russell is "a function whose truth or falsehood depends only upon the truth or falsity of its arguments. This covers all the cases with which we are ever concerned" (p. 184). The restriction of the theorems of *Principia* to formulae containing descriptions in truth-functional contexts comes out clearly in *14.3 below, which is meant to state that when ϕ is uniquely satisfied, the scope of $\iota x\phi$ does not matter to the truth-value of any truth-functional sentence in which it occurs:

*14.3 $\forall f [\{\forall p \forall q\,((p \equiv q) \supset f(p)) \equiv f(q))\} \cdot E!\iota x\phi)\} \supset$
$\{f([\iota x\phi]G\iota x\phi) \equiv [\iota x\phi]\,f(G\iota x\phi)\}].$

The variable f is meant to range over functions of propositions (I have added the quantifier $\forall f$ which is not present in *Principia*, but retained the authors' desire to keep the theorem short by ignoring issues of use and mention that will not lead to confusion).[10] Whitehead and Russell conclude their discussion of this topic with the observation that "the proposition in which $\iota x\phi$ has larger scope always implies the corresponding one in which it has the smaller scope, but the converse implication holds only if either (a) we have $E!\iota x\phi$ or (b) the proposition in which $\iota x\phi$ has the smaller scope implies $E!\iota x\phi$". (p. 186)

The restriction of scope permutation to truth-functional contexts expressed (or meant to be expressed) by *14.3 is something Quine (1980) accepts he overlooked in his 1953 and 1961 responses to Smullyan's (1948) postulation of truth-conditional ambiguity in sentences containing descriptions and modal operators. Mirroring Russell's postulation of an ambiguity of scope in sentences containing descriptions and psychological verbs – an ambiguity Russell (1905) milked to tackle puzzles about substitutivity – Smullyan sought to characterise what he saw as an example of the same type of ambiguity in modal contexts in terms of scope. Using "Px" for "x numbers the planets in our solar system" the surface sentences (21) and (22) can be given their respective a. and b. readings below:

(21) George IV wondered whether the number of planets in our solar system > 7
   a. George IV wondered whether $\exists x(\forall y(Py \equiv y{=}x) \cdot (x > 7))$
   b. $\exists x(\forall y(Py \equiv y{=}x) \cdot$ George IV wondered whether $(x > 7))$

(22) Necessarily the number of planets in our solar system > 7
    a. necessarily $\exists x(\forall y(Py \equiv y=x) \cdot (x > 7))$
    b. $\exists x(\forall y(Py \equiv y=x) \cdot$ necessarily $(x > 7))$.

As is well known, Quine (1943, 1947, 1953a, 1953c, 1960) has expressed the view that the interpretation of (21b) and (22b) pose great philosophical difficulties. I shall have a good deal to say about this and the Quine-Smullyan controversy in Part II, where I explain why the ambiguity Smullyan sees is, in fact, by itself powerless to deflect Quine's worries about quantified modal logic. For the moment, it will suffice to make a minor point on Smullyan's behalf. As Quine (1980) recognised, he erred in *From a Logical Point of View* in 1953 and 1961 when he accused Smullyan of "propounding, in modal contexts, an alteration of Russell's familiar logic of descriptions" by "allow[ing] difference of scope to affect truth value even in cases where the description concerned succeeds in naming [sic.]." (1961, p. 154). In a copy of the 1961 edition of the book Quine donated to the library at the Villa Serbelloni in Bellagio when he was there in 1975, Quine inscribed in the margin, "Kripke has convinced me that Russell shared Smullyan's position. See *Principia* pp. 184f. esp. *14.3: the explicit condition of extensionality"; and by the time of the 1980 reprinting of the book, the charge against Smullyan on p. 154 has been excised. In a new foreword, Quine points out that the page in question originally "contained mistaken criticisms of Church and Smullyan" (p. vii), and the relevant part of the page now reads as follows:

Then, taking a leaf from Russell ["On Denoting"], he [Smullyan] explains the failure of substitutivity [in modal contexts] by differences in the structure of the contexts, in respect of what Russell called the scopes of the descriptions. [*Footnote*: Unless a description fails to name [sic.], its scope is indifferent to extensional contexts. But it can still matter to intensional ones.]" (1980, p. 154)

I bring this up for two reasons. Firstly, it indicates Quine's openness to overtly acknowledging an error – a non-fatal error, as we shall see – in his early responses to Smullyan. Secondly, it highlights a misunderstanding that still pervades discussions of Smullyan's position on the failure of substitutivity in modal contexts: Smullyan does *not* claim that substitutivity is restored by appealing to the scopes of descriptions. His position is that on Russell's account (i) descriptions are not singular terms and so do not appear in primitive notation, and (ii) the false reading of (22), viz. (22a), cannot be derived from "necessarily 9 > 7" and "9 = the number of planets". The putative existence and truth (cf. Smullyan) or unintelligibility (cf. Quine) of (22b) is completely irrelevant to *this* point.

The situation with respect to propositional attitudes and logical modality appears to be mirrored elsewhere, as noted by Chisholm (1965), Føllesdal (1965), Sharvy (1969) and others. If, as a result of some astronomical cataclysm, Mercury gets sucked into the sun in the year 2002, the number of planets will be reduced from 9 to 8 – assuming no consequences for the other planets – but 9 would not be so reduced. And if "in 2001" functions semantically as some sort of one-place connective – there may be better treatments of course – and if (23a) and (23b) are intelligible, then in the year 2003 an utterance of (23) will be true when read as (23a), false when read as (23b), where ">" is shorthand for "exceeded":

(23) In 2001 the number of planets in our solar system > 8
    a. In 2001 $\exists x(\forall y(Py \equiv y=x) \cdot (x > 8))$
    b. $\exists x(\forall y(Py \equiv y=x) \cdot$ in 2001 $(x > 8))$.

(Of course difficult questions about $\exists$ and tense need to be answered before this sort of observation can be developed properly.)

Quine's initial interpretive error about applications of Russell's Theory of Descriptions in modal contexts has been repeated by others, and in some cases this has led to further error. For example, Hintikka (1968) says that

> ...when the Russellian theory [of descriptions] is put to work in modal and other complicated contexts, much of its success depends on a clever choice of scope conventions and on similar adjustments. With a clever choice of these, a suitable modification of the theory gives us a certain amount of mileage (cf. e.g. Smullyan [1948], Montague and Kalish [1959], Linsky [1966]), but the deeper reasons for the choice remain unaccounted for, and cry out for further analysis. (p. 5)

Thus far, Hintikka is amplifying Quine's error.[11] But from here, he goes straight on to say,

> Donnellan [1966a, 1966b] has even argued that definite descriptions are used in two essentially different ways in ordinary language. No matter whether these can be caught by juggling the scope conventions, the reasons for doing so will remain in the dark. (...) Thus it also seems to me that the theoretical significance (or lack thereof) of the ingenious use of the theory of descriptions to simplify one's "canonical notation" by Quine [1960, ch. v] and others is too incompletely understood to be evaluated here. (p. 5)

This passage betrays a confusion. It has seemed attractive to some philosophers to explicate the *de re* readings of sentences containing descriptions and non-extensional devices in terms of something like Donnellan's (1966) "referential" (rather than "attributive" $\backsimeq$ Russellian) interpretation of the relevant description. For example, Rundle (1965) makes this suggestion for descriptions in modal contexts; and Hintikka (1968), Partee (1972), Stalnaker (1972), Cole (1978) and others suggest that a referential interpretation can be used in characterising the *de re* readings of descriptions in non-extensional contexts more generally. This will not do. Cartwright (1968) and Kripke (1971, 1977) have pointed out very clearly that attempts to provide accounts of either the large scope-small scope distinction or the *de re–de dicto* distinction in terms of a referential/attributive distinction are misguided. A speaker may make a *de re* use of (22), or the variant (22′), without using the definite description "the number of planets" referentially:

(22′) The number of planets, whatever it is, is necessarily odd

The following passage from Kripke makes the point very clearly:

> Suppose I have no idea how many planets there are, but (for some reason) astronomical theory dictates that that number must be odd. If I say, "The number of planets (whatever it may be) is odd," my description is

used attributively. If I am an essentialist, I will also say, "The number of planets (whatever it may be) is necessarily odd," on the grounds that all odd numbers are necessarily odd; and my use is just as attributive as the first case (1977, p. 9).[12]

The point, quite simply, is that the proposition expressed by an utterance of (22′) is not *object-dependent*, even if the description is understood *de re*. The Russellian account is quite consistent with this fact. The proposition expressed is descriptive; the *de re* reading is obtained by giving the description large scope over the modal operator as in (22b) above. As Kripke goes on to point out, we find exactly the same situation with definite descriptions in attitude contexts. Suppose Quine has an object-dependent wonder concerning the man who lives upstairs: he wonders whether the man is a spy. I may correctly report this state of affairs by saying

(24) Quine wonders whether the man who lives upstairs is a spy

with the definite description "the man who lives upstairs" understood *de re*. But this does not mean that I have used the description referentially. I may have no relevant object-dependent thought about the man who lives upstairs and no intention of communicating such a thought. Russell captures this *de re* reading by giving the definite description maximal scope:

(24′) $\exists x(\forall y(My \equiv y=x) \cdot$ Quine wonders whether x is a spy).[13]

Furthermore, Kripke points out that no binary semantical distinction can *replace* Russell's notion of scope. A sentence like (25) is *three* ways ambiguous according as the description is given maximal, intermediate, or minimal scope:

(25) Watson doubts that Holmes believes that the murderer is insane
   a. $\exists x(\forall y(My \equiv y=x) \cdot$ (Watson doubts that (Holmes believes that x is insane))
   b. Watson doubts that $\exists x(\forall y(My \equiv y=x) \cdot$ (Holmes believes that x is insane))
   c. Watson doubts that (Holmes believes that $\exists x(\forall y(My \equiv y=x) \cdot x$ is insane)).

The reading given by (25b) is neither *de re* nor fully *de dicto*.[14]

These facts demonstrate quite conclusively that descriptions understood *de re* cannot, in general, be identified with descriptions understood referentially, and that a semantical ambiguity between Russellian and referential interpretations of descriptions cannot *replace* either the *de re–de dicto* distinction or the large scope-small scope distinction as they show up in non-extensional contexts. So even if one could provide good arguments for the existence of referential interpretations of the definite and indefinite descriptions in some of the examples we have been considering, *the large scope readings would still be needed*.[15]

If it were not for the particular aims of *Principia*, Whitehead and Russell could have pushed their abbreviatory conventions to their logical resting place. The square-bracketed occurrence of a description that is used to indicate scope is effectively marking the scope of an existential quantifier, and to that extent it is actually functioning rather like a quantifier. So why not replace the *original occurrence* of the description in the formula to which the square-bracketed occurrence is attached by a *variable* bound by the square-bracketed occurrence? After all, this is just an abbreviatory notation. On this simplification, (19′) and (20′) are reduced to (19″) and (20″) respectively, effectively yielding the notation of restricted quantification (see footnote 17):

(19″) $\sim[\iota x\phi]\, Gx$
(20″) $[\iota x\phi]\sim Gx$.

If this notational suggestion had been adopted, *14.01 and *14.02 would have looked like this:

*14.01′ $[\iota x\phi]Gx =_{df} \exists x(\forall y(\phi(y) \equiv y=x) \cdot Gx)$.
*14.02′ $[\iota x\phi]E!x =_{df} \exists x\forall y(\phi(y) \equiv y=x)$.

This would be real progress, but it is not what Whitehead and Russell do, for reasons to which I have already alluded.

(ii) a. If we were so disposed, we could add the *iota*-operator directly to L'' by way of other symbols of L''. There would be some work involved if *iota*-compounds were to continue to occupy singular term positions yet not be subject to reference axioms.[16] It would be much easier to use the more transparently quantificational notation above according to which expressions of the form $\iota x\phi$ are one-place variable-binding operators:

(26) If ϕ and ψ are formulae and x_k is a variable, then $[\iota x_k\phi]\psi$ is a formula;
(27) $\forall k\forall j\forall\phi\forall\psi\forall s$ {s satisfies $[\iota x_k\phi]\psi$ iff s satisfies $\exists x_k(\forall x_j(\phi(y) \equiv x_j=x_k) \cdot \psi)$}.

(ii) b. Alternatively, we could add the *iota*-operator to L'' by way of the same syntactic rule and a semantic axiom that does not involve any other primitive symbols of L'':

(28) $\forall k\forall\phi\forall\psi\forall s$ {s satisfies $[\iota x_k\phi]\psi$ iff ϕ is satisfied by exactly one sequence differing from s at most in the kth position and ψ is satisfied by every such sequence}.

The right hand side of (28) simply encodes Russell's truth conditions. It is a very short step from (28) to understanding how the Theory of Descriptions fits into a systematic account of quantification in natural language, quantified noun phrases functioning as restricted quantifiers of the form [DET x: ϕ], where ϕ is a formula and DET is (e.g.) "every", "some", "a", "no", "the", or "most"[17].

§4. The elimination of singular terms

As Quine sees it, Russell's Theory of Descriptions

… involved defining a term not by presenting a direct equivalent of it, but by what Bentham called *paraphrasis:* by providing equivalents of all desired sentences containing the term. In this way, reference to fictitious objects can be simulated in meaningful sentences without being committed to the objects. Frege and Peano had allowed singular description the status of primitive notation; only with Russell did it become an "incomplete symbol defined in use. [This sentence added in 1981 revision, p. 75 SRAN]

The new freedom that paraphrasis confers is our reward for recognizing that the unit of communication is the sentence and not the word. (1966, p. 659)[18]

And as Quine stresses repeatedly, Russell's theory ought to be very attractive to those who laud the virtues of "extensionalism" and first-order logic.[19] Besides its evident success as an account of the semantics of descriptive phrases, (i) the theory requires the postulation of no new entities, (ii) it avoids problematic existence assumptions and truth-value gaps, (iii) it provides a treatment of descriptions within first-order quantification theory with identity, and (iv) it captures a range of inferences involving descriptions as a matter of first-order logic, for example the fact that "the F is G" entails "there is at least one F", "there is at most one F", "there is at least one G", "some F is G", and "every F is G".

These evident virtues have led to numerous examinations of its potential application to noun phrases other than phrases of the from "the F", for example possessive noun phrases ("Socrates' wife", "Socrates' death"), ordinary proper names ("Socrates"), *that*-clauses ("that Socrates died in prison"), demonstratives ("that", "this vase"), indexical pronouns ("I", "you"), anaphoric pronouns ("it" as it occurs in, e.g., "John gave Mary a vase; he had purchased it at Sotheby's").[20]

Why is this? There are two reasons, I believe. The first stems from a somewhat *a priori* desire to eliminate singular terms, a desire that appears to be a reflex of the linguistic counterpart of a principle of ontological parsimony. The second is more empirical in nature. In the course of providing an adequate semantics of natural language and an account of many logical features of natural language, the possibility of analysing problematic noun phrases in terms of Russellian descriptions has promised instant solutions to nagging logical and ontological problems.

Quine has interesting ideas about the elimination of singular terms in connection with Russell's Theory of Descriptions. As an account of descriptive phrases, Quine sees only logical and philosophical good coming from Russell's theory.[21] Part of the appeal for Quine is that on this account descriptions are analysed in terms of the well-understood devices of first-order extensional logic. Second, Quine is happy with the idea that ordinary proper names can be "trivially" reconstrued as descriptions – and thereby analysed in accordance with Russell's theory. Names are "frills," says Quine, that can be omitted, "a convenient redundancy." (1970, p. 25) Apparently following Wittgenstein in the *Tractatus* (5.441, 5.47), he says that "Fa" is "equivalent" to (29):

(29) $\exists x(Fx \cdot a=x)$.

So the name "a" need never occur in a formula except in the context "$a=$"; but "$a=$" can be rendered as a simple one-place predicate "A", uniquely true of the object a; so "Fa" can, in fact, be rendered as

(30) $\exists x(Fx \cdot Ax)$

which contains no occurrence of "a"; indeed all occurrences of "a" – or any other name – are everywhere replaceable by combinations of quantifiers, variables, connectives, and predicates.

Two issues come up immediately here. The first concerns the nature of the predicate "A". It must hold uniquely of a, and it is not obvious that such a predicate can be drummed up without either reintroducing "a" or appealing to some principle of either linguistic or metaphysical essentialism. Like Church and Carnap, Quine

(1947, 1962) balks at the idea that coreferring names are synonymous; now suppose "$a = b$" is true; "a" and "b" should be intersubstitutable *salva veritate* (*s.v.*), at the very least in truth-functional contexts; thus "A" must be uniquely true of b (i.e. a). But what if "a" drops out of use leaving only "b" as a name of a? And what if "a" had never come into being in the first place, "b" being the only name of a? Unless the existence of "A" is ontologically dependent upon the existence of "a" (rather than a), in both scenarios "A" would still be true of b. Thus Quine's claim that (30) is equivalent to "Fa" seems to commit him to some form of either linguistic or metaphysical essentialism.

Secondly, the Quinean "paraphrase" of "Fa" might be questioned on the following grounds: it is in the nature of an occurrence of a name that it is understood as applying to a single object; but it is not in the nature of an occurrence of a predicate that it is understood as satisfied by a single object; so the "paraphrase deprives us of an assurance of uniqueness that the name afforded." (1970 p. 25) Quine's response to this point is straightforward: if we are worried about uniqueness, we can import it explicitly in the way Russell does in his analyses of sentences containing definite descriptions. That is, (30) can give way to the following:

(31) $\exists x(\forall y(Ay \equiv x=y) \cdot Fx)$

which is the Russellian spelling out of "$F\iota x Ax$" (i.e. "$F\iota x(x=a)$"). So anything that can be said using names, Quine assures us, can be said using formulae like (30) or (31) because the objects that names name are the values of variables. Moreover:

...names can even be restored at pleasure, as a convenient redundancy, by a convention of abbreviation. This convention would be simply the converse of the procedure by which we just now eliminated names" (1970, pp. 25–25).

A predication such as "Fa" is an abbreviation of (31) – or (30) if the uniqueness condition is cashed out elsewhere. "In effect," Quine adds, "this is somewhat the idea behind Russell's theory of singular descriptions" (1970, p. 26).

So, (i) Quine envisions a language in which the devices of quantification, variation, truth-functional connection, and predication do the work that we normally associate with names; and (ii) he sees this idea as essentially a refinement of, or a twist on, Russell's idea that ordinary proper names can be analysed in terms of definite descriptions. One point about Quine's use of descriptions should be noted, however. Quine (1969, pp. 326–7) uses Russell's *iota*-notation only if the description has minimal scope; when he wants to express something that requires larger scope for the description, he uses the unabbreviated form.

The "theoretical advantages" of analysing names as descriptions are "overwhelming" says Quine:

The whole category of singular terms is thereby swept away, so far as theory is concerned; for we know how to eliminate descriptions. In dispensing with the category of singular terms we dispense with a major source of theoretical confusion, to instances of which I have called attention in ... discussions of ontological commitment. (1953b, p. 167)[22]

As Russell stressed from the outset, endorsing the Theory of Descriptions has interesting and far-reaching consequences for logical issues involving substitutivity. But as we shall soon see, in two related settings Quine's use of examples involving the substitution of descriptions has had the unfortunate consequence of creating almost as much confusion as it has eradicated. In one case, the problem can be (and *was*) fixed; in the other it was not fixed because it *cannot* be fixed. There is a lot of groundwork to do before we get where we want to go.

§5. Extensions, connectives, and extensional entities

Let us be clear what is meant, or should be meant, by "extensional" and "extensionality". The occurrence of the name "Quine" in the sentence "Quine frowned voluntarily" might be said to refer to Quine. But pre-theoretically it is much less usual – indeed it is very strained – to say that the sentence itself or the occurrences of "frowned", "voluntarily", and "frowned voluntarily" refer to things. It might seem futile, then, to search for parts of the world to correspond to all the parts of sentences. However, many philosophers interested in constructing theories of meaning for natural languages have assigned entities to serve as the referents of expressions other than names and other devices of singular reference.

We avoid controversy here by adopting some well-defined theoretical vocabulary and stipulating that terms, predicates, sentences, and sentence connectives all have *extensions*. Consider a simple formal language L that contains just names (individual constants), predicates, and truth-functional connectives. The extensions of the various types of expressions are stipulated as follows: (i) The extension of a name is simply its referent (for immediate concerns let us agree to exile names that fail to refer, if there are such expressions). (ii) The extension of an n-place predicate is the set of n-tuples of which the predicate holds. (iii) The extension of a sentence is its truth-value.[23] (iv) The extension of an n-place connective is a function from n-tuples of truth-values to truth-values (such functions are *truth-functions*; hence the idea that connectives having truth-functions as their extensions are *truth-functional*). An extensional semantics for L will assign to every complex expression ζ of L an extension based on ζ's syntax and the extensions of ζ's parts.

As far as L is concerned, the class of so-called "extensional entities" is simply the class consisting of objects, sets of n-tuples of objects, the two truth-values, and functions from n-tuples of truth-values to truth values, i.e. the class consisting of all and only those entities capable of serving as the extensions of names, predicates, sentences, or connectives of L. There is nothing deep or mysterious about this class: it is determined *by stipulation* because the class of entities capable of serving as the extensions of terms, predicates, sentences and connectives of L has been determined by stipulation. Any entity that is not in this class is, by definition, a nonextensional entity as far as L is concerned. (I shall not use "non-extensional" and "intensional" interchangeably; it is useful to reserve the term "intensionality" for a restricted type of non-extensionality.)

§6. Scope (II)

Although the concept of scope is present in earlier work, the word "scope" itself is first used in the way that has become current by Whitehead and Russell in *Principia* (replacing Russell's earlier talk of primary, secondary, tertiary,... "occurrences" of symbols. But it is surprising how much confusion there is amongst philosophers about the *concept* of scope and about why we say that the scope of such-and-such is so-and-so. Anyone who has done some basic logic can tell you the scopes of the occurrences of "~", and "·" in (32), construed as a sentence of the language L used a moment ago:

(32) $\sim (\sim Fa \cdot Ga)$.

The scope of the first occurrence of "~" is the whole of (32); the scope of the second occurrence is "$\sim Fa$"; and the scope of "·" is "$(\sim Fa \cdot Ga)$". Let us reflect, for a moment, on why we say these things, on how scope is related to compositionality, and on what it would mean for expressions in natural language to have scopes.

Scope is a primitive syntactico-semantic notion: it concerns how semantic evaluation is driven by composition. In any interesting language there are molecular expressions, where the latter are expressions whose ultimate constituents are atomic expressions. It is the way a complex expression in a formal language is put together – its syntax – that tells us the scope of any part of the expression that may interest us. It betrays a fundamental misunderstanding of this concept when people claim, e.g., that "only operators have scopes".[24]

In L there is only one type of complex expression: the sentence. The scope of an n-place connective $\bullet$ ("ball") is simply the sentence (sanctioned by the syntax) that results from combining $\bullet$ with n sentences $\phi_1...\phi_n$, i.e. the smallest sentence containing both $\bullet$ and $\phi_1...\phi_n$ as constituents, i.e. the smallest sentence containing $\bullet$ as a constituent. If we stop here we fail to do justice to the fundamental simplicity of the concept of scope and also to its generality. The idea at the heart of the concept is simply that of two or more expressions combining to form a larger expression. In L, the scope of an expression – any expression – is the smallest sentence containing that expression as a proper constituent. So in "$\sim Fa$", not only is the scope of "~" the sentence "$\sim Fa$", so is the scope of "Fa". And in "Fa", the scope of "F" is "Fa", as is the scope of "a". Nothing changes when we examine more complex expressions. In "$(\sim Fa \cdot Ga)$" the scope of "·" is the whole sentence, and this is also the scope of both "$\sim Fa$" and "Ga". In "Rab", the scopes of "R", "a", and "b" are all "Rab". We now see that there was nothing *ad hoc* or arbitrary in our answers to the questions about the scopes of the occurrences of "~", and "·" in (32) above. To ask for the scope of an occurrence of an expression is to ask for information about any expressions with which it combined directly to form a larger expression, i.e. it is to ask for that larger expression.

Seen this way, it is evident that scope is a concept that applies to *any* compositional language, be it formal or natural, but a concept *whose semantic utility depends on whether the language being investigated is like standard formal languages in using scope to make semantic relations transparent.* And seen this way, it is evident that scope is semantically imperfect as far as the *surface* grammar of natural language is concerned, witness familiar "ambiguities of scope" ("a doctor examined every victim", "no-one has met the king of France", "bring your aunt or come alone and drink lots of Scotch"). It is, however, an interesting question, one being investigated by the more respectable strains of theoretical linguistics, whether there is a level of linguistic representation, the level relevant to semantic interpretation, at which scope does, in fact, play the desired role.[25]

There is one obvious mismatch between the syntax of the elementary formal languages just sketched and the syntax of English. Consider (33)

(33) $\sim(Fa \cdot \sim Gb)$.

The scope of the first negation sign "$\sim$" is the whole sentence, the scope of the conjunction sign "$\cdot$" is the sub-sentence "$(Fa \cdot \sim Gb)$", and the scope of the second negation sign is the sub-sub-sentence "$\sim Gb$". All of this is readily seen by looking at the phrase structure tree for (33), readily extractable from the standard syntax we would write for L:

(33′)

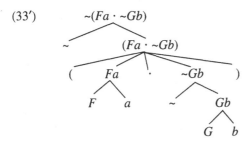

In terms of tree-geometry, the scope of an expression (as already defined) is simply the first node properly dominating it (similarly in the languages of propositional logic, first-order logic, and modal extensions thereof).[26] Following logical and philosophical tradition, in L a syntactic and semantic distinction was drawn between one- and two-place predicates. In terms of tree-geometry, the syntactic difference between a sentence containing a one-place predicate and a sentence containing a two-place predicate manifests itself clearly in (34) and (35):

Since the scope of an expression is simply the first node properly dominating it, in (34) "F" and "a" are within one another's scopes, just as "Alfie" and "fumbled" are within one another's scopes in an English analogue such as "Alfie fumbled". But when it comes to (35), our formal language and English come apart. In (35) "R", "a", and "b" are all within each other's scopes; in particular, "a" is within the scope of "R". But in an English analogue such as "Alfie respects Bertie", "Alfie" is not within the scope of the verb "respects" (though it is within the scope of the *verb phrase* "respects Bertie"):

(36)

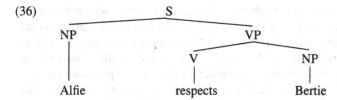

In (36) "respects" combines with the singular term "Bertie" to form an expression (a verb phrase) that combines with the singular term "Alfie" to form a sentence. That is, "respects" functions as a one-place, one-place-predicate-former. It does no harm to say it functions as a two-place predicate as long as it is kept in mind that the syntactic relations it bears to the two singular terms that function as its "arguments" are quite different.

The situation is strikingly similar when we turn to purported connectives, about which Quine (1953a, 1953c, 1960) and Davidson (1980) have worried. The English words "and" and "or" seem to be, on at least one of their uses, two-place connectives. Thus a sentence of the form of "φ and ψ" would seem to have the following tree:

(37)

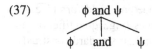

With the syntactic simplicity of philosophers" favourite formal languages in mind, very often expressions such as "if", "only if", "unless", "before", "after", "because", "although", "when", and "while" are treated as two-place connectives, and for purposes of this essay I shall adopt this (naïve) policy. From the perspective of providing a syntactic theory of English, these expressions should almost certainly be treated in a more complex fashion, perhaps as constituents of complex one-place connectives. On such an account, the basic syntactic structures of, say, "φ because ψ" and "because ψ, φ" are given by (38) and (38′) respectively:

(38)

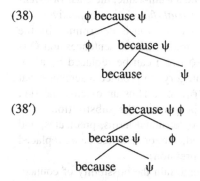

(38′)

Similarly for "φ if ψ" and "if φ, ψ", "φ before ψ" and "before ψ, φ", and so on.[27]

On this account, expressions such as "if Quine left", "before Quine leaves", "because Quine left" and so on are complex expressions that combine with a sentence to form another sentence. In other words they are complex sentence connectives. Much of what I have to say in the sequel is concerned with certain logical properties of connectives and of expressions that form *parts* of connectives such as "if", "only if", "before", "after", "because", "when", and "while". But to keep things simple, I shall partake of the common fiction that these expressions are two-place connectives on a par with "and" and "or" (just as we often partake of the common fiction that transitive verbs are two place predicates). This will make it easier to focus on the important logical and semantic issues at hand. Any accidental damage that results could easily be repaired if everything were transposed into a format that does justice to a more adequate syntax.

§7. *Extensional operators*

Extensional operators map *extensions* into extensions, i.e. they operate on the *extensions* of their operands. Consider an expression $❽(\phi_1...\phi_n)$ composed of an n-place operator $❽$ and operands $\phi_1...\phi_n$. $❽$ is an extensional operator if and only if the extension of $❽(\phi_1...\phi_n)$ depends upon no features of $❽$ and $\phi_1...\phi_n$ other than their *extensions* (the syntactic structure of $❽(\phi_1...\phi_n)$ supplying anything else that is required). Among those operators that operate on formulae and yield formulae, let us distinguish between those capable of binding variables (e.g., quantifiers) and those that are not (e.g. the truth-functional connectives, on their standard construal). Let us call the latter group *connectives*. Since they take us from the *extensions* of expressions to the *extensions* of larger expressions, the class of *extensional* connectives is a subclass of the class of extensional operators: an n-place connective $❽$ is extensional if and only if the extension of $❽(\phi_1...\phi_n)$ depends upon no features of $❽$ and $\phi_1...\phi_n$ other than their extensions. Thus a connective $❽$ is extensional if and only if any sentence with the same extension (i.e. truth value) as the sentence ϕ_k (where $1 \le k \le n$) can be substituted for ϕ_k in the sentence $❽(\phi_1...\phi_n)$ to produce a sentence with the same extension (i.e. truth value) as $❽(\phi_1...\phi_n)$. Since the extension of a sentence has already been stipulated to be a truth-value, the class of *extensional connectives* is the same thing as the class of *truth-functional* connectives.

A sentence ϕ is *extensional* if and only if its extension is determined by the extensions of its parts (and its syntax). If $\phi_1...\phi_n$ are extensional sentences and $❽$ is an extensional connective, then any part of $❽(\phi_1...\phi_n)$ can be replaced by a co-extensional expression (of the same syntactic category) to produce a sentence that has the same extension (i.e. truth-value) as $❽(\phi_1...\phi_n)$. Thus an extensional (i.e. truth-functional) connective $❽$ with large scope permits the substitution *salva veritate* (henceforth *s.v.*) within its scope of co-extensional terms, predicates, and sentences (assuming, of course, that the term, predicate, or sentence being replaced is not within the scope of any non-extensional expression).

The issues I want to address require that we entertain the possibility of connectives (in English or formalised languages designed to mirror parts of English) that are non-extensional. Quine (1953a, 1953c, 1960) has used slingshot arguments to cast doubt upon the viability of such connectives. I want to bracket such worries for a short while; all that is required for immediate purposes is a grasp of the *intended* difference between extensional and (purportedly) non-extensional connectives.

The following are sometimes treated as non-extensional connectives (on some of their uses): "necessarily" ($\Box$), "possibly" ($\Diamond$), "probably", "provably", "since", "because", "before", "after", "sometimes", "usually", "today", "yesterday", "intentionally", "voluntarily", "freely".[28] I will come to Quine's concerns about philosophers' uses of "necessarily" and "possibly" soon enough. For the moment it is sufficient to appreciate that if such expressions are connectives, their *non-extensional* nature is easily established: sentences with the same extension may not always be substituted for one another *s.v.* within their scopes. For example, if $\Box$ were extensional, then $\Box\phi$ and $\Box\psi$ would have the same truth-value whenever ϕ and ψ had the same truth-value. But this is not so. The sentences "9 > 7" and "9 has a name that rhymes with a name of Quine" have the same truth-value, but (39) and (40) do not:

(39) $\Box(9 > 7)$

(40) $\Box$(9 has a name that rhymes with a name of Quine).

Hence $\Box$ is not extensional. As we might say, the truth value of $\Box\phi$ depends upon features of ϕ other than its truth-value.[29] The intended difference between extensional and non-extensional connectives is clear: extensional connectives permit the substitution *s.v.* of co-extensional sentences; non-extensional connectives do not. (Alternatively, but equivalently, a connective $\otimes$ is extensional if and only if it has a complete truth table.)

In order to keep things as simple as possible and avoid digressions on semantical issues orthogonal to those being discussed here, let us bracket the existence of any purportedly non-extensional operators that are not connectives (e.g., adjectives and verbs such as "fake", "alleged", "fear", and "want"). Where X is a particular occurrence of an expression, we can now say that (i) X occupies an *extensional position*, and (ii) X occurs in an *extensional context* if and only if (iii) X is not within the scope of any non-extensional connective.

§8. *Principles of substitution*

There is an elegant strategy, essentially due to Quine (1953c, 1960), for investigating non-extensional constructions: (i) take an arbitrary n-place sentence connective $\otimes$, an arbitrary truth-functional sentence ϕ, and an arbitrary compound sentence $\otimes$ $(...\phi...)$; (ii) examine the deductive consequences of replacing the occurrence of ϕ in $\otimes(...\phi...)$ by another sentence ϕ' obtained directly from ϕ by using inference principles that are valid in truth-functional contexts.[30] Importantly, on this strategy the inference principles are applied *not* to $\otimes(...\phi...)$ itself but to the occurrence of the extensional sentence ϕ within such a sentence, i.e. the sentence ϕ within the scope of $\otimes$. With this fact in mind, I want to set out, in a rather unorthodox way and with some unorthodox terminology, various inference principles common in extensional logic. They will, perhaps, not be formulated quite as cleanly as one might like, but they will be clean enough for present purposes.

Two philosophically useful principles of inference that can be employed in extensional contexts concern the substitution of co-extensional sentences and co-extensional singular terms.

(i) *The Principle of Substitutivity for Material Equivalents* (PSME) can be put thus:

PSME $(\phi \equiv \psi)$
$$\frac{\Sigma(\phi)}{\Sigma(\psi).}$$

For present purposes this should be read as saying that if two sentences ϕ and ψ have the same truth-value and $\Sigma(\phi)$ is a true sentence containing at least one occurrence of ϕ, then $\Sigma(\psi)$, is also true where $\Sigma(\psi)$ is the result of replacing at least one occurrence of ϕ in $\Sigma(\phi)$ by ψ.

A *context* is extensional if and only if it permits the substitution *s.v.* of co-extensional terms, predicates, and sentences. So it is a truism that PSME is a principle that validly can be used on a sentence ϕ occurring in an extensional context. As shorthand for this, let us say that extensional contexts are +PSME (as opposed to –PSME).

By an unproblematic and extremely useful extension of terminology, let us say that any extensional connective ❽ is +PSME in the sense that ❽ permits the use of PSME on any sentence ϕ within its scope (assuming, of course, that ϕ does not occur within the scope of any non-extensional connective). To avoid confusion, let me spell out the idea precisely:

An *n*-place connective ❽ is +PSME iff in any sentence ❽$(...\Sigma(\phi)...)$ in which $\Sigma(\phi)$ is an extensional sentence occurring as an operand of ❽, if ϕ and ψ are sentences with the same truth-value then replacing the contained sentence $\Sigma(\phi)$ by $\Sigma(\psi)$ in the original sentence ❽$(...\Sigma(\phi)...)$ yields a sentence ❽$(...\Sigma(\psi)...)$ with the same truth-value as the original.

(ii) *The Principle of Substitutivity for Logical Equivalents* (PSLE). Following Tarski, and common practice, let us say that ϕ and ψ are logically equivalent if, and only if, ϕ and ψ have the same truth-value in every model. We can now state another rule of inference, the *Principle of Substitutivity for Logical Equivalents* (PSLE):

PSLE: $\phi \models \dashv \psi$
$$\frac{\Sigma(\phi)}{\Sigma(\psi)}$$

For present purposes this should be read as saying that if two sentences ϕ and ψ are logically equivalent and $\Sigma(\phi)$ is a true sentence containing at least one occurrence of ϕ, then $\Sigma(\psi)$, is also true where $\Sigma(\psi)$ is the result of replacing at least one occurrence of ϕ in $\Sigma(\phi)$ by ψ.

PSLE is, of course, a rule that validly can be used on a sentence φ occurring in an extensional context (i.e. a context that is +PSME). As shorthand for this, let us say that extensional contexts and extensional connectives are +PSLE (as opposed to −PSLE).

Continuing with the useful extension of terminology introduced earlier, let us say that any extensional connective ❽ is +PSLE in the sense that ❽ permits the use of +PSLE on any sentence φ within its scope (assuming, of course, that φ does not occur within the scope of any non-extensional connective). More precisely:

An n-place connective ❽ is +PSLE iff in any sentence ❽$(...\Sigma(\phi)...)$ in which $\Sigma(\phi)$ is an extensional sentence occurring as an operand of ❽, if φ and Ψ are logically equivalent sentences, then replacing the contained sentence $\Sigma(\phi)$ by $\Sigma(\Psi)$ in the original sentence ❽$(...\Sigma(\phi)...)$ yields a sentence ❽$(...\Sigma(\Psi)...)$ with the same truth-value as the original.

(iii) *The Principle of Substitutivity for Singular Terms* (PSST) can be put thus:

$$\text{PSST:} \quad \frac{\alpha=\beta}{\Sigma(\alpha)} \qquad \text{or:} \qquad \frac{\Sigma(\alpha)}{\sim\Sigma(\beta)}$$
$$\overline{\Sigma(\beta).} \qquad\qquad\qquad \overline{\alpha\neq\beta}$$

This just says that if two singular terms α and β have the same extension (i.e. if "$\alpha=\beta$" is a true identity statement) and $\Sigma(\alpha)$ is a true extensional sentence containing at least one occurrence of α, then $\Sigma(\beta)$ is also true where $\Sigma(\beta)$ is the result of replacing at least one occurrence of α in $\Sigma(\alpha)$ by β.

There is a manifest difficulty in applying PSST: it presupposes a clear answer to the question *which noun phrases that occur in the grammatical singular are singular terms?*[31] For the sake of having a provisional answer, let us suppose the class of singular terms to comprise just the following: (i) ordinary proper names; (ii) the simple demonstratives "this" and "that"; (iii) complex demonstratives such as "this man" and "that man"; (iv) the first- and second-person singular pronouns "I", "me", and "you"; and (v) at least some occurrences of the third-person singular pronouns "he", "him", "she", "her", and "it" (including those occurrences that Quine (1960) argues function as variables hooked up to quantifiers). In order to get things moving, I simply stipulate that a description such as "the prince" is Russellian, hence *not* a singular term but a quantificational noun phrase along with its syntactic siblings "a prince", "one prince", "no prince", "some prince", "each prince", etc., reserving the right to redraw the boundary of the class of singular terms should this provisional characterisation proves to be lacking in any way.

A context is extensional only if it permits the substitution *s.v.* of co-extensional singular terms. So it is a truism that PSST is a valid rule of inference for terms occurring in extensional contexts. As shorthand for this, let us say that extensional contexts are +PSST (as opposed to −PSST).

In accordance with the convention introduced in connection with PSME, let us say that any extensional connective ❽ is +PSST in the sense that ❽ permits the use of

PSST on any sentence ϕ within its scope (assuming, of course, that ϕ does not occur within the scope of any non-extensional connective). More precisely:

An n-place connective $\mathbf{O}$ is +PSST iff in any sentence $\mathbf{O}(...\Sigma(\alpha)...)$ in which $\Sigma(\alpha)$ is an extensional sentence occurring as an operand of $\mathbf{O}$, if α and β are co-extensional singular terms then replacing the contained sentence $\Sigma(\alpha)$ by the sentence $\Sigma(\beta)$ in the original sentence $\mathbf{O}(...\Sigma(\alpha)...)$ yields a sentence $\mathbf{O}(...\Sigma(\beta)...)$ with the same truth-value as the original.

Of course, a context or connective that is +PSME is also +PSST; but nothing on the table guarantees the converse (an argument would be needed to demonstrate it).

§9. An inference principle involving "exportation"

The inference principles examined in the previous section involved replacing one expression in a formula by another, though it is not necessary to characterise the principles in that way. I want now to set out an inference principle that cannot be characterised in terms of such a simple substitution, a principle often known as EG, "existential generalisation". In view of the unorthodox way I am setting out my versions of inference principles, some care must be taken here. For present purposes, I want to state what I shall call "EG" as follows,

EG: $$\frac{\Sigma(x/\alpha)}{\exists x \Sigma(x)}$$

where $\Sigma(x)$ is any extensional formula containing *at least one* occurrence of the variable x and $\Sigma(x/\alpha)$ is the result of replacing *every* occurrence of the variable x in $\Sigma(x)$ by the (closed) singular term α (there may be other occurrences of α in $\Sigma(x)$ as well).

EG is a valid inference principle when applied to extensional sentences. An extensional connective $\mathbf{O}$ is +EG in the sense that $\mathbf{O}$ permits the use of EG on any formula ϕ within its scope (assuming, of course, that ϕ does not occur within the scope of any non-extensional connective). More precisely:

An n-place connective $\mathbf{O}$ is +EG iff for any sentence $\mathbf{O}(...\Sigma(x/\alpha)...)$ in which $\Sigma(x/\alpha)$ is an extensional sentence occurring as an operand of $\mathbf{O}$, $\exists x \mathbf{O}(...\Sigma(x)...)$ has the same truth-value as $\mathbf{O}(...\Sigma(x/\alpha)...)$.

Questions about propositional attitude contexts in connection with EG are notoriously complex, as Quine (1943, 1953a, 1956, 1960, 1977) has stressed. For the sake of argument, assume that (41) is true and (42) is false:

(41) Philip is unaware that Tully denounced Catiline
(42) Philip is unaware that Cicero denounced Catiline.

To use Quine's terminology, the difference in truth-value indicates that attitude contexts are "referentially opaque" in the sense of being –PSST. (There is a wrinkle here because Quine sometimes uses examples involving definite descriptions to exemplify referential opacity, especially when considering modal contexts. But we can put this aside until Part II.) Quine's main worry is the intelligibility of an example such as (43), the truth of which should be inferable from (41) if attitude contexts are +EG:

(43) $\exists x$(Philip is unaware that x denounced Catiline).

Now Quine asks his famous question: "What is this object, that denounced Catiline without Philip yet having become aware of the fact? Tully, i.e., Cicero? But to suppose this would conflict with the fact that [42] is false." (1943a, p. 118)[32]

§10. An inference principle for definite descriptions

We turn now to two substitution rules that are less familiar. If definite descriptions are treated in accordance with Russell's theory – or any other theory that does not treat descriptions as singular terms – then, as Russell and others have pointed out, substitutions involving descriptions are not licensed directly by PSST.[33] This matter merits some attention as philosophers who appeal or profess allegiance to Russell's Theory of Descriptions often fail to do justice to the point and thereby run into logical difficulties of a type that will concern us very soon.

On Russell's account, what might look like an identity statement involving one or two descriptive phrases is really no such thing. An identity statement has the general form "$\alpha=\beta$", where α and β are singular terms. The way PSST was stated, it is the truth of a statement of this form that licenses its applications. But on a Russellian analysis of descriptive phrases, the logical forms of sentences of the superficial grammatical forms "a = the F" and "the G = the F" are given by the following quantificational formulae:

(44) $\exists x(\forall y(Fy \equiv y=x) \cdot \underline{x=a})$
(45) $\exists x(\forall y(Fy \equiv y=x) \cdot \exists u(\forall v(Gu \equiv v=u) \cdot \underline{u=x}))$.

And neither (45) nor (45) is an identity statement; each is a quantificational statement that *contains* an important identity statement (underlined) as a proper part.

The real force of this point emerges once we reflect on the nature of derivations in first-order logic with identity. The inference in (46) is obviously valid (on the standard definition of validity):

(46)   [1]   Cicero = Tully
           [2]   Cicero snored

           [3]   Tully snored.

In order to provide a formal derivation of the conclusion from the premises, we can use PSST, which sanctions a direct move from [1] and [2] to [3]:

1	[1]	$c = t$	premiss
2	[2]	Sc	premiss
1,2	[3]	St	1, 2, PSST.

Now consider (47), which looks like a very similar argument.

(47)   [1]   Cicero = the greatest Roman orator;
       [2]   Cicero snored;

       [3]   The greatest Roman orator snored.

Clearly this is valid. But – and this is the important point – *if* definite descriptions are Russellian, then they are *not* singular terms so we cannot use PSST to move *directly* from lines [1] and [2] to line [3] in the formal analogue of this argument in first-order logic with identity. Reading "Rx" as "x is greatest Roman orator" it might be tempting to set out a derivation as follows:

1	[1]	$c = \iota x R x$	premiss
2	[2]	Sc	premiss
1,2	[3]	$S \iota x R x$	1, 2, PSST.

The Russellian accepts that the conclusion follows from the premises, but rejects this particular derivation: it is illegitimate because PSST can be invoked only where we have an identity statement, and an identity statement has *singular terms* on either side of the identity sign. As Church (1942) points out (in the course of scolding Quine (1941) for employing this sort of direct inference in a purportedly non-extensional context – more on this later):

On the basis of *Principia* or of Quine's own *Mathematical Logic*…any formal deduction must refer to the unabbreviated forms of the sentences in question (p. 101).

Premise [1] is not an identity statement; on Russell's account it is, to use Quine's (1941) wording again, "mere shorthand" for a complex quantificational statement; indeed, the purported derivation is just shorthand for the following illegitimate derivation:

1	[1]	$\exists x (\forall y (Ry \equiv y{=}x) \cdot x{=}c)$	premiss
2	[2]	Sc	premiss
1,2	[3]	$\exists x (\forall y (Ry \equiv y{=}x) \cdot Sx)$	1, 2, PSST.

To say that the derivation is illegitimate is to say that PSST does not sanction a direct move from line [2] to line [3] on the basis of the truth of the entry on line [1]; it is *not* to say that the argument is invalid – it *is* valid – nor is it to say that one cannot *derive* the entry on line [3] from the entries on lines [1] and [2] using standard rules of inference, which include, of course, PSST. Indeed, it is a routine exercise – the logical and philosophical importance of which is stressed in better introductory logic texts – to provide the relevant derivation:

1	[1]	$c = \iota x R x$	premiss
2	[2]	Sc	premiss
1	[3]	$\exists x (\forall y (Ry \equiv y{=}x) \cdot c{=}x)$	1, def. of "ιx"
4	[4]	$(\forall y (Ry \equiv y{=}\alpha) \cdot c{=}\alpha)$	assumption

4	[5]	$c = \alpha$	4, $\cdot$-ELIM
2,4	[6]	$S\alpha$	2, 5, PSST[34]
4	[7]	$\forall y(Ry \equiv y=\alpha)$	4, $\cdot$-ELIM
2,4	[8]	$(\forall y(Ry \equiv y=\alpha) \cdot S\alpha)$	6, 7, $\cdot$-INTR
2,4	[9]	$\exists x(\forall y(Ry \equiv y=x) \cdot Sx)$	8, EG
1,2	[10]	$\exists x(\forall y(Ry \equiv y=x) \cdot Sx)$	3, 4, 9, EI
1,2	[11]	$S\imath x Rx$	10, def. of "$\imath x$".

Within a purely extensional system, it would be tedious to proceed in this way every time one wanted to prove something involving one or more descriptions, and it would be practical to have a fool-proof method of shortening such proofs. Whitehead and Russell recognised this and reduced their workload by demonstrating that, although descriptions are not genuine singular terms (in their system), if a predicate F applies to exactly one object (i.e. if it has exactly one thing in its extension), in truth-functional contexts the description "$\imath x F x$" can be treated *as if* it were a singular term for derivational purposes. As noted earlier, the following theorem to this effect is proved by them for truth-functional contexts:

*14.15 $(\imath x\phi = \alpha) \supset \{\Sigma(\imath x\phi) \equiv \Sigma(\alpha)\}$.

(Recall Whitehead and Russell's convention that absence of the scope indicator "[$\imath x\phi$]" signals that the description has minimal scope.) *14.15 says that if the individual that α stands for is the unique object satisfying a formula ϕ, then one can, as Whitehead and Russell sometimes put it, "verbally substitute" α for the description $\imath x\phi$, or vice-versa (in truth-functional contexts). (Obviously the derivation given above can be recast to provide a proof of the same thing.) If descriptions are treated in accordance with Russell's theory, it is a mistake to think that when one performs a "verbal substitution" of this sort, one is simply making a direct application of PSST. *14.15 is *not* PSST; it is a *derived rule of inference* that can be used in truth-functional contexts, a rule that licenses certain substitutions when the referent of a particular singular term is identical to the unique object satisfying a particular formula. Naturally, Whitehead and Russell prove the analogue of *14.15 where both noun phrases are descriptions:

*14.16 $(\imath x\phi = \imath x\psi) \supset \{\Sigma(\imath x\phi) \equiv \Sigma(\imath x\psi)\}$.

This says that if the unique object satisfying a formula ϕ is identical to the unique object satisfying a formula ψ, then one can "verbally substitute" the description "$\imath x\phi$" for the description "$\imath x\psi$", or vice versa (in truth-functional contexts).

On the basis of *14.15 and *14.16, we can add a fourth inference rule (actually, a triple of rules) to our collection, $\imath$-SUBSTITUTION:

$\imath$-SUBS:

$$\frac{\imath x\phi = \imath x\psi \qquad \Sigma(\imath x\phi)}{\Sigma(\imath x\psi)} \qquad \frac{\imath x\phi = \alpha \qquad \Sigma(\imath x\phi)}{\Sigma(\alpha)} \qquad \frac{\imath x\phi = \alpha \qquad \Sigma(\alpha)}{\Sigma(\imath x\phi)}.$$

$\imath$-SUBS is a valid rule of inference when the description "$\imath x\phi$" occurs in an extensional context. As shorthand for this, let us say that extensional contexts are $+\imath$-

SUBS (as opposed to $-\iota$-SUBS). Continuing with our useful extension of terminology, let us say that any extensional connective $\circledast$ is $+\iota$-SUBS in the sense that $\circledast$ permits the use of ι-SUBS on any sentence ϕ within its scope (assuming, of course, that ϕ does not occur within the scope of any non-extensional connective). More precisely:

An n-place connective $\circledast$ is $+\iota$-SUBS iff in any sentence $\circledast(...\Sigma(\iota x\phi)...)$ in which $\Sigma(\iota x\phi)$ is an extensional sentence occurring as an operand of $\circledast$, if $\iota x\phi = \iota x\psi$ then replacing the contained sentence $\Sigma(\iota x\phi)$ by the sentence $\Sigma(\iota x\psi)$ in the original sentence $\circledast(...\Sigma(\iota x\phi)...)$ yields a sentence $\circledast(...\Sigma(\iota x\psi)...)$ with the same truth-value as the original.

(There is no need to build in the second and third allomorphs of ι-SUBS here.) No residual issues concerning scope arise here in connection with those theories of descriptions for which matters of scope are important (e.g. Russell's): scopes remain constant with respect to $\circledast$. Of course, if descriptions are treated as singular terms, ι-SUBS is redundant, its work already done by PSST.

It is surely only *because* truth-functional (i.e. extensional) contexts are $+\iota$-SUBS that Whitehead and Russell introduce descriptive terms (and through them condensed expressions of the form "$R'x$" mentioned in note 7) in *Principia*: they simplify both formulae and proofs. Adding such rules to an extensional deductive system, we can now formally capture the inference from "Cicero = the greatest roman orator" and "Cicero snored" to "the greatest roman orator snored":

1	[1]	$c = \iota xRx$	premiss
2	[2]	Sc	premiss
1,2	[3]	$S\iota xRx$	1, 2, ι-SUBS.

As far as truth-functional contexts are concerned, there are no interesting reflexes of the formal differences between PSST and ι-SUBS. Only in so far as there are linguistic contexts and connectives that are $+$PSST but $-\iota$-SUBS does the distinction become interesting. Many philosophers *today* take the one-place modal connectives $\square$ and $\lozenge$ to be $+$PSST and $-\iota$-SUBS, where these connectives are understood as expressing necessity and possibility in some metaphysical sense, such as that expounded by Kripke (1971, 1972) – most certainly *not* the sense Quine (1941, 1943, 1947, 1953a, 1953c, 1960, 1961, 1962) was attacking.

§11. Intensions and intensionality

A good deal of contemporary philosophy involves manoeuvring within linguistic contexts governed by modal, causal, deontic, and other purportedly non-extensional operators. If this sort of manoeuvring is to be effective, it must respect the logical and other semantical properties of the contexts within which it takes place. It is an unfortunate fact about much of today's technical philosophy that the relevant logico-semantical groundwork is not properly done (if it is done at all), and this is one reason so much work in metaphysics, ethics, the philosophy of mind, and the philosophy of language reduces to utter nonsense. Great progress has been made in the last fifty years in understanding the logic, structure, and use of language; technical philosophy in the absence of logical grammar in this age barely deserves the name "philosophy."

If ❽ is an *n*-place extensional *S*-connective, then the extension of $❽(\phi_1...\phi_n)$ is determined by the extensions of ❽ and $\phi_1...\phi_n$. But suppose ❽ is a *non-extensional* *S*-connective; what properties of ❽ and $\phi_1...\phi_n$ determine the extension of ❽ $(\phi_1...\phi_n)$? Inspired by Frege's distinction between *sense* and *reference*, many philosophers have attempted to answer this question, for particular values of ❽, by postulating a second level of "semantic value" to supplement extensions. For example, Carnap (1947) and those he has influenced have suggested that each expression has an *intension* as well as an extension, and that for certain interesting non-extensional *S*-connectives ❽, the extension of ❽ together with the *intensions* of $\phi_1...\phi_n$ determine the extension of $❽(\phi_1...\phi_n)$.

Consider the modal *S*-connectives $\square$ and $\lozenge$ (where $\square\phi$ is understood as $\sim\lozenge\sim\phi$). Allowing ourselves talk of so-called "possible worlds" for a moment – we can, and will, avoid such talk soon enough – one common idea is to regard the intension of an expression as *a function from possible worlds to extensions*. On such an account, (i) the intension of a singular term is a (possibly partial) function from possible worlds to *objects*;[35] (ii) the intension of an *n*-place predicate is a function from possible worlds to *sets of ordered n-tuple of objects*; (iii) the intension of a sentence is a function from possible worlds to *truth-values*; and (iv) the intensions of $\square$ and $\lozenge$ are functions *from* functions from possible worlds to truth-values *to* functions from possible worlds to truth-values.

As far as $\square$ and $\lozenge$ are concerned, the answer to our original question – "if ❽ is a non-extensional *S*-connective, what properties of ❽ and $\phi_1...\phi_n$ determine the extension of $❽(\phi_1...\phi_n)$?" – is now within sight. The extension of, say, $\square\phi$ is determined (in part) by the extension of $\square$. And the extension of $\square$ must be a function from something ξ to the potential extensions of $\square\phi$, *i.e.* a function from ξ to truth-values. Clearly ξ cannot be the potential extensions (*i.e.* truth-values) of ϕ, for otherwise $\square$ would be an extensional *S*-connective. But if ξ is the potential *intension* of ϕ, everything fits together perfectly. The extension of $\square$ is simply a function from intensions to truth-values, *i.e.* a function *from* functions from possible worlds to truth-values *to* truth-values. Thus the extension of $\square\phi$ is determined by (a) the *extension* of $\square$, and (b) the *intension* of ϕ. We say that $\square$ and $\lozenge$ are *intensional S-connectives* because they operate on the *intensions* of their operands.

Where X is a particular occurrence of an expression we can say that (i) X occupies an *intensional position*, and (ii) X occurs in an *intensional context*, if and only if (a) X is within the scope of an intensional *S*-connective, and (b) any *S*-connective within whose scope X lies is either intensional or extensional.

Carnap (1947) was very clear in his use of "extensional" and "intensional", recognising that the class of non-extensional contexts need not collapse into the class of intensional contexts. However, many philosophers and linguists to use "intensional" and "intensionality" in ways that are much looser, thereby encouraging talk of the "intensionality of propositional attitude reports", and talk of attitude constructions involving "intensional operators," and so on. (This surely has something to do with the fact that "intensional" and "intentional" are homophonic.) Such talk muddies already cloudy waters and can lead to philosophical mistakes engendered by running together modal and psychological contexts. There is no *a priori* reason to think of the logical or metaphysical modalities and psychological attitudes as sharing a logic – they *don't* – nor is there any reason to think that talk of "possible worlds" is of any help in thinking about the semantics of propositional attitude reports – it *isn't*. If we need a technical word to use in connection with attitude reports and constructions we can settle for "atensional", applied uniformly to contexts, constructions, and operators (including connectives). To the extent that we want to group non-extensional contexts, constructions, and operators together, there is already a perfectly good word: "non-extensional". There is only misery in store for those who would too easily lump together constructions involving the attitudes and those involving logical or metaphysical modality.

At this point we do well to recall how possible worlds and intensions have been related to several other notions in the literature. (1) Whereas some philosophers have been inclined to view possible worlds as primitive, others have been tempted to view them as sets of (consistent) states-of-affairs. Still others, such as Fine (1982), have been tempted to see them as "very large facts." (2) A common idea is to equate the intension of a sentence with the *proposition* it expresses and characterise the notion of a proposition in terms of *possible worlds*. The basic idea is this: if the intension of a sentence is a function from possible worlds to truth-values, then (assuming an extensional characterisation of functions) the intension of a sentence can be viewed as a set of possible worlds, viz. those at which the sentence is true. And this set of worlds can be called a "proposition." This notion of a proposition corresponds to the common philosophical notion of the *truth-condition* of a sentence, *i.e.* the condition under which it is true. Thus we reach the familiar positions that (i) the intension of a sentence is its truth condition, and (ii) the truth-value of the intensional sentence $\Box\phi$ depends upon ϕ's *truth condition* (whereas the truth-value of the extensional sentence $\sim\phi$ depends only upon ϕ's *truth-value*). Obviously this will suggest to some – for example, those who view the notion of the truth-condition of a sentence as more basic than the notion of a possible world – that we can talk perfectly well about the intensions of expressions without talking about possible worlds.

It should be clear that intensional S-connectives (in the sense given above) are understood to be –PSME and +PSST. That they are –PSME is self-evident; that they are +PSST is readily proved: If α and β are terms that both refer to x, and $\Re$ is a one-place extensional predicate, then $\Re\alpha$ and $\Re\beta$ have the same truth-condition: $\Re\alpha$ and $\Re\beta$ are both true if and only if x is $\Re$. If $\mathbf{O}$ is an intensional operator, then by definition the truth value of $\mathbf{O}\Re\alpha$ depends only upon the truth condition of $\Re\alpha$; and the truth-value of $\mathbf{O}\Re\beta$ depends only upon the truth-condition of $\Re\beta$. But $\Re\alpha$ and $\Re\alpha$ have the same truth-condition (they are both true if and only if x is $\Re$). Hence $\mathbf{O}\Re\alpha$ is true if and only if $\mathbf{O}\Re\beta$ is true. Hence $\mathbf{O}$ is +PSST. (There is almost enough in this proof to undermine the idea that definite descriptions are singular terms unless free logic is assumed.)

§12. *Quine's connective slingshot*

We are now ready to begin investigating the logic of purportedly non-extensional connectives using the Quinean strategy mentioned earlier. The idea, recall, is to (i) take an arbitrary n-place connective ⦾, an arbitrary extensional sentence φ, and an arbitrary compound sentence ⦾(...φ...); and then (ii) examine the deductive consequences of replacing the occurrence of φ in ⦾(...φ...) by another sentence φ′ obtained directly from φ using inference principles known to be valid in truth-functional contexts. The inference principles are applied *not* to ⦾(...φ...) itself but to an occurrence of the extensional sentence φ within such a sentence, i.e. a sentence φ within the scope of ⦾.

I now ask the reader to put out of mind thoughts about ◊ and □, indeed thoughts about any particular non-extensional connective. I want to proceed in the most abstract way possible so as to avoid being distracted by prejudicial thoughts about the semantics of this or that connective, which may have something to do with one's views about, say, necessity, causation, time, facts, or states of affairs, or orthogonal semantic worries about such things as rigidity, direct reference, or semantic innocence.[36]

The Quinean proof I am about to set out is going to be the central component of a deductive argument meant to show that an inconsistency results when one posits a non-extensional connective that freely permits the use of ι-SUBS (or its analogue for class abstracts) together with some other inference principle within its scope. The inconsistency arises because (roughly) descriptions (and abstracts) contain *formulae* as proper parts; permitting the interchange of such devices when their contained formulae are satisfied by the same object, is tantamount to permitting the interchange of formulae themselves; and once some weak additional inference principle is assumed – and it is in the precise character of the additional principle that a proof hinted at by Gödel (1944) is more interesting than those proposed by Church and Quine – the formulae in question can be drawn out of their *iota*-governed contexts to make purportedly non-extensional connectives provably extensional.

Quine (1953a, 1953c, 1960) gives three versions of the argument, but they all three have the same essential ingredients and structure. For expository convenience, I take the 1960 version.[37] Recall Quine's (1941) hunch that any non-extensional connective will fail to admit "inference by interchanging terms that designate the same object." The argument we are about to examine is as close as Quine gets to confirming his hunch. With the modal connectives in mind – but forget this once you reach the end of this sentence – he proposes to show that any connective ⦾ that (i) permits what he calls "the substitutivity of identicals" and (ii) is +PSLE is, in fact, an extensional connective. The argument begins with an abbreviatory convention (1960, p. 148):

Where "p" represents a sentence, let us write "δp" (following Kronecker) as short for the description: the number x such that $((x = 1)$ and $p)$ or $((x = 0)$ and not $p)$.[38]

The central part of the argument can be set out as the following derivation:

1	[1]	$(p \equiv q)$	premiss
2	[2]	$\otimes p$	premiss
2	[3]	$\otimes(\delta p=1)$	2, $\otimes$+PSLE (ass. that $\delta p=1$ and p are log. equiv.)
1	[4]	$\delta p=\delta q$	1, def. of δ
1,2	[5]	$\otimes(\delta q=1)$	3, 4, "substitutivity of identicals"
1,2	[6]	$\otimes q$	5, $\otimes$+PSLE (ass. $\delta q=1$ and q are log. equiv.).

Gloss: on the assumptions that (i) p and q have the same truth-value, (ii) $\otimes p$ is true, (iii) $\otimes$ is +PSLE, and (iv) $\otimes$ permits the "substitutivity of identicals", it appears to be provable that $\otimes q$ is true. And since $\otimes q$ differs from $\otimes p$ just in the substitution of the mere material equivalents p and q, it would seem that $\otimes$ is, contrary to initial hypothesis, +PSME, i.e. it would seem that $\otimes$ is actually an extensional connective after all. Philosophical consequences of this argument can then drawn by interpreting $\otimes(...)$ as, e.g., "necessarily (...)", "ϕ because (...)", "it is a moral (/legal/ constitutional) requirement that (...), or "the statement that ϕ corresponds to the fact that (...)".

It will not do to object to Quine's argument on the grounds that a description $\iota x \phi$ is not well-formed or not interpretable unless every atomic formulae in its matrix contains at least one occurrence of x that ιx can bind. Even if it would normally be odd to use the analogue of a description tailing this condition in ordinary or theoretical talk, there is no formal difficulty involved in making sense of such a description.

The argument does have a weakness however. An important but frequently overlooked fact about the argument is that if it is to be of any interest it must be supplemented with a precise semantics for definite descriptions. The reason is simple: (i) the notion of *logical equivalence* is invoked in getting from line [2] to line [3], and from line [5] to line [6]; (ii) lines [3] and [5] both contain definite descriptions; (iii) on some treatments of descriptions the logical equivalences obtain, on others they do not; (iv) the treatment of descriptions assumed by the argument will determine whether it is PSST or ι-SUBS that is invoked in getting from lines [3] and [4] to line [5].[39]

The validity of the derivational component of this argument turns on two related matters then: (i) the interpretation of the expression "substitutivity of identicals," and (ii) the semantics ascribed to Kronecker's δ-functional. Quine uses the definite description "the number x such that $((x = 1)$ and $p)$ or $((x = 0)$ and not $p)$" when rendering δp in English; so the question naturally arises whether he is assuming the description to have a Russellian or a referential semantics. The power of any collapsing argument involving logical equivalence and the substitution of descriptions whose matrices are satisfied by the same object depends upon the precise semantics assumed.

If Quine is assuming that δp is treated in accordance with Russell's theory, then it is easy enough for him to justify the logical equivalences that his proof invokes. On a Russellian account, $\delta p=1$ is simply an abbreviation for the first-order sentence, (48), which is logically equivalent to p:

(48) $\exists x(\forall y(((y=1 \cdot p) \vee (y=0 \cdot \sim p)) \equiv y=x) \cdot x=1)$.

Using Quine's (1936) criterion, for all interpretations of their non-logical vocabulary, p and (48) have the same truth-value. (Equivalently, following Tarski, the sentences are true in exactly the same models; equivalently, they are interdeducible.) Perhaps it was this fact that prompted Quine to claim that $\delta p=1$ and p are logically equivalent.

But now a question arises. If δp is Russellian, the entry on line [4] of Quine's proof is not a genuine identity statement; it is just shorthand for (49):

(49) $\exists x(\forall y(((y=1 \cdot p) \vee (y=0 \cdot \sim p)) \equiv y=x) \cdot$
$\exists z(\forall w(((w=1 \cdot q) \vee (w=0 \cdot \sim q)) \equiv w=z) \cdot x=z))$.

So on the reading that interests us, in their unabbreviated forms the central lines of the derivation given above will be as follows:

[3] $\mathbf{❽}(\exists x(\forall y(((y=1 \cdot p) \vee (y=0 \cdot \sim p)) \equiv y=x) \cdot x=1))$ 2, $\mathbf{❽}$+PSLE
[4] $\exists x(\forall y(((y=1 \cdot p) \vee (y=0 \cdot \sim p)) \equiv y=x) \cdot$
$\exists z(\forall w(((w=1 \cdot q) \vee (w=0 \cdot \sim q)) \equiv w=z) \cdot x=z))$ 1, def of "δ"
[5] $\mathbf{❽}(\exists z(\forall w(((w=1 \cdot q) \vee (w=0 \cdot \sim q)) \equiv w=z) \cdot z=1))$ 3, 4, "subst".

But what is the inference rule "subst" ("substitutivity of identicals") that licenses the move from lines [3] and [4] to line [5]? On a Russellian treatment of descriptions, "subst" cannot be PSST, which is a rule of inference governing the substitution of coreferring singular terms; it has to be ι-SUBS. So on such a treatment Quine's proof shows decisively that $\mathbf{❽}$ cannot have the following combination of features:

(50) +PSLE   −PSME   +ι-SUBS.[40]

But this result, incontrovertible as it is, will not worry all proponents of non-extensional logics. *Individual systems and their interpretations need to be investigated.*

It is doubtful that today's modal logician, who works with Kripke's metaphysical modality, will be troubled. It is the combination given in (51) that most logicians seem to want to ascribe to ◊ and □:

(51) +PSLE   −PSME   +PSST.

And Quine's proof has no bearing on the viability of *this* combination if descriptions are provided with a Russellian treatment. Second, today's modal logicians are antecedently predisposed to think that □ and ◊ are −ι-SUBS, largely because of some of the examples brought up by Quine in connection with analytic necessity, examples which, as Kripke (1971, 1972) saw, carry over to a metaphysical conception. Consider the following argument, where the description "the number of planets in our solar system" is substituted for "nine" within the scope of □:

(52) $\square(9 > 7)$;
          9 = the number of planets in our solar system
          _____
          $\square$(the number of planets >7).

The fact that (52) is invalid (when $\Box$ has large scope) shows that $\Box$ is not +ι-SUBS. (The purported existence of a separate (true) reading of the conclusion upon which the description has large scope is irrelevant to this point.) The upshot of all this, then, is that if Quine treats Kronecker descriptions as Russellian, (a) the logical equivalences he needs for his slingshot are guaranteed, but (b) the conclusion of the argument is one that today's modal logicians will endorse, viz., that $\Box$ and $\Diamond$ do not possess the set of features given in (50). None of this has any bearing on whether or not $\Diamond$ and $\Box$ are +PSST, which many of today's modal logicians think they are. So no argument against the possibility of treating today's $\Diamond$ and $\Box$ as non-extensional sentence connectives emerges from Quine's slingshot (which is not to say, of course, that Quine does not have other good reasons for being suspicious of $\Diamond$ and $\Box$).

The only way that Quine could convert his slingshot into a general argument against non-extensional sentence connectives would be to articulate a treatment of descriptions as singular terms and ensure that it licenses the logical equivalences to which he appeals. Of course this would be a strikingly odd move for Quine to make as he has been a champion of *Russell's* theory for ontological, semantic, and logical reasons (its ontological parsimony, its utility in eliminating singular terms, its simple first-order nature, and so on). Moreover, as I shall illustrate in §14, it is far from clear that there is an antecedently plausible referential theory that simultaneously (a) licenses the logical equivalences Quine needs, (b) provides a plausible account of descriptions whose matrices are unsatisfied, and (c) respects the intuitive meaning of the description operator accorded it by the referentialist: if a matrix $\Sigma(x)$ containing at least one occurrence of the variable x (and no free occurrence of any other variable) is uniquely satisfied by A, then the description $\iota x\Sigma(x)$ must refer to A.[41]

§13. *Gödel's slingshot in Quinean format*

It is now easy to set out in Quinean connective format, a slingshot proof based on suggestions made by Gödel's (1944).[42]

In natural language there appear to be ways of reorganising a sentence, or converting it into a related sentence, without changing meaning, or at least without changing meaning in any philosophically interesting sense. For example, double negation, passivisation, and topicalisation will convert (53) into sentences that seem to have the same truth condition:

(53) Cicero denounced Catiline
   a. It is false that Cicero did not denounce Catiline
   b. Catiline was denounced by Cicero
   c. It was *Cicero* who denounced Catiline.

Reorganisations and conversions not entirely dissimilar to these grammatical processes are sometimes employed in logic and semantics, perhaps the most common being the inference rule DNN (double negation) and λ-CONV (*lambda* conversion) in roughly the sense of Church (1940, 1944). On Church's account, $\lambda x\phi$,

$\iota x \phi, \hat{x}\phi, \mu x \phi$, etc. are all singular terms. For purposes of exposition and to avoid distracting side-issues about the status of complex singular terms, I want to adopt a use of λx that is found in much contemporary work in semantics and differs in a harmless way from Church's usage: where ϕ is a formula $\lambda x \phi$ will be a one-place predicate; and where α is a singular term, $(\lambda x \phi)\alpha$ will be a formula. However, I follow Church in introducing λ-conversion by way of inference rules.

For certain purposes the extensional sentence $(Fa \cdot Ga)$ might be rendered as $(\lambda x(Fx \cdot Gx))a$, which, depending upon one's taste, can be read as (a) "a is something that is both F and G"; (b) "the class of things that are both F and G contains a"; or (c) "the property of being both F and G is a property a has." For concreteness, let us think of λ-conversions as sanctioned by two rules of inference, λ-INTRODUCTION and λ-ELIMINATION, where α is a singular term and x is a variable:

$$\lambda\text{-INTR:} \quad \frac{T[\Sigma(x/\alpha)]}{T[(\lambda x \Sigma(x))\alpha]}$$

$$\lambda\text{-ELIM:} \quad \frac{T[(\lambda x \Sigma(x))\alpha]}{T[\Sigma(x/\alpha)].}$$

Here $\Sigma(x)$ is an extensional formula containing *at least one* occurrence of the variable x, $\Sigma(x/\alpha)$ is the result of replacing *every* occurrence of the variable x in $\Sigma(x)$ by the (closed) singular term α (there may be other occurrences of α in $\Sigma(x)$ as well), $T[\Sigma(x/\alpha)]$ is an extensional sentence containing $\Sigma(x/\alpha)$ (which may be just $\Sigma(x/\alpha)$ itself), and $T[(\lambda x \Sigma(x))\alpha]$ is a sentence in which $(\lambda x \Sigma(x))$ has minimal scope. λ-INTR and λ-ELIM are valid rules of inference in extensional contexts. As shorthand for this, let us say that extensional contexts are $+\lambda$-INTR and $+\lambda$-ELIM. Continuing with the useful extension of terminology introduced earlier in connection with principles of substitution, let us say that any extensional connective $\circledast$ is $+\lambda$-INTR and $+\lambda$-ELIM in the sense that $\circledast$ permits the use of λ-INTR and λ-ELIM on any sentence ϕ within its scope (assuming, of course, that ϕ does not occur within the scope of any nonextensional connective). More precisely:

An n-place connective $\circledast$ is $+\lambda$-INTR iff in any sentence $\circledast(...T[\Sigma(x/\alpha)]...)$ in which $T[\Sigma(x/\alpha)]$ is an extensional sentence occurring as an operand of $\circledast$ and containing $\Sigma(x/\alpha)$, replacing $T[\Sigma(x/\alpha)]$ by $T[(\lambda x \Sigma(x))\alpha]$ in the original sentence $\circledast(...T[\Sigma(x/\alpha)]...)$ yields a sentence $\circledast(...T[(\lambda x \Sigma(x))\alpha]...)$ with the same truth-value as the original. (*Mutatis mutandis* for $+\lambda$-ELIM.)

No residual issues concerning scope arise: scopes remain constant with respect to $\circledast$. When a connective is both $+\lambda$-INTR and $+\lambda$-ELIM, let us say that it is $+\lambda$-CONV. On the weakest reading of λ-expressions, i.e. (a) above, those who make use of such expressions will view extensional connectives as $+\lambda$-CONV, just as they will view them as +PSST, +PSME, and +ι-SUBS.

With Gödel's (1944) thoughts about the tight relationship between Fa and $a = \iota x(x=a \cdot Fx)$ in mind, I want now to draw up two similar inference principles involving the description-operator, principles I shall call ι-INTRODUCTION and ι-ELIMINATION:

ι-INTR: $\dfrac{T[\Sigma(x/\alpha)]}{T[\alpha=\iota x(x=\alpha \cdot \Sigma(x))]}$

ι-ELIM: $\dfrac{T[\alpha=\iota x(x=\alpha \cdot \Sigma(x))]}{T[\Sigma(x/\alpha)].}$

Here $\Sigma(x)$ is an extensional formula containing *at least one* occurrence of the variable x, $\Sigma(x/\alpha)$ is the (result of replacing *every* occurrence of the variable x in $\Sigma(x)$ by the (closed) singular term α (there may be other occurrences of α in $\Sigma(x)$ as well), $T[\Sigma(x/\alpha)]$ is an extensional sentence containing $\Sigma(x/\alpha)$ (it may be just $\Sigma(x/\alpha)$ itself), and $T[\alpha=\iota x(x=\alpha \cdot \Sigma(x))]$ is a sentence in which $\iota x(x=\alpha \cdot \Sigma(x))$ has minimal scope.

Notice that this wording ensures that (54a) and (54b) can *both* be inferred from $\sim Fa$ (and vice versa):

(54) a. $a = \iota x(x=a \cdot \sim Fx)$
      b. $\sim a = \iota x(x=a \cdot Fx).$

This is because $\Sigma(x/\alpha)$ is extensional and can be taken as either Fa or $\sim Fa$.

ι-INTR and ι-ELIM are valid rules of inference in extensional contexts. As shorthand for this, let us say that extensional contexts are +ι-INTR and +ι-ELIM. (Any adequate theory of descriptions must be compatible with this fact, as Russell's theory is. On Russell's account, the fact that extensional contexts are +ι-INTR and +ι-ELIM follows immediately from the fact that extensional contexts are +PSLE). Continuing with our useful extension of terminology, let us say that any extensional connective ❽ is +ι-INTR and +ι-ELIM in the sense that ❽ permits the use of ι-INTR and ι-ELIM on any sentence φ within its scope (assuming, of course, that φ does not occur within the scope of any nonextensional connective). More precisely:

An n-place connective ❽ is +ι-INTR iff in any sentence ❽$(...T[\Sigma(x/\alpha)]...)$ in which $T[\Sigma(x/\alpha)]$ is an extensional sentence occurring as an operand of ❽ and containing $\Sigma(x/\alpha)$, replacing the $T[\Sigma(x/\alpha)]$ by $T[\alpha=\iota x(x=\alpha \cdot \Sigma(x))]$ in the original sentence ❽$(...T[\Sigma(x/\alpha)]...)$ produces a sentence ❽$(...T[\alpha=\iota x(x=\alpha \cdot \Sigma(x)]...)$ with the same truth-value as the original. (*Mutatis mutandis* for +ι-ELIM.)

Again no residual issues concerning scope arise: scopes remain constant with respect to ❽. When a context or connective is both +ι-INTR and +ι-ELIM, let us say that it is +ι-CONV (it permits ι-conversion *s.v.*).[43]

In just a moment I am going to convert Gödel's hints for constructing a slingshot into a proof that has interesting consequences for the semantics of natural language and for any system that is meant to contain definite descriptions and identity. But first I want to provide a short proof of something that should already be worrying.

Let ❽ be an *arbitrary* one-place connective that is +ι-CONV and +ι-SUBS (and also +DNN). Let "❽+ι-CONV" be shorthand for "the assumption that ❽ is +ι-CONV", and so on. From the premises $\sim Fa$ and ❽(Fa) we can now derive the startling conclusion that ❽$(\sim a=a)$:

1	[1]	$\sim Fa$	premiss
2	[2]	$●(Fa)$	premiss
2	[3]	$●(\sim\sim Fa)$	2, $●$+DNN
2	[4]	$●(\sim a=\iota x(x=a \cdot \sim Fx))$	3, $●$+ι-CONV
1	[5]	$a=\iota x(x=a \cdot \sim Fx)$	1, ι-CONV
1,2	[6]	$●(\sim a=a)$	4, 5, $●$+ι-SUBS.

I say "startling" because this holds of *any* connective that is +ι-CONV and +ι-SUBS. Where $●$ is an extensional connective this creates no difficulty (e.g., if $●$ is $\sim$ then all that is proved is $a=a$). But the proof should worry anyone contemplating treating a non-extensional connective as +ι-CONV and +ι-SUBS.

The magnitude of the problem comes out clearly in a proof $\Im$ based on Gödel's slingshot. $\Im$ is a proof, in four parts, $\Im_1$–$\Im_4$, from premises $Fa \equiv Gb$ and $●(Fa)$ to the conclusion $●(Gb)$. The general idea should be clear by the end of $\Im_1$.

$\Im_1$: From premises Fa, $a \neq b$, Gb, and $●(Fa)$ to conclusion $●(Gb)$.

1	[1]	Fa	premiss
2	[2]	$a \neq b$	premiss
3	[3]	Gb	premiss
1	[4]	$a=\iota x(x=a \cdot Fx)$	1, ι-CONV
2	[5]	$a=\iota x(x=a \cdot x \neq b)$	2, ι-CONV
2	[6]	$b=\iota x(x=b \cdot x \neq a)$	2, ι-CONV
3	[7]	$b=\iota x(x=b \cdot Gx)$	3, ι-CONV
1,2	[8]	$\iota x(x=a \cdot Fx) = \iota x(x=a \cdot x \neq b)$	4,5, ι-SUBS
2,3	[9]	$\iota x(x=b \cdot Gx) = \iota x(x=b \cdot x \neq a)$	6,7, ι-SUBS
10	[10]	$●(Fa)$	premiss
10	[11]	$●(a=\iota x(x=a \cdot Fx))$	10, $●$+ι-CONV
1,2,10	[12]	$●(a=\iota x(x=a \cdot x \neq b))$	11,8, $●$+ι-SUBS
1,2,10	[13]	$●(a \neq b)$	12, $●$+ι-CONV
1,2,10	[14]	$●(b=\iota x(x=b \cdot x \neq a))$	13, $●$+ι-CONV
1,2,3,10	[15]	$●(b=\iota x(x=b \cdot Gx))$	14,9, $●$+ι-SUBS
1,2,3,10	[16]	$●(Gb)$	15, $●$+ι-CONV

$\Im_2$: *Mutatis mutandis* where premise [2] is '$a=b$' rather than '$a \neq b$', though of course a shorter proof can be constructed when '$a=b$' is used.[44]

Putting $\Im_1$ and $\Im_2$ together we have a proof $\Im_1\Im_2$: if $●$ is +ι-SUBS and +ι-CONV, then it permits the substitution *s.v.* of (atomic) truths for (atomic) truths within its scope (e.g. Fa and Gb).

The attraction of $\Im_1$ and $\Im_2$ is twofold. Firstly, they are in Quinean connective format and have as their entries sentences that are meant to be equivalent only in truth-value with their predecessors, not equivalent in any richer respect (e.g., logically equivalent or synonymous). Secondly, in all crucial respects they are semantically neutral. Most importantly, unlike Quine's connective proof, $\Im_1$ and $\Im_2$ *require no supplementation with a precise semantics for definite descriptions*. This is because (a) they nowhere appeal to logical equivalences involving descriptions, and (b) the apparent identities involving descriptions – e.g. at lines [8] and [9] of $\Im_1$ are obtained by appeal to ι-SUBS not by any *specific* assumption about the semantics of descriptions.

It should also be noted that $\Im_1$ and $\Im_2$ nowhere assume "direct" (or "indirect") reference, semantic "innocence" (or "guilt"), or any facts about the semantics of ⊗ beyond the hypothesis being investigated, viz. that it is a one-place connective that is +ι-SUBS and +ι-CONV. What the proofs shows, without any qualification whatsoever, is that if ⊗ is +ι-SUBS and +ι-CONV, then it permits the substitution *s.v.* of (atomic) truths for (atomic) truths within its scope (e.g. *Fa* and *Gb*).

Immediate philosophical consequences of $\Im_1\Im_2$ can be drawn by interpreting ⊗ (...) as, e.g., "the fact that ϕ = the fact that (...)", "the statement that ϕ corresponds to the fact that (...)", "the fact that ϕ caused it to be the case that (...)" or any other connective that, at least *prima facie*, someone might be tempted to view as +ι-SUBS and +ι-CONV. The friend of facts needs a theory according to which these connectives are either −ι-SUBS or −ι-CONV.[45]

It ought to be clear intuitively that we have, in fact proved something stronger, viz. that if ⊗ is +ι-SUBS and +ι-CONV, then it permits not only the substitution *s.v.* of truths for truths within its scope, but also falsehoods for falsehoods, i.e. that if ⊗ is +ι-SUBS and +ι-CONV it is extensional, as I claimed in 1995, and as Dever and I claimed in 1997. Most commentators have accepted the stronger conclusion without further ado; but as some have not, let me complete everything with two more sub-proofs $\Im_3$ and $\Im_4$.

$\Im_3$: We replace the premises *Fa* and *Ga* in $\Im_1$ by ~*Fa* and ~*Gb* to produce a proof of ⊗*Gb* from ~*Fa*, ~*Gb*, $a\neq b$, and ⊗*Fa*:

1	[1]	~*Fa*	premiss
2	[2]	$a\neq b$	premiss
3	[3]	~*Gb*	premiss
1	[4]	$a=\iota x(x=a \cdot \sim Fx)$	1, ι-CONV
2	[5]	$a=\iota x(x=a \cdot x\neq b)$	2, ι-CONV
2	[6]	$b=\iota x(x=b \cdot x\neq a)$	2, ι-CONV
3	[7]	$b=\iota x(x=b \cdot \sim Gx)$	3, ι-CONV
8	[8]	⊗(*Fa*)	premiss
8	[9]	⊗(~~*Fa*)	8, ⊗+DNN
8	[10]	⊗($\sim a=\iota x(x=a \cdot \sim Fx)$)	9, ⊗+ι-CONV
1,8	[11]	⊗($\sim a=a$)	4, 10, ⊗+ι-SUBS
1,2,8	[12]	⊗($\sim a=\iota x(x=a \cdot x\neq b)$)	5, 11, ⊗+ι-SUBS
1,2,8	[13]	⊗($\sim a\neq b$)	12, ⊗+ι-CONV
1,2,8	[14]	⊗($\sim b=\iota x(x=b \cdot x\neq a)$)	13, ⊗+ι-CONV
1,2,3,8	[15]	⊗($\sim b=b$)	6, 14, ⊗+ι-SUBS
1,2,3,8	[16]	⊗($\sim b=\iota x(x=b \cdot \sim Gx)$)	7, 15, ⊗+ι-SUBS
1,2,3,8	[17]	⊗(~~*Gb*)	16, ⊗+ι-CONV
1,2,3,8	[18]	⊗(*Gb*)	17, ⊗+DNN

$\Im_4$: *Mutatis mutandis* where premise [2] is '$a=b$' rather than '$a\neq b$', though again a shorter proof can be constructed when '$a=b$' is used.

Putting $\mathfrak{I}_1\mathfrak{I}_2$ and $\mathfrak{I}_3\mathfrak{I}_4$ together, we have our proof $\mathfrak{I}$: if a connective ❸ is +ι-CONV and +ι-SUBS then it is *extensional*. (Again, this assumes, with Gödel, that every sentence can be brought into subject-predicate form. If this is problematic, the argument demonstrates the technically weaker – but equally important – conclusion that if ❸ is +ι-SUBS and +ι-CONV, then it is extensional with respect to *atomic* sentences.)

It is important to distinguish this formal consequence of $\mathfrak{I}$ from any philosophical implications – concerning, say, facts, causation, necessity, or action – that it may have. It ought to be clear that metaphysical issues about (e.g.) whether there are facts or whether there are any general facts, and semantic issues about whether descriptions are singular terms, whether singular terms are directly referential, or whether semantic innocence can be maintained, are *not* issues concerning the proof (which is not to say, of course, that interesting implications for such issues might not emerge from reflecting on the proof in the context of this or that philosophical theory).

The relation between the sub-proof $\mathfrak{I}_1\mathfrak{I}_2$ and the proof Gödel originally suggested comes into focus if ❸(...) is interpreted as "the fact that Fa = the fact that (...)" or as "the sentence 'Fa' corresponds to the fact that (...)".

How worrying is $\mathfrak{I}$ for friends of facts, non-extensional logics, and purportedly non-extensional connectives in natural language? That is something I cannot take up here. But one formal point needs to be made. $\mathfrak{I}$ does *not* show that a connective is extensional if it is both +PSST and +ι-CONV, i.e. it does *not* directly demonstrate that the following combination of features is inconsistent:

(60) +ι-CONV   +PSST   −PSME.

But the inconsistency of (60) *would* be shown if it were possible to prove that any connective that is +PSST is also +ι-SUBS, or prove that any connective with the three features in (60) must also be +ι-SUBS. I know of no attempts to construct such proofs; indeed, if descriptions are Russellian, such proofs cannot be constructed. So anyone hoping to use $\mathfrak{I}$ as part of an argument against the consistency of (60) needs to (i) provide a viable treatment of descriptions according to which they are (a) singular terms (and hence subject to PSST), and (b) still the plausible inputs and outputs of ι-CONV, and then (ii) construct a proof exactly like the one above except that it appeals to PSST (rather than ι-SUBS) in the obvious places.

§14. Descriptions and equivalence

Although there are good reasons for thinking that descriptions should be analysed in accordance with Russell's theory, as devices of quantification rather than reference, there are a number of superficially attractive referential treatments of descriptions on the market. For the purposes of this essay, it will not be necessary to pronounce on most of the virtues and vices of this or that referential theory: it is necessary only to examine the commitments of such theories in connection with logical and Gödelian equivalence, for the aim is to uncover the consequences of employing referential theories (satisfying a minimal condition any such theory must satisfy) in connection with the Quinean connective versions of slingshot arguments already considered. Quine was seeking a stronger conclusion than the one that can be obtained if descriptions are Russellian, and so was Davidson (1980). We owe it to them at least to see if the stronger conclusion is forthcoming on a referential treatment; and we owe it to ourselves to see if a stronger result can be obtained from a connective version of Gödel's slingshot. The main issue is whether the logical and Gödelian equivalences the respective proofs invoke can be sustained if descriptions are referential.

If the intuitive meaning of the description operator is to be honoured, one definite condition must be satisfied by any referential treatment: if a formula $\Sigma(x)$ containing at least one occurrence of the variable x (and no free occurrence of any other variable) is uniquely satisfied by A, then the description $\iota x\Sigma(x)$ must refer to A. But the wording of this condition brings out questions that any referential treatment must answer, questions that can be sharpened by reflecting on the logical simplicity afforded by Russell's treatment.

Russell's theory provides straightforward accounts of sentences containing descriptions whose matrices are not uniquely satisfied, so-called "improper" descriptions. Refining our terminology, let us say that a description $\iota x\Sigma(x)$ is *proper* (according to some model M) if, and only if, its matrix $\Sigma(x)$ is (on M's interpretation) true of exactly one item in the domain over which the variables of quantification range, and *improper* otherwise.[46] Let us assume, then, that we are dealing only with referential treatments that generate no serious problems for proper descriptions (if there are such theories): a model M interprets a proper description $\iota x\Sigma(x)$ as referring to the unique object satisfying $\Sigma(x)$.

But what of improper descriptions? An adequate referential treatment must say something about them, and there are various competing approaches in the literature, or else constructible on the basis of existing informal suggestions. A good deal of the relevant work here has been done by Carnap (1947) and Taylor (1985), from whom I shall draw liberally.

(a) Hilbert and Bernays. In the system of Hilbert and Bernays (1934), a description $\iota x\Sigma(x)$ can be used only after it has been proved proper, *i.e.* only after (61) has been proved:

(61) $\exists x(\forall y(\Sigma(x/y) \equiv y=x)$

Although this treatment may be useful for certain mathematical purposes, as Quine (1940), Carnap (1947), and Scott (1967) have pointed out, there are insurmountable problems involved in viewing it as a treatment of descriptions in any interesting fragment of natural language. Firstly, since the class of well-formed formulae will not be recursive, the question of whether a string of symbols containing the sub-string 'ιx' is a formula depending not only upon a set of syntactical rules but also on, for example, matters of logic and the "contingency of facts". And secondly, utterances of many sentences of natural language containing improper descriptions (or descriptions not known to be proper) are straightforwardly true or false (e.g. 'last night Quine dined with the king of France') or straightforwardly used to conjecture, as even Strawson was finally forced to concede (see below). There would appear to be no prospect, then, of using Hilbert and Bernays' treatment in connection with descriptions belonging to any interesting fragment of natural language.

Furthermore, since the use of a description of the form of (62)

(62) $\iota x(x=a \cdot \Sigma)$

is permitted on Hilbert and Bernays' treatment only if either Σ or $\Sigma(x/a)$ is provable, the adoption of this treatment will not bring about stronger results in connection with the connective proofs based on the slingshots of Church-Quine and Gödel. Certainly the reinterpreted proofs would not show that there can be no connective with the combinations of features given in (63) (the Church version) or (64) (the Gödel version):

(63) +PSST   −PSME   +PSLE
(64) +PSST   −PSME   +ι-CONV.

The proofs would show only that connectives satisfying (63) and (64) also permit the substitution s.v. of logical truths for logical truths.[47]

As far as the proof based on the Church-Quine slingshot is concerned, this result is already a direct consequence of one of its assumptions, viz. that the connective +PSLE, so the new proof is of no interest.

The connective version of Gödel's slingshot would demonstrate something of mild interest: if ❽ is +PSST and +ι-CONV then it also permits the inter-substitution s.v. of logical truths. This is not a direct consequence of any particular assumption made by the argument. On standard Tarskian conceptions of logical truth, logical consequence, and logical equivalence, if ❽ permits the inter-substitution of logical truths s.v., then it must also be +PSLE. Thus it would seem that a mildly interesting result emerges on Hilbert and Bernays' treatment of descriptions, viz. that the slingshots of Church-Quine and Gödel are basically equivalent (every Gödelian equivalence being a logical equivalence).

To sum up, on the plausible assumption that any connective that permits the inter-substitution s.v. of logical truths is +PSLE, all that can be demonstrated by adopting Hilbert and Bernays' (antecedently implausible) treatment of definite descriptions is that no connective can have the following combination of features:

(65) +ι-CONV   +PSST   −PSLE.

There would still be no collapse of non-extensional connectives into the class of extensional connectives.[48]

(b) Fregean Theories (I): Chosen Objects. Rather different approaches to improper descriptions have been inspired by Frege, who thought it an imperfection of languages that they contain apparent singular terms that fail to refer. Frege (1892) suggests that a description is a "compound proper name" and as such,

---

… must actually always be assured a reference, by means of a special stipulation, e.g. by the convention that it shall count as referring to 0 when the concept applies to no object or to more than one (p. 71).

Elsewhere, Frege (1893) suggests an alternative treatment according to which an improper description refers to *the class of entities satisfying its matrix* (thus all empty descriptions refer to the empty class). Within the context of Frege's overall theory of reference, this looks like an improvement as far as compositionality and extensionality are concerned.

The matrix of a definite description, on Frege's account, is a concept expression and, as such, it is paired – though not directly through Frege's semantics – with a class of entities, its extension. As required by minimally respecting the intuitive meaning of the definite article, where the extension of a matrix ϕ is one-membered, that one member qualifies as the referent of the resulting description $\iota x\phi$; if the class in question is anything other than one-membered, *the class itself* serves as the referent. So, on this account, there is a straightforward extensionality constraint governing *all* definite descriptions: if $\Sigma(x)$ and $\Sigma'(x)$ are satisfied by the same elements, then $\iota x\Sigma(x)$ and $\iota x\Sigma'(x)$ have the same reference.[49] Once the suggestion has been made that empty descriptions refer to the empty class, it would be misleading to say, with Quine (1940, p. 149) that there is something "arbitrary" about Frege's suggestion that those "uninteresting" descriptions whose matrices are satisfied by more than one entity refer to the class of things satisfying the matrix: for this is *exactly* the suggestion for empty descriptions.

Frege's suggestions have been developed in a number of promising ways, notably by Carnap (1947), Scott (1967), and Grandy (1972). Carnap's position on improper descriptions might be summarised model-theoretically as follows: in each model M, some arbitrary element $*_M$ in the domain (over which the variables of quantification range) serves as the referent (in M) of *all* descriptions that are improper with respect to M.[50]

On this treatment of descriptions, about the only thing that Gödel's slingshot demonstrates is just how bad a treatment of descriptions it is (it was hardly a plausible treatment of descriptions in natural language anyway). Consider a model M in which Fa is false and the singular term 'a' refers to $*_M$. (The existence of such a model presupposes that *which* element of the domain is functioning as the referent of improper descriptions is one feature (*i.e.* assignment) used in individuating models. This is surely the way to make sense of the idea that an "arbitrarily" chosen entity in the domain serves as the referent of improper descriptions.) In M, (Γ) is false while (Γ') is true:

(Γ) Fa
(Γ') $a=\iota x(x=a \cdot Fx)$.

So on this model-theoretic treatment of descriptions, neither +ι-INTR nor +ι-ELIM is truth-preserving in *truth-functional* contexts. Surely this finishes off Carnap's account once and for all; but even if some die-hard Carnapian semanticist insists on the theory, the fact that it permits (Γ) and (Γ') to differ in truth-value means it cannot be used in conjunction with a connective version of Gödel's slingshot to demonstrate any conclusion more interesting than the one we have already established.

The problem just raised concerning +ι-CONV could be eradicated by a special stipulation to the effect that only those singular terms that are also definite descriptions can be assigned $*_M$ as their reference. If the resulting treatment of descriptions turns out to be the correct one – remember descriptions are singular terms on this treatment – then the connective version of Gödel's slingshot demonstrates that no S-connective can have the following combination of features:

(66) +ι-CONV   +PSST   –PSME.[51]

The situation is slightly different when it comes to the connective version of Church's slingshot. As pointed out by Taylor (1985), on the Carnapian treatment of descriptions, ($\triangle$) and ($\triangle'$) are *not* logically equivalent:

($\triangle$) ϕ
($\triangle'$) $a=\iota x(x=a \cdot \phi)$.

Consider a model M in which ϕ is false, and the singular term 'a' refers to $*_M$. ($\triangle'$) will be true in M; and since ($\triangle$) and ($\triangle'$) have different truth values in M, they do not have the same truth-value in all models, hence they are not logically equivalent. So the connective version of Church-Quine slingshot itself collapses if descriptions are treated in the strict Carnapian way. Again the required logical equivalence could be regained by stipulating that only descriptions can refer, in M, to $*_M$; and with this stipulation the connective version of Church's slingshot would now produce a result on the modified Carnapian treatment, viz. that no connective can have the combination of features given in (67):

(67) +PSLE   +PSST   –PSME.[52]

The failure of the desired logical equivalence on the original Carnapian treatment of descriptions suggests to Taylor a modified slingshot. The idea, put into connective format, is to tack "$a \neq \iota x(x \neq x)$" onto ϕ and derive "$❽(\psi \cdot a \neq \iota x(x \neq x))$" from "$❽(\phi \cdot a \neq \iota x(x \neq x))$" and "$(\phi \equiv \psi)$" in exactly the same way as "$❽(\psi)$" is meant to be derived from "$❽(\phi)$" and "$(\phi \equiv \psi)$" using the original proof. The beauty of Taylor's version of the slingshot is that it avoids any special stipulation concerning which terms can refer to $*_M$, *and* it guarantees the logical equivalence of ($\triangle$) and ($\triangle'$).[53]

Taylor's strategy for avoiding this slingshot is to define a notion of "tight" logical equivalence, and then maintain that tight logical equivalents stand for the same state-of-affairs whereas mere logical equivalents need not. This involves defining a class of expressions that might be called the "tight" logical constants, a class that includes the quantifiers and truth-functional connectives but not the description operator or the identity sign. Whilst I have sympathy with Taylor's view that the standard notion of logical equivalence is still somewhat murky, I am not sure that he really improves matters by bringing tight equivalence into the picture. More importantly, Taylor's manoeuvres do not, by themselves, allow him to avoid *Gödel's* slingshot, which makes no appeal to logical equivalence whether standard or tight.[54]

(c) *Fregean Theories (II): Pseudo-domains.* Fregean treatments of descriptions have also been proposed by Scott (1967) and Grandy (1972). On these treatments, bound variables range over a domain D, but the values of singular terms and free variables may lie in a so-called "pseudo-domain" $D°$, stipulated to be disjoint from D and non-empty. An improper description is given a value in $D°$, "thereby emphasising its impropriety" as Scott says.[55] The situation with respect to Gödel's slingshot is much as before. Consider a model M in which 'Fa' is false and the singular term 'a' refers to $*_M°$, the pseudo-object selected from $D°$ to be the referent of descriptions that are improper with respect to M. In such a model, (Γ) is false while (Γ') is true. So, again, we have treatments of descriptions according to which truthfunctional contexts are neither +ι-ELIM nor +ι-INTR. So while these treatments certainly solve some of the problems with which Scott and Grandy are grappling, the fact that they permit (Γ) and (Γ') to differ in truth-value ensures that they cannot be used in conjunction with Gödel's slingshot to show anything new about connectives and also suggests strongly that they are inadequate as treatments of descriptions in natural language.

Church's slingshot fares no better because ($\triangle$) and ($\triangle'$) are *not* logically equivalent. Scott's treatment declares ($\triangle$) false and ($\triangle'$) true in any model M in which φ is false and 'a' refers to $*_M°$; Grandy's declares ($\triangle$) false and ($\triangle'$) true in any model M in which φ is false and 'a' refers to the referent of descriptions whose matrix has the *intension* of '$(x=a \cdot φ)$'. Thus Church's slingshot falls apart if descriptions are treated in the ways Scott and Grandy suggest.

Again, tinkering with the class of expressions that can take $*_M°$ (or anything else in $D°$) as a value would alter things, but such a move would constitute a clear departure from the theories of Scott and Grandy. However, such tinkering may be what Olson (1987) has in mind when he suggests that ($\triangle$) and ($\triangle'$) are logically equivalent upon a "Fregean" theory of descriptions according to which an improper description refers to "some object outside the universe" (p. 84, footnote 9). Assuming that Olson has not simply overlooked models in which 'a' refers to $*_M°$, he must have in mind a semantics quite different from those envisaged by Scott and Grandy (not to mention Frege). It is a feature of the Scott-Grandy systems that the values of singular terms (but not bound variables) may lie in $D°$, and it is this feature that legitimises the selection of an element in $D°$ to serve as the value of a description – a singular term on this proposal – whose matrix is not uniquely satisfied by something in D. So if Olson has in mind a referential semantics according to which ($\triangle$) and ($\triangle'$) *are* logically equivalent, then he must postulate two distinct classes of singular terms, those that can take values in $D°$ and those that cannot, and he must put descriptions into the former class and proper names into the latter. Treating definite descriptions so differently from other singular terms would make the resulting theory less attractive than the Scott-Grandy theories.

(d) Strawsonian Theories. The final treatment of descriptions I want to consider is the one Taylor (1985) produces (but does not endorse) by recasting some of Strawson's (1950b) views in model-theoretic terms.[56] The key features of this account are (a) the rejection of bivalence: a sentence containing a description that is improper with respect to a model M will lack a truth-value in M; and (b) a refinement of the notion of logical equivalence to take into account cases in which

sentences lack truth-values: two sentences are *logically equivalent* if and only if they have the same truth-value *in every model in which they both have a truth-value*. (It would be very odd for *Quine* to pursue such an approach as he has consistently opposed truth-value gaps and praised Russell's Theory of Descriptions for eliminating them where descriptions and names are concerned.) On this account, ($\triangle$) and ($\triangle'$) are logically equivalent: in any model in which ($\triangle$) is true, so is ($\triangle'$); in any model in which ($\triangle$) is either false or lacks a truth-value, the description $\iota x(x{=}a \cdot \phi)$ is improper and so ($\triangle'$) lacks a truth value; so every model in which ($\triangle$) and ($\triangle'$) both have a truth-value is a model in which they are both true; thus they are ("Strawsonian") logical equivalents.

On this treatment of descriptions, the Church-Quine slingshot appears to demonstrate that no connective can have the following combination of features:

(69) +PSST   +PSLE   −PSME.

However, the rejection of bivalence and subsequent refining of logical equivalence bring up important questions. Firstly, logical equivalence is standardly taken to be tightly, if not definitionally, connected to other notions, for example *logical consequence*, *logical truth*, and *material equivalence*. On the proposed refinement, is there pressure to redefine the notion of logical truth (from the standard (i) "$\models \phi$ if, and only if, ϕ is true in all models," to (ii) "$\models \phi$ if and only if ϕ is true in all models in which ϕ has a truth-value")? Given that standardly $\phi \models\dashv \psi$ if and only if $\phi \models \psi$ and $\psi \models \phi$, is there pressure to redefine the notion of logical consequence? And given that standardly if $\phi \models\dashv \psi$ then $\models (\phi \equiv \psi)$, should the truth-table for '$\equiv$' be the one given by Halldén (1949) and Körner (1960) for certain logics in which bivalence is rejected – "($\phi \equiv \psi$) is true if, and only if, ϕ and ψ are either both true or both false; without (standard) truth-value otherwise" – or should it differ in some way? And what of the truth-table for negation? I do not mean to be insisting that all of these (and related) questions cannot be answered together to produce a consistent and attractive package; I simply want to point out that such questions need to be answered by anyone who wants to give up bivalence and refine logical consequence in the way Taylor suggests.

Secondly, and more importantly, even if there is no formal problem with the account, it appears to be inadequate to the task of providing an account of descriptions in natural language. Put bluntly, there are just too many (utterances of) sentences of natural language that seem to have clear truth-values despite containing improper descriptions. I have discussed such cases at length elsewhere, so I will be brief here. Utterances of (70)-(72) made today would surely be true, false, and false respectively precisely because there is no king of France:

(70) The king of France does not exist
(71) Quine is the King of France
(72) The king of France is not bald since there is no king of France.

Perhaps clever theories of negation, existence, identity, and predication could help the Strawsonian here, but they could not help with a sentence like (73):

(73) Last night Quine dined with the king of France.

And appeals to a semantically relevant asymmetry between singular terms in subject position and those that form part of a predicate phrase might help with (73), but they will not help with (74) and (75):

(74) The king of France borrowed Quine's car last night.
(75) The king of France shot himself last night.

Descriptions occurring in non-extensional contexts create similar problems. I may say something true or false by uttering (76) or (77):

(76) The first person to land on Mars in 1990 might have been from Akron
(77) Kurt thinks the largest prime lies between 10^{27} and 10^{31}.

At the very least, then, we must reject the view that the use of an empty description *always* results in an utterance without a (standard) truth-value. Strawson (1964, 1972, 1986) came to realise this; and in an attempt to reduce the number of incorrect predictions made by his earlier theory, he suggests that sometimes the presence of an improper description renders the proposition expressed false and at other times it prevents a proposition from being expressed at all. Since nothing appears to turn on structural or logical facts about the sentence used, Strawson suggests restricting the "truth-value gap" result by appealing to the *topic of discourse*. Once the Strawsonian model-theorist makes this concession, even if a workable semantics can be salvaged, the logical equivalence that is being sought must surely drift away.

   The status of slingshot arguments is clearly a complex matter if definite descriptions are treated as singular terms. On Hilbert and Bernays' treatment, Gödel's slingshot demonstrates something of mild interest, but the Church-Quine version proves only the truth of one of its own premises. On Fregean treatments, according to which improper descriptions refer, by stipulation, to some entity in the domain D of quantification or to some entity in a nonempty and disjoint "pseudo-domain" $D°$, both slingshots demonstrate something of significance only if descriptions are treated differently from other singular terms, a move which robs the placement of descriptions into the class of singular terms of some of its appeal and has formal consequences that still need to be explored. (But as we saw, a modified slingshot, due to Taylor, appears to hit its target without such a contortion.) The full range of consequences of the model-theoretic Strawsonian treatment (which abandons bivalence) also needs to be explored. On the assumption that the treatment is coherent, both slingshots hit their targets, but the treatment itself, even if coherent, does not come at all close to succeeding as an account of descriptions in natural language.

   At this juncture it is worth reminding ourselves of the force of slingshot arguments on a standard Russellian analysis of descriptions. The Church-Quine slingshot succeeds in showing only that any *S*-connective that is +PSLE and +ι-SUBS is also +PSME. Gödel's, by contrast, shows something more worrying: any connective that is +ι-CONV and +ι-SUBS is also +PSME (this is more worrying on the obvious assumption that every "Gödelian equivalence", as given by ι-CONV, is also a logical equivalence, but not vice versa). This fact will be of interest if we find connectives or contexts that are −PSLE, +ι-SUBS, and +ι-CONV, because defenders of such connectives will have no recourse to the most common rejoinder to slingshot arguments: denying that the connective in question is +PSLE, a rejoinder made explicitly by defenders of facts and situations such as Barwise and Perry (1981, 1983), Searle (1995), and Taylor (1985).

§15. Concluding remarks

A few final remarks about $\mathfrak{J}$. As already noted, Whitehead and Russell prove that if ❽ is +PSME, then it is also +ι-SUBS. But what about the converse, which Whitehead and Russell do *not* prove? As far as I am aware, the proof just presented is as close as anyone has come to confirming it, but there is still a way to go. One way to get all the way would be to show that if ❽ is +ι-SUBS, then it is also +ι-CONV. But I hazard this cannot be proved in a way that is sufficiently neutral on the semantics of definite descriptions.

Proofs based on slingshots are the central components of watertight deductive arguments; they show that an inconsistency results when one posits a non-extensional connective that freely permits the use of inference principles involving descriptions within its scope. This much we now know from examining Quinean connective versions of slingshots. The inconsistency arises, roughly speaking, because descriptions and abstracts (as standardly understood) contain *formulae* as proper parts; by permitting the interchange of such devices when their contained formulae are satisfied by the same object, one is essentially permitting the interchange of formulae; and once a weak additional inference principle is assumed − and as I have argued it is in the precise character of the second principle that proofs based on Gödel's slingshot are superior to those based on Church's − the formulae in question can be drawn out of their *iota*-governed contexts to make the purportedly non-truth-functional connectives provably truth-functional.

As Quine (1941) first pointed out, such substitutions are not truth-preserving in modal contexts. In one respect, to notice this is to notice very little; but in another, it is to catch sight of something important: an argument based on the assumptions that (i) □ is a non-truth-functional connective, and (ii) □ is +ι-SUBS. It is not exactly an inconsistency that flows from *this* argument but a sentence that seems false ('□(the number of planets > 7)') from premises that are true ('□(9 > 7)' and '9 = the number of planets'). What Quine conjectured was that if one could saddle □ with one more plausible inference principle, perhaps one could derive an outright inconsistency; and if one could make the additional principle sufficiently innocuous, perhaps the inconsistency would generalise to *all* non-truth-functional connectives, not just the modal ones. Following Church (1943), Quine (1953a, 1953c, 1960) appealed to an inference principle for substituting logically equivalent sentences. The failure of a particular non-truth-functional connective to be +ι-SUBS comprised most of what was needed to produce the full argument because descriptions contain formulae as proper parts, a fact that the second inference principle (PSLE) exploits. And the full argument has two obvious advantages: (a) it produces an inconsistency, rather than just counterintuitive truth conditions, and (b) the result is fully general, producing a constraint on *all* non-truth-functional operators, not just modal ones. However the constraint is not the eliminativist one that Quine was seeking. There *can* be non-truth-functional sentence connectives that are +PSLE and +PSST, but not ones that are +PSLE and +ι-SUBS. We have learned a great deal from Quine's bold excursions in philosophical logic.

PART II

QUINE'S CRITIQUE OF QUANTIFIED MODAL LOGIC, THE ORIGINS OF
QUINE'S SLINGSHOT IN THIS CRITIQUE AND IN THE WORK OF CHURCH
AND GÖDEL, AND APPEALS TO THEORIES OF DESCRIPTIONS IN
DISCUSSIONS OF MODALITY

§16. Preliminaries to the discussion of quantified modal statements

Much of what I said about Quine's attack on modality in my 1990 book *Descriptions* was wrong. My elaboration of Smullyan's (1947, 1948) position and my harsh critique of Quine's (1953, 1961, 1969) response to Smullyan were founded on an anachronism: the false belief that by the late 1940s or early 1950s, passing talk of Leibniz and "possible worlds", and suggestions for systems to model temporal, physical, causal, and other modalities were all indicative of a background conception of what we are today inclined to call *metaphysical* necessity, a conception not driven purely by the logico-linguistic considerations underlying *strict* necessity, which was, in fact, Quine's principal target.[57]

In discussing the history of modal logic (as I am about to) matters are made excruciating by terminological distinctions that are common coin today but were not in the 1940s and 1950s, and by terms which are still not used in uniform ways. The main problems concern the use of the adjectives "strict", "logical", "analytic", "necessary", "*a priori*" (and any corresponding adverbs). I will attempt to clarify some of the issues as I proceed. I am going to make three main points in connection with Quine's attack on logical modality. Firstly, the interesting observations about Russellian descriptions in modal contexts made by Church (1942), Carnap (1947), Smullyan (1947, 1948), Marcus (1948), and Fitch (1949, 1950) are just interesting observations: by themselves they are incapable of robbing Quine's (1943, 1947, 1953a, 1953c) critique of its real force. Secondly, an alternative way of dealing with Quine's worries, also suggested by Smullyan and Fitch, presupposes a highly controversial thesis about the semantics of proper names, a thesis explicitly rejected by Quine, Church, and Carnap. Thirdly, Quine (1953a, 1953c) made a *tactical* error in responding to Smullyan the way he did, compounded matters by including in his responses inaccurate charges about the use of Russell's Theory of Descriptions, and further compounded the issues in 1969 remarks and again in 1980 when he revised his response to Smullyan for a new printing of *From a Logical Point of View*.

Some of Smullyan's *observations* – which are essentially Russell's observations about descriptions in attitude contexts carried over to modal contexts – can be reinterpreted as useful *today* because, thanks largely to Kripke (1971, 1972), a *non-linguistic* conception of necessity (a "metaphysical" conception) has come to dominate, and thanks mainly to Kripke and Kaplan (1989) we have theories of reference that, in connection with such a conception of necessity, are supported by arguments showing that a genuine singular term (e.g., a name or a simple demonstrative or indexical) refers to the same thing in every metaphysical possibility in which that thing exists (they are "rigid designators", to use Kripke's label).[58]

It is vital not to confuse Quine's *general* (slingshot) argument against non-extensional connectives with any of his *particular* arguments against the viability of *modal* connectives expressing logical necessity and possibility, although the general argument certainly has its origins in the particular arguments and also in arguments first presented by Church (1943a) and Gödel (1944). I propose to take things chronologically, stopping at key points to mention important developments in logic (due to Gödel, Church, Tarski, Kripke and others) that have a direct bearing on a proper understanding of the nature of the debate about the intelligibility of quantified modal logic and the nature of slingshot arguments that bear on the construction of non-extensional logics quite generally. From 1930 to 1953, we will need to proceed a year at a time (more or less). It will become clear, I think, that Carnap, Church, Fitch, Føllesdal, Gödel, Kripke, Lewis, Marcus, Quine, and Smullyan have all either made subtle errors or else been misread in subtle ways that have engendered deep confusions in today's literature.

As we go back in time, I ask the reader to assume a certain naïveté, to suspend certain distinctions we take for granted today. (However, I will add occasional parenthetical remarks pointing to future developments where useful.) It will be important to bear in mind that for half a century many philosophers and logicians will use two or more of the following adjectives interchangeably in connection with statements: "logical", "tautological", "analytic", "necessary", "*a priori*". Furthermore, some will use two or more of the following interchangeably when discussing modality and necessity: "strict", "logical", and "analytic". All of this can lead to exegetical difficulties. It will be useful to begin in 1910.

§17. 1910–1940: The logical background

1910. In volume I of *Principia Mathematica*, Whitehead and Russell use the English word "implies" in connection with their symbol for the material conditional, $\supset$, which is defined contextually – $(\phi \supset \psi) =_{df} (\sim\phi \vee \psi)$ – as are definite descriptions and class abstracts.

(An important aside. In the terminology of the present essay, Whitehead and Russell's proofs of *14.15 and *14.16 (see Part I) for truth-functional contexts amount to a proof of the following (where ❽ is an *n*-place connective):

(I)   If ❽ is +PSME, then ❽ is also +ι-SUBS.

But what about the converse, which Whitehead and Russell do *not* prove?

(II)   If ❽ is +ι-SUBS, then ❽ is +PSME.

Quine (1941, 1960) will suggest it is true, and Gödel (1944) will come tantalisingly close to making the same suggestion. In due course, someone is sure to come up with a fiendishly simple proof of (ii). Its truth *seems* obvious, particularly if descriptions are treated in accordance with Whitehead and Russell's *14.01 and *14.02.)

1912–17. In various articles, C. I. Lewis argues that the material implication ($\supset$) of *Principia* does not capture the meaning of "if...then", understood as expressing an intuitive relation of *implication* or *consequence*, because it gives rise to some highly counterintuitive results (e.g., ($\phi \supset \psi$) $\vee$ ($\psi \supset \phi$) comes out as a theorem). Lewis proposes a two-place connective (which I shall render as $\Rightarrow$) in the object language that expresses a "stricter" notion of implication, something with the force of "necessarily if ϕ then ψ", where the notion of necessity is to be understood in a "logical" or "semantic" sense that is not well articulated, but which is meant, at least in the first instance, to give substance to a notion of implication that coincides with deducibility: ($\phi \Rightarrow \psi$) can be read as "the conclusion that ψ is logically deducible from the premise that ϕ". (A similar idea was suggested by MacColl in 1906.)

Norbert Wiener (1916) argues that Lewis's criticisms of Russell's logic and interpretation of $\supset$ are both baseless and irrelevant to the construction of Lewis's own systems:

> Mr Lewis's arguments against Mr Russell have little logical cogency, and one feels that they were developed to give an excuse for his constructive work in the definition of his system of "strict implication," which really requires no such apology for its existence....[the logical worth] of this exceedingly valuable and interesting piece of work is utterly independent of an acceptance of Mr Lewis's contentions against Mr Russell....It is...a supplement to the Russellian logic, and not a refutation of it. (p. 662)

Wiener's basic point is that Lewis has simply expanded classical logic by adding to the object language a symbol that expresses a concept previously expressible only in the metalanguage, viz. deducibility/provability. (It is precisely this point that will lead Quine, in future decades) to say that the enterprise of modal logic was founded on a use-mention confusion.)

1918. In his *Survey of Symbolic Logic*, Lewis spells out his notion of strict implication in more detail and provides axiomatizations of various propositional modal systems. Lewis also wants in the object-language a one-place connective $\lozenge$ meaning something like "it is not a contradiction in terms that" or "it is possible that": $\lozenge\phi$ is true if and only if ϕ is possible in this (still vague) sense. Strict implication and $\lozenge$ are connected in an obvious way. If $\lozenge$ is taken as primitive (as it will be by Lewis and Langford in 1932), then ($\phi \Rightarrow \psi$) $=_{df}$ $\sim\lozenge(\phi \cdot \sim\psi)$. (As it will be put in the future, ($\phi \Rightarrow \psi$) $=_{df}$ $\square(\phi \supset \psi)$.)

Whereas earlier work on necessity and possibility (with the possible exception of MacColl's) was primarily *metaphysical* in character (in its evocation of a realm of possibility distinct from the actual) Lewis is interested in *logical* possibility, a notion that is considerably more linguistic than metaphysical. Strict implication itself gives rise to unintuitive results: e.g. $\sim\!\Diamond\phi \supset (\phi \Rightarrow \psi)$ is a theorem (i.e. a logically impossible proposition strictly implies every proposition). Lewis bites the bullet and sees this result as a discovery. This is a marked shift from his earliest attack on Russell, in 1912, in which he suggested there must be some "connection in meaning" between ϕ and ψ in order for ϕ to imply ψ. (This idea will be taken up by Nelson in 1930, by Ackerman in 1956, and by Anderson and Belnap in 1962.)

1922. In his *Tractatus Logico-Philosophicus*, Wittgenstein analyses logical truth in terms of truth-tables, argues (6.375) that the only necessity that exists is *logical* necessity, and claims (5.441, 5.47) that Fa "says the same as" $\exists x(a{=}x \cdot Fx)$. (Quine will sometimes exploit the latter idea, at other times Russell's Theory of Descriptions, in eliminating all singular terms apart from variables.)

1930. Gödel publishes his proof of the (weak) completeness of the first-order functional calculus (the question of (weak) completeness was posed clearly by Hilbert and Ackerman in 1928). Becker publishes "Zur Logic der Modalitäten". Nelson begins criticising strict implication on the grounds that it validates unintuitive inferences, and he pushes for a return to the spirit of Lewis's 1912 original critique of Russell, which seemed to demand a stricter notion of implication than the one Lewis settled for in 1918.

1931. Gödel publishes his incompleteness proofs: (i) arithmetic is incomplete in the sense that for any formal system S adequate for elementary number theory there is a proposition Φ such that neither the formula ϕ expressing Φ, nor its negation $\sim\!\phi$ is provable in S; (ii) the formula expressing the consistency of a consistent system S is not provable in S itself. The second incompleteness proof has implications for systems of modal logic as, at least on one understanding, the notion of provability is itself expressed in such systems.

1932. Lewis and Langford's *Symbolic Logic* contains a treatment of modality that is a marked improvement on Lewis's *Survey*. The (non-modal) quantificational system adopts a version of Russell's Theory of Descriptions.[59] The modal propositional system uses Lewis's $\Diamond$, where $\Diamond\phi$ is meant to be interpreted as "ϕ is possible" or "ϕ is self-consistent", by which is meant that it is not possible to deduce $\sim\!\phi$ from ϕ (p. 153.) The authors also say,

It should be noted that the words "possible", "impossible", and "necessary" are highly ambiguous in ordinary discourse. The meaning here assigned to $\Diamond p$ is a *wide* meaning of "possibility" – namely, logical conceivability or the absence of self-contradiction. (pp. 160–1)

It is not at all obvious that ϕ's *being logically conceivable* and ϕ's *not being self-contradictory* (i.e. *being self-consistent*) amount to the same thing.

Importantly, the Lewis systems are formulated *sententially*. In Chapter IX, Lewis and Langford do discuss quantifiers in connection with modal operators; but such operators attach only to closed sentences (not matrices), hence there is no discussion of quantifiers binding variables across modal connectives. It is not the case, however, that modal connectives always have maximal scope: Lewis and Langford often use formulae in which ◊ occurs within the scope of ~. Their uncertainty about the logical status of meaning relations between predicates probably explains their looseness in characterising possibility. *Philosophically,* the basic idea seems to be that ◊φ is true if and only if φ is *not a contradiction in terms, not false in virtue of meaning, not analytic.* At present, this is cashed out *logically* as the idea that it is not possible to deduce ~φ from φ, the justification for this being that in propositional systems the only vocabulary items assigned "meanings" (in any interesting sense) are the logical constants – the atomic sentences are just assigned truth-values; thus ◊φ is true if and only if φ is *not false by virtue of the meanings of the logical constants,* i.e. *not logically false.* (*Mutatis mutandis* for what will in the future be rendered as □φ.) (In a few years the notions of logical consequence and logical truth will be given precise characterisations by Tarski and Quine, making it easier to see the relevant issues here.)

(The Lewis systems that will concern us most are those now called T, S4, and S5. The axiom characteristic of T (also known as the basic Gödel-Feys-von Wright system) is □φ ⊃ φ. S4 can be obtained from T by adding the characteristic S4 axiom, □φ ⊃ □□φ. S5 can be obtained from S4 in numerous ways, e.g. by adding the characteristic S5 axiom, ◊φ ⊃ □◊φ.)

Church publishes the first of two papers (the second to be published next year) in which he introduces the λ-operator: $\lambda x\Sigma(x)$ stands for that function which has the value denoted by Σ( ) for each value of x. (The overall logical systems that Church presents in 1932 and 1933 will turn out to be inconsistent, but the λ-calculus itself will be proved consistent by Church and Rosser in 1936.)

1933. Tarski publishes (in Polish) his definition of truth for formalised languages. He solves the problem of the truth-relevant contribution of matrices (open formulae, such as *Fx* as it occurs in ∃*xFx*) by defining truth in terms of the satisfaction of formulae by sequences of objects. (It will be the possibility of extending this technique to modal matrices such as □*Fx* that will lead Quine to criticise quantified modal logic in ten years' time.)

Gödel publishes a short piece (which will be misleadingly cited by future proponents of quantified modal logic). He provides an interpretation of Heyting's intuitionistic propositional calculus using (a) the ordinary calculus, (b) the notion of "φ is provable" (written *B*φ), (c) the axioms (i) *B*φ ⊃ φ, (ii) *B*(φ ⊃ ψ) ⊃ (*B*φ ⊃ *B*ψ), and (iii) *B*φ ⊃ *BB*φ, and a rule of inference: (RL) if φ is a theorem, so is *B*φ. (In the hands of modal logicians, (RL) will become the rule of *necessitation,* which will make for elegant formulations of modal systems.) If *B*φ is interpreted as "necessarily φ", Gödel's system is equivalent to Lewis's S4 – axiom (iii) is just the characteristic S4 axiom. When it is interpreted as "φ is provable", this must be understood as *provable by some means,* not provable in a given formal system, otherwise axiom (iii) will clash with Gödel's second incompleteness theorem.[60] (In the early 1960s, Kaplan and Montague will show that using a broad notion of provability still leads to inconsistency in connection with quite weak modal postulates.) The construction of Gödel's system underscores Wiener's (1913) point that Lewis's systems are not alternatives to non-modal calculi, but extensions that import into the object language a notion of necessity hitherto assumed in the metalanguage.

1934. Hilbert and Bernays publish the first volume of their *Grundlagen der Mathematik.* Borrowing an idea from Peano, in their system, a definite description $\iota x\phi$ can be used only after it has been proved that $\exists x(\forall y(\phi(x/y) \equiv y=x))$.

In *Logische Syntax der Sprache,* Carnap expresses worries about Russell's use of "implication" in connection with $\supset$, as it has engendered confusion with talk of "consequence". He introduces *L-implication*: "ϕ *L*-implies ψ" is true if and only if ($\phi \supset \psi$) is valid. For necessity, possibility, and impossibility, Carnap has N, P, and I: N(ϕ) for "ϕ is analytic", P(ϕ) for "ϕ is not contradictory", and I(ϕ) for "ϕ is contradictory". Intensional sentences like N(ϕ) are said to be "quasi-syntactical" in the sense that they can be translated into "extensional syntactical sentences."

1936. Church proves the undecidability of general first-order logic (the decidability of monadic first-order logic was proved by Löwenheim in 1915). Church and Rosser prove the consistency of the λ-calculus.

Tarski's 1933 paper is published in German, as is a newer paper in which he provides what he sees as fully general, model-theoretic characterisations of the "intuitive" consequence relation and the corresponding property of logical truth.

In "Truth by Convention", Quine characterises logical truth as truth under all uniform reinterpretations of the non-logical vocabulary. (The work of Quine, and Tarski in this area bears affinities to work by Bolzano last century.[61]) Where predicate logic is concerned, Quine now draws a sharp distinction between *logical* truths and *analytic* truths, the former constituting a proper subclass of the latter (assuming sense can be made of "analytic"). If a sentence is *analytically* true just in case it is "true in virtue of meaning" (or something like that), then by such a criterion there will be analytic truths that are not logical truths, e.g. "No bachelor is married" (assuming this to be "true in virtue of meaning"). Quine will address this matter in 1947 and in more detail in 1951.)

1937. Carnap's 1934 book is published in English with additions. In "Les Logiques Nouvelles des Modalités", Feys uses the axiom system presented by Gödel, minus (iii) – the characteristic S4 axiom – to yield what will come to be called Gödel's "basic" system. Parry, Fitch, McKinsey, Smullyan, Vredenduin, and others are beginning to construct modal systems. In "New Foundations for Mathematical Logic", Quine lays out the importance to logic of Russell's Theory of Descriptions.

1939. In "Designation and Existence", Quine stresses the relationship between bound variables, reference, and ontology. Hilbert and Bernays publish the second volume of their *Grundlagen der Mathematik,* which contains their ε-operator.

1940. In *Mathematical Logic,* Quine provides contextual definitions of definite descriptions and class abstracts in accordance with Russell's Theory of Descriptions. He then argues that *all* singular terms except variables bound by quantifiers can be eliminated in this way and are therefore dispensable. In his review of Quine's book, Church (1940a) expresses his preference for adding a device of class abstraction to the primitive notation and for a Fregean rather than a Russellian approach to names and descriptions. Where Quine appeals to Russell's (1905) arguments in "On Denoting" for treating descriptions as incomplete symbols, Church draws the Fregean conclusion that co-referential singular terms need not be synonymous. (Church also publishes "A Formulation of the Simple Theory of Types".)

§18. 1941–1947: The formal assault on quantified modal logic

1941. Church's monograph *The Calculi of Lambda Conversion* is finally published, as are Becker's *Einführung* and Tarski's *Introduction to Logic*.

In "Whitehead and the Rise of Modern Logic", Quine again endorses Russell's Theory of Descriptions and goes on to express doubts about the viability of *any* form of non-truth-functional logic:

> In *Principia*, as in Frege's logic, one statement is capable of containing other statements truth-function-ally only; i.e., in such a way that the truth value (truth or falsehood) of the whole remains unchanged when a true part is replaced by any other truth, or a false part by any other falsehood. Preservation of the principle of truth-functionality is *essential* [emphasis added] to simplicity and convenience of logical theory. In *all* [emphasis added] departures from this norm that have to my knowledge ever been pro-pounded, moreover, a sacrifice is made not only with regard to simplicity and convenience, but with regard even to the admissibility of a certain *common-sense mode of inference* [emphasis added]: infer-ence by interchanging terms that designate the same object. (1941a, pp. 141–2).

(In the terminology of the present essay, one of the things Quine seems to be claim-ing at the end of this passage is that any connective he has encountered that is –PSME is also –PSST.)

In a footnote appended to the passage just quoted, Quine mentions two concrete cases of departures from the "norm" of truth-functionality that appear to "sacrifice" the "common-sense mode of inference":

> C.I. Lewis and C.H. Langford (Symbolic Logic, New York, 1932), e.g., use a non-truth-functional oper-ator "◊" to express logical possibility. Thus the statements:
>
> ◊ (number of planets in solar system < 7)
> ◊ (9 < 7)
>
> would be judged as true and false respectively, despite the fact that they are interconvertible by inter-changing the terms "9" and "number of planets in solar system", both of which designate the same object. Similar examples are readily devised for the early Whitehead system discussed in §III.] On grounds of technical expediency and on *common-sense* grounds as well [emphasis added], thus, there is a strong case for the principle of truth-functionality. (1941a, pp. 141–2)

This is interesting for a number of reasons.

(i) We have here Quine's first published attempt to demonstrate what he will later call the *referential opacity* of modal contexts. There is no talk about quantifying into modal contexts here, just talk of a substitution failure (Quine will not discuss this until 1943). However, in his review this year of Russell's "Inquiry into Meaning and Truth", Quine expresses concern that Russell does not explain how to interpret sentences in which a quantifier binds a variable across verbs of propositional attitude like "believe", sentences Russell has been using since 1905 in discussions of scope ambiguities involving definite and indefinite descriptions.

(ii) Although Quine's example makes use of the definite description "the number of planets", it will soon become clear to him (if it is not already) that *the use of a description is not essential to his main point,* which can in fact be made using ordinary names (*modulo* the rejection of a controversial theory of names), and in a more technical and ultimately more revealing way by using variables.

(iii) In hindsight, by putting together Quine's footnote and the passage to which it is appended, we dimly discern a general "slingshot" argument forming in Quine's mind – as it must have been forming in the minds of Church and Gödel at around the same time – an argument that *does* require an example involving a definite description (or a class abstract). For one of Quine's claims here seems to be that, as far as he can ascertain, giving up truth-functionality requires giving up a "common-sense mode of inference" – which he will later call *substitutivity* – the implication being that holding onto the common-sense mode of inference requires holding on to truth-functionality. (When the formal slingshots of Church (1943a) and Quine (1953a, 1953c, 1960) finally appear, they will involve not just *inference by interchanging terms that designate the same object* but also *inference by interchanging logical equivalents*; Gödel's (1944) slingshot will involve *interchanging Gödelian equivalents*.)

For immediate purposes, we can simply note that, in 1941, Quine is pointing out that the following argument is invalid:

(1)   ◊(the number of planets in our solar system < 7)
        the number of planets = 9
        ——————————————————————————————
        ◊(9 < 7).

(In the terminology of the present essay, Quine is pointing out that Lewis's ◊, if used in a system of predicate logic, would be –ι-SUBS. (In papers that will be published in 1971 and 1972 Kripke will show that ◊ understood *metaphysically* is also –ι-SUBS.) If Quine continues with the policy of treating definite descriptions in accordance with Russell's theory, then (a) the first premise of (1) would seem to be ambiguous (as will be pointed out by Smullyan in 1948) in the same way as "George IV wondered whether the number of planets in our solar system < 7", and (b) the invalidity of (1) does not, as it stands, show that ◊ is also –PSST. Point (a) is considerably less important than future work will contend; a change of example will fix point (b) in 1943. As is customary, let's stop in 1942 on the way to 1943.)

1942. Carnap publishes *Introduction to Semantics* in which he breaks with Frege by taking the designation of a sentence to be a proposition rather than a truth-value.

In his logical vocabulary, Carnap again includes symbols for what he calls the "logical modalities", and mentions the possibility of future systems containing "physical and causal modalities."

Church will review Carnap's book next year. This year he reviews Quine's 1941 paper and says,

> [Quine's] statement on p. 146 that Russell was the first to define the notation or device of *description* in terms of more basic notions, or indeed that Russell made such a definition at all, seems to involve a misleading emphasis. In fact Russell's well-known method of dealing with descriptions is not a definition of a notation for descriptions, but a proposal to do without descriptions (to accomplish certain purposes without using them). This aspect of contextual definition is, it seems to the reviewer, too often overlooked, or at least underemphasized...Russell's contextual definition of a description...does not mean that the primitive notation provides an expression having the same meaning, but only that it provides, for every sentence containing this expression, *a substitute sentence* – i.e., a sentence which in a certain way serves the same purpose without containing anything which could be called a description. (1942, pp. 100–1)

Thus far, Church appears to be complaining that Quine, like others, has not sufficiently appreciated an important feature of Russell's Theory of Descriptions: on Russell's account, descriptions are not singular terms that can be defined in terms of expressions already in the primitive notation; *they are not singular terms at all and are defined away in primitive notation*. A careful examination of the relevant parts of Quine's 1941 article and his 1940 book *Mathematical Logic* reveal there is some truth to Church's complaint of "misleading emphasis". (Unfortunately, this will have dialectical repercussions for the next forty years. It will derail not only Quine but also his critics, including Church. It will lead Quine to use "singular term" in misleading ways, it will lead Carnap (1947), Smullyan (1947, 1948), Marcus (1948, 1962), Fitch (1949, 1950), and even Church (1950) himself into error in connection with Quine's attack on quantified modal logic, and it will lead Quine (1953a, 1953c, 1961, 1969, 1980) into error in dealing with the errors of his critics. But only once will any of this have a major consequence for Quine, and this will be in connection with a *general* argument against non-extensional connectives in 1953 and 1960. It will have no serious consequences for his main objection to quantified modal logic, it will simply get him into an unnecessary tangle dealing with points raised by Smullyan (1947, 1948), who will pick up on Church's next point). Church claims the "misleading emphasis" effectively undermines Quine's attack on Lewis's use of ◊:

> This point is relevant to an objection made by Quine in another passage (pp. 141–142, cf. also p. 148) to "non-truth-functional statement composition," in particular to Lewis's use of ◊. It is objected that the sentences "◊(the number of planets < 7)" and "◊(9 < 7)" would be judged to have opposite truth-values, whereas "the number of planets" and "9" denote the same number, and the two sentences should therefore be interdeducible by substituting one term for the other. On the basis of *Principia* or of Quine's own *Mathematical Logic* (...), however, *the reply to this is immediate* [emphasis added]. The translation into symbolic notation of the phrase "the number of planets" would render it either as a description or as a class abstract, and in either case it would be construed contextually; any formal deduction must refer to the unabbreviated forms of the sentences in question, and the unabbreviated form of the first sentence is found actually to contain no name of the number 9. (1942, p. 101)

This is where things start to get tricky. Church's remarks also seem to involve misleading emphasis, although he isn't exactly saying anything *false*. First, there are proponents of modal logic who treat descriptions as singular terms, e.g. Carnap and Church. Second, it is true that in its unabbreviated form the simple non-modal sentence "the number of planets < 7" will contain no "name" of 9:

(2) $\exists x(\forall y(y \text{ numbers the planets} \equiv y{=}x) \cdot x < 7)$.

Now what of the modalised sentence "◊(the number of planets < 7)"? Attaching ◊ to a sentence ϕ in Lewis's system is equivalent to saying that ~ϕ is not an analytic truth, so the sentence should be understood as (3), itself understood as (4):

(3) $\Diamond\exists x(\forall y(y \text{ numbers the planets} \equiv y{=}x) \cdot x < 7)$
(4) "~$\exists x(\forall y(y \text{ numbers the planets} \equiv y{=}x) \cdot x < 7)$" is not an analytic truth.

So far so good. Church is right that from (3) Quine will not be able to obtain "◊(9 < 7)" using just an inference principle for interchanging terms that designate the same object. But to stop at this point is to suggest that Quine's objection cannot be salvaged, which is not the case. *Given Lewis's interpretation of ◊, Quine can make the same point using two ordinary names of the same object* (in fact he will do this, but not until next year). Note: if Church were to deny that Quine's case can be made using simple names, then he would have to hold "Hesperus = Phosphorus" is just as analytic as "Hesperus = Hesperus", i.e. he would have to hold that "Hesperus" and "Phosphorus" were *synonymous*, which would not sit well with his Fregean leanings.

Church adds that he would prefer a system in which class abstracts and descriptions are construed as "names", and hence not contextually defined. Within such a system, "Sentences, if names at all, are perhaps best taken as names of truth-values, and must be taken as *expressing* rather than *denoting* propositions" (p. 101). He concludes, rather cryptically, that "Quine's argument...shows that a non-truth-functional operator, such as ◊, if it is admitted, must be prefixed to names of propositions rather than to sentences." (p. 101.) The fact that some logicians wish to treat descriptions as singular terms may explain why Quine, in future works, will continue to use "singular term" in connection with descriptions despite his endorsement of Russell's theory. But the explanation might also lie in the "misleading emphasis" noted by Church, i.e. in the fact that Quine appears to view contextual definition as a way of defining rather than eliminating descriptions in primitive notation.

1943. Quine publishes "Notes on Existence and Necessity". Church reviews Quine's paper and also Carnap's *Introduction to Semantics*.

In his review of Carnap's book, Church (1943a) presents what will later be called a "slingshot" argument to undermine Carnap's proposal that a sentence designates a proposition. The argument is meant to show that in any system in which singular terms and sentences designate, there must be some unique entity that all true sentences designate, and some distinct unique entity that all false sentences designate. Church attributes the argument to Frege, but amongst his premises are the following:

(i) class abstracts are complex singular terms; (ii) interchanging singular terms that designate the same object cannot alter the designation of the containing sentence; and (iii) logically equivalent sentences have the same designation.[62] Church rejects Russell's Theory of Descriptions and endorses Frege's distinction between *Sinn* and *Bedeutung*. He adds that if we reject Fregean senses we will almost certainly have to treat all names in accordance with the Theory of Descriptions, and hence the notion of designation will largely disappear, remaining only for variables and (perhaps) also for sentences. As Church notes, if all names are treated in accordance with Russell's Theory of Descriptions, there is no longer any barrier to treating sentences as designating propositions as the slingshot will lose its ammunition, viz. constant singular terms that designate.

In his 1943 article, Quine lays out his first main attack on the notion of necessity employed by modal logicians. The article contains much that we need to be clear about.

(i) Quine begins by characterising what he calls *substitutivity*:

One of the fundamental principles governing identity is that of substitutivity...It provides that, *given a true statement of identity, one of its two terms may be substituted for the other in any true statement and the result will be true*. (p. 113)

(ii) After pointing to the obvious failure of substitutivity in overt and covert quotational contexts, Quine then characterises what he calls *purely designative* occurrences of names:

The relation of name to the object whose name it is, is called *designation*; the name "Cicero" designates the man Cicero. An occurrence of the name in which the name refers simply to the object designated, I shall call *purely designative*. (p. 114)

(In 1953, when much of this 1943 paper is incorporated verbatim into "Reference and Modality", Quine will replace "purely designative" by "purely referential".)

(iii) Quine then tells us that,

Failure of substitutivity reveals merely that the occurrence to be supplanted is not purely designative, and that the statement depends not only upon the object but on the form of the name. (p. 114)

(It will be forty years before Kaplan (1986) points out that, technically, in the context of Quine's argument, failure of substitutivity reveals something weaker: that at least one of the supplantee or the supplanter is not purely designative/referential.[63])

(iv) Quine argues by example that occurrences of singular terms in contexts governed by psychological verbs are not purely designative: the inference from (5) to (6) is not valid despite the truth of "Tully = Cicero":

(5) Philip is unaware that Tully denounced Catiline
(6) Philip is unaware that Cicero denounced Catiline.

(As Quine will put it in 1953 in "Reference and Modality", this shows that the context in question is *referentially opaque*.)

(v) He reiterates his 1939 point about the "intimate connection between designation and existential quantification...implicit in *existential generalization*" (p. 116) and argues that the principle is "unwarranted" in connection with the occurrences of "Tully" and "Cicero" (similarly "Catiline") in (5) and (6). Applied to the former, it yields,

(5′) $\exists x$(Philip is unaware that x denounced Catiline).

We now get the first published statement of a Quinean question, posed in the form of a paradox (broadly construed), that will become famous:

What is this object that denounced Catiline without Philip yet having become aware of the fact? Tully, i.e., Cicero? But to suppose this would conflict with the fact that [(6)] is false. (p. 118)

This application of existential generalisation is "unwarranted" because in (5) "Tully" does not occur purely designatively.[64] The point can be restated thus: on the standard objectual interpretation of quantification, i.e. when the variables that quantifiers bind are ranging over ordinary objects, $\exists x \Sigma(x)$ is true if and only if there is at least one object, however specified, that satisfies $\Sigma(x)$. So (5′) should be true if and only if there is at least one object, however specified, that satisfies "Philip is unaware that x denounced Catiline". But an *object's* satisfying this condition is precisely what Quine has shown to be nonsensical: the contrast between (5) and (6) shows that "is unaware that" ("believes that", "hopes that", "doubts that", etc.) renders the contexts it governs sensitive to the *linguistic manner* in which an object is specified.

(vi) Quine now pushes a logical and ontological point. Existential generalisation is an inference principle "only by courtesy" (p. 118) and "anomalous as an adjunct to the purely logical theory of quantification" (p. 118, n. 3). Hence the importance, he claims, of the fact that all singular terms, except variables bound by quantifiers, can be contextually eliminated: "...contact between language and object comes to be concentrated in the variable, or pronoun." (p. 118, n. 3). Existential generalisation holds good only when a term occurs purely designatively. "It is simply the logical content of the idea that a given occurrence is designative." (p. 118).

(vii) Quine's next point is Fregean:

To say that two names designate the same object is not to say that they are *synonymous* ...in order to determine whether a statement of the form "$a=b$" is true "it is commonly necessary to investigate the world. The names "Evening Star" and "Morning Star", for example, are not synonymous, having been applied each to a certain ball of matter according to a different criterion...The identity

Evening Star = Morning Star

is a truth of astronomy, not following merely from the meanings of the words. (p. 119)

(viii) Quine now begins to set out his worries about any "calculus of necessity", such as that of Lewis. First, he makes clear precisely what he is attacking:

Among the various possible senses of the vague adverb "necessarily" we can single out one – the sense of *analytic* necessity – according to the following criterion: the result of applying "necessarily" itself to a statement is true if, and only if, the original statement is analytic." (p. 121).

Why does Quine talk about *analytic* rather than *logical* or *strict* necessity? Abstracting away from the axioms of this or that system, what has thus far been *formally* characteristic of modal logic is the introduction of symbols whose inter- pretation has typically been defined in terms of *deducibility* (*provability*). But is important to recall that *philosophically* the motivation for Lewis and those he has influenced is the construction of systems in which something close to the "intuitive relation" of *implication* or *consequence* is expressible, a relation that holds not just between "Heath is tall and Sneath is tall" and "Heath is tall", but between "Heath is a bachelor" and "Heath is unmarried". In propositional systems the atomic sen- tences lack structure, so there are no semantic relations between predicates to con- sider, so there is no intuitively valid implication of the main propositional systems that cannot be captured by appeal to deducibility. But modal logicians also want predicate systems in which, say, "Heath is a bachelor" implies "Heath is unmar- ried", and in which "all bachelors are unmarried" is a theorem. This requires going beyond *logical* necessity, narrowly construed, and invoking a broader notion that subsumes meaning relations between predicates, viz. *analytic* necessity.

It is "necessarily" so understood that is under attack. One of Quine's 1943 exam- ples makes use of the description "the number of planets" and the name "9", whilst another makes use of two names, "(the) Morning Star" and "(the) Evening Star" (thereby side-stepping Church's (1942) complaint). Before looking at any argu- ments, it is worth asking why Quine has chosen to use *these* names rather than "Cicero" and "Tully", which he used when examining psychological contexts. My hunch – and that's all it is – is this. Borrowing some future terminology, we can remind ourselves, so to speak, that Whitehead and Russell proved that in exten- sional contexts the difference between PSST and ι-SUBS is immaterial; Quine oper- ates on extensional turf, and so he does not sharply distinguish names and descriptions; Russell's (1905) first application of the Theory of Descriptions was to *descriptions*; however, he subsequently enlarged its scope to take in ordinary proper names, leaving just a handful of *logically proper names* (at least that is how he will be understood); Quine (1940) pushed this idea to its limit: all singular terms, except variables, can be eliminated by means of a form of Russellian contex- tual definition, so that only variables occupy singular term positions in unabbrevi- ated notation; so for Quine in 1941, unlike Russell in 1905, a sharp distinction between names and descriptions seemed unnecessary; but as a result of Church's review last year, in which Church pushed for Fregean, referential treatments of both names and descriptions, Quine has been reminded that, in principle, one might opt to treat names and descriptions differently from one another, and he has decided that he should not restrict himself to examples that make use of descriptions; the failure of two sentences $\Sigma(\alpha)$ and $\Sigma(\beta)$ to be analytically equivalent is certainly more *blatant* when one or both of α and β is a description; and it still seems fairly obvious when one or both of α and β is a name that contains the definite article or appears to have descriptive content or heritage, as in, say, "(the) Morning Star" or "(the) Evening Star". (It will be nearly three decades before Kripke argues that ordinary proper names cannot be analysed descriptively.)

Basically, Church was right to *complain* last year, but wrong to think the complaint ultimately mattered: Quine's charge was thrown out on a technicality. More precisely, given Quine's (1941) endorsement of Russell's Theory of Descriptions, Church was right to object to Quine's (1941) use of an example containing a description in a non-extensional context to show that "a common-sense mode of inference" fails in that context – remember, (a) Whitehead and Russell proved only that ι-SUBS holds in extensional contexts, and (b) quantification wasn't the issue in 1941. But this year Quine shows us that Church's complaint has no real bite. The point made in 1941 with a name and a description is now made with two names – albeit names with descriptive heritage. (The point stands even with names that seem to function more like arbitrary labels. Actually, one has to think for a while to come up with a handful of good examples where neither name has an obvious descriptive heritage. Pairs involving "Hesperus", "Constantinople", and "Karlsbad" are no good for anyone exposed to any Greek or German.) Let's use "Phosphorus" and "Hesperus" (pretending we don't see the etymologies). Quine's point is that strictly speaking (pun intended) substitution of co-referential names fails in modal contexts, in so far as "necessarily" is understood *analytically*. (If, in the future, the claim is made that some alternative, non-linguistic, non-semantic, interpretation of "necessarily" is common at the moment, such a claim will be false.) "Phosphorus = Phosphorus" is analytic, but "Phosphorus = Hesperus" is not; so (7) is true while (8) is false:

(7) $\Box$(Phosphorus = Phosphorus)

(8) $\Box$(Phosphorus = Hesperus).

It would seem that if the common-sense mode of inference is to be maintained in modal logic, co-referential names will have to be *synonymous*. No-one has rushed to embrace such a view yet. Perhaps they will.

(ix) But Quine has more for us. He takes himself to have demonstrated that existential generalisation is "unwarranted" in connection with occurrences of names that are not purely designative. The basic problem was that of making sense of the quantified sentence (5′) given the truth of (5) and the falsity of (6). Since names occurring within the scope of $\Box$ are not purely designative, the same problem will arise in connection with modal contexts. Existentially generalising on (7) we get

(7′) $\exists x \Box (x = \text{Phosphorus})$.

But this is an export with no import. Mirroring the question asked in connection with (5′), Quine can ask us the following: What is this object that as a matter of analytic necessity is identical to Phosphorus? Phosphorus, i.e., Hesperus? But to suppose this would conflict with the fact that (8) is false. Quine's main point can be stated thus: on the standard objectual interpretation of quantification, i.e. when the variables that quantifiers bind range over ordinary objects, $\exists x \Sigma(x)$ is true if and only if there is at least one object, however specified, that satisfies $\Sigma(x)$. So (7′) should be true if and only if there is at least object, however specified, that satisfies "$\Box(x =$ Phosphorus)". But an *object's* satisfying this matrix is precisely what Quine has shown to be nonsensical: the contrast between (7) and (8) shows that $\Box$ renders the contexts it governs sensitive to the *linguistic manner* in which an object is specified.

(x) It should be noted that it is in this 1943 paper that Quine first talks about other modalities:

These observations apply, naturally, to the prefix "necessarily" only in the explained sense of analytic necessity; and correspondingly for possibility, impossibility, and the necessary conditional. As for other notions of necessity, possibility, etc., for example, notions of physical necessity and possibility, the first problem would be to formulate the notions clearly and exactly. Afterwards we could investigate whether such notions involve non-designative occurrences of names and hence resist the introduction of pronouns and exterior quantifiers. This question concerns intimately the practical use of language. It concerns, for example, the use of the contrary-to-fact conditional within a quantification; for it is reasonable to suppose that the contrary-to-fact conditional reduces to the form "necessarily, if p then q" in some sense of necessity. Upon the contrary-to-fact conditional depends in turn, for example, this definition of solubility in water: To say that an object is soluble in water is to say that it would dissolve if it were in water. In discussions of physics, naturally we need quantifications containing the clause "x is soluble in water", or the equivalent in words; but, according to the definition suggested, we should then have to admit within quantifications the expression "if x were in water then x would dissolve", that is, "necessarily if x is in water then x dissolves". Yet we do not know whether there is a suitable sense of "necessity" that admits pronouns referring thus to exterior quantifiers. (p. 124)

The moral that Quine appears to be drawing from his discussion is that if modal logics are to serve any useful purpose, they must include quantifiers that can attach to formulae to form new formulae as in extensional logic; in the modal logics most clearly formulated – those based on analytic necessity – names cannot be interchanged *s.v.* within the scope of $\Box$, so variables bound from outside would appear to defy interpretation. In short, the prospects for the significant use of modal operators look dim.

In his review of Quine's 1943 paper, Church (1943b) appears tacitly to accept the point that the appearance of a description in Quine's 1941 example was not essential to the main point. (Note that Church (1943a) and Carnap (1942) both take descriptions to be singular terms.) Church questions Quine's conclusion that a variable within the scope of a modal connective may not be bound by a quantifier outside and suggests an alternative conclusion: "...variables must have an intensional range – a range, for instance, composed of attributes rather than classes." (p. 46) This, he believes, leads naturally to Frege's position that a singular term in "oblique" contexts designates (or *denotes* as Church is now putting it) its customary sense. He borrows some of Quine's own words to portray the sense-designation distinction: to determine that two names have the same sense it is sufficient to understand them; but to determine that two names have the same designation "it is commonly necessary to investigate the world" (p. 47).

1944. In "Russell's Mathematical Logic", Gödel provides the material for a more interesting slingshot than Church's, relying not on the interchange of mere logical equivalents but on the interchange of sentences such as *"Fa"* and *"a = ɿx(Fx ·* *x=a)"*, which are (a) logically equivalent (b) contain the same non-logical vocabulary (assuming "=" to be part of the logical vocabulary), and (c) are interconvertible another using an elementary syntactic transformation. (For details, see Part I.) Towards the end of the paper, Gödel discusses "whether (and in what sense) the axioms of *Principia* can be considered to be analytic." (p. 150).[65]

1946. Bernays reviews Gödel's (1944) paper and claims (mistakenly) that Gödel reaches his conclusion because he fails to separate the sense (*Sinn*) and reference (*Bedeutung)* of a sentence despite mentioning Frege in his discussion of Russell's Theory of Descriptions. Bernays" point seems to be that Fa and $a=ɿx(x=a · Fx)$ have different senses. But this is irrelevant. Gödel's argument concerns the relationship between descriptions and facts, not descriptions and *modes of presentation* of facts (which Russell did not have in any case). (Bernays' mistake will be repeated *ad nauseam* by Oppy in 1997.)[66]

In an abstract of "A Formulation of the Logic of Sense and Denotation" Church sketches a theory according to which singular terms undergo reference shifts in modal contexts (the worked out theory, somewhat modified, will be published in 1951).

"A Functional Calculus of First Order Based on Strict Implication" by Barcan (Marcus) contains the first published axiomatization of a first-order modal system (containing variables but no individual constants), an extension of Lewis's S2.[67] (If, in the future, is seen the idea that names are rigid designators in Marcus's formal work, it will be the product of fantasy or poor scholarship.) Interestingly, in Barcan's system "$\exists x \Diamond Fx$" and "$\Diamond \exists x Fx$" are provably equivalent.

In "Modalities and Quantification", Carnap (1946) constructs a first-order extension of S5 for which he provides a semantics in terms of "state-descriptions" and a restricted completeness proof. The system contains denumerably many individual constants and variables, but no descriptions or abstracts. Every individual is denoted by some individual constant, and different constants denote different individuals. The primitive symbol "N" expresses logical necessity; I shall follow Marcus in using Fitch's $\Box$. Lewis's symbol $\Diamond$ for logical possibility is defined in the usual way: $\Diamond\phi = $ df $\sim\Box\sim\phi$.

Carnap's brief remarks on the aims and underpinnings of modal logic are noteworthy. He begins by saying that

[t]he guiding idea in our constructions of systems of modal logic is this: a proposition p is logically necessary if and only if a sentence expressing p is logically true. That is to say, the modal concept of the logical necessity of a proposition and the semantical concept of the logical truth or analyticity of a sentence correspond to each other (p. 34).

He proposes the following "convention" governing the interpretation of his basic modal operator: $\Box\phi$ is true if and only if ϕ is L-true (logically true). This is not be a *definition*; syntactically and semantically □ is a primitive symbol in the object language. But the semantical rules of Carnap's system are framed so that the condition specified by the convention is satisfied.

The explanation of L-truth comes from Carnap's semantical concept of *state-description*, which he glosses informally as

... a class of sentences which represents a possible specific state of affairs by giving a complete description of the universe of individuals with respect to all properties and relations designated by predicates in the system (p. 50).

More precisely, a class of sentences Σ is a state-description = $_{df}$ for every atomic sentence ϕ, Σ contains either ϕ or $\sim\phi$, but not both. (Thus every complete, non-contradictory assignment of truth-values to the atomic sentences of a system corresponds to a state-description and represents some possible state of the universe.) Since a sentence ϕ "is usually regarded as logically true or logically necessary if it is true in every possible case" (p. 50), Carnap proposes that ϕ is L-true true if it holds in every state-description. Informally, to say that a sentence ϕ holds in a given state description Σ – which is just a class of sentences – is to say that Σ "entails" ϕ, that ϕ "would be true if this state-description were the description of the actual state of the universe" (p. 50).

A further convention governing the understanding of □ emerges in connection with the quantifiers: it is to be "interpreted in such a way that" $\forall x \Box \phi$ and $\Box \forall x \phi$ are L-equivalent, i.e. they hold in exactly the same state-descriptions (p. 37).

Carnap gives a restricted completeness proof for his quantified S5: if ϕ is semantically valid, ϕ is deducible in the accompanying calculus as long as ϕ contains no occurrence of "=" and no formula of the form "$\sim\Box(...)$". (In a 1948 review, Bernays will explain why he thinks a full completeness proof can be provided. Such proofs will be given by Bayart and Kripke, independently, in 1959.)

In closing, Carnap claims (p. 64) that in his forthcoming book (*Meaning and Necessity*) he will deal with the difficulties raised by Quine (1943) for quantified modal logic by distinguishing between the extension of an individual expression (singular term) and its intension, which is "a concept of a special kind which we may call an individual concept" (p. 64).

1947. A big year. Carnap publishes *Meaning and Necessity,* in which he includes a two-page criticism from Quine. Quine himself publishes "The Problem of Interpreting Modal Logic", which Smullyan reviews. Barcan (Marcus) publishes "The Identity of Individuals in a Strict Functional Calculus of First Order", Reichenbach publishes *Elements of Symbolic Logic.*

The discussions surround analytic or logical necessity, but Reichenbach wants to distinguish logical and physical necessity and to introduce the notion of *relative* modality. For Reichenbach, ϕ is logically necessary $=_{df}$ ϕ is a tautology; ϕ is impossible $=_{df}$ ϕ is a contradiction. He uses "possible" in such a way as to exclude what *is* the case, i.e. "it is logically possible that ϕ" entails $\sim\phi$, and "ϕ is logically necessary" does not entail "ϕ is logically possible". What is logically necessary is physically necessary (but not vice versa), what is logically impossible is physically impossible (but not vice versa), and what is physically possible is logically possible (but not vice versa). He adopts Russell's Theory of Descriptions as an account of both descriptions and names.

Quine (1947) launches a blistering attack on quantified modal logic. He begins with the statement that

...the ideas of modal logic (e.g. Lewis's) are not intuitively clear until explained in non-modal terms...When modal logic is extended (as by Miss Barcan [Marcus])...to include quantification theory... serious obstacles to *interpretation* are encountered. [emphasis added] (p. 43).

The non-modal notion of *analyticity*, Quine suggests, is what modal logicians are appealing to: the result of prefixing $\Box$ to a sentence ϕ is true if and only if ϕ is analytic. The class of analytic truths, he points out, is broader than the class of logical truths, adding that,

The notion of analyticity...appears, at present writing, to lack a satisfactory foundation. Even so, the notion is clearer to many of us, and obscurer surely to none, than the notions of modal logic; so we are still well advised to explain the latter in terms of it. This can be done...as long as modal logic stops short of quantification. (p. 45)

I suppose some such conception underlies the intuition whereby axioms are evaluated and adopted for modal logic. The explanation of modal logic thus afforded is adequate so long as modalities are not used inside the scopes of the quantifiers; i.e. so long as $\Box$ is applied only to statements and not to matrices. (p. 46)

The points Quine raised in 1943 about quantification into modal contexts are then sharpened. To say that $\Box\phi$ is true if and only if ϕ is analytic and that $\Diamond\phi$ is true if and only if $\sim\phi$ is not analytic is fine as long as $\Box$ and $\Diamond$ are applied to closed sentences and not matrices (open sentences). The failure of names to permit interchange *s.v.* in modal contexts seems to be merely a symptom of a deeper problem raised by quantified modal matrices, a problem that can be articulated as a challenge. If $\Box$ and $\Diamond$ are understood analytically, how are we to make sense of the matrices $\Box Fx$ and $\Diamond Fx$ as they occur in $\exists x\Box Fx$ and $\exists x\Diamond Fx$? It is senseless to ask, e.g., whether Fx is analytically true of an object. There is an important use-mention issue here – but not a crude confusion – that comes into relief when the interpretation of the proposed necessity operator, "$\Box$" is contrasted with that of the negation operator, "$\sim$".

Whether "$\sim Fa$" and "$\Box Fa$" are true depends upon semantic properties of the embedded sentence "Fa". What properties are these? Standardly,

"$\sim Fa$" is true if and only if "Fa" is *not true*.

And, according to the theory of modality under consideration.

"$\Box Fa$" is true if and only if "Fa" is *analytically true*".

So far so good. Assuming a clear criterion of analyticity, to ask whether "*Fa*" has the property of being *analytically true* seems to make as much formal sense as asking whether it has the property of being *not true*. But once quantifiers and variables are introduced, matters are more complex. We can see this by comparing the evaluation of "∃*x~Fx*" and "∃*x*□*Fx*". Whether "∃*x~Fx*" is true depends upon a semantic property of the negative matrix "*~Fx*"; whether "∃*x*□*Fx*" is true depends upon a semantic property of the modal matrix "□*Fx*"; and whether these matrices have the relevant semantic properties in each case depends upon some semantic property of the atomic matrix "*Fx*".

Whatever the relevant properties are, they are not *truth*: "*Fx*" is not the right sort of thing to be true or false as it contains a free variable. Tarski (1933) has shown us how to proceed here. Or, rather, Tarski has shown us how to get what we want for "∃*x~Fx*", but not for "∃*x*□*Fx*". The basic idea is to use the notion of a formula (a closed or open sentence) being *true of (satisfied by)* an object – actually a *sequence* of objects because of formulae containing occurrences of two or more distinct variables. That is, the Tarskian strategy is to work up to the one-place notion expressed by "is true" from the two-place notion expressed by "if true of". A closed sentence is *true* if and only if it is *true of* every sequence of objects. To simplify exposition of the crucial respect in which "*~Fx*" and "□*Fx*" differ, let us suppose our sentences contain occurrences of at most one variable, "*x*". This allows us to talk of a sentence being true of an object. Under this harmless simplifications,

"∃*x~Fx*" is true if and only if "*~Fx*" is true of at least one object.

At this point, the interpretation of the one-place connective "~" enters the picture. Given what we take the connective to mean, the matter is straightforward:

"*~Fx*" is true of an object ∗ if and only if "*Fx*" is *not true* of ∗.

It is clear what it means for "*Fx*" to be *not true* of ∗: it is for the object ∗ to fail to be *F*. There is no barrier, then, to making sense of a quantifier binding a variable across the negation operator.

By contrast, when we turn to binding a variable across a modal operator, as in "∃*x*□*Fx*", we face an apparently insurmountable obstacle. The first step is straightforward enough:

"∃*x*□*Fx*" is true if and only if "□*Fx*" is true of at least one object.

But the situation deteriorates once the interpretation of the one-place connective "□" enters the picture. Given what we take "□" to mean, we seem to need the following:

"□*Fx*" is true of an object ∗ if and only if "*Fx*" is *analytically true* of ∗.

But what is it for "*Fx*" to be *analytically true* of ∗? It is for ∗ to be *F analytically*. But what does *that* mean?

Here we must be careful. The idea of analytic satisfaction is not itself incoherent. We can see this by considering certain non-atomic matrices, e.g. "*Fx* ⊃ *Fx*" or "*Fx* ∨ *~Fx*". Surely these matrices are true of everything; moreover, surely they are *analytically* true of everything (by virtue of being *logically* true of everything). The real problem that Quine is raising concerns *atomic* matrices: it seems to make no sense to ask whether an atomic matrix is analytically true of something (except, of course, in the trivial case where the matrix is "$x = x$" (or, perhaps, "x exists")). Consider a concrete example: of which objects is "$x =$ Phosphorus" *analytically true*? The question seems to lack sense. The closed sentence "Phosphorus = Phosphorus" is analytically true, but the closed sentences "Hesperus = Phosphorus" and "Venus = Phosphorus" are not, despite the fact that "Phosphorus", "Hesperus", and "Venus" are all names of the same object. Quine is leaving the modal logician with a challenge: explain the semantics of modalised atomic matrices.

If every matrix containing a modal connective could be converted systematically into a closed sentence with the modal connective at the beginning, i.e. if every *de re* modal statement could be reduced to a *de dicto* modal statement, then the problem of how to understand, say, $\exists x \Diamond Fx$ would seem to disappear.

For Marcus (1946) and Carnap (1946) $\exists x \Diamond Fx$ and $\Diamond \exists x Fx$ are provably equivalent, so part of the job is done, at least if one is prepared to accept this equivalence (which some are not). But this conversion (reduction) will not help, Quine points out, with a sentence such as the following,

(9) $\exists x(x$ is red $\cdot \Diamond(x$ is round))

which the advocate of quantified modal logic will surely hold true.

Quine then throws the modal logician a bone, a supplementary *substitutional thesis*, a generalisation based on the following idea: $\exists x Fx$ is true if there is some individual constant whose substitution for x in Fx would yield a true sentence; so $\exists x \Box Fx$ is true if there is some individual constant whose substitution for x in Fx would yield a sentence that is an *analytic* truth, while $\exists x \Diamond Fx$ is true if there is some individual constant whose substitution for x in Fx would yield a sentence that is not an *analytic* falsehood. (This suggestion will be taken up next year by Smullyan.) But, says Quine, the substitutional thesis has two drawbacks: (a) It would provide at best a *partial* solution to the problem of interpreting modalised matrices because of unnamed (and unnameable) objects; (b) it would have "queer ontological conse-quences" (p. 47). The first point is self-evident; the second requires examination.

Sentence (10) is true, so by the substitutional thesis, (10′) is also true:

(10) Phosphorus = Hesperus $\cdot \Box$(Phosphorus = Phosphorus)
(10′) $\exists x(x =$ Hesperus $\cdot \Box(x =$ Phosphorus)).

But (11) is also true, says Quine, so by the substitutional thesis, (11′) is also true:

(11) Hesperus = Hesperus $\cdot \sim\Box$(Hesperus = Phosphorus)
(11′) $\exists x(x =$ Hesperus $\cdot \sim\Box(x =$ Phosphorus)).

Since the quantified matrices in (10′) and (11′) contain mutual contraries, there must be two distinct entities x such that $x =$ Hesperus! If we were to introduce the term "Venus" we could infer a third such object in similar fashion.

Thus it is that the contemplated version of quantified modal logic is committed to an ontology which repudiates material objects (such as the Evening Star properly so-called) and leaves only multiplicities of distinct objects (perhaps the Evening-Star concept, the Morning-Star concept, etc.) in their place. For the ontology of a logic is nothing other than the range of admissible values of the variables of quantification. (p. 47) Quine concludes that Church's (1943b, 1946) suggestion to restrict the range of the variables to intensional objects in modal contexts might be the best option for the modal logician at this point. (In 1953 and 1960, he will argue that such a restriction will not solve the problem.)

Carnap (1947) appears to have accepted Church's (1943a) slingshot as he now takes sentences to designate truth-values, though they also have intensions. He is explicit in his preface that his "L-terms" are to be understood in terms of analyticity; e.g., on page v he says that "ϕ is L-true" is to be understood as ϕ is "logically true" or "analytic".

As in his 1946 paper, Carnap works up to his discussion of modal operators in the object language via analyses of the L-concepts (L-truth, L-equivalence, etc.). "The concept of *L-truth* is here defined as an explicatum for what philosophers call logical or necessary or analytic truth" (p. 7). On the following page, he adds that "'*L-true*' is meant as an explicatum of what Leibniz called necessary truth and Kant analytic truth" (p. 8). The *L*-concepts Carnap defines

> ... are meant as explicata for certain concepts which have long been used by philosophers without being defined in a satisfactory way. Our concept of L-truth is ... intended as an explicatum for the familiar but vague concept of logical or necessary or analytic truth as explicandum (p. 10).

Again, the L-concepts are introduced using the concepts of *state-dsecription*. A class of sentences Σ is a state-description = $_{df}$ for every atomic sentence ϕ, Σ contains either ϕ or $\sim\phi$, but not both. A state-description

> ...obviously gives a complete *description* [emphasis added] of a possible state of the universe of individuals with respect to all properties and relations expressed by predicates of the system. Thus the state-descriptions *represent* [emphasis added] Leibniz' possible worlds or Wittgenstein's possible states of affairs (p. 9).

These allusions to Leibniz' possible worlds and Wittgenstein's possible states of affairs in informally characterising state-descriptions should not be seen as inconsistent with Carnap's general aversion to metaphysics. Formally, state-descriptions are just complete, non-contradictory assignments of extensions and, as such, they carry no metaphysical baggage. The modal notions that interest Carnap are logical or analytic, certainly not metaphysical.

Carnap imposes the following informal condition on any proposed definition of L-truth: A sentence ϕ is L-true in a system S if and only if "its truth can be established on the basis of the semantical rules of the system S alone, without any reference to (extra-linguistic) facts" (p. 10). A definition that satisfies this condition is suggested, says Carnap "... by Leibniz' conception that a necessary truth must hold in all possible worlds" (p. 10). And since "our state-descriptions represent the possible worlds" (p. 10) – every complete, non-contradictory assignment of extensions is a state-description representing a possible world – we arrive at the following definition: A sentence ϕ is L-true $=_{df}$ ϕ holds in every state-description.

Carnap sets up his semantical rules so that $\Box\phi$ is true if and only if ϕ is L-true. However, certain complications arise because Carnap now has individual descriptions and abstracts in his system and insists on treating them as singular terms – in his 1946 system, the only singular terms were individual constants and variables.

Apparently drawing on Church's (1942) observation about contextual definition and primitive notation, Carnap (1947) and Smullyan (1947) both claim (mistakenly, as it turns out) that appeals to Russell's Theory of Descriptions will dispose of Quine's paradox; however, Carnap rejects Russell's theory in favour of a "Fregean" treatment according to which descriptions are singular terms, and so has to reject this purported solution. His own solution is similar to Church's and boils down to restricting the range of variables – at least in modal contexts – to intensional entities. However, in order to avoid the validity of ($\Box\phi \equiv \phi$) and thereby the collapse of modal distinctions, Carnap has to make an ad hoc stipulation in connection with descriptions: their matrices may not contain $\Box$ (p. 184). Thus, e.g., $\imath x\Box(x=a \cdot Fx)$ has to be ruled ill-formed. Carnap includes in his book a letter from Quine in which the latter suggests that restricting the range of variables to intensions is "an effective way of reconciling quantification and modality" (p. 197).

Smullyan (1947) by contrast – who says that by $\Box$ he means "what is ordinarily meant by 'it is necessary that'," (pp. 139–140) – embraces both Russell's Theory of Descriptions (at least as an account of *descriptions*) and the substitutional thesis, which he calls the *principle of existential generalisation*, presumably because of its connection to the inference principle that goes by the same name (it is important not to confuse the two). In attempting to dispose of Quine's ontological paradox, Smullyan first appeals to the idea that *if* ordinary names are *not* analysed in accordance with the Theory of Descriptions, then they function like Russell's *logically proper names* and, as such, when two of them have the same bearer they are *synonymous* (p. 140 and p. 141). He then claims that the correct deployment of Russell's theories of descriptions and logically proper names renders modal contexts referentially transparent and thereby disposes of the paradox facing the modal logician who subscribes to the substitutional thesis as a way of explicating the semantics of modalised matrices. Next he points to a fallacy involving existential generalisation – the inference principle – in connection with descriptions occurring within the scopes of modal connectives. Finally, he suggests that the intensional contortion to which Church and Carnap are subjecting themselves in order to get around Quine's objections to quantification in modal systems is simply the price to be paid for abandoning Russell's Theory of Descriptions in favour of a referential treatment. Some of these points need to be spelled out to see where they lead to trouble.

(i) According to Smullyan, if "Hesperus" and "Phosphorus" are logically proper names, then the second conjunct of (11) is false, so the truth of (11') cannot be inferred by the substitutional thesis, and Quine's paradoxical conclusion is avoided. This is correct of course, *as long as one is prepared to accept that "Hesperus" and "Phosphorus" are synonymous*. On such an account modal contexts – indeed, *all* non-quotational contexts – will be referentially transparent, and "$(a=b) \supset \Box(a=b)$" will be true, whenever a and b are names, as Smullyan recognises. But Quine, Church, and Carnap all explicitly *reject* the synonymy thesis, and so do many others. The systems of Marcus (1946, 1947) do not contain proper names – or, more precisely, they do not contain the nearest thing to proper names found in formal systems, viz. individual constants – so it is an open question whether she will side with Smullyan or Quine *et al.* on this matter. (In 1962 she will side explicitly with Smullyan.)

(ii) If "Phosphorus" and "Hesperus" are disguised Russellian descriptions, then the situation is more complicated than Smullyan makes out. It is important to bear in mind that Smullyan is trying to defend quantified modal logic, explicated in part by appeal to the substitutional thesis, from paradox. Suppose "Phosphorus and "Hesperus" abbreviate "the uniquely brightest star before sunrise" and "the uniquely brightest star after sunset", respectively. Smullyan (a) declares the second conjunct of (10) false on the grounds that it is not logically necessary that there exist exactly one brightest star before sunrise (as required by employing Russell's Theory of Descriptions within the scope of $\Box$), and (b) points out that if that conjunct is false, the truth of (10′) cannot be inferred by the substitutional thesis. Thus, he claims, Quine's paradoxical conclusion is again avoided.

This is too quick. First, a minor point. Surely Smullyan is just pointing to a slight deficiency in the choice of example by bringing up the fact that it is not logically necessary that there exist exactly one brightest star before sunrise. Quine could have stated the paradox using conditionals as the second conjuncts, as in (12) and (13):

(12) Phosphorus = Hesperus ·

    $\Box$((Phosphorus exists · Hesperus exists) $\supset$ Phosphorus = Phosphorus)

(13) Hesperus = Hesperus ·

    ~$\Box$((Phosphorus exists · Hesperus exists) $\supset$ Hesperus = Phosphorus).

(12′) and (13′) could then have been inferred from the substitutional thesis, displaying the same paradox:

(12′) $\exists x(x = $ Hesperus · $\Box$((Phosphorus exists · Hesperus exists) $\supset x = $ Phosphorus)).

(13′) $\exists x(x = $ Hesperus · ~$\Box$((Phosphorus exists · Hesperus exists) $\supset x = $ Phosphorus)).

So Smullyan's (correct) observation that it is not logically necessary that there exist exactly one brightest star before sunrise does not dispel the paradox. But, as I said, this is only a minor point. There are two aspects to the important point: (1) *Quine's paradox cannot even be stated if "Hesperus" and "Phosphorus" are analysed as Russellian descriptions; but (2) far from helping Smullyan, such analyses render the substitutional thesis unworkable, so no paradox is needed to undermine it anyway.* The dedicated reader is welcome to use $\imath x Px$ and $\imath x Hx$ as shorthand for the descriptions that "Phosphorus" and "Hesperus" abbreviate and then spell out (12) and (13) in primitive notation using Whitehead and Russell's *14.01 and *14.02 for all possible scopes. Fortunately, my main points can be made without doing that. (a) Regardless of scope assignments, the results will contain no singular terms except variables, so there is no way of using the substitutional thesis to produce, from these results, additional existential generalisations that create a paradox, but also no way of using the substitutional thesis to explain the semantics of modalised matrices in terms of the replacement of variables by individual constants. (b) The fact that there are possible readings of the second conjuncts of (12) and (13) upon which $\Box$ appears within the scopes of all four descriptions in the second conjunct, will not reinstate the paradox because the positions occupied by "Hesperus" and "Phosphorus" in the consequents of the conditionals will contain different variables, and only in (12) can any link be forged between them (through the truth of the *first* conjunct). So the net result is the same as before: no hope of paradox, but also no hope of using the substitutional thesis to respond to Quine's *interpretive* challenge, which still stands.[68]

It is worth noting that if ιxPx and ιxHx lie within the scope of $\Box$ in the second conjuncts of (12) and (13), then both are true. By contrast, if $\Box$ lies within the scopes of the relevant descriptions, then Smullyan and others who permit quantification into modal contexts will want to say that (12) is true and (13) false. But this observation will not address the paradox; indeed, it only serves to underscore the point that Quine is still owed an account of how to interpret a modalised matrix $\Box Fx$ if names are treated as Russellian descriptions.

Smullyan would seem to have shown, then, that the advocate of the substitutional thesis can avoid Quine's ontological paradox in just one way: by drawing a sharp distinction between names and descriptions, treating ordinary names as logically proper names, and contextually defining definite descriptions. In dealing with the ontological paradox in this way, no appeal is made to a reading of a modal sentence in which a description has large scope over (and hence binds a variable within the scope of) a modal operator, so Smullyan is begging no question if he subscribes to the substitutional thesis, as he suggests he does. *All of the work is done by a controversial theory of names, a theory that Quine, Church, and Carnap reject.*

Although the ontological paradox appears to be avoidable by adopting the controversial theory of names, Quine's main question – how are we to *understand* examples in which $\Box$ is attached to a matrix rather than a closed sentence? – is still not *completely* answered. Recall Quine brought up the purported ontological paradox as one of *two* drawbacks of a particular response to that question, the response based on the *substitutional thesis*. At best, Smullyan has shown that invoking the substitutional thesis does not lead to a certain paradox if one is prepared to treat names as logically proper names. But there is still the problem of unnamed (and unnameable) objects to address. Surely our quantifiers don't range over only the things we have named. Perhaps someone will propose to so restrict them, or to allow them to range only over names in some way.

Taking stock. By the end of the 1947, it would seem Quine has demonstrated that quantified modal logic is intelligible only on the assumption that the entities over which variables range in modal contexts are not the same entities over which they typically range in extensional contexts. Following Church and Carnap, one can opt

for intensional entities (in intensional contexts and, if so desired, also for extensional contexts); following Smullyan, one can opt for entities that have names (or,
in some way yet to be elaborated, for names themselves) and treat co-referring
names as synonymous. All of the important points relevant to the interpretation of
logical necessity and possibility are now in print, although not always in the clearest
form. From 1948 onwards, we will find (a) interesting formal observations about
scope and deducibility made by Smullyan (1948) and Fitch (1949), (b) interpretive
mistakes by Smullyan, Fitch, Marcus, Church, and Carnap, (c) mistaken rejoinders to
Smullyan *et al.* by Quine in 1953, 1961, 1969, and 1980, (d) talk about a commitment
to "Aristotelian essentialism" by Quine (1953c, 1960), which *seems* to presage a
metaphysical interpretation of $\Box$ and $\Diamond$, (e) arguments by Quine (1960, 1961) that
restricting the range of the variables of quantification to intensional entities will not
avoid modal paradox, (f) the use of slingshots by Quine (1953a, 1953c, 1960) against
non-extensional logics generally, (g) mathematical work on models for modal logic
in the late 1950 and early 1960s – work that does not bear as directly as some would
make out on the matter of intelligibility – and (h) formal work on modalities other
than the purely logical culminating in Kripke's (1972) discussion of a metaphysical
conception, perhaps latent in some of his 1963 work.

§20. 1948–1953: The reign of confusion

1948. In "Modality and Description", Smullyan observes that sentences containing
descriptions and modal connectives (e.g. "necessarily the number of planets in our
solar system = 9") seem to be ambiguous in the same way as sentences containing
descriptions and verbs of propositional attitude (e.g. "George IV doubts that the
number of planets in our solar system = 9"), i.e. ambiguous in respect of the scope
of the description, as predicted by Russell's Theory of Descriptions. Smullyan also
echoes a point made by Church (1942): if descriptions are treated in accordance
with Russell's theory, the false (16) is not deducible from the true premises (14)
and (15), (where "P" is shorthand for "numbers the planets in our solar system):[69]

(14) $\Box(9 = 9)$
(15) $9 = \iota x P x$
(16) $\Box([\iota x P x]\ 9 = \iota x P x)$.

For (15) and (16) are "mere shorthand" for (15′) and (16′):

(15′) $\exists x (\forall y (P y \equiv y = x) \cdot x = 9)$
(16′) $\Box \exists x (\forall y (P y \equiv y = x) \cdot x = 9)$.

So far, then, Smullyan is simply elaborating Church's point that *an example involving a Russellian description* does not, by itself, demonstrate the failure of *substitutivity of singular terms* in modal contexts, because Russellian descriptions are not
singular terms. He then makes a *new* point: the purportedly true (17), which is
shorthand for (17′) *is* derivable from (14) and (15):

(17) $[\iota x P x]\ \Box(9 = \iota x P x)$
(17′) $\exists x (\forall y (P y \equiv y = x) \cdot \Box(x = 9))$.

More carefully, Smullyan should say that *if co-referring singular terms, including variables, are intersubstitutable in modal contexts,* (17′) is derivable from (14) and (15), as we can see from the following derivation:

1	[1]	$9 = \iota x P x$	premiss
2	[2]	$\Box(9 = 9)$	premiss
1	[3]	$\exists x(\forall y(Py \equiv y = x) \cdot x = 9)$	1, def. of ιx
4	[4]	$\forall y(Py \equiv y = \alpha) \cdot \alpha = 9$	assumption
4	[5]	$\alpha = 9$	4, ·-ELIM
2,4	[6]	$\Box(\alpha = 9)$	2, 5, $\Box$+PSST
4	[7]	$\forall y(Py \equiv y = \alpha)$	4, ·-ELIM
2,4	[8]	$\forall y(Py \equiv y = \alpha) \cdot \Box(\alpha = 9)$	6, 7, ·-INTR
2,4	[9]	$\exists x(\forall y(Py \equiv y = x) \cdot \Box(x = 9))$	8, $\Box$+EG
1,2	[10]	$\exists x(\forall y(Py \equiv y=x) \cdot \Box(x = 9))$	3, 4, 9, $\Box$+EI

Crucially, the move from lines [2] and [5] to line [6] presupposes that $\Box$ is +PSST, (i.e. that modal contexts are referentially transparent).

Contrary to what people have claimed (on both sides), the reading of "necessarily the number of planets in our solar system = 9" given by (17)/(17′) in which the description has large scope is *completely irrelevant to the main debate*. Quine does not need it reject it at any point in his argument for the referential opacity of modal contexts, in stating his challenge to provide an interpretation of modal matrices, or in setting out the alleged paradox facing those who endorse the substitutional thesis; and Smullyan does not appeal to it in attempting to refute the argument for referential opacity or defuse the paradox. In connection with (17′), however, it is legitimate for Quine to ask the same old question from 1943 and 1947: how is such a sentence to be *interpreted*, given that it contains a matrix "$x = 9$" governed by $\Box$?

Given what he said in his review of Quine's 1947 paper, presumably Smullyan will appeal to a suitably elaborated version of the substitutional thesis: $\exists x(\dots \cdot \Box(x=9))$ is true if there is some individual name (constant) c such that (i) "$\dots$" is true when c replaces all occurrences of x it contains and (ii) "$c = 9$" analytic. (Of course it would be more complicated in the case at hand because of the universal quantifier within the scope of the existential.) He would then offer "9" as such a name. But as it is easy to introduce a new name for 9, e.g. "Planeto", so again Smullyan would need to appeal to the thesis that co-referential names are synonymous in order to avoid paradox. (In the case of names for numbers, it might, perhaps, be feasible to hold that some coreferring names are synonymous, but surely not those whose references are (to use Kripke's (1972) terminology) fixed by descriptions like "the number of planets" or "my favourite number".)

To sum up, then, Smullyan's 1948 paper does not advance the case for an interpretation of modal matrices and thereby quantified modal statements. Everything Smullyan had to say about *that* was said in his 1947 review.

Marcus (1948) in her review of Smullyan's (1948) paper claims that,

Smullyan is justified in his contention that the solution of Quine's dilemma does not require any radical departure from a system such as that of *Principia Mathematica*. Indeed, since such a solution is available it would seem to be an argument in favor of Russell's method of introducing descriptions and abstracts. (p. 150).

This remark seems to betray a misunderstanding of the central issue. Endorsing "Russell's method of introducing descriptions" will certainly prevent Quine from demonstrating that substitutivity of singular terms fails in modal contexts by *using an example involving a description*. But it does not prevent him from demonstrating the same point with ordinary names that co-refer, or from setting out his paradox, unless Smullyan or Marcus can provide an argument for the highly controversial thesis that ordinary proper names, when co-referential, are synonymous, a thesis explicitly rejected by, *inter alia*, Quine, Carnap, Church, Russell, and Frege. The paradox remains. The central question is, I repeat, how to interpret modal matrices, and no appeal to Russell's Theory of Descriptions, glorious as it may be, can provide an answer to *that question*.

1949. In "The Problem of the Morning Star and the Evening Star", Fitch recapitulates several points made by Smullyan in 1947 and then adds one of his own.

(i) Fitch claims (with Smullyan) that the second conjunct of (11) is false if "Hesperus" and "Phosphorus" are names:

(11) Hesperus = Hesperus $\cdot$ $\sim\square$(Hesperus = Phosphorus).

(ii) Like Smullyan, he provides no argument for the synonymy of "Hesperus" and "Phosphorus" upon which this claim is based.

(iii) Fitch then claims (with Smullyan) that the second conjunct of (10) is false if "Hesperus" and "Phosphorus" are treated as Russellian descriptions within the scope of $\square$:

(10) Phosphorus = Hesperus $\cdot$ $\square$(Phosphorus = Phosphorus).

The reason (again with Smullyan) is that the second conjunct entails the existence of a unique brightest star before sunrise. We have already seen that Quine can circumvent this minor irritation. To keep things simple, let us forget about it for the remaining points.

(iv) Fitch then considers the case in which the descriptions in the second conjuncts have larger scope than $\square$. This time it is the second conjunct of (11) that is said to be false, for in primitive notation that conjunct will have the following form:

(11*) $\exists x(\forall y(Hy \equiv y=x) \cdot \exists z(\forall w(Pw \equiv w=z) \cdot \sim\square(x=z)))$.

And this, says Fitch, is (by Barcan's theorem 2.32*) equivalent to (11**), in which $\square$ has disappeared: [70]

(11**) $\exists x(\forall y(Hy \equiv y=x) \cdot \exists z(\forall w(Pw \equiv w=z) \cdot \sim(x=z)))$.

And this conflicts with the truth of the first conjunct of (10). But again Fitch is simply exploiting a contingent feature of the example rather than addressing the real issue. An appeal to Barcan's 2.32* can be made only when the second conjunct of the original sentence contains an identity statement, which is not a necessary feature of the example (unlike the identity statement in the first conjunct). Quine can, and does, make his point perfectly well by using as his second conjuncts the following:

(10#) □(there is life on Phosphorus ⊃ there is life on Phosphorus)
(11#) ~□(there is life on Hesperus ⊃ there is life on Phosphorus).

Once (11#) is put into primitive notation with the descriptions assigned larger scope than □, as Fitch requires, Barcans's 2.32* will not provide a □-free sentence that is equivalent. And so the same old question reappears: how do we interpret the modal matrix?

(v) Fitch's final point is a solid one. By *Principia* *14.3, if E!ιxφ then altering the scope of ιxφ does not alter truth-value in extensional contexts:

*14.3 $\forall f[\{\forall p \forall q ((p \equiv q) \supset f(p) \equiv f(q)) \cdot E!\iota x\phi)\} \supset$
$\quad \{f[\iota x\phi]G\iota x\phi) \equiv [\iota x\phi] f(G\iota x\phi)\}].$

But as Smullyan observed, this is not so for modal contexts. Fitch points out that an analogue of *14.3 for modal contexts holds if "E!ιxφ" is replaced by "□E!ιxφ".

There has been no progress in answering Quine's question for two years now.

1950. In "On Referring", Strawson launches a full-scale assault on Russell's Theory of Descriptions. In a review of Fitch's (1949) paper, Church says:

It is pointed out that if the phrases "the Morning Star" and "the Evening Star" are construed as descriptions in Russell's contextual sense, then Quine's argument fails to show any difficulty in modal logic (compare the review VII 100(2)). In particular, no objections appear from this point of view against the [quantified extensions of the Lewis] systems S2 and S4 of Ruth Barcan's XII 95(4) (p. 63)

(VII 100(2) is Church's 1942 review of Quine's 1941 paper.) On the assumption that "points out" is a factive verb, the same false claim has now been made by Church, as well as by Carnap, Smullyan, Marcus and Fitch. While treating "Phosphorus" and "Hesperus" as Russellian descriptions will undermine a *direct* argument for the referential opacity of modal contexts, it cannot dispel the ontological paradox facing the advocate of the substitutional thesis, and it leaves Quine's main question unanswered: how are we to understand modal matrices when □ expresses logical (i.e. analytic) necessity?

As we have seen, a partial solution to the paradox and a partial answer to Quine's question emerge if "Phosphorus" and "Hesperus" are Russellian logically proper names and hence, by virtue of being co-referential, *synonymous*. (On such an account, as long as descriptions and abstracts are contextually defined, modal contexts are referentially transparent.) But Church (1950) dismisses this idea. Echoing Quine, he says that,

...as ordinarily used, "the Morning Star" and "the Evening Star" cannot be taken to be proper names in this sense; for it is possible to understand the meaning of both phrases without knowing that the Morning Star and the Evening Star are the same planet. Indeed, for like reasons, it is hard to find any clear examples of a proper name in this sense. (p. 63)

1951. Von Wright (1951) publishes *An Essay on Modal Logic* in which he distinguishes and characterises *alethic, deontic, dynamic,* and *epistemic* modalities; stressing the similarities between quantifiers and modalities, he takes up the idea that quantification expresses an *existential* modality. In "The Logic of Causal Propositions", Burks sets out a system of quantified *causal* logic. In "On the Need for Abstract Entities in Semantic Analysis" and "A Formulation of the Logic of Sense and Denotation" Church advances his intensional agenda and attaches the necessity operator not to sentences but to names of propositions (construed as senses). Rasiowa proves the completeness of a system of quantified S4. In "Three Dogmas of Empiricism", Quine presses his worries about analyticity and claims that Carnap's (1947) appeal to state descriptions yields, at best, a reconstruction of logical truth, not of analyticity. He suggests that Carnap's idea of a sentence being *L*-true if it is true in every state description is an adaptation of Leibniz's "true in all possible worlds". Quine is not here attributing a metaphysical, rather than a logical or analytic conception of possibility to Carnap. Talk of "possible worlds" carries no such commitment.

1952. Three interesting logic books appear, two of which articulate quantified modal systems. In his *Untersuchungen*, Becker focuses on questions of interpretation, especially statistical and normative, in connection with the system he presented in 1941. In *Symbolic Logic*, Fitch's preferred system is a modification of S4 that does not assume the Law of Excluded Middle and in which $\Diamond\exists x\phi$ can be derived from $\exists x\Diamond\phi$ (but not *vice versa*). In *Introduction to Logical Theory*, Strawson points out that it is usual to use "necessary", "logically necessary", "logically true", and "analytic" interchangeably, mounts an assault on the idea that the logical particles map cleanly onto their natural language counterparts, and argues that standard extensional logic is "inadequate for the dissection of most ordinary types of empirical statement." (p. 216)

1953. Quine publishes *From a Logical Point of View*, which contains "Reference and Modality", largely a fusion of his 1943 and 1947 papers, but with interesting discussions of Carnap, Church, Fitch, and Smullyan. (The paper will be revised for the 1961 and 1980 printings; page numbers below are to the original 1953 edition.)

(i) Quine introduces the phrase *referential opacity*, characterises the notion cleanly in terms of the interpretation of variables (rather than constant singular terms, which he still sees as contextually eliminable "frills"), and amplifies his 1947 remarks concerning the possibility of making sense of quantified modal logic by taking the course suggested by Church (1943) of restricting the range of variables to intensional entities, a course he also sees in the work of Carnap (1947) and (perhaps) Fitch (1951). On such a view, the universe of discourse is restricted to objects x such that any two conditions uniquely determining x are analytically equivalent.

(ii) Such a restriction, Quine argues, leads to the following theorem:

(18) $\forall x \forall y ((x = y) \supset \Box (x = y))$.

He notes that Fitch proves the following (his 23.6):

(19) $(a = b) \supset \Box(a = b)$

and suggests it is unclear whether (19) should be read as equivalent to (18) because it is unclear whether Fitch construes "a" and "b" as bindable variables or as schematic letters for available names. (It is crucial to understand that neither (18) nor (19) entails any claim about what Kripke (1971, 1972) will later call the "rigidity" of proper names. (18) talks about *objects* (not names or statements): for all *objects* x and y, if x and y are the same object, then necessarily x and y are the same object. So (18) is silent on the semantic properties of proper names and on the modal status of particular true statements of identity. It is compatible with all sorts of theories of names, including theories that treat *no* names as rigid designators. It is also a mistake a to think that Fitch's (19) entails that names are rigid designators (taking "a" and "b" as schematic letters for names). (19) is compatible with two names α and β co-referring *at* every possible world and neither preserving reference *across* possible worlds (e.g., α and β both referring to Venus in world W_1, both referring to Mars in W_2, to Quine in W_3, and so on). Kripke's forthcoming thesis that names are rigid designators entails (19), but not *vice versa*.)

(iii) Smullyan, Quine notes, has taken a different course by challenging the reasoning that leads to the alleged "ontological burden" of quantified modal logic:

[Smullyan's] argument depends on positing a fundamental division of names into proper names and (overt or covert) descriptions, such that proper names which name the same object are always synonymous....He observes, quite rightly on these assumptions, that any examples which...show failure of substitutivity of identity in modal contexts, must exploit some descriptions rather than just proper names. (p. 155)

So far so good: Quine accepts that if co-referential names are synonymous, examples involving "Phosphorus" and "Hesperus" cannot be used to demonstrate the referential opacity of modal contexts. But then Quine makes a mistake:

Then, [Smullyan] undertakes to adjust matters by propounding, in connection with modal contexts, an alteration of Russell's familiar logic of descriptions. [*Footnote*: Russell's theory of descriptions, in its original formulation, involved distinctions of so-called "scope". Change in the scope of a description was indifferent to the truth value of any statement, however, unless the description failed to name. This indifference was important to the fulfillment, by Russell's theory, of its purpose as an analysis or surrogate of the practical idiom of singular description. On the other hand, Smullyan allows difference of scope to affect truth value even where the description concerned succeeds in naming.] (p. 155)

In fact Smullyan makes no "alteration" of Russell's theory: he applies it exactly as *14.01 and *14.02 dictate, exactly as Russell applied it in connection with e.g., sentences containing negation or verbs of propositional attitude. (This error will remain in Quine's 1961 revision of the paper, but will be expunged in a further revision in 1980, in which Quine will represent Smullyan as "taking a leaf from Russell".)

(iv) Quine claims that substitutivity fails in what we can call "entity identity" contexts.[71] The truths (20) and (21) become the falsehoods (20') and (21') if "9" is replaced by "the number of "planets", says Quine

(20) The proposition that $9 > 7$ = the proposition that $9 > 7$

(20') The proposition that $9 > 7$ = the proposition that the number of planets > 7

(21) The attribute of exceeding 9 = the attribute of exceeding 9

(21') The attribute of exceeding 9 = the attribute of exceeding the number of planets.

And existential generalisations on positions occupied by "9" in these examples, says Quine, yield unintelligible sentences.

Here Quine has overstepped the mark. Firstly, anyone who (a) thinks there are no propositions or attributes, and (b) endorses Russell's Theory of Descriptions, will take (20) and (21) to be false. Secondly, if there are propositions and attributes, it is far from clear that the case for substitution failure can be made using only names. Thirdly, it depends upon the details of *particular theories of propositions and attributes* whether (20') and (21') are false, or whether statements obtained from (20) and (21) by replacing "9" with a co-referential name are true or false. The same goes for fact-identity statements like (22') and (22"):

(22) The fact that Cicero denounced Catiline = the fact that Cicero denounced Catiline

(22') The fact that Cicero denounced Catiline = the fact that Tully denounced Catiline

(22") The fact that Cicero denounced Catiline = the fact that the author of *De Fato* denounced Catiline.

A Russellian about facts and descriptions will take (22') to be true and (22") to be false.[72]

(v) In the final paragraph, Quine presents an argument (based squarely on Church's (1943a) slingshot) against the viability of *any* non-truth-functional logics. (His 1960 version of essentially the same argument was the one analysed in detail in Part I).

The 1953 International Congress of Philosophy includes several papers on modal logic including "Three Grades of Modal Involvement", in which Quine sharpens his critique by focussing on referential opacity in connection with variables rather than constant singular terms – which is vital if the latter are all to be contextually defined – but makes what seem to be important concessions to Smullyan by accepting that quantified modal logic simply "complicates the logic of singular terms" (p. 81). The concession is quite unnecessary and appears to be based on the same misunderstanding of Smullyan's invocation of Russell's Theory of Descriptions on display in "Reference and Modality" earlier this year. This matter needs to be examined. (In the passages quoted below I have replaced Quine's "nec" by □.) The referential opacity of modal contexts, says Quine,

…has been shown by a breakdown in the operation of putting one constant singular term for another which names [sic.] the same object. But it may justly be protested that constant singular terms are a notational accident, not needed at the level of primitive notation....nothing in the way of singular terms is needed except the variables of quantification themselves. Derivatively all manner of singular terms may be introduced by contextual definition in conformity with Russell's theory of singular descriptions.... (p. 78)

Here Quine seems finally to realise that eliminating all constant singular terms contextually makes it impossible even to *formulate* his simplest (1941) argument for the referential opacity of modal contexts. So far so good. Quine continues,

Now the modal logician intent on quantifying into $\Box$ sentences may say that $\Box$ is not referentially opaque, but that it merely interferes somewhat with the contextual definition of singular terms. He may argue that "$(\exists x)\Box(x > 5)$" is not meaningless but true, and in particular that the number 9 is one of the things of which "$\Box(x > 5)$" is true. He may blame the real or apparent discrepancy in truth value between (4) and (18)

(4) $\Box(9 > 5)$
(18) $\Box(\text{the number of planets} > 5)$

simply on a queer behavior of contextually defined singular terms. Specifically he may hold that (18) is true if construed as:

(49) $(\exists x)[\text{there are exactly } x \text{ planets} \cdot \Box(x > 5)]$

and false if construed as

(50) $\Box(\exists x)[\text{there are exactly } x \text{ planets} \cdot x > 5]$

and that (18) as it stands is ambiguous for lack of a distinguishing mark favoring (49) or (50). [*Footnote referring to Smullyan (1948).*] No such ambiguity arises in the contextual definition of a singular term in extensional logic (as long as the named [sic.] object exists), and our modal logician may well deplore the complications which thus issue from the presence of $\Box$ in his primitive notation. Still he can fairly protest that the erratic behavior of contextually defined singular terms is no reflection on the meaningfulness of his primitive notation, including his open $\Box$ sentences and his quantification of them. (p. 78)

The claim that no ambiguity mirroring the one in Quine's (18) is to be found in extensional contexts (when the description's matrix is uniquely satisfied) is interesting. Strictly speaking, the point should be that no *truth-conditional* ambiguity arises. *Every description has to be given some scope or other* in order for *14.01 to be applied in any particular case to obtain a formula in primitive notation. The English sentence "the king of France kissed the queen of France" is *structurally* ambiguous in just the same way as "some boy kissed some girl" is. It is a trivial mathematical fact about these examples (unlike, say, "most boys kissed most girls') that the pairs of readings are logically equivalent. Furthermore, there is no guarantee that every description in use has a matrix that is uniquely satisfied – unless one adopts Hilbert and Bernays' stringent condition – so truth-conditional ambiguities will arise in extensional contexts, as in "No U.S. president has talked to the present king of France".

Quine does not cite *Principia* *14.3. (It will not be until some time between 1961 and 1980 that Quine, at Kripke's prompting, will realise the relevance of this theorem to the debate.) This naturally raises the issue of other non-extensional connectives, in particular those that express non-logical modalities, where analogous "complications" and "queer" or "erratic" behaviour will also arise. An example discussed earlier is helpful here. If Mercury gets sucked into the sun in the year 2002, the number of planets will be reduced from 9 to 8 – assuming no consequences for the other planets – but 9 would not be so reduced. And if "in 2001" functions semantically as some sort of one-place connective – there may be better treatments of course – and if (23a) and (23b) are intelligible, then in the year 2003 an utterance of (23) will be true when read as the former, false when read as the latter:

(23) In 2001 the number of planets in our solar > 8
  a. In 2001 $\exists x(\forall y(Py \equiv y=x) \cdot (x > 8))$
  b. $\exists x(\forall y(Py \equiv y=x) \cdot$ in 2001 $(x > 8))$.

Indeed, such an example raises the spectre of Church's (1942) complaint about the legitimacy of using an example containing a description in any argument for referential opacity. In the year 2003, the truth of (23) and of "8 = the number of planets" will not licence the truth of "In 2001 $(8 > 8)$". Indeed this seems to be part of Quine's next point:

> Looking upon quantification as fundamental, and constant singular terms as contextually defined, one must indeed concede the inconclusiveness of a criterion of referential opacity that rests on interchanges of constant singular terms. The objects of a theory are not properly describable as the things named by the singular terms; they are the values rather of the variables of quantification. [*Footnote:* See *From a Logical Point of View*, pp. 12ff., 75f., 102–110, 113ff., 148ff.] Fundamentally, the proper criterion of referential opacity turns on quantification rather than naming, and is this: a referentially opaque context is one that cannot be properly *quantified into*...However, to object to necessity as sentence operator on the grounds of referential opacity [thus construed] would be simply to beg the question. (p. 79)

But then Quine suddenly appears to retract the claim that quantifying into modal contexts is *unintelligible*:[73]

> ...necessity in quantificational application...is not *prima facie* absurd if we accept some interference in the contextual definition of singular terms. The effect of this interference is that constant singular terms cannot be manipulated with the customary freedom, even when their objects exist. (p. 80)

Here Quine has lost his thread. His misunderstanding of Smullyan's deployment of the Theory of Descriptions has distracted him from the points he made so clearly in 1947. The ambiguity that Smullyan, Fitch, and Marcus see in, e.g. "□(the number of planets > 7)" is *irrelevant*. More precisely, just as the existence of a reading upon which the description has large scope was of no use to, and was not used by Smullyan to make his perfectly valid point that *the small scope reading cannot be derived in the way Quine suggests*, so the alleged existence of that same reading is of no use to Smullyan in explicating the semantics of a modal matrix. What is still needed is an account of the rôle of $\Box Fx$ in fixing the truth-value of $\exists x \Box Fx$. An appeal to scope alone cannot hope to provide *that*. Quine seems to have succumbed to the same misunderstanding as his critics.

 But there is, perhaps, something else lurking behind Quine's remarks. In the same paper, a brand new problem is presented for quantified modal logic: a commitment to "Aristotelian essentialism".[74] It is just about conceivable that Quine has begun to suspect that some modal logicians – but certainly not Lewis and Carnap – are groping for a non-linguistic interpretation of $\square$ according to which it is objects, rather than the expressions used to specify them, in which the attributes relevant to understanding necessity reside.

§21. Post-1953

1956. In a thoroughly revised and expanded edition of *Introduction to Mathematical Logic*, Church presents an informal version of his 1943 slingshot. Ackerman presents his theory of *strenge Implikation* ("rigorous" implication). In "Quantifiers and Propositional Attitudes", Quine follows up on the worry he expressed in 1941 about Russell's idea of explaining perceived ambiguities in, e.g. "Ralph believes someone is spying" and "Ralph believes the author of *Waverley* is spying" in terms of scope, on the grounds that the *de re* readings obtained by giving the quantifiers large scope involve quantifying into opaque contexts. Instead of a structural ambiguity he suggests a *lexical* ambiguity: "believe" has both a *relational* sense which permits both substitution and quantification, and a *notional* sense that permits neither. He does not consider extending this idea to "necessarily" and "possibly". The justification for this seems to be that quantifying into modal contexts, unlike quantifying into attitude contexts, is (even if intelligible) dispensable.

1957. Mostowski inaugurates work on generalised quantifiers. (It will be some years before Russell's Theory of Descriptions is neatly recast in this framework.) Craig publishes papers that include his interpolation lemma; Prior publishes *Time and Modality*. Myhill follows Smullyan in claiming that modal paradoxes are avoided if singular terms are contextually defined. Hintikka and Kanger publish work in which they explore modality using "model sets of formulas" and "accessibility relations between possible worlds" to capture differences between different modal systems, i.e. by considering a set of worlds that are not possible *simpliciter* but possible (or not) relative to one another – if the accessibility relation is reflexive we get T, if it is also transitive we get S4, if it is reflexive, transitive, and symmetric we get S5. Bayart, Guillaume, Kripke, Montague, and possibly others are working along similar lines. The "possible worlds" central to this work (and subsequent work until a paper by Kripke in 1963) are identified with, or at least based on *linguistic assignments*, i.e. they are the progeny of Carnap's state-descriptions, models or complete assignments of extensions to the expressions of the object language. As such, possible worlds are well suited to the task of providing semantical analyses of languages containing an operator meant to express *logical* possibility (or necessity). And while talk of possibility generally does seem to be shorthand for talk of *logical* possibility, the mathematical systems themselves are consistent with quite an array of intuitive interpretations of modal operators. Contrary to what some people suggest, this mathematical work does not bear directly on Quine's worries about the intelligibility of quantified modal logic. A mathematical model can refute a technical claim that a particular system is inconsistent, but it cannot refute a philosophical claim about *intelligibility*.[75]

1959. Bayart and Kripke independently publish first-order extensions of S5 for which they prove completeness (the former drawing on Henkin's 1949 completeness proof for extensional logic). Kripke points out that his theorems can be formalised in an extensional metalanguage. In *The Place of Language,* Wilson tries to resolve the Quine-Smullyan debate and concludes that Smullyan has not provided a fully general rebuttal of Quine's argument. Let $\imath x W x$ stand for the author of *Waverley*, and $\imath x M x$ stand for "the author of *Marmion*". According to Wilson, on Smullyan's Russellian account, it is now possible to derive a false conclusion (24c) from two true premises (24a) and (24b), where scope assignments have been fixed in advance:

(24) a. $\imath x W x = \imath x M x$
 b. $\Box([\imath x W x] G \imath x W x \equiv [\imath x W x] G \imath x W x)$
 c. $\Box([\imath x W x] G \imath x W x \equiv [\imath x M x] G \imath x M x)$.

He concludes, with Church and Carnap, that "the individuals of a modal language are individual concepts rather than ordinary concrete individuals." (p. 43)

There is a serious error here. (The error will be repeated by Føllesdal in 1961, 1966.) If descriptions are Russellian (24c) is *not* derivable from (24a) and (24b), even with the indicated scopes. Using $\imath u M u$ in place of $\imath x M x$ so as to avoid obvious confusion, on a Russellian analysis of descriptions, the sentences in (24) abbreviate their counterparts in (25):

(25) a. $\exists x(\forall y(Wy \equiv y=x) \cdot \exists u(\forall v(Mv \equiv v=u) \cdot x=u))$
 b. $\Box[\exists x(\forall y(Wy \equiv y=x) \cdot Gx) \equiv \exists x(\forall y(Wy \equiv y=x) \cdot Gx)]$
 c. $\Box[(\exists x(\forall y(Wy \equiv y=x) \cdot Gx) \equiv \exists u(\forall v(Mv \equiv v=u) \cdot Gu)]$.

And (25c) simply does *not* follow from (25a) and (25b). Wilson's error seems to be attributable to the assumption that Smullyan's proposal reduces to the following two propositions: (i) a sentence containing a description and a modal operator is ambiguous, and (ii) once scope has been fixed, one can substitute for the description a co-denoting description having the same scope. Smullyan certainly holds (i), but he certainly does *not* subscribe to (ii), which is obviously incorrect.

Montague and Kalish (1959) point out that quantifying into contexts governed by $\Box$ interpreted as "it is provable that" (following Gödel's 1933 suggestion) must be intelligible for us to make sense of elementary facts about arithmetic, for example the fact expressed by (26)

(26) For each prime number x it is provable in arithmetic that x is prime.

In *Thought and Object,* Hampshire alludes to a potential ambiguity between Russellian and referential interpretations of definite descriptions, a topic discussed by Grice and others in lectures and tutorials at Oxford. Hampshire takes the Gricean position that (roughly) a distinction between what a speaker *says* and what he *means* explains the phenomenon in question without appeal to ambiguity. In "Proper Names", Searle presents a clear statement of the view that proper names should be understood in terms of descriptions.

1960. In the preface to an abridged printing of his 1918 book, Lewis says, "I wish the system S2, as developed in *Symbolic Logic*, Chapter V and Appendices II and III to be regarded as the definitive form of Strict Implication." (p. vii). Montague presents semantics for logical, physical, and deontic modalities based on S5, providing model-theoretic interpretations for each.

In *Word and Object*, Quine renews and summarises his assault on logical (strict) modality; he presents another slingshot using descriptions (the one discussed in Part I) and a further argument – a bad one – designed to show that under minimal assumptions that ought to be granted by modal logicians, every true sentence will be necessarily true.[76]

1961. In the foreword to the second edition of *From a Logical Point of View*, Quine says that

The principal revision affects pages 152–159, on the controversial topic of modal logic. A point that was made in those pages underwent radical extension on page 198 of my *Word and Object* (New York, 1960); and lately the situation has further clarified itself, thanks in part to a current doctoral dissertation by my student Dagfinn Føllesdal. These revised pages embody the resulting assessment of the situation.

The relevant pages are in "Reference and Modality" (pp. 139–159) where Quine discusses the positions of Carnap, Church, Fitch, and Smullyan. He adds that he has made "substantive emendations" also to pages 148 and 150 of the same article.

(i) The discussion of Fitch is now entirely deleted; the discussion of Smullyan is virtually unaltered; the matter of essentialism (mentioned in 1953 in "Three Grades of Modal Involvement" but not in the original version of "Reference and Modality") is raised again.

(ii) Quine repeals his concession to Church and Carnap that restricting the range of the variables to intensional entities is sufficient to restore substitutivity and hence interpretability. The idea would be to purge the universe over which the variables range of those entities that can be specified in ways that fail of necessary equivalence. What *ought* to happen after such a purge is that whenever two singular terms α and β designate one and the same entity, $\Box\Sigma(\alpha)$ and $\Box\Sigma(\beta)$ have the same truth-value, thus restoring substitutivity and opening the door to the long arm of the quantifier. But in the revision, Quine argues that the required purge cannot be carried out: there will always be intensional entities – for this is all the purged universe can contain – that fail of necessary equivalence of specification. If a is any intensional entity named by "a", and "ϕ" is a synthetic/contingent truth (these notions are not yet sharply distinguished) then "$a = \iota x(x=a \cdot \phi)$" is true and "$\iota x(x=a \cdot \phi)$" also designates a. But "a" and "$\iota x(x=a \cdot \phi)$" are not intersubstitutable *s.v.* in modal contexts, so substitutivity has not been restored and no door opened to quantification. (Smullyan and Fitch will doubtless say that Church and Carnap are again victims of their rejection of Russell's Theory of Descriptions in favour of a singular term treatment.)

In his 1961 dissertation, Føllesdal explores various ways of rendering quantified modal logic intelligible, including (i) analysing all singular terms in accordance with Russell's Theory of Descriptions, and (ii) restricting the class of singular terms to those that are "genuine" in the sense of referring to the same object in every possible world (i.e. those that Kripke will later call "rigid designators").

(i) Føllesdal concludes, with Quine (1953a, 1961) and Wilson (1959), that an alteration of Russell's contextual definitions would be required in order to take into account facts about scope. This is wrong, as we have already seen.

(ii) Føllesdal suggests that ordinary names are genuine singular terms because they refer to the same object in every possible world, but *only certain definite descriptions* have this property, e.g., "the positive square root of 81". PSST and EG can be applied only in connection with genuine singular terms. There are two main issues to take up here. Firstly, the idea that proper names refer to the same object in every possible world ensures that a true identity statement involving proper names such as "Cicero = Tully" is a *necessary* truth. So either (a) Føllesdal is disagreeing with Church, Carnap, and Quine, who all hold that "Cicero" and Tully" are not synonymous and hence that "Cicero = Tully" is not *analytically* or *logically* necessary; or else (b) he is unconsciously equivocating between logical necessity and some alternative conception under which "Cicero = Tully" is necessary. If (b), then Føllesdal's dissertation marks not only a major turning point in the theory of reference but a turning point in the interpretation of modal logic. Secondly, because of the misunderstanding of Russell's Theory of Descriptions Føllesdal has inherited from Quine and Wilson, he fails to see that he can treat descriptions in accordance with Russell's theory and retain a logically cleaner class of "genuine singular terms" without appealing to modal facts, viz. the class of proper names (and variables).

1962. In *An Introduction to Logic*, Mitchell argues for a distinction between Russellian and referential interpretations of descriptions in natural language. Marcus alludes to a similar ambiguity in her "Modalities and Intensional Languages", presented, with a response by Quine, at a Boston Colloquium for the Philosophy of Science. Marcus mostly rehearses points made by Smullyan and Fitch, and seems to suggest that Quine thinks he has shown the formal inconsistency of quantified modal systems. Quine points out that he has never argued for such inconsistencies, that his objections have concerned *interpretation* and *ontology*. The ensuing discussion involving Føllesdal, Kripke, McCarthy, Marcus, and Quine is inconclusive.[77]

1963. Acta Philosophica Fennica publishes papers on modal logic presented by Hintikka, Kripke, Prior, Montague, Rasiowa, and others at a colloquium on modal and many-valued logics held in Helsinki in August last year. Hintikka demonstrates how his notion of "model sets of formulas" can be used to build various systems of modal logic; Kripke provides completeness proofs for first order extensions of B, S4, and S5; Montague, drawing on work by Gödel (1932, 1933), demonstrates the hopelessness of treating "necessarily" as a predicate of sentences in even weak modal systems; Prior alludes to an ambiguity between Russellian and referential interpretations of descriptions and suggests implementing Russell's Theory of Descriptions using restricted quantifiers, $[\iota xFx]Gx$ representing "the F is G" (see Part I).

Besides his Helsinki paper ("Semantical Considerations on Modal Logic"), Kripke publishes "Semantical Analysis of Modal Logic, I", which marks a formal and perhaps also a philosophical turning point in modal logic: the treatment of possible worlds as *primitive points* (rather than *models*). Possible worlds are no longer identified with (or based on) linguistic assignments (i.e. complete assignments of extensions to the expressions of the object language). Rather they are antecedently stipulated possibilities. One might be excused for seeing Kripke's "points" as *metaphysical primitives* and "Semantical Analysis" as dangling the idea that $\square$ and $\lozenge$ can be understood as expressing *metaphysical* notions of necessity and possibility. (It is surely more than coincidental that in "Semantical Considerations", Kripke introduces a "varying domain" of individuals. On this matter, see Almog, 1986.)

1965. In "Modality and Quantification", Rundle suggests that co-referential names are intersubstitutable *s.v.* in modal contexts. After drawing a distinction between Russellian and referential interpretations of descriptions in extensional contexts, he then suggests that invoking this ambiguity dispels Quine's concerns about quantified modal statements. On the Russellian interpretation of "the number of planets", the sentence "$\square$the number of planets > 7" is false but not derivable from "$\square(9 > 7)$" and "9 = the number of planets". Presumably, Rundle has in mind the reading upon which the description has small scope; in which case he is agreeing whole-heartedly with Smullyan and Fitch (though curiously he mentions neither.) But on Rundle's referential interpretation of descriptions, "$\square$(the number of planets > 7)" is true – the description serves as just another name of 9 – and indeed follows from the aforementioned premises. So, unlike Smullyan and Fitch, Rundle sees the purported ambiguity in "$\square$(the number of planets > 7)" as *lexical* rather than scope-based. (Herein lies a problem with the account, which will be exposed by Cartwright (1968): unlike Smullyan's account of *de re* modal statements, Rundle's is not sufficiently general because of the possibility of non-referential *de re* uses of descriptions.) More importantly, as a response to Quine, Rundle's suggestion is beside the point unless, on his account, a referential description and a co-referential name are *synonymous*, an idea Quine explicitly rejects.

In "Quantification into Causal Contexts", Føllesdal sets out the details of an arguments for the referential opacity of contexts expressing *causal* necessity. His examples all involve definite descriptions. Suppose there is a well such that anyone who drinks from it gets poisoned; and suppose exactly one man has drunk from that well (and, of course, got poisoned). Now consider the following argument:

(27) It is causally necessary that the man who drank from that well got poisoned

(28) the man who drank from that well = the first Finn to memorise Homer

---

(29) It is causally necessary that the first Finn to memorise Homer got poisoned.

Since it is possible for the premises of this argument to be true while the conclusion is false, it would appear that the context governed by "it is causally necessary that" is referentially opaque. Consequently, it would appear to follow that quantification into such a context is unintelligible Suppose (30) is obtained from (27) by EG:

(30) $\exists x$(it is causally necessary that x got poisoned).

Who would be an example of someone of whom it would be true to say that it is causally necessary that he or she got poisoned? According to (27), the man who drank from that well is such a person; but the man who drank from that well is the first Finn to memorise Homer, so the the first Finn to memorise Homer is such a person; but to suppose this would conflict with the purported falsity of (29). Quantification into contexts of causal necessity appears to make no sense.

Føllesdal's preferred solution follows the one proposed in his dissertation: quantification into causal/modal contexts can be rescued by treating as "genuine singular terms" only proper names and those descriptions that refer to the same object in every physically/logically possible world. He now seems to accepts that he, Quine, and Wilson were wrong about descriptions, claiming that "undesirable results disappear from modal logic when descriptions are not treated as names, but contextually defined" (p. 271), as Church, Smullyan, Fitch, Marcus, and Myhill suggest. But we have already seen that this does not seem to be completely right as it leaves open the question of how to interpret modalised matrices when *logical* or *analytic* necessity is the issue. Indeed, Føllesdal is being too generous to Church, Smullyan, *et al.* A supplementary thesis is required: *essentialism*, construed as the thesis that *if* $\Box Fx$ is true of an object, it is true of it "regardless of the way in which this object is referred to." (p. 272) And this thesis, Føllesdal points out is not one Lewis and Carnap can accept. With an ironic twist he suggests that "by insisting on the "primacy of predicates" and the eliminability of *all* singular terms, Quine can thus be said to have levelled the road for modal logic." (p. 274, n. 14)

In a response paper, Chisholm (1965) recognises that Føllesdal has floated the possibility that proper names are "genuine referring expressions", although he states the point incorrectly: "an object in this world might be *identical* with an object named "De Gaulle" in some other possible world despite the fact that the first-named De Gaulle *differs* in various respects (there is a slight difference in weight, say) from the second" (p. 275) – but finds the idea "unintelligible" (*ibid.*)

Chisholm notes the impact of tense leading to failure of substitutions involving descriptions within the scope of simple causal verbs:

The assassination of Kennedy caused Johnson to become the U.S. President

Johnson = the U.S. President

The assassination of Kennedy caused Johnson to become Johnson.

1966. In "Reference and Definite Descriptions", Donnellan argues for a distinction between *attributive* and *referential* interpretations of descriptions, where the former are (roughly) Russellian in respect of their contributions to truth-conditions, and the latter more like demonstratives. Føllesdal's monograph *Referential Opacity and Modal Logic* – a lightly edited version of his dissertation – is published as a monograph.

1968. "In Some Remarks on Essentialism", Cartwright argues that we have to accept that there is an intuitive *de dicto-de re* ambiguity in,

(31) the number of planets is necessarily greater than 7

mirroring those in examples obtained by putting, say, "provably", "probably", "certainly", or "obviously" in place of "necessarily". He claims the *de dicto-de re* distinction in (31) is not characterised adequately by Smullyan's appeal to the scope of the description because the distinction shows up when the description is replaced by a name, as in

(32) 9 is necessarily greater than 7.

(It would seem, then, that Cartwright must hold that (31) and (32) are both *four* ways ambiguous, as each contains two singular noun phrases, neither, either, or both of which may be read *de re*.) To the potential rejoinder that Smullyan could treat names as disguised descriptions, Cartwright replies that Smullyan countenances names that are not descriptions, so the point can be made by using examples that contain such names. (In fact, Smullyan is a bit vague about whether he *does* countenance names that are *not* disguised descriptions.) Following Quine (1960), Cartwright claims that in bringing out an ambiguity of scope, Smullyan has used an "expanded version" of Russell's Theory of Descriptions. (p. 617) He also makes the important point that neither Rundle's (1965) Russellian-referential distinction nor Donnellan's (1966) attributive-referential distinction maps onto the *de dicto-de re* distinction: one may use a description attributively to make a *de re* statement as in an utterance of "the number of planets, whatever it may be, is necessarily odd". (p. 618)

In "Modal Logic", Marcus claims that Smullyan resolved the problem of substitutivity of identity and "to a considerable extent" the problem of quantifying into modal contexts by employing Russell's theories of names and descriptions. (p. 92) It is tempting to read the qualification as an acknowledgement that the substitutional thesis cannot fully explain quantifying in because of unnamed (and unnameable) objects.

1969. Publication of *Words and Objections*. In "Vacuous Names", Grice explains his position (dating to the 1950s) that the distinction between what a speaker *says* and what he *means* obviates the need for a semantically distinct non-Russellian sense of definite descriptions of the sort Marcus (1962), Mitchell (1962) Prior (1963), Rundle (1965), and Donnellan (1966) have been talking about. In "Quine on Modality", Føllesdal again suggests that the course taken by Smullyan *et al.* will enable the modal logician to avoid paradox. It is beginning to look as though Føllesdal is thinking about necessity in more of a metaphysical than an analytic way.

In his "Replies" Quine again takes on Smullyan:

My objection to quantifying into non-substitutive positions dates from 1942.[78] In response Arthur Smullyan invoked Russell's distinction of scopes of descriptions to show that the failure of substitutivity on the part of descriptions is no valid objection to quantification...

Still, what answer is there to Smullyan? Notice to begin with that if we are to bring out Russell's distinction of scopes we must make two contrasting applications of Russell's contextual definition of description. But when the description is in a non-substitutive position, one of the two contrasting applications of the contextual definition is going to require quantifying into a non-substitutive position. So the appeal to scopes of descriptions does not justify such quantification, it just begs the question.

First, note that making "two contrasting applications of Russell's contextual definition of descriptions" is not itself the issue. That is the case even in some extensional sentences, e.g., $\sim G\iota x Fx$, $(Fa \supset G\iota x Fx)$, and $\forall y (R y \iota x Fx)$. Quine's point seems to be that no appeal to ambiguity in, e.g., "$\Box$(the number of planets > 7)", *emdash* the purportedly true reading of which is meant to be represented by giving the description large scope *emdash* will help Smullyan explicate the interpretation of modal matrices. This is correct. But in fact Smullyan never claimed it would. Flushed with success after rightly pointing out that the false reading upon which the description has small scope cannot be derived by appeal to PSST or any other standard set of inference rules, Smullyan went on to miss the point that an account of modal matrices was still needed. Quine goes on,

Anyway, my objection to quantifying into non-substitutive positions can be made without use of descriptions. It can be made using no singular terms except variables...

... let us ban singular terms other than variables. We can still specify things; instead of specifying them by designation we specify them by conditions that uniquely determine them. On this approach we can still challenge the coherence of (4),

[(4) $(\exists x)$ necessarily x is odd]

by asking that such an object x be specified. One answer is that

(5) $(\exists y)(y \neq x = yy = y + y + y)$

But that same number x is uniquely determined also by this different condition: there are x planets. Yet (5) entails "x is odd" and thus evidently sustains "necessarily x is odd", while "there are x planets" does not.

The point I have just now tried to make is this: (I) If a position of quantification can be objected to on the score of failures of substitutivity of identity involving descriptions, it remains equally objectionable when no singular terms but variables are available. (1969, p. 338–9)

1971. In "Identity and Substitutivity", Cartwright urges a sharp distinction between the Principle of Identity (PI) and the Principle of Substitution (PS). The former is a metaphysical principle expressible thus:

(PI) $\forall x \forall y ((x = y) \supset (Fx \supset Fy))$.

By contrast, (PS) is a syntactico-semantic principle with essentially the content of PSST. Cartwright claims that logical muddles are in store for those who run together PI and PS.

In "Identity and Necessity", Kripke introduces the notion of a *rigid designator*, an expression that designates the same object in every possible world in which that object exists, where "possible" is to be understood in a *metaphysical* sense. Ordinary proper names are said to be rigid. (Detailed arguments for this position will be published next year in "Naming and Necessity"). One consequence of Kripke's position is that co-referential names are intersubstitutable in modal contexts, *if □ is read metaphysically*: if (i) Phosphorus = Hesperus and (ii) "Phosphorus" and "Hesperus" are names, then (iii) □(Phosphorus = Hesperus).[79] Kripke rehearses and endorses the argument From (PI) and the Principle of Necessary Self-identity (SI),

(SI) $\forall x \Box (x = x)$

to the conclusion that all identities are necessary (NI), a conclusion that many philosophers have regarded as paradoxical because of the existence of apparently contingent identity statements:

(NI) $\forall x \forall y((x = y) \supset \Box(x = y))$.

Kripke then stresses the point that (NI)

by itself does not assert, of any particular true statement of identity, that it is necessary. It does not say anything about *statements* at all. It says for every *object* x and *object* y, if x and y are the same object, then it is necessary that x and y are the same object. (p. 137)

But, from [NI] one may apparently be able to *deduce* [emphasis added] various particular statements must be necessary and this is then supposed to be a very paradoxical consequence. (p. 138)

How, then, are we to reconcile the truth of the apparently contingent identity (35) with the truth of (NI)?

(35) The first Postmaster General of the United States = the inventor of bifocals.

Kripke says he is going to be "dogmatic" about the issue: "It was I think settled quite well by Bertrand Russell in his notion of the scope of a description." (p. 138) The basic idea is that attaching a modal operator to (35) results in various readings, according to how scopes are assigned. Since there are 3 (relevant) operators there are 3! (=6) possibilities. Three non-equivalent readings will suffice for present purposes. Let "*Px*" and "*Ix*" abbreviate the matrices of the two descriptions in the obvious way:

(36) □The first Postmaster General of the United States = the inventor of bifocals.
    a. $\Box \exists x[\forall y(Py \equiv y=x) \cdot \exists u(\forall w(Iw \equiv w=u) \cdot u=x)]$
    b. $\exists x[\forall y(Py \equiv y=x) \cdot \Box \exists u(\forall w(Iw \equiv w=u) \cdot u=x)]$
    c. $\exists x[\forall y(Py \equiv y=x) \cdot \exists u(\forall w(Iw \equiv w=u) \cdot \Box u=x)]$.

Kripke then says that,

Provided that the notion of modality *de re*, and thus of quantifying into modal contexts, makes any sense at all, we have quite an adequate solution to the problem of avoiding paradoxes if we substitute descriptions for the universal quantifiers in [NI] because the only consequence we will draw, [*footnote omitted*] for example, in the bifocals case, is [36c] (p. 139)

And to the extent that quantifying into modal contexts makes sense, (36c) would seem to be true, for it says that,

> There is an object x such that x invented bifocals, and as a matter of contingent fact an object y, such that y is the first Postmaster General of the United States, and finally it is necessary, that x is y. What are x and y here? Here, x and y are both Benjamin Franklin, and it can certainly be necessary that Benjamin Franklin is identical with himself. So there is no problem in the case of descriptions if we accept Russell's notion of scope. (p. 139)

Appended to the last remark is a fascinating footnote, the first half of which reads as follows:

> An earlier distinction with the same purpose was, of course, the medieval one of *de dicto-de re*. That Russell's distinction of scope eliminates modal paradoxes has been pointed out by many logicians, especially Smullyan.
>
> So as to avoid misunderstanding, let me emphasize that I am of course not asserting that Russell's notion of scope solves Quine's problem of "essentialism'; what is does show, especially in conjunction with modern model-theoretic approaches to modal logic is that, is that quantified modal logic need not deny the truth of all instances of $\forall x \forall y((x = y) \supset (Fx \supset Fy))$, nor all instances of "$\forall x(Gx \supset Ga)$" (where "$a$" is to be replaced by a nonvacuous definite description whose scope is all of "Ga"), in order to avoid making it a necessary truth that one and the same man invented bifocals and headed the original Postal Department. Russell's contextual definition of descriptions need not be adopted in order to ensure these results; but other logical theories, Fregean or other, which take descriptions as primitive must somehow express the same logical facts. (139–40, note 5)

Now, precisely what "modal paradoxes" is Kripke referring to in the first paragraph of this note? Like Quine in *The Ways of Paradox*, Kripke seems to be using "paradox" in a perfectly acceptable way to mean something broader than "formal inconsistency".[80] He uses the plural "paradoxes", and obviously the context suggests that one of the modal paradoxes he has in mind is the alleged clash between (NI) and the contingency of (35). But Kripke's use of the plural suggests he also has in mind the other modal paradoxes discussed by Smullyan, for example Quine's semantic paradox involving "9" and "the number of planets", and his ontological paradox involving "Phosphorus" and "Hesperus". It should be clear by now that if *this* is Kripke's intention, then the point is mistaken. In the second half of the footnote, Kripke says that

> Some logicians have been interested in the question of the conditions under which, in an intensional context, a description with small scope is equivalent to the same one with large scope. One of the virtues of a Russellian treatment of descriptions in modal logic is that the answer (roughly that the description be a "rigid designator" in the sense of this lecture) then often follows from the other postulates of modal logic.

1972. Kripke publishes "Naming and Necessity", in which (*inter alia*) metaphysical necessity is expounded and the rigidity of ordinary names is introduced to the philosophical public.

1975. Whilst in residence at the Villa Serbelloni in Bellagio, Quine donates to the library some of his books, including a copy of the 1961 edition of *From a*

Logical Point of View. In the margin of page 154 next to the footnote in which he chastises Smullyan, Quine inscribes "Kripke has convinced me that Russell shared Smullyan's position. See *Principia* pp. 184f. esp. *14.3: the explicit condition of extensionality".

1980. Quine publishes a revised second edition of *From a Logical Point of View*. Only one page is changed from the 1961 edition, page 154, which, Quine says in a new foreword, "contained mistaken criticisms of Church and Smullyan" (p. vii). The charge against Smullyan has been excised and the relevant part of the page now reads as follows:

Then, taking a leaf from Russell ["On Denoting"], he [Smullyan] explains the failure of substitutivity by differences in the structure of the contexts, in respect of what Russell called the scopes of the descriptions. [*Footnote*: Unless a description fails to name [sic.], its scope is indifferent to extensional contexts. But it can still matter to intensional ones.]

This retraction hardly constitutes an admission that Smullyan has solved the fundamental *interpretive* issue and thereby rendered quantified modal logic intelligible. At the very least, Smullyan and his followers still need the thesis that co-referring names are synonymous and the substitutional thesis (both of which are highly problematic), or some alternative way of interpreting modal matrices if names are eliminated contextually. To date, no plausible interpretation of quantified modal logic has been provided where $\Box$ and $\Diamond$ express the strict notions of necessity and possibility that Quine was attacking.[81]

Rutgers University

NOTES

[*]   It is a great honour to present this essay to Professor Quine. It will not go unnoticed that I develop or push hard on points I find unclear or underdeveloped in his powerful and provocative work; but I hope it is as transparent as a modal context – non-analytically conceived – how stimulating, illuminating and plain enjoyable I find Quine's work. I thank John Burgess, Liz Camp, William Craig, Josh Dever, Donald Davidson, Dagfinn Føllesdal, Leon Henkin, Ernie LePore, Hans Kamp, James Levine, Jerry Katz, Michael Levin, Paul Boghossian, Saul Kripke, Simon Blackburn, Zsofia Zvolensky, and Alex Orenstein for comments and editorial assistance, and Quine for his patient and instructive reply in Karlovy Vary as well as his written comments. The issues addressed here are treated in more detail in *Facing Facts*. I gratefully acknowledge the support of the National Endowment for the Humanities and the Rockefeller Foundation.
[1]   I partake of the sin – *sloth*, rather than confusing *use* and *mention* – of not bothering with quotes around free-standing occurrences of $\Box$, $\Diamond$, and other symbols; not always, but sometimes.
[2]   That there might *be* such a general argument lurking seems to be suggested as early as his 1941 paper, "Whitehead and the Rise of Modern Logic". I have recently seen it claimed that what I am calling the *general* argument is aimed only at modal connectives, and then only in connection with quantified modal logic. The falsity of such a claim is easily seen by reading the last paragraph of Quine's 1953 paper, "Reference and Modality".
[3]   I notice that my anachronistic and incorrect 1990 critique of Quine has convinced some of my peers, e.g., Marti (1997) and Della Rocca (1996a, 1996b). I put the anachronism down to our youth and relative unfamiliarity (at the time) with the enormous literature on modality

– Marti and I were graduate students at Stanford together in the mid 1980s – and the mis-reading of Quine down to ... well, carelessness. What lies behind the under-estimations of Quine's critique made by Church (1942, 1987), Smullyan (1947, 1948), Fitch (1949, 1950), and Marcus (1948, 1962, 1990, 1993) is not so obvious as they are or were Quine's "chronies" (rhymes with "Ronnie's").

For a powerful, if rather polemical, analysis of some of the central issues, see Burgess's 1997 paper, from which I have learned a great deal. Anyone with an interest in the recent history of the theory of reference and its connection to theories of modality and modal logic – let us not confuse the two! – should read Burgess's paper. To say this is *not* to say I agree with everything in that paper. In particular, I disagree with some of what Burgess says about Smullyan and "Smullyanites", as will become clear.

[4]   I set out a preliminary version of this argument in my 1995 paper mentioned in the bibli-ography. It is set out more cleanly and in more detail in the present paper and in *Facing Facts*.

[5]   See Russell (1905) and Whitehead and Russell (1925), *14.

[6]   Throughout, I shall use $\iota x\phi$ where Russell uses $(\iota x)(\phi x)$. This policy will be applied in all contexts, even when quoting from *Principia*.

[7]   The *iota*-notation is rarely used in *Principia* after *14, "being chiefly required to lead up to another notation" (1925, p. 67), namely the inverted comma *of*-notation:

(i) $R'z$

is used as shorthand for "the object that bears R to z" and is introduced by a further contex-tual definition:

(ii) $R'z =_{df} \iota x R x z$.

Russell calls both (ii) and *14.01 "contextual definitions," and he also says that "$\iota x\phi$" and "$R'z$" are both "defined in use". Notice that in *14.01 we get whole formulae on both the right and left of "$=_{df}$", whereas in (ii) we do not. On Russell's account, then, technically it is not in the nature of a contextual definition that it involve whole formulae (Quine's shift from terms to sentences is not mandatory). Of course, (ii) can easily be recast in terms of formulae – just attach a one-place predicate to either side – but this would buy Russell nothing. According to (ii), (iii) is analysed as (iv) which, according to *14.01 is, in turn, analysed as (v):

(iii) $G(R'z)$

(iv) $G(\iota x R x z)$

(v) $\exists x(\forall y(Ryz \equiv y=x) \cdot Gx)$.

*14.01 differs from (ii) in that it involves a *logical reparsing*. It is only in the context of a whole formula that the analysis of "$\iota x\phi$" can be stated. Of course, *ultimately* this is true of "$R'z$" too.

[8]   For Russell, all of this is intimately tied up with his epistemology and psychology. Just as one can grasp the proposition expressed by an utterance of a sentence of the form "every F is G" or "no F is G" without knowing who or what satisfies the matrix Fx, indeed indepen-dently of whether or not anything does satisfy it, so one can perfectly well grasp the proposi-tion expressed by an utterance of a sentence of the form "the F is G" without knowing who or what satisfies Fx, and independently of whether or not anything does satisfy it. As it is sometimes put, one can perfectly well grasp the proposition expressed without knowing who or what is "denoted" by "the F", indeed independently of whether or not anything actually is "denoted" by it. To this extent, it makes no sense to say that the existence of the proposition depends upon the identity of the "denotation" of "the F"; so the proposition expressed is not singular.

[9]   There is an unfortunate tendency in the literature on Russell to construe the notion of an incomplete symbol as driven by notational considerations. See, e.g. Evans (1982), Linsky (1992, forthcoming). For discussion, see *Facing Facts*.

[10]   Whitehead and Russell do not appeal to *14.3 in subsequent proofs because of its use of propositions as values of variables "an apparatus not required elsewhere" (p. 185.) They proceed by individual cases, as they come up. Notice that *14.3 does not entail that the scope of a *relativized* description – i.e. a description containing a free variable such as $\iota x W x y$ (rep-resenting, say, "the woman sitting opposite him" as it occurs in "every man talked to the woman sitting opposite him" – is irrelevant in truth-functional contexts. In order for $E!\iota x\phi$ to be true $\iota x\phi$ cannot contain a free variable.

[11]   Less severe versions of this error can be found in remarks made by Carnap (1947), Føllesdal (1961), Hintikka (1989), Hintikka and Kulas (1985), Kalish *et al.* (1973), Lambert

(1991), Scott (1967), Thomason (1969), Wallace (1969), Wedberg (1966, 1984), and Wilson (1959). The fact that the error is so widespread naturally leads one to speculate why, especially as Russell is explicit as early as 1905 in "On Denoting" that altering the scope of a description – even one whose matrix is uniquely satisfied – *can* alter truth value in non-extensional constructions: that is how he deals with the puzzle involving "George IV wondered whether Scott was the author of Waverley"! (The point is made again in *Principia*.) I suspect the answer lies in the contrast between, on the one hand, the discussion of descriptions in the introduction to *Principia* and the informal remarks at the end of *14, and, on the other, the formal presentation of the theory and the relevant theorems. For example, in the introduction, Whitehead and Russell say that "when E!ιxφ, we may enlarge or diminish the scope of ιxφ as much as we please without altering the truth-value of any proposition in which it occurs." (p. 70); and the end of *14 they say "when E!ιxφ, the scope of ιxφ does not matter to the truth-value of any proposition in which ιxφ occurs. This proposition cannot be proved generally, but it can be proved in each particular case." (p. 184). But then they go on to add that "The proposition can be proved generally when ιxφ occurs in the form χιxφ and χιxφ occurs in what we may call a "truth function", i.e. a function whose truth or falsehood depends only upon the truth or falsehood of its argument or arguments." (p. 184) This, of course, is what theorem *14.3 says.

[12] See also Cartwright (1968, p. 618), who seems to have been the first person to make this point in print.

[13] As Kripke (1977) points out, the situation is the same with indefinite descriptions. Consider the following:

    (i)   Ralph thinks that a member of the board has been selling industrial secrets
    (ii)   Each teacher overheard the rumor that a student of mine cheated
    (iii)   A man from Ohio might have patented the zipper
    (iv)   Next year a man from Ohio will be President.

The indefinite descriptions in these sentences can be read as existentially quantified phrases with maximal scope. But contrary to what is assumed by (e.g.) Fodor and Sag (1982), it would be quite incorrect to account for these readings in terms of referential interpretations of the indefinite descriptions they contain. Kripke's point effectively undermines *all* of Fodor and Sag's arguments for a referential interpretation of indefinite descriptions. Fodor and Sag's general strategy is (*a*) to uncover readings of sentences in which indefinite descriptions appear to take maximal scope, and then (*b*) argue that plausible syntactical constraints on quantifier scope in natural language – so-called "island constraints" – are violated if the indefinite descriptions in question are given the required scope assignments. Their "solution" is to treat the offending indefinite descriptions as referring expressions. But, of course, this type of argument is powerless: indefinite and definite descriptions can take maximal scope without being used referentially, and in such cases referential interpretations are quite wrong. So even if one could demonstrate conclusively (in some other way) that indefinite descriptions have referential interpretations, an appeal to Russell's notion of scope is still needed in order to capture the readings on which indefinite descriptions take maximal scope. For detailed discussion of Fodor and Sag's arguments, see Ludlow and Neale (1991).

[14] Again, the situation is the same with indefinite descriptions. As Kripke (1977) observes, the following example is three ways ambiguous according as "a high American official" is given large, intermediate, or narrow scope:

(i)   Hoover charged that the Berrigans plotted to kidnap a high American official.

Similar examples can be constructed using iterated modalities.

[15] In the light of the publication of Kripke's (1977) therapeutic remarks on scope, the tendency to confuse referential and large scope readings appears to have abated, though see Fodor and Sag (1982), Hornstein (1984), and Stich (1986). It should be pointed out that Hornstein's accounts of both definite and indefinite descriptions are undermined by the behaviour of descriptions in non-extensional contexts. (This has been stressed by Soames (1987) and Neale (1990), but in the context of the present discussion certain points need to be stressed.) Firstly, Hornstein's account predicts (falsely) that only *large* scope readings are available for the definite descriptions in sentences such as the following:

    (i)   Ralph believes that the man who lives upstairs is a spy
    (ii)   John claims to have proved that the largest prime number lies somewhere between 10^{27} and 10^{31}
    (iii)   The first man in space might have been American
    (iv)   The number of planets is necessarily odd.

Secondly, since Hornstein claims (*a*) that *indefinite* descriptions (by virtue of belonging to his *Type II* category of quantified noun phrase) cannot take large scope over certain sentence-embedding operators, then to the extent that it is possible to get clear predictions from his account, it predicts (falsely) that only *small* scope readings are available for the indefinite descriptions in the sentences discussed in the previous two footnotes. Hornstein attempts to rescue his accounts of definite and indefinite descriptions by appealing to largely unspecified non-linguistic considerations. With respect to definite descriptions, Hornstein claims (falsely) that any ambiguities turn on whether or not the descriptions succeed in denoting. Something that does the work of the (linguistically prohibited) narrow scope reading is allowed to materialise at a level of semantical representation external to "the language faculty." With indefinite descriptions, the situation is apparently reversed: something that does the work of the (linguistically prohibited) *large* scope reading is generated by obscure non-linguistic factors. It is quite unclear to me how such plasticity is compatible with a firm and genuine distinction between his Type I and Type II quantifiers..

[16] In fact, there is even work to be done if *iota*-compounds are to occupy singular term positions and be subject to reference axioms, as in non-Russellian theories that treat descriptions as singular terms. The work is caused by the fact that descriptions contain formulae (sometimes quantified formulae) but at the same time occupy term positions in formulae; this makes it impossible to first define the class of terms, and then the class of formulae, as one does standardly; the classes must be defined together. It is surprising how often this fact is overlooked by those proposing referential theories of descriptions (or of class and functional abstracts for that matter).

[17] Without doubt, Russell's own formal implementation of the Theory of Descriptions suggests a fairly significant mismatch between surface syntax and "logical form," but this has little to do with descriptions *per se*. In order to characterise the logical forms of quantified sentences such as "every human is wise" or "some human is wise" in standard first-order logic we have to use formulae containing sentence connectives, no counterparts of which occur in the surface forms of the sentences. And when we turn to a sentence like "just two men are wise", we have to use many more expressions that do not have counterparts in surface syntax, as well as repetitions of a number that do:

(i) $\exists x \exists y ((x \neq y \cdot \text{man } x \cdot \text{man } y \cdot \text{wise } x \cdot \text{wise } y) \cdot$
        $\forall z ((\text{man } z \cdot \text{wise } z) \supset (z = x \vee z = y)))$.

So there is no real problem of fidelity to surface syntax that is specific to *descriptions*. The case involving descriptions is a symptom of – and also helps us to see the severity of – a larger problem involving the use of standard first-order logic to characterise the logical forms of sentences of natural language. Similarly, if Russell's theory predicts that ambiguities of scope arise where there actually *is* ambiguity in natural language, this is a virtue rather than a vice; and if there is any "problem", it concerns only the fact that the use of Russell's abbreviatory conventions may, on occasion, require the insertion of scope indicators in order to make it clear which of two (or more) unambiguous formulae in primitive notation is being abbreviated by a particular pseudo-formula.

For the purposes of providing a systematic semantics for natural language we can capture Russell's insights about the logic and semantics of descriptions without using Russell's own notation (or even the notation of standard first-order logic). Indeed, a more perspicuous notation is not hard to construct and already widely used. We modify our simple quantificational language L'' by throwing out the two unrestricted quantifiers (and associated rules of syntax and semantics) and bringing in two quantificational determiners *every* and *some*, devices which are used to create restricted quantifiers. We take a determiner D to combine with a variable x_k and a formula ϕ to form a restricted quantifier $[D\ x_k\colon \phi]$ such as (ii) or (iii)

(ii) $[every\ x_1\colon man\ x_1]$
(iii) $[some\ x_1\colon man\ x_1]$.

And we take a restricted quantifier to combine with a formula ψ to form a formula $[D\ x_k\colon \phi]$ ψ, such as (iv) or (v):

(iv) $[every\ x_1\colon man\ x_1]\ snores\ x_1$
(v) $[some\ x_1\colon man\ x_1]\ snores\ x_1$.

Adding axioms such as the following will suffice for defining truth:

(vi) $\forall s \forall k \forall \phi \forall \psi$ (s satisfies "$[every\ x_k\colon \phi]\ \psi$" iff every sequence satisfying ϕ and differing from s at most in the kth position satisfies ψ)
(vii) $\forall s \forall k \forall \phi \forall \psi$($s$ satisfies "$[some\ x_k\colon \phi]\ \psi$" iff there is some sequence satisfying ϕ and differing from s at most in the kth position that satisfies ψ).

In this system, we could represent "no man snores" as (viii) or (ix):

(viii) $\neg$ [*some* x_1: *man* x_1] *snores* x_1

(ix) [*every* x_1: *man* x_1] $\neg$ *snores* x_1.

But of course we could make for a more direct mapping between sentences of English and our new formalism by adding to the latter a new quantificational determiner "*no*" and an appropriate axiom. The formula in (x) would be subject to the axiom in (xi):

(x) [*no* x_1: *man* x_1] *snores* x_1

(xi) $\forall s \forall k \forall \phi \forall \psi$ (*s* satisfies "[*no* x_k: ϕ] ψ" iff no sequence satisfying ϕ and differing from *s* at most in the kth position satisfies ψ).

Now what about definite descriptions? The particular formalism and notation of *Principia* are not essential to the Theory of Descriptions itself. Indeed, it was presented in "On Denoting" without this particular formalism and notation; in *Principia* it was presented much more clearly with it. If we want, we can use the formal language containing restricted quantifiers. In English, the word "the" is a one-place determiner just like "every", "some", "no", "most", and so on; so we could add to our new formalism yet another quantificational determiner "*the*" and an appropriate axiom (see (e.g.) Grice (1969), Barwise and Cooper (1981), Evans (1977), Higginbotham and May (1981), Westersta[o]hl (1989), Neale (1990)). The formula in (xii) would be subject to the axiom in (xiii), which captures Russell's insights perfectly:

(xii) [*the* x_1: *king* x_1] *snores* x_1

(xiii) $\forall s \forall k \forall \phi \forall \psi$ (*s* satisfies [*the* x_k: ϕ] ψ" iff the sequence satisfying ϕ and differing from s at most in the kth position also satisfies ψ).

I have used an English determiner "the" in the metalanguage so as to make (xiii) congruent with the axioms for the other determiners in (vi), (vii), and (viii) above. The right-hand side of (xiii) is to be understood as equivalent to "there is exactly one sequence satisfying ϕ and differing from *s* at most in the kth position and every such sequence also satisfies ψ." (See Neale (1992)). The axiom can be simplified; however, questions about the syntax, semantics, and systematicity of the axioms (and the metalanguage in which they are stated), as well as questions about the distinction between semantics and analysis have a considerable bearing on the proper form of any truth definition that is to play a serious rôle in a semantical theory for natural language and also on the characterisation of that rôle itself.)

The viability of a formal language containing restricted quantifiers shows that the language of *Principia* is not an essential ingredient of a theory of quantification and logical form; in particular, it is not an essential ingredient of the Theory of Descriptions construed as a component of a systematic semantics for natural language

The restricted quantifier implementation of the theory has a great deal to recommend it. For one thing, it draws out the syntactical and semantical similarities between "every", "some", "a", "the", and so on; for another, it makes the scope of a description utterly transparent in the formal notation. For example, Russell's (xiv) and (xv) will be rendered as (xiv′) and (xv′) respectively:

(xiv) $\neg[\iota x F x] \, G \iota x F x$

(xv) $[\iota x F x] \, \neg G \iota x F x$

(xiv′) $\neg$ [*the x*: *Fx*] *Gx*

(xv′) [*the x*: *Fx*] $\neg$ *Gx*.

To use a restricted quantifier notation in connection with descriptions is to make a move that Russell failed to make when he did not simplify his abbreviatory notation in the way I mentioned earlier, according to which (xiv) reduces to (xiv″) and (xv) to (xv″):

(xiv″) $\neg[\iota x F x] \, Gx$

(xv″) $[\iota x F x] \, \neg Gx$.

Russell's failure to make this elementary simplification has already been explained. It should be clear that using a formal language that deploys devices of restricted quantification to render a sentence containing a description does not conflict with Russell's claim that a description is an "incomplete symbol." A description no more *stands for something* if restricted quantifiers are used than it does if unrestricted quantifiers are used. If there are viable non-extensional sentence connectives in natural language – something we do not want to assume at this point – analogues of (xiv′) and (xv′) can be used to represent the notorious ambiguities that are claimed to arise in natural language when such connectives occur with definite descriptions. For example, the addition of $\square$ and an appropriate axiom – no straightforward matter, as difficult choices have to be made which have repercussions for the rest of the axiomatization and the philosophical status of the axioms themselves – ought to make it possible to characterise the so-called *de dicto* and *de re* readings of "the number of planets > 7" as (xvi) and (xvii) respectively:

(xvi) $\Box[the\ x\colon Px_1]\ (x > 7)$

(xvii) $[the\ x\colon Px]\ \Box(x > 7)$

(Of course, (xvii) assumes that quantification into modal contexts is intelligible, something we might not want to assume at this point.)

Although Russell did not have the resources of restricted quantifier theory at his disposal, and although he had philosophical aims that went beyond the semantics of natural language, the restricted quantifier analysis of descriptions given above just *is* Russell's theory, semantically speaking, but stated in a way that allows us to see the relationship between surface syntax and logical form more clearly. By virtue of being Russellians about descriptions, we are not committed to the view that the only way to represent the "logical form" of a sentence *S* containing a description is to translate *S* into a formula of the language of *Principia* (or a similar language). As far as explicating the logical structure of sentences containing descriptions is concerned, treating them in terms of restricted quantifiers results not in a departure from, or a falling out with Russell but in the beginnings of an elegant explanation of where his theory fits into a more general theory of quantification, a theory in which determiners like "every", "some", "all", "most", "a", "the", and so on, are treated as members of a unified syntactical and semantical category.

The fact that Russell's theory can be so easily implemented in this way shows the hollowness of claims to the effect Russell's theory has difficulties that arise because of matters of scope, logical form, fidelity to surface syntax, and so on. In view of the need to discuss certain "derived" rules of inference employed by Whitehead and Russell, I will revert to standard logical notation supplemented with the *iota*-operator in much of the sequel. But the fact that Russell's theory can be implemented in a system of restricted quantification should quell any fears about the degree of mismatch between logical and grammatical form and also defuse a worry of Gödel's (that will emerge) by indicating how Russell's theory can function as a component of a general and systematic theory of quantified noun phrases in natural language.

[18]   In the 1981 revision (p. 75) "the" becomes "this" and "and" disappears.

[19]   See, e.g., Quine (1937, 1940, 1941a, 1948, 1953b, 1957, 1960, 1966, 1970, 1981, 1982). But see also Quine's (1995) brief, and I think aberrant, foray into free logic.

[20]   Russell (1905) treated possessive noun phrases as subject to the Theory of Descriptions. He also argued that from certain perspectives, ordinary proper names should be analysed in terms of definite descriptions (a handful of logically proper names (basically, "this" and "that", and in some moods "I") resisting analysis). The precise content of this claim and its relevance to semantics, as opposed to pragmatics, are matters of debate; but in the light of Kripke's (1972) work it is now widely held that it is not possible to provide an adequate semantical analysis of ordinary proper names by treating them as *synonymous* with definite descriptions or as having their *references fixed* by description.

[21]   For references, see note 19.

[22]   Let us put aside, for the moment, the fact that Kripke (1972) has argued that it is not possible, in general, to replace every name α in every context by a definite description that is true of the referent of α.

[23]   Given (ii), (iii) is not entirely arbitrary: two *n*-place predicates $\mathfrak{R}$ and $\mathfrak{R}'$ have the same extension if and only if "$(\forall x_1 \ldots x_n)(\mathfrak{R}(x_1 \ldots x_n) \equiv \mathfrak{R}'(x_1 \ldots x_n))$" is true; if a sentence can be viewed as a 0-place predicate (i.e. an expression that combines with zero terms to form a sentence), then two sentences ϕ and ψ have the same extension if and only if "$(\phi \equiv \psi)$" is true; so, as Carnap (1947, p. 26) points out, on such an account it seems "natural" to regard the truth-values of sentences as their extensions

[24]   See, e.g., Patton (1997) who uses precisely these words. Of course, with some ingenuity any expression can be made to function as an operator, in which case it would be true to say that "only operators have scopes". But this is not what Patton has in mind. He is chastising Kripke and Dummett: "People, Kripke and Dummett for two, freely ascribe *scopes* to singular terms, which is incoherent since only operators have scopes." (p. 251). Unlike Dummett and Kripke, Patton hasn't reflected upon what scope *is*. I don't mean to suggest he is alone in this; indeed, I hesitated to pick on Patton here because the failure is so widespread, but his statement has the virtue of being exceptionally clear.

[25]   With respect to multiple quantifications, the original motivation in logic for allowing permutations of quantifier scope was the desire to capture readings with distinct truth condi-

tions. But as is well known, permuting quantifiers does not always result in such a difference. For example, the truth conditions of neither (i) nor (ii) are sensitive to which quantifier
has larger scope:

(i)   every philosopher respects every logician
(ii)   some logician respects some philosopher.

But there is no getting around the fact that scopes must still be assigned in providing translations of (i) and (ii) in standard first-order notation, and all intelligible work in that strain of
theoretical linguistics which is attempting to characterise a syntactical level of "logical form"
respects this fact, declaring each of (i) and (ii) the surface manifestation of two distinct but
logically equivalent underlying structures. This might strike some as introducing an unnecessary redundancy, but the perception is illusory. Firstly, theorists should be striving after the
most general and aesthetically satisfying theory, and the fact that no truth-conditional differences result from scope permutations in *some* simple sentences is of no import by itself.
Secondly, contrary to what some people have claimed, in order to produce such examples, it
is neither necessary nor sufficient to use the same quantificational determiner twice. This is
easily seen by adopting restricted quantifier notation. That sameness of determiner is not
necessary is clear from the fact that the pairs of readings for (iii) and (iv) are equivalent:

(iii)   the queen owns a bicycle
        [*the x: queen x*] [*a y: bicycle y*] *x owns y*
        [*a y: bicycle y*] [*the x: queen x*] *x owns y*
(iv)   every outlaw talked to the sheriff
        [*every x: outlaw x*] [*the y: sheriff y*] *x talked to y*
        [*the y: sheriff y*] [*every x: outlaw x*] *x talked to y.*

(In effect, this was pointed out by Whitehead and Russell in *Principia*.) That sameness of
determiner is not *sufficient* follows from the fact that (e.g.) "most" is not self-commutative;
the two readings of (v) are not equivalent (a point I believe was first made by Nicholas
Rescher):

(v)   most outlaws shoot most sheriffs
        [*most x: outlaw x*] [*most y: sheriff y*] *x shoots y*
        [*most y: sheriff y*] [*most x: outlaw x*] *x shoots y.*

The real moral that emerges from reflecting on (i)–(v) is that a theory of logical form is
rather more than a theory that associates a sentence of a well-behaved formal language with
each sentence of a natural language. If the best syntax and semantics we have both say (or
jointly entail) that there are two distinct "logical forms" associated with some particular
string, then it would be preposterous to claim that the string in question is not the surface
form of two distinct sentences just because the two purported "logical forms" are logically
equivalent. My point here is not the familiar one that truth conditions are not fine-grained
enough to serve as propositions or meanings. This matter is irrelevant to the point at hand.
(Notice that although the pairs of readings of (i)-(iv) are equivalent, the axioms of a truth
definition will apply in a different order, and to that extent there is still room for the truth-
conditional semanticist to say that the sentences (construed, as pairs of "logical" and
"surface" forms) differ in an interesting *semantic* respect. My point is much simpler. We all
accept that the string "visiting professors can be a nuisance" is the surface manifestation of
two distinct sentences with distinct truth conditions, and we don't mind saying this even
though the two sentences are written and sound alike. Equally, we all accept that "Bill sold
Mary a car" and "Mary bought a car from Bill" are the surface manifestations of two distinct
sentences with the same truth conditions. So neither the "surface sameness" nor the "truth-
conditional sameness" of two purported sentences is sufficient to demonstrate that a single
sentence is actually under scrutiny. And as far as I can see, there is no reason to think that
the *combination* of surface sameness and truth-conditional sameness demonstrates it either.
So there is no reason to reject the view that each of (i)-(iv) is the surface manifestation of a
pair of sentences. At times we must let the theory decide. If the best syntax and semantics we
have say there are two distinct sentences corresponding to a single string, so be it.

It has been argued by Hornstein (1984) that the absence of a difference in truth conditions
for the pairs associated with (iii) and (iv) lends support to his view that descriptions are *not*
regular quantified noun phrases that admit of various scope assignments but are always interpreted as if they took maximal scope, something he sees as explicable on the assumption that
descriptions are more like referential than quantificational noun phrases. As argued in detail
by Soames (1987) and Neale (1990), Hornstein's position is riddled with philosophical and
technical problems so severe that it is unintelligible where it is not plain wrong. To gain a

glimpse of where it is (if intelligible then) wrong in connection with descriptions, notice that (as Russell observed back in 1905), scope matters truth-conditionally in (vi) just as much as it does in (vii):

(vi)   Ralph thinks that the person who lives upstairs is a spy
(vii) Ralph thinks that someone who lives upstairs is a spy.

As noted by Smullyan (1948), Chisholm (1965), and Sharvy (1969), the point extends from attitude contexts to modal and temporal contexts as in (viii) and (ix):

(viii) the number of planets is necessarily odd
(ix)   the president used to be a republican
(x)   the death of Kennedy caused the vice-president to become president.

Furthermore, even within the relative safety of extensional constructions, the scope of a description is important. Russell makes the point with the sentence "the king of France is not bald", which he claims has two readings according as the description or the negation has larger scope. We can bolster Russell's point by considering sentences that contain a description together with a quantifier that is monotone decreasing:

(xi)   few men have met the king of France.

[26]   When considering scope in natural language, the scope of an expression is the first *branching* node properly dominating it. This is because of the possibility (in some theories) of non-branching nodes in connection with, e.g., nouns and intransitive verbs. This need not concern us here.

[27]   It is as if such structures may "swivel" at the dominant sentence node, at least in English. Interestingly, infants master "φ before ψ" and "after φ, ψ" before they master "before ψ, φ" and "ψ after φ" (see Clark (1971) and Johnson (1975)). This would appear to comport with – and perhaps ultimately underpin – Grice's (1989) views about utterances of "φ and ψ" giving rise, *ceteris paribus*, to suggestions that the event described by φ preceded the event described by ψ (where φ and ψ describe events that can be temporally ordered), a fact Grice attributes to a (contextually defeasible) sub-maxim of conversation enjoining orderliness.

[28]   Of course it is possible to treat (e.g.) "necessarily" and "possibly" as quantifiers (over "possible worlds") rather than connectives in my sense.

[29]   On the now common *metaphysical* interpretation of $\Box$, the truth value of $\Box\phi$ depends upon φ's truth *conditions*. This makes $\Box$ a special kind of non-extensional connective, one that sticklers (myself included) call an *intensional* connective. Not all non-extensional connectives are intensional in this sense.

[30]   For applications, see Neale (1995), Neale and Dever (1997), and *Facing Facts*.

[31]   There is no *syntactical* criterion for being a singular term. "Quine", "he", "that prince", "the prince", "a prince", "one prince", "only one prince", "no prince", "no one prince", "some prince", "each prince", and "every prince" are all grammatically singular: we need to make a decision based on *semantic* arguments or intuitions.

[32]   Quine notes that we should not confuse the seemingly unintelligible sentence "∃x(Philip is unaware that x denounced Catiline)" with the false sentence "Philip is unaware that ∃x(x denounced Catiline)". This remark raises an interesting issue. Notice that the following inference is valid:

(i)   <u>Philip believes that Cicero denounced Catiline</u>
       Philip believes that ∃x(x denounced Catiline).

But attempting to capture its validity in terms of an inference principle meant to function as a "smaller scope" version of EG has obvious pitfalls, not least of which is the existence of elements that import negation, as Quine's (1943, p. 148) observation about the invalidity of the following brings out:

(ii)   <u>Philip is unaware that Tully denounced Catiline</u>
       Philip is unaware that ∃x(x denounced Catiline).

The contrast between the validity of (i) and the invalidity of (ii) is obviously connected to the negation imported by "is unaware" – we find the same with "doubts", "is sceptical", "does not believe", etc. And of course the invalidity of (ii) mirrors that of (iii):

(iii) <u>~ Tully admired Catiline</u>
       ~ ∃x(x admired Catiline).

Similarly, in modal systems one wants to be able to infer from $\Diamond Fa$ to $\Diamond\exists xFx$, but not from $\sim\Diamond Fa$ to $\sim\Diamond\exists xFx$. There are difficult issues here which, to my mind, have not been satisfactorily resolved in the literature.

[33]   See in particular Russell (1905, esp. p. 47 and pp. 51–52); Whitehead and Russell (1925, *14); Church (1942); Smullyan (1947, 1948); and Fitch (1949, 1950).

[34] This particular application of PSST assumes that variables and temporary names function as genuine singular terms. I am fully at ease with this assumption, as, in effect, were Whitehead and Russell. It is not obvious how it might be contested, but it is an assumption nonetheless.

[35] If all singular terms are "rigid designators" in Kripke's (1972) sense, then the extension of a singular term will be a *constant* (but, again, possibly partial) function.

[36] On the value of keeping apart logical and non-logical issues in explorations of the non-extensional case, see Kaplan (1986).

[37] Quine chooses to use an attitude construction when formulating his 1960 version of the argument but (a) nothing in the 1960 version turns on any particular non-extensional construction, (b) Quine's *formal* target is non-extensional connectives quite generally, (c) his main *philosophical* targets were modal connectives, and (d) Quine's (1953a, 1953c) versions do not use an attitude construction and are, moreover, formulated in ways that are meant to show that they generalise to *all* non-extensional connectives. For explicit confirmation of the generality, see the last paragraph of "Reference and Modality", and also Quine's (1941) earliest worries about the very idea of non-truth-functional logics (discussed in detail below). Indeed Quine was the first person to propose a general, connective-based slingshot, his intention being to argue on purely formal grounds without getting embroiled in the semantics of this or that connective. It would be a poor scholar who complained that Quine used a slingshot only against logical modality, and a poor philosopher who saw an illicit running together of modal and attitude contexts in any attempt to set out a general slingshot using Quine's 1953 and 1960 versions.

[38] Note the similarity with Carnap's (1937) K-operator: for Carnap "$(Kx)m(x > 7)$" is read as "the smallest positive integer x up to and including m such that "$x > 7$" is true, and 0 if there is no such integer".

[39] As Church (1943a) notes, if arguments of this general type are restated using class abstracts rather than descriptions, exactly analogous questions must be answered concerning logical equivalence and the precise semantics for class abstracts. The fact that Church (1943) uses abstracts where Gödel (1944) uses descriptions in the first published slingshots of the early 1940s should not obscure the fact that Gödel and Church are in complete harmony on the matter of what happens when there is contextual elimination of purported term-forming devices such as $\iota x, \hat{x}, \lambda x, \mu x, Kx$ etc. If ιx and $\hat{x}$ do not belong to the primitive symbols and are defined contextually, then it will not be possible to use their slingshots to demonstrate that if true sentences stand for facts, all true sentences stand for the same fact (Gödel) or that if sentences designate propositions, all true sentences designate the same proposition (Church). Church has in mind the following contextual definition for class abstracts:

(i) $\hat{x}Fx = \hat{x}Gx =_{df} \forall x(Fx \equiv Gx)$.

Quine seems to have overlooked these points made by Church and Gödel in connection with their original slingshots, something I put down to the "misleading emphasis" Church (1942) mentioned in his review of Quine's 1941 paper (See Part II). An alternative to the contextual definition in (i) would be to view "$\hat{x}Fx$" as a definite description ("the class of things that are F") that can be analysed in accordance with Russell's theory. This suggestion has been made by Quine (1941), who offers (ii), and by Smullyan (1948) who offers (iii):

(ii) $\hat{x}Fx =_{df} \iota\alpha(\forall x(Fx \equiv x \in \alpha))$
(iii) $[\hat{x}Fx]G^\wedge xFx =_{df} \exists\alpha(\forall x(Fx \equiv x \in \alpha) \cdot G\alpha)$.

(In Smullyan's (iii), "$[\hat{x}Fx]$" is a scope marker just like Whitehead and Russell's "$[\iota xFx)]$".) It makes no difference whether descriptions or abstracts are used in setting up the basic slingshot I examine here. I have a preference for (first-order definable) description over abstraction.

[40] It will not do to object to this argument on the grounds that its use of "singular terms" is incompatible with the Quinean elimination of such devices (recall Quine wants to reduce sentences containing singular terms to sentences containing quantifiers, variables, the identity sign, predicates, and truth-functional connectives). The reinterpretation of the argument when singular terms are eliminated is straightforward: (i) "substitutivity" is interpreted as shorthand for an inference principle in which the descriptions it contains are given their Russellian expansions, in accordance with *Principia* *14.01, *14.02, *14.15, and *14.16.

[41] See also my 1995 paper "On the Philosophical Significance of Gödel's Slingshot".

[42] I first set out a version of this proof in my 1995 paper. Dever and I set out a cleaner version in our 1997 paper. Everything is done in more detail and still more cleanly here and in *Facing Facts*. The version presented here should be viewed as superseding the 1995 and 1997 versions.

[43] Inferences of the following forms involving modal operators seem intuitively valid to many people when $\Box$ receives a contemporary metaphysical interpretation (the absence of scope markers indicating smallest scope for the descriptions):

$$\frac{\Box Fa}{\Box a=\iota x(x=a \cdot Fx)} \qquad \frac{\Box a=\iota x(x=a \cdot Fx)}{\Box Fa}$$

If this is right, one task for the modal logician will be to provide a semantics for descriptions that captures these facts. Notice that Russell's theory will do the job perfectly.

[44] For example, by appealing to the fact that if ❽ is +ι-SUBS it is also +PSST (see §10), the following will suffice:

1	[1]	Fa	premiss
2	[2]	$a=b$	premiss
3	[3]	Gb	premiss
1	[4]	$a=\iota x(x=a \cdot Fx)$	1, ι-CONV
3	[5]	$b=\iota x(x=b \cdot Gx)$	3, ι-CONV
2,3	[6]	$a=\iota x(x=b \cdot Gx)$	5, 2, PSST
1,2,3	[7]	$\iota x(x=a \cdot Fx) = \iota x(x=b \cdot Gx)$	4,6, ι-SUBS
8	[8]	❽(Fa)	premiss
8	[9]	❽$(a=\iota x(x=a \cdot Fx))$	8, ❽+ι-CONV
8	[10]	❽$(a=\iota x(x=b \cdot Gx))$	9,7, ❽+ι-SUBS
1,2,3,8	[11]	❽$(b=\iota x(x=b \cdot Gx))$	10,2 ❽+PSST
1,2,3,8	[12]	❽(Gb)	11, ❽+ι-CONV

[45] In *Facing Facts*, I explore this matter in connection with various theories of facts, drawing on work published with Josh Dever in 1997.

[46] Terminology aside, I here follow Taylor (1985). For the sake of simplicity I propose (again with Taylor) to ignore the irrelevant complexities raised by relativized descriptions such as "the woman sitting next to him" where "him" is bound by a higher quantifier. Nothing of any bearing upon the point at hand turns on the existence or interpretation of such descriptions. Russell's theory both predicts the existence of, and provides an automatic and successful interpretation of, such descriptions without any special stipulation or additional machinery. With some work, presumably some referential accounts of descriptions can also supply what is necessary here; so I propose to ignore any potential problems that relativization creates for the non-Russellian.

[47] Following Tarski, let us say that a sentence ϕ of first-order logic is *logically true* if, and only if, it is true in every first-order model.

[48] Carried back over to the discussion of facts, the proofs would show only that logically equivalent sentences stand for the same fact. There would still be no collapse of all facts into one.

[49] I am here indebted to Mark Sainsbury and Barry Smith.

[50] It is unclear whether it makes a great deal of sense to attribute to Carnap, as part of his overall account of descriptions – which, as he points out (1947, p. 8), "deviates deliberately from the meaning of descriptions in the ordinary language" – even the informal analogue of this model-theoretic account of improper descriptions, given the way he characterises state descriptions. As Carnap points out, his system requires a prohibition on descriptions containing modal operators.

[51] This would be devastating for any theory of facts requiring "the fact that Fa = the fact that $(\ldots)$" to have this set of features.

[52] While this will herald the demise of certain theories of facts, it will not bother people like Barwise and Perry (1983) who see logical equivalence as no guarantee of factual equivalence, *i.e.* those who deny that "the fact that ϕ = the fact that $(\ldots)$" is +PSLE.

[53] It could be viewed as damaging to some theories of facts and states-of-affairs because it demonstrates the truth of a statement such as (i):
   (i)   the fact that $(\phi \cdot a \neq \iota x(x \neq x))$ = the fact that $(\psi \cdot a \neq \iota x(x \neq x))$.

[54] In order to avoid Gödel's slingshot, Taylor would have to deny, in addition, that "Fa" and "$a = \iota x(x=a \cdot Fx)$" stand for the same fact.

[55] According to Grandy, "Not all objects in the pseudo-domain are possible objects for one of them will be the denotation of $(\iota x)(x \neq x)$" (1972, p. 175).

[56] Taylor is well aware that his reformulation cannot capture Strawson's own intentions and that these intentions are not important for the purposes at hand. On Strawson's account, it is *speakers* rather than singular terms that refer; and his assault on Russell's Theory of Descriptions is part of a general campaign against the ideas that terms refer and sentences are true or false; thus some distortion of Strawson's views is inevitable in any attempt to recast them model-theoretically; important choices where Strawson is unclear or inconsistent are also necessary.

[57] In this I have been convinced (slowly) by conversations with Donald Davidson, Leon Henkin, James Levine, W.V. Quine, and John Searle. The conviction has been reinforced by ploughing through the relevant literature and by Burgess's (1997) paper, to which I have already directed the reader. The mistake I believe I made in 1990 also infects self-contained portions of several subsequent articles – most obviously those published in 1993 and 1995 and mentioned in the bibliography. Lest one not read too much into my admission: the main claims of these earlier works are unaffected by the error induced by the anachronism.

[58] That clear metaphysical conceptions of $\Box$ and $\Diamond$ were certainly not part of the background culture of modal logic and the theory of reference until after Kripke's (1971, 1972) seminal work is argued by Burgess (1997).

[59] Lewis and Langford make an interesting suggestion for complex demonstratives that has been rediscovered of late: "that F is G" is equivalent to "that is F and G" where "that" functions as a logically proper name.

[60] For detailed discussion of modal logic in application to the concept of provability in formal theories, see Boolos (1993).

[61] See Bar-Hillel (1950).

[62] On whether such an argument can be attributed to Frege, see my "Colouring and Composition".

[63] On this topic, see also Kazmi (1987), Richard (1987), and Soames (1995).

[64] As Quine notes, there is no analogous problem for "Philip is unaware that $\exists x(x$ denounced Catiline)".

[65] In Gödel's *Nachlass*, an annotated page of an offprint of his 1944 paper (p. 150) has the remark "Th. der natürlichen Zahlen nachweislich nicht analytisch im Kantschen Sinn" ("The theory of natural numbers [is] demonstrably not analytic in Kant's sense"). See his *Collected Works*, Vol. II., p. 314. For a penetrating analysis of Gödel's discussions of analyticity and their relation to Quine's, see Parsons (1990).

[66] In Gödel's *Nachlass*, an annotated page of an offprint of his 1944 paper bears the heading "Bernays Rev. Meiner Arbeit über Russell." (Bernays' review of my paper on Russell.) On the same page, Gödel wrote the following: "Das Probl. der Beschreibung ist durch "Sinn" und "Bedeutung" in befriedingender weise gelöst." (The problem of description is solved in a satisfactory way by "sense" and "signification" – Gödel makes it clear in the article that he wishes to render "Bedeutung" as "signification" in English.) He is not here conceding Bernays' point; he is simply stating it. If there is any more to Gödel's annotation it is likely an expression of sympathy with Church's view that the problems about names and descriptions *raised by Frege and Russell* are handled better by Fregean than by Russellian treatments.

[67] In 1959, Lewis wrote an appendix for the second edition of *Symbolic Logic*, in which he says that "...Dr Ruth Marcus has shown that the system S2 can be extended to first-order propositional functions. Though it happens I hold certain logical convictions in the light of which I should prefer to approach the logic of propositional functions in a different way, I appreciate this demonstration that there is a calculus of functions which bears to the calculus of Strict Implication a relation similar to that which holds between the calculus of functions in *Principia Mathematica* (*9–*11) and the calculus of propositions (*1–*5)." (p. 508)

[68] Dragging in "Venus" or any other name of the heavenly body in question will not help as Quine's example can be restated using any two co-referential names of Venus.

[69] I have used "$\Box(9 = 9)$" rather than "$\Box(9 > 7)$" in order to avoid the distracting side issue of whether arithmetical truths are analytic, which as Quine (1943) points out is required for the truth of "$\Box(9 > 7)$".

[70] Marcus's (1947) theorem 2.32* says that $\forall x \forall y[(x=y)$ $\forall x \forall y[(x=y) \Leftrightarrow \Box(x=y)]$, where $\Leftrightarrow$ expresses strict equivalence and = is understood as strict identity. Her proof of 2.32* assumes the Barcan and converse Barcan formulae as well as a principle to the effect that $\Box\phi \Leftrightarrow \Box\Box\phi$. On this matter, see Soames (1995), p. 209, note 4.

[71] On this notion, see Prior (1963), Neale (1995), and Neale and Dever (1997).

[72] On this matter, see Neale (1995), Neale and Dever (1997), and *Facing Facts*.

[73] I am not alone in reading this as a retraction. See Kaplan (1986, p. 249).

[74] For discussion, see, e.g., Marcus (1967), Cartwright (1968), Parsons (1969), Føllesdal (1986), Kaplan (1986), and Burgess (1995).

[75] As Quine (1972) observes: "Models afford consistency proofs; also they have heuristic value; but they do not constitute explication. Models, however clear they be in themselves, may leave us still at a loss for the primary, intended interpretation." (p. 492). On this matter, see also Burgess (1997).

[76] On this argument, see Føllesdal (1961, 1965, 1966, 1983) and Marti (1995, 1997).

[77] For an edited version of the discussion, see Marcus, Quine, *et al.* (1962).

[78] The date Quine gives must be when he *wrote* his 1943 paper "Notes on Existence and Necessity". He alludes to the existence of an unstated difficulty in connection with quantifying into attitude contexts in his 1941 review of Russell's *Inquiry into Meaning and Truth*.

[79] It is important to distinguish the valid argument (i) $\alpha=\beta$, (ii) α and β are rigid designators, therefore (iii) $\Box(\alpha=\beta)$, from the invalid argument (i) $\Box(\alpha=\beta)$, therefore (ii) α and β are rigid designators See the discussion of (18) and (19) above.

[80] As John Burgess has pointed out to me, this is consistent with Kripke's (1982) use of "skeptical paradox".

[81] Quine (1953c, 1961) appears to have seen that the interpretive problem I have been discussing is avoided if modal operators are taken to express *metaphysical* notions. Limitations of space prevent me from addressing what Quine means when he says that a "reversion to Aristotelian essentialism is required ... if quantified modal logic is to be insisted on" (1961, p. 155). My position may be summarised thus. Once Quine's remarks about essentialism are stripped of certain confusions, his basic position is correct and amounts to the following: (i) substitutivity and quantification fail where strict necessity is concerned because the notion of necessity in question is linguistic; (ii) restoring substitutivity and quantification requires construing modal operators such a way that objects themselves – rather than objects relative to modes of specifying them – have traits necessarily or contingently; and (iii) this means moving from a linguistic to a *metaphysical* construal. To this extent, Quine is pointing to (but not endorsing) the road ultimately taken by Kripke.

REFERENCES

Ackerman, W. 1956: Begründung einer Strengen Implikation", *Journal of Symbolic Logic*, 21, 113–28.

Almog, J. 1986: "Naming Without Necessity", *Journal of Philosophy*, 86, 210–242.

Anderson, A.R. 1957: Review of Prior's "Modality and Quantification in S5", *Journal of Symbolic Logic*, 29, 79–87.

Anderson, A.R. and Belnap, N.D. 1962a: "Tautological Entailments", *Philosophical Studies*, 13, 9–24.

—— 1962b: "The Pure Calculus of Entailment", *Journal of Symbolic Logic*, 27, 19–52.

Apostel, L. 1953: "Modalités Physiques et Techniques", *Actes du XIᵉ Congres International de Philosophie, Bruxelles, 1953, Vol. XIV*, Amsterdam: North-Holland, 97–104.

Barwise, J and Perry, J. 1981: "Semantic Innocence and Uncompromising Situations", *Midwest Studies in Philosophy* VI, 387–403.

Bar-Hillel, Y. 1950: "Bolzano's Definition of Analytic Propositions", *Theoria*, 16, 91–117.

Bayart, A. 1958: "La Correction de la Logique Modale de Premier et Second Ordre S5", *Logique et Analyse*, 2, 28–44.

—— 1959: "Quasi-adéquation de la Logique Modale de Second Ordre S5 et Adéquation de la Logique Modale de Premier Ordre S5", *Logique et Analyse*, 1, 99–121.

Becker, O. 1930: "Zur Logic der Modalitäten", *Jahrbuch für Philosophie und Phänomenologische Forschung*, 11, 497–548.

—— 1941: *Einführung in die Logistik, Verzüglich in der Modalitäten*, Meisenheim am Glan: Anton Heim.

—— 1952: *Untersuchungen über den Modalkalkül*, Meisenheim am Glan: Anton Heim.

Bernays, P. 1946: Review of Gödel's "Russell's Mathematical Logic." *Journal of Symbolic Logic* 11, 75–9.

—— 1948: Review of Carnap's "Modalities and Quantification", *Journal of Symbolic Logic*, 13, 218–19.

Boolos, G. 1993: *The Logic of Provability*, Cambridge: Cambridge University Press.

Burks, A. 1951: "The Logic of Causal Propositions", *Mind* 60, 363–382.

Burgess, J. 1997: "Quinus ab Omni Naevo Vindicatus", in A. Kazmi (ed.), *Meaning and Rererence. Canadian Journal of Philosophy*, suppl. vol. 23, 25–65.

Burks, A. 1951: "The Logic of Causal Propositions", *Mind*, 60, 363–82.

Carnap, R. 1928: *Der Logische Aufbau der Welt*. Vienna: Springer.

—— 1934: *Logische Syntax der Sprache*. Vienna: Springer.

—— 1937: *The Logical Syntax of Language*, London: Kegan Paul (translation, with revisions, of *Logische Syntax der Sprache*).

—— 1942: *Introduction to Semantics*, Studies in Semantics, vol. I. Cambridge, Mass.: Harvard University Press.

—— 1946: "Modalities and Quantification" *Journal of Symbolic Logic*, 11, 33–64.

—— 1947: *Meaning and Necessity, a Study in Meaning and Modal Logic*. Chicago: University of Chicago Press (2nd edition, with additions, 1956).

Cartwright, R. 1968: "Some Remarks on Essentialism", *Journal of Philosophy*, 65, 615–26.

—— 1971: "Identity and Substitutivity" in M.K. Munitz (ed.), *Identity and Individuation*. New York: New York University, 119–33.

Chisholm, R. 1965: "Query on Substitutivity: Comment on Føllesdal", in R.S. Cohen and M.W. Wartofsky (eds.), *Boston Studies in the Philosophy of Science, Vol. II: In Honor of Philipp Frank*. New York: Humanities Press, 275–78.

Church, A. 1932/33: "A Set of Postulates for the Foundation of Logic", *Annals of Mathematics*, vol. 33, 346–66 and vol. 34, 839–64.

—— 1936: "A Note on the Entscheidungsproblem", *Journal of Symbolic Logic*, 1, 40–1 and 101–2.

—— 1940a: Review of Quine's *Mathematical Logic*, *Journal of Symbolic Logic*, 5, 163–4.

—— 1940b: "A Formulation of the Simple Theory of Types", *Journal of Symbolic Logic*, 5, 56–68.

—— 1941: *The Calculi of Lambda Conversion*. Princeton: Princeton University Press.

—— 1942a: "On Sense and Denotation" (abstract), *Journal of Symbolic Logic*, 7, 47.

—— 1942b: Review of Quine's "Whitehead and the Rise of Modern Logic", *Journal of Symbolic Logic* 7, 100–1.

—— 1943a: Review of Carnap's *Introduction to Semantics*, *Philosophical Review*, 52, 298–304.

—— 1943b: Review of Quine's "Notes on Existence and Necessity", *Journal of Symbolic Logic*, 8, 45–7.

—— 1944: *Introduction to Mathematical Logic*, Part I. Princeton: Princeton University Press.

—— 1946: "A Formulation of the Logic of Sense and Denotation" (abstract), *Journal of Symbolic Logic*, 11, 31.

—— 1950: Review of Fitch's "The Problem of the Morning Star and the Evening Star", *Journal of Symbolic Logic*, 15, 63.

—— 1951a: "A Formulation of the Logic of Sense and Denotation", in P. Henle *et al.* (eds.), *Structure, Method and Meaning: Essays in Honor of Henry M. Scheffer*. New York: Liberal Arts Press, 3–24.

—— 1951b: "The Need for Abstract Entities in Semantic Analysis", *Proceedings of the American Academy of Arts and Sciences*, 80, 110–12.

—— 1951c: Review of Lewis and Langford's *Symbolic Logic*, 2nd ed., *Journal of Symbolic Logic*, 225.

—— 1956: *Introduction to Mathematical Logic*, exp. and rev. ed. Princeton: Princeton University Press.

Church, A., and Rosser, J.B. 1936: "Some Properties of Conversion", *Transactions of the American Mathematical Society*, 39, 472–82.

Clark, E. 1971: "On the Acquisition of the Meaning of 'before' and 'after'", *Journal of Verbal Learning and Verbal Behavior*, 10, 266–75.

Craig, W. 1957: "Linear Reasoning: A New Form of the Herbrand-Gentzen Theorem, *Journal of Symbolic Logic*, 22, 250–68, and "Three Uses of the Herbrand-Gentzen Theorem in Relating Model Theory and Proof Theory", *ibid.*, 269–85.

—— 1960: "Bases for First-Order Theories and Subtheories", *Journal of Symbolic Logic*, 25, 97–142.

Davidson, D. 1963: "The Method of Extension and Intension", in P.A. Schilpp (ed.), *The Philosophy of Rudolph Carnap.* La Salle: Open Court, 311–50.

—— 1984: *Inquiries into Truth and Interpretation*, Oxford: Clarendon.

Della Rocca, M. 1996a: "Essentialists and Essentialism", *Journal of Philosophy*, 96, 186–202.

Della Rocca, M. 1996b: "Essentialism: Part 1", *Philosophical Books*, 37, 1–13.

Donnellan, K.S. 1966: "Reference and Definite Descriptions", *Philosophical Review*, 75, 281–304.

—— 1966b: "Substitution and Reference", *Journal of Philosophy*, 63, 685–8.

Dreben, B. 1952: "On the Completeness of Quantification Theory", *Proceedings of the National Academy of Sciences*, 38, 1047–52.

Dummett, M.E. 1981a: *Frege: Philosophy of Language*, 2nd edition. London: Duckworth.

—— 1981b: *The Interpretation of Frege's Philosophy*. London: Duckworth.

Dummett, M.A.E. and Lemmon, E.J. 1959: "Modal Logics Between S4 and S5", *Zeitschrift für Mathematische Logik und Grundlagen der Mathematik*, 5, 250–64.

Evans, G. 1977: "Pronouns, Quantifiers and Relative Clauses (I)", *Canadian Journal of Philosophy*, 7, 467–536.

—— 1982: *The Varieties of Reference*. Oxford: Clarendon Press.

Feys, R. 1937/38: "Les Logiques Nouvelles des Modalités", *Revue Néoscolastique de Philosophie*, vol. 40, 517–53 and vol. 41, 217–52.

—— 1950: Les Systèmes Formalisés des Modalités Aristotéliennes, *Revue Philosophique de Louvain*, 48, 478–509.

—— 1965: *Modal Logics* (edited by J. Dopp) Louvain: Nauwelaerts; Paris: Gauthier-Villars.

Field, H. 1989: *Realism, Mathematics, Modality*, Oxford: Oxford University Press.

—— 1991: "Metalogic and Modality", *Philosophical Studies*, 62, 1–22.

Fine, K. 1978a: "Model Theory for Modal Logic. Part I: The *De Re/De Dicto* Distinction", *Journal of Philosophical Logic*, 7, 125–56.

—— 1978b: "Model Theory for Modal Logic. Part II: The Elimination of the *De Re*", *Journal of Philosophical Logic*, 7, 277–306.

—— 1981: "Model Theory for Modal Logic. Part III: Existence and Predication", *Journal of Philosophical Logic*, 10, 293–307.

Fine, K. 1982: "First-order Modal Theories III – Facts". *Synthese* 53, 43–122.

—— 1986: "Modality *De Re*", in Almog *et al.* (eds.), *Themes from Kaplan*, Oxford: Oxford University Press 197–272.

—— 1991: "Quine on Quantifying in", in C. Anderson and J. Owens (eds.). *Propositional Attitudes*, Stanford: CSLI, 1–25.

Fitch, G., 1937: "Modal Functions in Two-valued Logics", *Journal of Symbolic Logic*, 2, 125–8.

—— 1939: "Note on Modal Functions", *Journal of Symbolic Logic*, 4, 115–6.

—— 1948: "Intuitionistic Modal Logic with Quantifiers", *Portugaliae Mathematica*, 7, 113–18.

—— 1949: "The Problem of the Morning Star and the Evening Star", *Philosophy of Science*, 16, 137–41.

—— 1950: "Attribute and Class", in M. Farber (ed.), *Philosophic Thought in France and the United States*. Buffalo: University of Buffalo, 545–63.

—— 1952: *Symbolic Logic*. New York: Ronald Press.

—— 1960: "Some Logical Aspects of Reference and Existence", *Journal of Philosophy*, 57, 640–7.

Fodor, J. D. and Sag, I. 1982: "Referential and Quantificational Indefinites." *Linguistics and Philosophy*, 5, 355–398.

Føllesdal, D. 1961: *Referential Opacity and Modal Logic*. Doctoral thesis, Harvard University.

—— 1965: "Quantification into Causal Contexts", in R.S. Cohen and M.W. Wartofsky (eds.), *Boston Studies in the Philosophy of Science, Vol. II: In Honor of Philipp Frank*. New York: Humanities Press, 263–74.

—— 1966: "Referential Opacity and Modal Logic", *Filosofiske Problemer,* vol. 32, Oslo: Universitetforslaget.

—— 1967: "Knowledge, Identity, and Existence", *Theoria* 33, 1–27.

—— 1969: "Quine on Modality", in D. Davidson and J. Hintikka (eds.) *Words and Objections,* Dordrecht: Reidel, 175–85.

—— 1983: "Situation Semantics and the Slingshot", *Erkenntnis,* 19, 91–8.

—— 1986: "Essentialism and Reference", in L.E. Hahn and P.A. Schilpp (eds.), *The Philosophy of W.V. Quine.* La Salle: Open Court, 97–113.

Frege, G. 1879: *Begriffsschrift.* English translation in J. van Heijenoort (ed.), *From Frege to Gödel.* Cambridge, Mass.: Harvard University Press, 1967.

—— 1892: "Über Sinn und Bedeutung", *Zeitschrift fur Philosophie und Philosophische Kritik* 100, 25–50. Translated as "On Sense and Reference", in Geach P. and Black, M., eds. 1952: *Translations from the Philosophical Writings of Gottlob Frege.* Oxford: Blackwell, 56–78.

Geach, P. 1963: "Quantification Theory and the Problem of Identifying Objects of Reference", *Acta Philosophica Fennica,* 16, 41–52.

Gödel, K. 1930: "Die Vollständigkeit der Axiome des Logischen Funkionenkalküls", *Monatschefte für Mathematik und Physik,* 37, 349–60. Reprinted with an English translation ("The Completeness of the Axioms of the Functional Calculus") in S. Feferman *et al.* (eds.), *Kurt Gödel, Collected Works,* vol. I, Oxford: Oxford University Press, 1986, 102–23.

—— 1931: "Über Formal Unentscheidbare Sätze der *Principia Mathematica* und Verwandter Systeme I", *Monatschefte für Mathematik und Physik,* 38, 173–98. Reprinted with an English translation ("On Formally Undecidable Propositions of *Principia Mathematica* and Related Systems I") in S. Feferman *et al.* (eds.), *Kurt Gödel, Collected Works,* vol. I, Oxford: Oxford University Press, 1986, 144–95.

—— 1933: "Eine Interpretation des Intuitionistischen Aussagenkalküls", *Ergebnisse eines Mathematischen Kolloquiums,* 4, 39–40. Reprinted with an English translation ("An Interpretation of Intuitionistic Propositional Calculus") in S. Feferman *et al.* (eds.), *Kurt Gödel, Collected Works,* vol. I, Oxford: Oxford University Press, 1986. In S. Feferman *et al.* (eds.), *Kurt Gödel, Collected Works,* vol. I (1929–1936), Oxford: Oxford University Press, 1986, 300–3.

—— 1944: "Russell's Mathematical Logic", in P.A. Schilpp (ed.), *The Philosophy of Bertrand Russell,* Evanston and Chicago: Northwestern University Press, 125–53. Reprinted in S. Feferman *et al.* (eds.), *Kurt Gödel, Collected Works,* vol. II (1938–1974), Oxford: Oxford University Press, 1990, pp. 119–41.

Grice, H.P. 1969: "Vacuous Names." In D. Davidson and J. Hintikka (eds.), *Words and Objections.* Dordrecht: Reidel, 118–145.

Grice, P. 1989: *Studies in the Way of Words.* Cambridge: Harvard University Press.

Guillaume, M. 1958: "Rapports Entre Calculs Propositionnels Modaux et Topologie Impliqués par Certaines Extensions de la Méthode des Tableaux Sémantiques. Système S4 de Lewis", *Comptes Rendues des Séances de l'Academie des Sciences,* Paris, 246, 11–40 and 2207–10, "Système de Lewis S5", *ibid.,* 247, 1282–3.

Halldén, S. 1949: *The Logic of Nonsense,* Uppsala: Uppsala Universitets Arsskrift.

—— 1951: "On the Semantic Non-completeness of Certain Lewis Calculi", *Journal of Symbolic Logic,* 16, 127–9.

—— 1963: "A Pragmatic Approach to modal Theory", *Acta Philosophica Fennica,* 16, 53–64.

Hampshire, S. 1959: *Thought and Action.* New York: Viking Press.

Henkin, L. 1949: "The Completeness of the First-order Functional Calculus", *Journal of Symbolic Logic,* 14, 159–66.

Heyting, A, 1931: "Die Intuititionische Grundlegung der Mathematik", *Erkenntnis,* 2, 106–15.

—— 1934: *Mathematische Grundlagenforschung: Intuititionismus. Beweistheorie.* Berlin: Springer.

Hilbert, D, and Ackerman, W. 1928: *Grundzüge der Theoretischen Logik.* Berlin: Springer, 2nd edition, 1938.

Hilbert, D. and Bernays, P. 1934: *Grundlagen der Mathematik,* vol. I. Berlin: Springer., 2nd edition, 1968.

—— 1939: *Grundlagen der Mathematik,* vol. II. Berlin: Springer, 2nd edition, 1970.

Hintikka, J. 1957a: "Modality as Referential Multiplicity", *Ajatus,* 20, 49–64.

—— 1957b: "Quantifiers in Deontic Logic", *Societas Scientiarum Fennica, Commentationes Humanarum Litterarum,* vol. 23, no. 4. Helsinki.

—— 1959: "Towards a Theory of Definite Descriptions", *Analysis,* 19, 79–85.

—— 1961: "Modality and Quantification", *Theoria,* 27, 119–28.

—— 1962: *Knowledge and Belief.* Ithaca: Cornell University Press.

—— 1963: "Modes of Modality", *Acta Philosophica Fennica,* 16, 65–82.

—— 1967: "Individuals, Possible Worlds, and Epistemic Logic", *Nous* 1, 33–62.

—— 1968: "Logic and Philosophy", in R. Klibansky (ed.), *Contemporary Philosophy.* Florence: La Nuova Italia Editrice, 3–30.

—— 1975: *The Intentions of Intentionality and Other New Models for Modalities.* Dordrecht: Reidel.

—— 1982: "Is Alethic Modal Logic Possible?", *Acta Philosophica Fennica,* 35, 89–105.

Hintikka, J. and Kulas J. 1985: *Anaphora and Definite Descriptions.* Dordrecht: Reidel.

Hornsby, J. 1996: "The Identity Theory of Truth", *Proceedings of the Aristotelian Society,* 1–24.

Hornstein, N. 1984: *Logic as Grammar.* Cambridge, Mass.: MIT Press.

Johnson, H. 1975: "The Meaning of "before" and "after" for Preschool Children", *Journal of Experimental Child Psychology,* 19, 88–99.

Kalish, D., Montague, R., and Mar D. 1980: *Logic: Techniques of Formal Reasoning,* 2nd ed. New York: Harcourt, Brace, Jovanovich.

Kanger, S. 1957a: *Provability in Logic.* Stockholm: Almqvist and Wiksill.

—— 1957b: "The Morning Star Paradox", *Theoria,* 23, 1–11.

—— 1957c: "A Note on Quantification and Modalities", *Theoria,* 23, 133–4.

—— 1957d: "On the Characterization of Modalities", *Theoria,* 23, 152–5.

Kaplan, D. 1968: "Quantifying In", *Synthese,* 19, 178–214.

—— 1972: "What is Russell's Theory of Descriptions?" in D.F. Pears (ed.), *Bertrand Russell: A Collection of Critical Essays.* Garden City, NY: Doubleday Anchor, 227–44.

—— 1975: "How to Russell a Frege-Church", *Journal of Philosophy* 72, 716–29.

—— 1986: "Opacity", in L.E. Hahn and P.A. Schilpp (eds.), *The Philosophy of W.V. Quine.* La Salle: Open Court, 229–89.

—— 1989a: "Demonstratives", in J. Almog *et al.* (eds.), *Themes from Kaplan,* Oxford: Oxford University Press, 481–563.

—— 1989b: "Afterthoughts", in J. Almog *et al.* (eds.), *Themes from Kaplan,* Oxford: Oxford University Press, 565–614.

Kaplan, D. and Montague, R. 1960: "A Paradox Regained", *Notre Dame Journal of Formal Logic,* 1, 79–90.

Kazmi, A. 1987: "Quantification and Opacity", *Linguistics and Philosophy,* 10, 77–100.

Kemeny, J.G. 1954: Review of Quine's "Reference and Modality", *Journal of Symbolic Logic,* 19, 137–8.

Körner, S. 1960: *The Philosophy of Mathematics,* London: Hutchison.

Kripke, S.A. 1959a: "A Completeness Proof in Modal Logic", *Journal of Symbolic Logic* 24, 1–14.

—— 1959b: "Semantical Analysis of Modal Logic" (abstract), *Journal of Symbolic Logic,* 24, 323–4.

—— 1962: "The Undecidability of Monadic Modal Quantification Theory", *Zeitschrift für Mathematische Logik und Grundlagen der Mathematik,* 8, 113–6.

—— 1963a: "Semantical Considerations on Modal Logic", *Acta Philosophica Fennica,* 16, 83–94.

—— 1963b: "Semantical Analysis of Modal Logic: I, Normal Modal Propositional Calculi", *Zeitschrift für Mathematische Logik und Grundlagen der Mathematik,* 9, 67–96.

—— 1965a: "Semantical Analysis of Modal Logic: II, Non-normal, Modal Propositional Calculi", in J. Addison *et al.* (eds.), *The Theory of Models,* Amsterdam: North Holland, pp. 206–20.

—— 1965b: "Semantical Analysis of Intuitionistic Logic I", in J.N. Crossley and M.A.E. Dummett (eds.), *Formal Systems and Recursive Functions.* Amsterdam: North Holland, 92–130.

—— 1971: "Identity and Necessity", in M.K. Munitz (ed.), *Identity and Individuation.* New York: New York University, 135–64.

—— 1972: "Naming and Necessity", in D. Davidson and G. Harman (eds.), *Semantics of Natural Language*. Dordrecht: Reidel, 253–355, and 763–9.

—— 1977: "Speaker Reference and Semantic Reference", in P.A. French *et al.* (eds.), *Contemporary Perspectives in the Philosophy of Language*. Minneapolis: University of Minnesota Press, 6–27.

—— 1979: "A Puzzle about Belief", in A. Margalit (ed.), *Meaning and Use*. Dordrecht: Reidel, 239–83.

—— 1982: *Wittgenstein on Rules and Private Language,* Cambridge: Harvard University Press.

—— 1980. Preface to *Naming and Necessity*, Cambridge, Mass.: Harvard University Press.

Lambert, K. 1991: "A Theory of Definite Descriptions", in Lambert, K. (ed.), *Philosophical Applications of Free Logic*, Oxford: Clarendon Press, 17–27.

Lemmon, E.J. 1957/58: "Quantifiers and Modal Operators", *Proceedings of the Aristotelian Society,* 58, 244–68.

—— 1963: "A Theory of Attributes Based on Modal Logic", *Acta Philosophica Fennica,* 16, 95–122.

Lewis, C. I. 1912: "Implication and the Algebra of Logic", *Mind,* 21, 522–31.

—— 1913a: "Interesting Theorems in Symbolic Logic", *Journal of Philosophy,* 10, 239–42.

—— 1913b: "A New Algebra of Implications and Some Consequences", *Journal of Philosophy,* 10, 428–38.

—— 1914: "The Matrix Algebra for Implications", *Journal of Philosophy,* 11, 589–600.

—— 1917: "The Issues Concerning Material Implication", *Journal of Philosophy,* 14, 350–6.

—— 1918: *Survey of Symbolic Logic.* Berkeley: University of California Press.

—— and Langford, C.H. 1932: *Symbolic Logic.* New York: The Century Company.

Lindström, P. 1966: "First-order Predicate Logic with Generalized Quantifiers", *Theoria* 32, 186–195.

Linsky, L. 1966: "Substitutivity and Descriptions", *Journal of Philosophy,* 63, 673–83.

—— 1969: "Reference, Essentialism, and Modality", *Journal of Philosophy,* 66, 687–700.

—— 1977: *Names and Descriptions.* Chicago: University of Chicago Press.

—— 1983: *Oblique Contexts.* Chicago: University of Chicago Press.

Löwenheim, L. 1915: "Über Möglichkeitenim Relativkalkül", *Mathematische Annalen,* 76, 447–70.

Ludlow, P.J. and Neale S.R.A. 1991: "Indefinite Descriptions: In Defence of Russell." *Linguistics and Philosophy,* 14, 171–202.

Lukasiewicz, J., 1953: "A System of Modal Logic", *Actes du XIᵉ Congres International de Philosophie Bruxelles, 1953, Vol. XIV.* Amsterdam: North-Holland, 111–49.

MacColl, H., 1906: *Symbolic Logic and its Applications*. London: Longmans.

Marcus, R. 1946a: "A Functional Calculus of first Order Based on Strict Implication", *Journal of Symbolic Logic,* 11, 1–16.

—— 1946b: "The Deduction Theorem in a Functional Calculus of First Order Based on Strict Implication", *Journal of Symbolic Logic,* 11, 115–8.

—— 1947: "The Identity of Individuals in a Strict Functional Calculus of First Order", *Journal of Symbolic Logic* 12, 12–15.

—— 1948: Review of Smullyan's "Modality and Description", *Journal of Symbolic Logic,* 13, 149–50.

—— 1953: "Strict Implication, Deducibility, and the Deduction Theorem", *Journal of Symbolic Logic,* 18, 234–6.

—— 1962: "Modal Logics I: Modalities and Intensional Languages", in M. W. Wartofsky (ed.), *Proceedings of the Boston Colloquium for the Philosophy of Science,* 1961/1962, 77–96.

—— 1967: "Essentialism in Modal Logic", *Nous,* 1, 91–6.

—— 1968: "Modal Logic", in R. Klibansky (ed.), *Contemporary Philosophy.* Florence: La Nuova Italia Editrice, 87–101.

—— 1990: "A Backward Look at Quine's Animadversions on Modalities", in R. Barrett and R. Gibson (eds.), *Perspectives on Quine*, Oxford: Blackwell, 230–243.

—— 1993: *Modalities,* Oxford: Oxford University Press.

Marcus, R., Quine, W.V., *et al.,* 1962: "Discussion", in M.W. Wartofsky (ed.), *Proceedings of the Boston Colloquium for the Philosophy of Science,* 1961/1962, 105–16.

Martí, G., 1995: "Do Modal Distinctions Collapse in Carnap's System?" *Journal of Philosophical Logic,* 23, 575–93.

—— 1997: "Rethinking Quine's Argument on the Collapse of Modal Distinctions", *Notre Dame Journal of Formal Logic*, 38, 276–94.

Mates, B. 1968: *Elementary Logic*. Berkeley: University of California Press.

—— 1973: Descriptions and Reference. *Foundations of Language* 10, 409–18.

Mill, J. S. 1872: *A System of Logic*. 8th edition. London: Longmans, 1949.

Mitchell, D. 1962: *An Introduction to Logic*. London: Hutchison.

Montague, R. 1960: "Logical Necessity, Physical Necessity, Ethics, and Quantifiers", *Inquiry*, 3, 259–69.

—— 1963: "Syntactical Treatments of Modality, with Corollaries on Reflexion Principles and Finite Axiomatizability", *Acta Philosophica Fennica*, 16, 153–68.

Montague, R. and Kalish, D. 1959: "That", *Philosophical Studies*, 10, 54–61.

Mostowski, A. (1957). "On a Generalization of Quantifiers", *Fundamenta Mathematicae*, 44, 12–36.

Myhill, J. 1958: "Problems Arising in the in the Formalization of Intensional Logic", *Logic et Analyse*, 1, 74–83.

—— 1963: "An Alternative to the Method of Extension and Intension", in P.A. Schilpp (ed.), *The Philosophy of Rudolph Carnap*. La Salle: Open Court, 229–310.

Neale, S.R.A. 1990: *Descriptions*, Cambridge: MIT Press.

—— 1992a "Binary Quantifiers and Unary Quantifier-formers" (abstract), *The Journal of Symbolic Logic* 57, p. 318.

—— 1993a: "Term Limits." *Philosophical Perspectives* 7, 89–124.

—— 1993b: "Grammatical Form, Logical Form, and Incomplete Symbols", In A.D. Irvine and G.A. Wedeking (eds.), *Russell and Analytic Philosophy,* Toronto: University of Toronto Press, 97–139.

—— "The Philosophical Significance of Gödel's Slingshot." *Mind* 104 (1995), 761–825.

—— 1999: "Colouring and Composition." In R. Stainton (ed.) *Philosophy and Linguistics*. MIT Press, 35–82.

—— *Facing Facts*, Oxford: Oxford University Press, forthcoming.

Neale, S.R.A. and Dever, J. 1997: "Slingshots and Boomerangs." *Mind* 106, 143–68.

Nelson, E.J. 1930: "Intensional Relations", *Mind*, 39, 440–53.

—— 1933: "On Three Logical Principles in Intension", *Monist*, 43, 268–84.

Omelyantchik, V. 1997: "Constructive Belief vis-à-vis the Slingshot", Italian Society for Analytic Philosophy.

Oppy, G. 1997: "The Philosophical Insignificance of Gödel's Slingshot". *Mind*, 106, 121–141.

Parry, W.T. 1939: "Modalities in the *Survey* System of Strict Implication", *Journal of Symbolic Logic,* 4, 137–54.

Parsons, C.A. 1990: Introduction to Gödel's "Russell's Mathematical Logic", in S. Feferman *et al.* (eds.), *Kurt Gödel, Collected Works,* vol. II (1938–1974), Oxford: Oxford University Press, 112–18.

—— 1995: "Quine and Gödel on Analyticity", P. Leonardi and M. Santambrogio (eds.), *On Quine,* Cambridge: Cambridge University Press, 297–313.

Parsons, T. 1967: "Grades of Essentialism in Quantified Modal Logic", *Nous,* 1, 181–200.

—— 1969: "Essentialism and Quantified Modal Logic", *Philosophical Review,* 78. 35–52.

Patton, T.E. 1997: "Explaining Referential/Attributive", *Mind,* 106, 245–61.

Prior, A. 1956: "Modality and Quantification in S5", *Journal of Symbolic Logic,* 21, 60–62.

—— 1957: *Time and Modality*. Oxford: Clarendon.

—— 1962: "Possible Worlds", *Philosophical Quarterly,* 36–43.

—— 1963: "Is the Concept of Referential Opacity Really Necessary?" *Acta Philosophica Fennica*, 16, 189–200.

—— 1967: "Modal Logic", in P. Edwards (ed.), *The Encyclopedia of Philosophy,* vol. 5, New York: Macmillan, 5–12,

Quine, W.V. 1935: Review of Carnap's *Logische Syntax der Sprache*, *Philosophical Review*, 44. 394–7.

—— 1936: "Truth by Convention", in O. H. Lee (ed.), *Philosophical Essays for A.N. Whitehead*. New York: Longmans. Reprinted with revisions in W.V. Quine, *The Ways of Paradox and Other Essays,* 2nd ed., rev. and enl., 1976, 77–106.

—— 1937: "New Foundations for Mathematical Logic", *American Mathematical Monthly,* 44, 79–80.

—— 1939: "Designation and Existence", *Journal of Philosophy,* 36, 701–9

—— 1940: *Mathematical Logic*. Cambridge, Mass.: Harvard University Press, rev. ed., 1951.
—— 1941a: "Whitehead and the Rise of Modern Logic", in P.A. Schilpp (ed.), *The Philosophy of Alfred North Whitehead*. Evanston, Illinois: Northwestern University Press, 127–63.
—— 1941b: Review of Russell's *Inquiry into Meaning and Truth*, *Journal of Symbolic Logic*, 6, 29–30.
—— 1943: "Notes on Existence and Necessity", *Journal of Philosophy*, 40, 113–27.
—— 1946: Review of Marcus's "A Functional Calculus of First Order Based on Strict Implication", *Journal of Symbolic Logic*, 12, 95.
—— 1947: "On the Problem of Interpreting Modal Logic", *Journal of Symbolic Logic* 12, 43–8.
—— 1948: "On What There is", *Review of Metaphysics*, ...! Reprinted in W.V. Quine, *From a Logical Point of View*, Cambridge: Harvard University Press, 1–19.
—— 1950: *Methods of Logic*, 1st ed., New York: Henry Holt.
—— 1951: "Two Dogmas of Empiricism", *Philosophical Review*, 60, 20–43.
—— 1953a: "Reference and Modality", in *From a Logical Point of View*. Cambridge: Harvard University Press, 139–59.
—— 1953b: "Meaning and Existential Inference", in *From a Logical Point of View*, 160–67.
—— 1953c: "Three Grades of Modal Involvement", *Actes du XIe Congrès International de Philosophie Bruxelles, 1953*, Vol. XIV. Amsterdam: North-Holland, 65–81. Reprinted in W.V. Quine, *The Ways of Paradox and Other Essays*, 2nd ed., rev. and enl., 1976.
—— 1956: "Quantifiers and Propositional Attitudes", *Journal of Philosophy*, 53, 177–87.
—— 1957: "Logic, Symbolic", *Encyclopedia Americana*. Reprinted in *Selected Logic Papers*, enlarged edition, Cambridge: Harvard University Press, 37–51.
—— 1960: *Word and Object*. Cambridge, Mass.: MIT Press.
—— 1961: "Reference and Modality" (revised version) in *From a Logical Point of View*, 2nd ed., Cambridge: Harvard University Press, 139–59.
—— 1962: "Reply to Professor Marcus" M.W. Wartofsky (ed.), *Proceedings of the Boston Colloquium for the Philosophy of Science*, 1961/1962, 97–104.
—— 1966: "Russell's Ontological Development", *Journal of Philosophy*, 63, 657–667. Reprinted with revisions in *Theories and Things*, Cambridge: Harvard, 1981, 73–85.
—— 1969: "Replies", in D. Davidson and J. Hintikka (eds.), *Words and Objections*. Dordrecht: Reidel, 292–352.
—— 1970: *Philosophy of Logic*. 2nd ed., Cambridge: Harvard University Press.
—— 1972: Review of M.K. Munitz (ed.), *Identity and Individuation*. *Journal of Philosophy*, 69, 488–97.
—— 1977: "Intensions Revisited", in *Midwest Studies in Philosophy*, 2, Reprinted in *Theories and Things*, Cambridge: Harvard, 1981, 113–123.
—— 1980: "Reference and Modality" (2nd revised version) in *From a Logical Point of View*, 2nd ed., rev. printing, New York: Harper and Row, 139–59.
—— 1981: "Five Milestones of Empiricism", in *Theories and Things*, Cambridge: Harvard, 1981, 67–72.
—— 1982: *Methods of Logic*. Cambridge Mass.: Harvard University Press, 4th ed.
—— 1993: "Comment on Marcus", in R. Barrett and R. Gibson (eds.), *Perspectives on Quine*, Oxford: Blackwell, 244.
—— 1995: "Free Logic, Description and Virtual Classes", in *Selected Logic Papers*, rev. and enl. ed., Cambridge: Harvard University Press, 278–285,
Rasiowa, H. 1951: "Algebraic Treatment of the Functional Calculi of Heyting and Lewis", *Fundamenta Mathematicae*, 38, 99–126.
—— 1963: "On Modal Theories", *Acta Philosophica Fennica*, 16, 201–14.
Rasiowa, H. and Sikorski, R. 1953: "Algebraic Treatment of the Notion of Satisfiability", *Fundamenta Mathematicae*, 40, 62–95.
—— 1954: "On Existential Theorems in Non-classical Algebraic Functional Calculi", *Fundamenta Mathematicae*, 41, 21–8.
Reichenbach, H. 1947: *Elements of Symbolic Logic*. New York: Macmillan.
Rescher, N. 1962: "Plurality Quantification". Abstract. *Journal of Symbolic Logic* 27, 373–4.
Richard, M. 1987: "Quantification and Leibniz's Law", *Philosophical Review*, 96, 555–78.

Rundle, B. 1965: "Modality and Quantification", in R. J. Butler (ed.), *Analytical Philosophy, Second Series*, Oxford: Blackwell, 27–39.

Russell, B. 1905: "On Denoting", *Mind*, 14, 479–93.

Scott, D. 1967: "Existence and Description in Formal Logic", in R. Schoenman (ed.), *Bertrand Russell, Philosopher of the Century*, London: Allen and Unwin, 181–200.

Searle. J. "Proper Names", *Mind*, 67, 166–73.

Sharvy, R. 1969: "Things", *The Monist* 53, 488–504.

—— 1970: "Truth Functionality and Referential Opacity", *Philosophical Studies* 21, 5–9.

—— 1972: "Three Types of Referential Opacity", *Philosophy of Science* 39, 153–61.

—— 1980: "A More General Theory of Definite Descriptions", *Philosophical Review* 89, 607–24.

Smullyan, A.F., 1941: "Entailment Schemata and Modal Functions", abstract, *Journal of Symbolic Logic*, 6, p. 40.

—— 1947: Review of Quine's "The Problem of Interpreting Modal Logic", *Journal of Symbolic Logic*, 12, 139–41.

—— 1948: "Modality and Description", *Journal of Symbolic Logic* 13, 31–7.

Soames, S. 1987: Review of Hornstein's *Logic as Grammar*, *Journal of Philosophy*, 84, 447–455.

Soames, S. 1995: "Revisionism about Reference", *Synthese*, 104, 191–216.

Stalnaker, R. 1972: "Pragmatics", in D. Davidson and G. Harman (eds.), *Semantics of Natural Language*, Dordrecht: Reidel, 380–97.

Stich, S. 1986: "Are Belief Predicates Systematically Ambiguous?" In R.J. Bogdan (ed.), *Belief*. Oxford: Clarendon Press, 119–147.

Strawson, P.F. 1950: "On Referring", *Mind*, 59, 320–44.

—— 1952: *Introduction to Logical Theory*. London: Methuen.

Tarski, A. 1933: "Pojecie Prawdy w Jezykach Nauk Dedukcynch". Translated as "Der Wahrheitsbegriff in den Formalisierten Sprachen", *Studia Philosophica*, 1 (1936), 438–460. Translated as "The Concept of Truth in Formalized Languages", in A. Tarski, *Logic, Semantics, and Metamathematics*. 2nd edition, edited by J. H. Woodger. Indiana: Hackett, 1983, 152–277.

—— 1936: "O Pojciu Wynikania Logicznego", Przeglad Filozoficzny, 39, 58–68. Translated as "Über den Begriff der Logischen Folgerung", *Actes du Congrès International de Philosophie Scientifique, Paris, 1935*, vol. 7, 1–11. Translated as "On the Concept of Logical Consequence", in A. Tarski, *Logic, Semantics, and Metamathematics*. 2nd edition, edited by J.H. Woodger. Indianapolis: Hackett, 1983, 409–20.

—— 1941: *Introduction to Logic and to the Methodology of the Deductive Sciences*. Oxford: Oxford University Press.

Taylor, B. 1985: *Modes of Occurrence*. Oxford: Blackwell.

Thomason, R. 1969: "Modal Logic and Metaphysics", in K. Lambert (ed.), *The Logical Way of Doing Things*. New Haven: Yale University Press, 119–46.

Vredenduin, P.G.J. 1939: "A System of Strict Implication", *Journal of Symbolic Logic*, 4, 73–6

Wallace, J. 1969: "Propositional Attitudes and Identity", *Journal of Philosophy*, 66, 145–152.

Wiener, N, 1913: "Mr Lewis and Implication", *Journal of Philosophy*, 13, 656–62.

Wedberg, A. 1966: Filosofins Historia: Fran Bolzano till Wittgenstein, Stockholm: Bonniers.

—— 1984: *A History of Philosophy. Volume 3: From Bolzano to Wittgenstein*, Oxford: Clarendon.

Whitehead, A.N. and Russell, B. 1925: *Principia Mathematica*, vol. I, 2nd ed. Cambridge: Cambridge University Press.

Wilson, N. 1959: *The Concept of Language*. Toronto: University of Toronto Press.

Wittgenstein, L. 1922: *Tractatus Logico-Philosophicus*. London: Kegan Paul.

von Wright, G.H. 1951: *An Essay in Modal Logic*, Amsterdam: North-Holland.

—— 1953: "A New System of Modal Logic", *Actes du XIe Congres International de Philosophie Bruxelles, 1953, Vol. V*. Amsterdam: North-Holland, 59–63.

GREG RAY

DE RE MODALITY: LESSONS FROM QUINE

INTRODUCTION

The aim of this paper is twofold: i) to give a logically explicit formulation of a slight generalization of Quine's master argument about *de re* modality – an argument which imposes important constraints on modal semantics, ii) to briefly present my favored account of modal locutions (especially locutions of the *de re* metaphysical flavor) and show how it successfully copes with Quine's argument. Let me apologize in advance for spending a good deal of time, as I do in this paper, making explicit an argument that Quine laid out so many years ago and that has been very often discussed and alluded to since. However, I have come to the conviction that this argument is still widely misunderstood, and so the careful attention to detail seems warranted. In espousing my favored view of modal locutions in various venues, a broad appeal to "Quinean considerations" has often been made by way articulating a worry for me. Sometimes the objection is simply: "But what about Quine's argument?" Since I think of the view I am promoting as very nearly Quinean, this response has always puzzled me. From what I have seen, philosophers' attitudes towards Quine's master argument fall into two kinds: i) there are those that think that the argument has no force, because it is based on some mistake (usually, something about definite descriptions), and ii) there are those that think that the argument poses some insuperable barrier to any kind of *de re* modality. Neither of these attitudes is justified. So, I hope to make plain along the way that a) the original version of Quine's argument is sound, b) there is a version of this same basic argument which imposes very definite constraints on any proposed account of *de re* "metaphysical" modality in particular, and c) there is an account that satisfies these constraints. Part 1 of this paper will be concerned with laying out and discussing three versions of Quine's argument, in the service of establishing points (a) and (b).

In Part 2 of the paper, I will briefly sketch what I take to be a very promising, and also very Quinean account of *de re* modality – one that respects the constraint on modal semantics that Quine's argument reveals and comports well with the few positive remarks Quine makes, for example, in *Word and Object* (1960) regarding our use of modal locutions. This will put us in a position to see that the proposed account does not fly in the face of Quine's master argument.

347

Alex Orenstein and Petr Kotatko (eds.), Knowledge, Language and Logic, 347–365.
© *2000 Kluwer Academic Publishers. Printed in Great Britain.*

PART 1. QUINE'S MASTER ARGUMENT ON DE RE MODALITY

1.1 Introduction

The basic form of Quine's argument is by now familiar enough.

1. Modal contexts are referentially opaque.
2. Quantification into referentially opaque contexts does not make sense. That is, $Qx\varphi x$ is a meaningful expression only if every free occurrence of "x" in φx is purely designative.
3. Therefore, quantification into modal contexts does not make sense.

Thus, quantified modal logic is in trouble. Likewise for modal languages with *de re* constructions (of the sort that semantically justify an unrestricted rule of substitution).[1]

It is useful to think of several notions that come into play in Quine's basic argument as free parameters of the argument. We will want to consider the successfulness and implications of the argument that results from fixing these parameters in a number of alternative ways. The notions I have in mind are as follows.

First, the argument makes essential reference to a certain class of referring *terms*. In particular, the argument for (1)

1a) A substitution principle for transparent contexts holds.
1b) There are modal contexts in which substitution of co-referring terms does not preserve truth value.

involves the application of a certain substitution principle, and an appeal to certain examples involving sentences differing only by the substitution of key terms. So, one parameter of the argument which could be varied would be that which determines the class of terms which are involved here. In my formulation of Quine's argument, I call the target class of terms 's-terms'. For example, we might fix the s-terms to be the class of grammatical singular terms, which would include proper names and descriptions. On the other hand, we might limit the s-terms just to proper names.

Second, part of Quine's argument depends on the provision of an example. In particular, the existential claim, (1b), above is justified by Quine by appeal to examples. Obviously, the argument will not depend on any particular choice of example. Quine offers two, and I will suggest a third. Any one will do, but different examples may serve different ends and different examples may be differentially successful. So, the choice of example is another parameter of the argument which we will want to consider varying. Notice that this second parameter of the argument and the first we identified are not independent of each other. One of Quine's examples famously involves a definite description, and the use of this example requires that we fix the s-terms to include definite descriptions. However, Quine's second example and my third example do not involve any definite descriptions, and if we use these examples to establish (1b), then it will turn out that the s-terms can be fixed narrowly to include just proper names. Part of the motivation for varying this parameter will be clear from the next parameter we consider.

Third, the argument concerns modal contexts of some sort, but there may be different kinds of modal contexts with importantly different semantic features, and which of these is taken as a target of the argument may make a difference. Thus, another parameter might be the kind of modal context which is at issue in the argument. Now there are two basic ways in which a particular kind of modal context could enter as a target for Quine's argument. Firstly, the target modal contexts might be taken to be those associated with a recognized sense of the term "necessary" for which an acceptable analysis is at hand. Secondly, the target contexts might be taken to be those associated with a recognized, but only vaguely circumscribed, sense of the term "necessary" for which no good analysis is known, but which sense is distinguished by a core set of modal claims whose truth values are widely agreed upon. Quine's original argument appears to be of the former sort, since Quine is explicitly concerned with a notion of necessity which he identifies with *analytic necessity*.

Notice again, that this third parameter of the argument is not independent of the others. The identification of a class of modal contexts which is the target of the argument makes an obvious difference to the suitability of the examples which may be used in support of (1b). Those examples make claims about the truth values of various modal sentences, and so the veracity of these claims will depend on what sense of 'necessary" or 'possible" is at issue.

Thus, we have identified three degrees of freedom for the argument: i) the range of s-terms, ii) the choice of example appealed to in the argument, and iii) the identification of the modality in question. There are two reasons I am interested in these particular parameters. First, I am interested in the second, because the availability of examples for the argument which do not involve definite descriptions moots a long-standing line of criticism and discussion of Quine's argument. I think this discussion distracts from the central lesson of the argument, and such objections as are based on descriptions do not, in the final analysis, have much force against Quine's argument. Shifting the example appealed to (parameter ii) enables us to narrow the range of the s-terms (parameter i) and so avoid a potential objection.

My second reason for interest in these parameters is that I am interested in seeing the way in which Quine's argument impinges on the current discourse on modality. The current discourse is almost wholly centered on so-called "metaphysical modality", but Quine's argument was originally pitched against a notion of analytic necessity. Shifting the modality in question (parameter iii) in the argument requires a change of example (parameter ii). Neither of Quine's original examples in support of (1b) is apt for saying something about *de re* metaphysical modality. Thus, in the course of giving my formulation of Quine's argument, I will introduce a third example which is apt for that purpose.

The remainder of Part One of the paper is given over to formulating Quine's argument. In Section 1.2, (1a) is proven from Leibniz' Law. A second way of proving (1a) from a second version of Leibniz' Law is given in an appendix. Neither represents an argument explicitly given by Quine, though he does indicate that (1a) is justified by Leibniz Law. I believe it is worth presenting the argument explicitly both for the sake of completeness, and because the possibility of such an argument has been questioned in connection with Quine's argument.[2] In section 1.3, claim (2) is argued for on the basis of a simple assumption about the semantics of the sort of quantification which is at issue. Finally, in section 1.4, I turn to discuss two examples given by Quine to justify (1b), as well as a third example of my own which will feature largely in Part 2.

1.2 A Formulation of Quine's Master Argument

Definitions

NOTATION: Throughout, we will use "φ" to range over formulas (of a given language L), and "α" will range over the s-terms (of L) – which form a (possibly improper) subclass of the singular terms (of L). For simplicity we will generally consider only open formulas with exactly one free occurrence of the variable "x". We sometimes write "φx" to emphasize the free occurrence of "x" in such a formula. "$\varphi \alpha$" shall designate the result of substituting for the free "x" in φ an occurrence of the s-term α.

CONVENTION: To conserve typography, I will use a somewhat relaxed quotational convention. Where a string consists of at least one Greek letter and some other symbols, I will suppress the Quinean corner-quotes. Thus, in place of "$\ulcorner \Box \varphi \urcorner$" I may write "$\Box \varphi$". Either expression is to be understood as an abbreviation for the definite description "the result of concatenating '$\Box$' with φ".

TERM: φx expresses a property iff there is a determinate fact of the matter whether an object per se does or does not satisfy φx, i.e. iff φx has a definite extension.

DEF: For all y, φx is true of y iff y satisfies φx.[3]

It is useful (especially in Lemma 3) to talk about formulas expressing properties and of a formula being true of some object. Note however that both these notions are defined here in terms of the uncontroversial notion of satisfaction.

DEF: The occurrence of an s-term (or variable), α, in $\varphi \alpha$ is purely designative iff for any s-term γ, $\varphi \gamma$ is true iff the denotatum of γ has the property expressed by φx.[4]

DEF: Modal contexts will be said to be referentially transparent (in a given language L) just in case for every formula, φ, (of L), every occurrence of a constant or free variable in φ that is purely designative is also purely designative in $\Box \varphi$, and in $\Diamond \varphi$. Otherwise, modal contexts will be said to be referentially opaque.

Main Argument: It does not make sense to quantify into modal contexts.

1. Modal contexts are referentially opaque. [Lemma 1]
2. Quantification into referentially opaque contexts does not make sense. That is, $Qx\varphi x$ is a meaningful expression only if every free occurrence of "x" in φx is purely designative. [Lemma 2]
3. Therefore, quantification into modal contexts does not make sense.

Lemma 1: Modal contexts are not referentially transparent.

1. The restricted substitution principle is true, i.e. if the occurrence of "x" in φx is purely designative, then if $\alpha=\beta$ is true and $\varphi\alpha$ is true, then $\varphi\beta$ is true. [Lemma 3]
2. There are modal contexts in which substitution of co-referring s-terms does not preserve truth value. In particular, there are cases where "x" is purely designative in φx, but where $\alpha=\beta$ is true and $\square\varphi\alpha$ is true, but $\square\varphi\beta$ is false. [Lemma 4]
3. Therefore, modal contexts are referentially opaque.

Lemma 3: The Restricted Substitution Principle is True.

Quine adverts to Leibniz' Law (LL) in justifying the restricted substitution principle in question. Here, we give an argument based on the property version of Leibniz' principle of the indiscernibility of identicals to Quine's restricted substitution principle which illustrates how the assumption that the locus of substitution is purely designative gets involved. The version of LL to which Quine adverts is what we might call the "satisfaction version" of LL,[5] since it relies on the locution "true of" which we have here characterized in terms of satisfaction. Arguments connecting the property version of LL to the Satisfaction Principle, and from that principle to the restricted substitution principle, are given in an appendix.

I hope I may be excused for going into this matter in what may seem like excessive detail. It seems to me that some of the discussion of substitution in the literature suggests that the availability and integrity of this argument has not been made sufficiently clear. Evidence (Marti, 1989), where it is straightforwardly denied that Quine's substitution principle can be obtained from Leibniz' Law. It should be noted though that Marti seems to hold that the substitution principle Quine endorses is the *unrestricted* one.[6] While I decline to think that Richard Cartwright (1971) makes the same mistake in reading Quine, nonetheless the only substitution principle which he mentions in the same breath with Quine's name is the unrestricted one as well. Cartwright also spends some time arguing that this principle does not follow from Leibniz Law. Rightly understood, both of these authors are just recapitulating the point Quine (1943) made at the outset. At the very least, such discussions as Marti's and Cartwright's are likely to muddy the waters, since, while both are critical of Quine's master argument, neither is concerned with the only matter in connection with Leibniz Law which could be of *any* moment for Quine's argument. Thus, it seems to me worthwhile to actually present the argument from Leibniz' Law (in either of two versions) to the *restricted* substitutivity principle that Quine really *does* endorse.

The Argument for Restricted Substitution from Leibniz' Law

1. $\alpha=\beta$ is true iff the denotatum of α is identical to the denotatum of β.
2. For any x and y, if x is identical to y, then for any property P, x has P iff y has P. [property version of Leibniz' Law]
3. Suppose, the occurrence of "x" in φx is purely designative.[7]
4. Suppose $\alpha=\beta$ is true.[8]

5. Suppose $\varphi\alpha$ is true.
6. Suppose the denotatum of α is a, and the denotatum of β is b.
7. Suppose φx expresses the property P.
8. The denotatum of α has the property expressed by φx. [3,5]
9. a has P. [8,6,7]
10. a is identical to b [4,1,6]
11. b has P. [9,10,2]
12. The denotatum of β has the property expressed by φx. [11,6,7]
13. $\varphi\beta$ is true. [12,3]

So, assuming the following two uncontroversial principles

1. For any s-terms α and β, $\alpha=\beta$ is true iff the denotatum of α is identical to the denotatum of β. [semantics of identity statements]
2. If a is identical to b, then for any property P, a has P iff b has P. [indiscernibility of identicals]

We have the following result:
    For s-terms α, β, and formula φx,
IF
        the occurrence of "x" in φx is purely designative
THEN
        if $\alpha=\beta$ and $\varphi\alpha$ are true, then $\varphi\beta$ is true.

Lemma 2: Quantification into referentially opaque contexts does not make sense.

1. Suppose $\exists x\varphi x$ is a meaningful existential sentence.
2. Then, that expression has the truth conditions appropriate to an existential sentence, namely
        $\exists x\varphi x$ is true iff something satisfies φx.
3. This truth condition presupposes that there is a determinate fact of the matter whether an object *per se* does or does not satisfy φx. In the terminology introduced earlier, it presupposes that φx expresses a property.
4. (*) But if the context φx expresses a property, then if α is a s-term, $\varphi\alpha$ is true iff the denotatum of α has the property expressed by φx. That is, if the context φx expresses a property, then "x" is purely designative in φx.
5. Therefore, if $\exists x\varphi x$ is a meaningful existential sentence, then "x" is purely designative in φx.

Our estimation of premise (4) may depend on what the class of s-terms encompasses. We will discuss this below.

Lemma 4: There are modal contexts in which substitution of co-referring
s-terms does not preserve truth value.

This claim is argued by example. Requirements on an example which will stand witness for the existential claim are as follows. We require an open sentence, φx, and s-terms α and β such that a) "x" is purely designative in φx, b) $\alpha=\beta$ is true, and c) either i) $\Box\varphi\alpha$ is true, but $\Box\varphi\beta$ is false, or ii) $\Diamond\varphi\alpha$ is true, but $\Diamond\varphi\beta$ is false.

Example One

0. "x" is purely designative in "H is identical to x"
1. "$H=P$" is true.
2. "H is necessarily identical to H" is true.
3. Yet, "H is necessarily identical to P" is false.

(1) and (2) are uncontroversial. Easily, the controversial claim will be (3). Recall, however, that Quine's original argument concerns a sense of "necessary" which can be characterized in terms of analyticity. Thus, (3) is justified just in case "$H=P$" is not an analytic truth. There is good reason to think that "$H=P$" is not an analytic truth, and, I dare say, there would be widespread agreement on this (scruples about the notion of analyticity aside).

But I am interested in Quine's argument as it might be applied to a different notion of necessity, namely so-called "metaphysical necessity". The sense of "necessary" in question there is not one for which there is a handy analysis. The sense of the term is picked out by a core set of cases about which there is widespread agreement as to the truth values of those cases. In regard to this sort of necessity, there is considerable consensus among philosophers in favor of the thesis that true identities are metaphysically necessary. Thus, if we make metaphysical necessity the target of the argument, the truth value assignment in (3) is not justified (or at least not uncontroversial).

Now, it is sometimes thought that Quine can be shown wrong simply by triumphing against the next example we shall consider. Nothing could be farther from the truth, and that should be very clear at this point. Doing so only undermines that one example. What an opponent needs in order to show Quine wrong is that there be NO examples of the sort Quine needs. Prepatory to a defense of Smullyan's (1948) objection, Stephen Neale (1990) boldly states that, outside of the description-involving cases, there is not even a *prima facie* case of substitution failure of the sort Quine needs. This is far too hasty, of course, since Quine offered the one we have just considered at the get-go.

Example Two (the famous one)

0. "x" is purely designative in "x is greater than 7"
1. "9 = the number of the planets" is true.
2. "9 is necessarily greater than 7" is true.
3. Yet, "the number of the planets is necessarily greater than 7" is false.

The third claim is once again the controversial one. However, it is easy to see here, too, how that claim is justified when the target is analytic necessity. (3) is justified just in case "the number of the planets is greater than 7" is not an analytic truth. No one will want to argue with that, I suppose.[9] But again, as it seems to me, when we shift focus from analytic necessity to metaphysical necessity, (3) does not appear to be justified (or at least is not uncontroversial).

Several remarks are in order. First, it seems to me that example one is effective for establishing what Quine needs to establish here, so long as we stick to analytic necessity, and so the second example is not really necessary anyway, whether it is problematic or not. In fact, I think the second example is not so secure. One thing about this example that appears to have gone unnoticed is that it requires that "9 is greater than 7" be an analytic truth. That is a substantive and controversial assumption after all. If it isn't an analytic truth, then the example is defeated with never a hackle raised about definite descriptions.

A second reason that this second example is less secure does have to do with the status of definite descriptions. In order for this example to work for Quine, it must be the case that definite descriptions are among the s-terms. This was made explicit when we identified the necessary features of an example. Now, a number of claims are made in the larger argument about s-terms, and at least one of these claims is probably false if the s-terms include descriptions. The claim I have in mind is premise (4) of Lemma 2.

4. But if the context φx expresses a property, then if α is an s-term, $\varphi\alpha$ is true iff the denotatum of α has the property expressed by φx. That is, if the context φx expresses a property, then "x" is purely designative in φx.

If there is a problem about the use of an example involving descriptions in Quine's argument, then it seems to me to be here (and not in the defense of Lemma 4, as Smullyan's essay would seem to suggest).[10,11]

Example Three
Following Allan Gibbard (1975), let us suppose that an artisan has fashioned two lumps of clay, and, in bringing them together brings into being a new statue, and incidentally a new lump of clay as well. So as not to beg any identity questions, let us name the statue "Goliath", after its intended likeness, and let us name the new lump of clay, "Lump1". We may suppose that Goliath and Lump1 are destroyed later after the clay has dried when the artist, dissatisfied with his work, dashes it to pieces. Intuitively, before the clay dried, the artist could have squeezed Lump1 into a ball without thereby destroying it, but not so, we think, for the statue. This sort of modal intuition is the ground for our next example.

0. "x" is purely designative in "x is squeezed harmlessly [i.e. non-terminally] into a ball"
1. "Goliath = Lump1" is true.
2. "Lump1 could have been squeezed harmlessly into a ball" is true.
3. "Goliath could have been squeezed harmlessly into a ball" is false.

This third example is not one that Quine has offered, and is not one which it would make sense to offer in the context of an argument which had analytic necessity as a target. We already have an appropriate example for that purpose (example one, above). I am interested in this example, because it is an appropriate example if we wish to shift the ambit of the argument to target the notion of metaphysical modality. The example also shares with example one the virtue of not involving any descriptions. In contradistinction to example one, however, when considered as a case for arguing about metaphysical modality, the present example does not fly in the face of the doctrine of necessity of identity, subscribed to by many advocates of metaphysical modality.

Now, someone (such as Quine) might be dubious about using an example which required subscribing to the truth value claims given in (2) and (3) above. One might be concerned about this either because one was a modal skeptic in general, or because one did not want to rely on the vagaries of metaphysical modal "intuitions". To allay such concern, it is important to note the following little-noted fact about the role of the examples in Quine's argument. So far as the argument's critical purposes go, it need be no part of Quine's (or any advocate's) commitments that the truth value claims made in the examples be correct (or even sensible). It is enough that these claims be the deliverance of the analysis, if such there be, of the modal context in question. In Quine's case, all that is needed to justify his examples is that the embedded sentences are understood to be analytic (or non-analytic, as the case may be). If there is no such analysis assumed, as in the present case, then it is enough that these claims are drawn from a core set of modal claims of the targeted type whose truth values are widely agreed upon (even if not subscribed to). In the present case, all that is needed to justify our example is that the truth values assigned to the modal claims in the example are of a kind with those that are supposed to typify the sort of modality in question. The modal claims involving a clay statue chosen above are undoubtedly "canonical" in just the needed sense.

1.3 Lessons from Quine

Like any significant argument, Quine's argument has a number of significant premises. And, as with any such argument in philosophy, various philosophers may object to one or another of the premises along the way. More often, in the case of Quine's argument, it seems that philosopher's have sought to find fault in the reasoning.[12] I think that the foregoing shows that this is not a fruitful avenue. The argument formulated above is certainly valid, or so I claim, and you may easily verify for yourself.

Moreover, if we take the s-terms to range only over proper names, use example one, and take the target modality to be analytic necessity, then, I believe, we get a sound argument to the conclusion that quantifying into modal contexts of that sort does not make sense.[13] This is, of course, just what Quine argued.

As far as I can see, the only good option open to defenders of *de re* modality is to claim for themselves a sense of "necessary" which is not properly captured by the notion of analytic necessity. The notion of metaphysical necessity is supposed to be just such a beast, and Quine's original argument does not address it *per se*. But the form of argument is not powerless to speak to this kind of response either, as my discussion above has aimed to show.

If we take the s-terms to range only over proper names, use example three, and take the target modality to be metaphysical modality, then we get what is at least a *prima facie* sound argument to the conclusion that quantifying into modal contexts of the metaphysical sort does not make sense.

Given that items (2) and (3) are uncontroversial in the context of metaphysical modality, the friend of modality would appear forced by the Quinean argument to either i) accept that quantifying into these modal contexts does not make sense, or ii) deny that these modal contexts are after all referentially opaque on the grounds that many prima facie true identities are actually false. Accepting the first option would count as a defeat for modern advocates, insofar as they are wedded to the idea that a genuinely metaphysical necessity must admit genuinely *de re* necessities. So it looks as though the second option is forced. In the case of our example, this would mean denying the identity of the statue and the lump of clay, i.e. item (1) of example three. Plenty of contemporary philosophers have taken this move to be indeed forced. Perhaps the most well-known of these is Saul Kripke. For such philosophers, a statue is never identical to any lump of clay, a lump of clay is never identical to any sum of molecules, a ship is never identical to any sum of planks, pain sensations are never identical to any neurological event, etc.[14]

Evidently, these metaphysicians pay for their modality in the coin of ontology. This is an under-advertised cost of much contemporary modal metaphysics. It would not be hard to imagine supplementing Quine's argument with argumentation against this sort of move simply on the grounds of its extraordinary ontological burden. Of course, this would be a bitter pill for modal metaphysics, if it sealed off the only remaining live option. Fortunately, there remains a option – and an attractive and simple one. If I understand rightly, the way out of this bind is a treatment of modality of a sort that Quine himself suggests in various places.[15]

PART 2. A VERY QUINEAN ACCOUNT OF METAPHYSICAL MODALITY

2.1 Introduction

When I have presented, in various venues, my favored view of metaphysical modality, philosophers have often thought that Quine's famous argument would undermine my view. Now that we have a careful formulation of a version of that famous argument which is pertinent to metaphysical modality, I aim to show why the view of modality I favor is not disturbed by this argument. In fact, as I said earlier, the view I favor appears to be close to Quine's own, judging by the few positive remarks Quine makes on this subject.[16]

Accordingly, let us focus our attention on the version of Quine's argument in which: i) the s-terms are just proper names, ii) the target modality is metaphysical modality, and iii) example three is used. What this version of Quine's argument shows is that the following nine propositions are not compatible.

1. For any s-terms α and β, $\alpha=\beta$ is true iff the denotatum of α is identical to the denotatum of β. [semantics of identity statements]
2. If a is identical to b, then for any property P, a has P iff b has P. [indiscernibility of identicals]
3. If $\exists x \varphi x$ is a meaningful existential sentence, then

$\exists x \varphi x$ is true iff something satisfies φx. [objectual quantification]

4. If the context φx expresses a property, then if α is an s-term, $\varphi \alpha$ is true iff the denotatum of α has the property expressed by φx. That is, if the context φx expresses a property, then "x" is purely designative in φx.

5. "x" is purely designative in "x is squeezed harmlessly [i.e. non-terminally] into a ball"

6. "Goliath = Lump1" is true.

7. "Lump1 could have been squeezed harmlessly into a ball" is true.

8. "Goliath could have been squeezed harmlessly into a ball" is false.

9. Quantifying into [metaphysical] modal contexts makes sense.

Claims (1)–(8) are just the premises of the argument, and (9) is the denial of its conclusion. Now, no one should want to deny (1)–(5). In fact, some of these claims are non-negotiable. (2) is non-negotiable because it is a conceptual truth. (3) is non-negotiable here for a different reason: it expresses a fact which is partially definitive of objectual quantification, and so (3) is justified by what are the terms of the debate, since, I take it, it is objectual quantification into modal contexts that is at issue – it is objectual quantification that the defenders of (9), defenders of *de re* modality, have been interested in.[17] I also think that one should not deny (6), because of the enormous ontological cost of doing so. Incurring that cost is unjustified since there is a simple and attractive, ontology-free alternative, namely, the view I will now very briefly describe. Before doing so, we should note that, since part of the object of this game is to uphold (9) in the face of Quine's argument, and since we have ruled out rejecting any of (1)–(6), the view we will spell out must have the consequence that one of (7) or (8) is false. Also, while this is a success condition for a view of the sort we seek, it is important to note that the view is motivated in a way that is *independent* of this consideration.

2.2 Restricted modalities

The leading idea of the account of modal locutions I have in mind is one according to which our statements of *de re* modality are best understood as implicitly restricted (in a way to be specified). This means that the modal sentences we utter (what we strictly and literally say) generally do not specify all that is needed to understand what claim we have given voice to (what we pragmatically implicate).[18] To motivate these claims about our use of modal terms, let me briefly review some familiar facts about our use of *quantifier* terms.

There is a well-acknowledged parallel between the logical behavior of modal terms and that of ordinary quantifiers. The same pattern of logical relations between basic quantifier sentences (e.g. "Everything is a G" implies "Something is a G") that is exhibited on the classical square of opposition is likewise exhibited by the corresponding modal sentences (e.g. "Necessarily P" implies "Possibly P"). This parallel between ordinary quantifiers and modal operators is the foundation of modal model-theoretic semantics as well as (so-called) *possible worlds semantics*.

Our ordinary quantificational talk is often *implicitly restricted*. When we say things like

Not a building was left standing after the fire.
All the seats were occupied.
Everyone came down with the flu.

it is abundantly clear (in the circumstances) that we do not mean to be making a claim about *absolutely every* building, seat, or person that there is. Rather, we meant to claim something about, say, all the buildings *in that town*, all the seats *on that plane*, and every person *in that class*. What if modal terms, which otherwise act so much like quantifiers, paralleled them in this respect too? I think our use of modal terms is best understood as implicitly restricted.

There is plenty of evidence from (honest-to-goodness) ordinary language to support this idea. Here is a simple-minded example based on an ordinary (albeit fictitious) phone conversation:

A: I just called to remind you one last time that we are going to the show tonight. You know, you could come with us, if you liked.
B: I know, I'd really like to, but I can't.
A: Why not?
B: I must finish preparing my lecture for tomorrow.
A: Too bad. Well, maybe next time.

There is an obvious sense in which it would be silly to try to track this conversation by fixing on a *single, unrestricted modality* and attending only to what these inter-locutors *strictly and literally say*. If we did that, we would, for example, be puzzled by A's assertion of the utterly obvious point that B *could* go to the show. We would also then have to wonder at A's saying that he *can't* go, since it is clear in the con-versation that neither speaker takes this as in tension with what A just said. And finally, though it seems strictly-speaking false that B *must* finish his preparations (we would allow that he might well not finish it, should some untimely accident befall him), A does not object on those grounds, but acquiesces to the assertion. The point of this exercise is to suggest that in perfectly ordinary uses of modal locutions either i) we are not tracking what is strictly and literally said, or ii) there are a lot of different senses to modal terms. Any reader who is attracted to the latter alternative is referred to the *Schulman tour de force* in William Lycan's (1994, Chapter 8), where, in a passage from a cheap novel, Lycan identifies what, on assumption (ii), would have to be thirty or so different senses of simple modal locu-tions. It is abundantly clear from that exercise that the multiplication of senses does not end there, since many of the sense of modality which would be required to make sense of the passage are so specialized to the context, that it is easy to see that other passages will generate ever more senses for modal terms. The net result is to quickly erode any plausibility that affirming (ii) could be the right way to respond to these examples.

On the other hand, these ordinary uses of modal terms begin to line up in an orderly fashion as soon as we begin to think of them as implicitly restricted. It is not hard to identify implicit conditions for each modal claim which helps make sense of the phone exchange above. Likewise for the Schulman passage which is Lycan's target (and this is also Lycan's point with that passage). More examples are to be found in (Kratzer, 1977), and, of course, much the same idea underlies David Lewis' (1979; 1986) multiple-counterpart theory.[19]

Finally, it is allowed by many philosophers that certain broad categories of modal claim can be treated as relativised modal claims. For example, rather than have a distinct sense of modality which we might term "physical necessity" in order to make sense of the claim

It is not possible to suspend this building by a human hair
(i.e. Necessarily, this building is not suspended by a human hair)

one can make sense of this claim by assimilating it to

Necessarily, if N then this building is not suspended by a human hair

where "N" is to be replaced either by a statement of the laws of nature, or by "the laws of nature are as they actually are". The sense of "necessarily" in this second claim now no longer need be taken to express some specialized notion of physical necessity, but may express some more generalized notion, e.g. conceptual necessity.

So, there is ample evidence that ordinary modal claims are quite generally implicitly restricted. If we treat metaphysical modal claims also as implicitly restricted, then, just as in the physical case, we will no longer need to suppose that there is a special sense of necessity, "metaphysical necessity", but will be able to make sense of our modal claims with just implicit restriction and some broader notion of necessity. Thus, it is an upshot of the view I favor that the idea that there is a distinct metaphysical modality turns out to be based on a kind of pragmatic illusion. It is the product of not recognizing that pragmatic factors are at work, and so making wrong semantic judgments, e.g. judgments of truth value.[20] One of these is the judgment which gives us (8) on our list in Section 2.1.

2.3 Regarding the object

To what are modal terms restricted and what do such restrictions do? Now, the implicit restriction on (an utterance of) a *quantifier phrase* partially determines *which objects* are at issue. "Which seats? Oh, the ones in the business class section of the plane." I say, the implicit restriction on (an utterance of) a *modal* operator determines *what it is about an object* that is at issue. So, the restriction on a modal operator will select out some *feature* of an object that will be relevant to the truth of a claim.

If universal claims are understood as implicitly restricted, then the evaluation of such claims (insofar as we are interested not so much in the sentences, but in what the speaker is committed to) will be partially determined by imputing the right restriction in a given case. The same must be said for the modal case, except what must be settled on here is a way of regarding the object(s) under consideration. Presumably, an appropriate determination will not generally be independent of the intentions of a claimant, of what a listener may be expected to understand in that context, etc. Thus, our evaluations will be unavoidably tied up with how we (as individual speakers, hearers) *regard* those objects under consideration, i.e. what feature(s) of those objects we consider relevant to the truth of our modal claims.[21]

To bring these considerations to bear on our own concerns, we need to determine what are the relevant restrictions, in the case of our modal claims about the statue, Goliath, and the lump of clay, Lump1. First, though, we need a principled way of determining such things. Again, we can take a tip from quantifier cases. Suppose K says "Every seat is already occupied", and we want to fix on an appropriate restriction on the quantifier in his utterance, then we might well look to *which* things K considers relevant to justifying his claim. Does he need to inspect both the fore and aft sections of the plane? Does he need to inspect the interior of all the planes at the airport? Or what?

For the modal case, suppose J says "Obleo is necessarily a G", and we want to fix on a appropriate restriction on the operator in her utterance. Then we should look to *what* it is *about* Obleo that J considers relevant to justifying her claim. We may ask of J, what it is about an object she would need to know, in order to know that that object is necessarily G (like Obleo is). J's answer to this question gets us at least in the neighborhood of that way of regarding Obleo which implicitly restricts her modal claim.[22] So, this gives us a kind of "evidential test" for imputing the right restrictions on a modal claim, though it is admittedly a rather crude guide. I hasten to add that I am *not* suggesting here that what possibilities Obleo has depends in any way on what J believes about Obleo or on how J conceives of Obleo. Rather, what possibilities J *claims* Obleo has depends in some way on her beliefs or conceptions. That is unsurprising, of course.

Let's use this test with an example we care about. What is it about Goliath that makes you think that it could not be squeezed harmlessly into a ball? What would you need to know about a thing in order to know that it could not have been squeezed into a ball harmlessly (just like Goliath could not). The obvious answer is: that it is a statue. We *know that* about Goliath from the story told earlier, and it is certainly that which has us believing that Goliath could not have been squeezed up.[23] If you knew Obleo was a statue, you would think the same of Obleo. Now, what is it about Lump1 that makes you think that it *could* have been squeezed into a ball harmlessly? Well, it's a lump of clay, right? Like any lump of clay it could have been so squeezed. I think this indicates that our two modal claims (about Goliath and Lump1) are restricted in different ways.

The factors which influence how we regard some objects under consideration are many.[24] One way we may be influenced in our conception is the employment of *suggestive names* for objects. This happens in the statue puzzle, as we heard it told earlier by Gibbard. One may also indicate an object via different descriptions, as David Lewis does when relating the argument. Sometimes we are drawn to change our way of regarding in yet more subtle ways. This is especially evident when we are engaging in hypothetical reasoning. We must sometimes regard an object in ways that we believe are merely hypothetical.[25]

2.4 The upshot

Consider again our list from Section 2.1. We cannot consistently maintain all (9) of the propositions listed there, and we ruled out rejecting (1)–(6) and (9). This left us with (7) and (8).

7. "Lump1 could have been squeezed harmlessly into a ball" is true.
8. "Goliath could have been squeezed harmlessly into a ball" is false.

One of these must be rejected. The account of modal locutions sketched above gives us a well-motivated and painless way to do that, however. According to that view, when we affirmed

10. Lump1 could have been squeezed harmlessly into a ball

there was an implicit restriction involved, so that we are, after all, only really committed to a restricted version of that claim – something like

10b. It could have been the case that Lump1 is a lump of clay and is squeezed harmlessly into a ball.

and when we affirmed

11. Goliath could *not* have been squeezed harmlessly into a ball

there was an implicit restriction involved, so that we are, after all, only really committed to a restricted version of that claim – something like

11b. It could not have been the case that Goliath is a statue and is squeezed harmlessly into a ball.

Now, since (10b) logically implies (10), being committed to (10b) likewise commits us to (10), and hence we should accept (7). However, (11b) does not logically or conceptually entail (11), so our commitment to (11b) does not require the acceptance of (8). In fact, of course, with everything else we have already accepted, including (6) and (7), Quine's argument can be used to prove that (8) must be rejected. What the view of modal locutions that we sketched above does is give us independently motivated grounds for doing so.

There is, of course, much more to be said about the view that metaphysical necessity is just implicitly restricted conceptual necessity, but this must wait for another occasion. I will close this section by listing a set of features which I count as virtues of the view.

Modal locutions get a standard modal logic.	[D1]
Goliath and Lump1 may be said to be identical.	[D2]
Identity may be said to be necessary.	[D3]
Our discrepant *de re* modal intuitions about Lump1 and Goliath are respected.	[D4]
Leibniz's Law is not constrained in any way.	[D5]
Modal *logical* intuitions are preserved.[26]	[D6]
Names can have the same semantic function in and out of modal contexts, namely just to refer.	[D7]
De re modal claims turn out genuinely *de re,* i.e. they attribute modal properties to objects *per se.*	[D8]
The view employs no linguistic, semantic, or pragmatic machinery that we did not already need on separate account.	[D9]

The *epistemology* of so-called metaphysical modality becomes
  as unmysterious as is knowledge of conceptual truths.   [D10]

3. Conclusion

In conclusion, I hope to have accomplished two things. First, I hope to have given a
clarifying and definitive form to Quine's master argument. Second, I hope to have
shown how an attractive account of *de re* modal locutions copes successfully with
the version of that argument which presses most directly on it.

4. Appendix

It is also possible to argue for Quine's restricted substitution principle from another
principle, sometimes also called "Leibniz' Law", the Satisfaction Principle (if what
are nominally two things are in fact one and the same, then whatever is true of one
if true of the other). Quine makes explicit reference to this principle as an undeni-
able principle from which it is sometimes thought that an unrestricted substitution
principle follows.

    What we can show is that the Satisfaction Principle follows from Leibniz' Law
proper, and the restricted substitution principle follows from the Satisfaction
Principle.

A) Argument for the Satisfaction Principle from Leibniz' Law

1. For any x and y, if x is identical to y, then for any property P, x has P iff y has
    P.
2. If φx expresses a property P, then for all y, y satisfies φx iff y has P.
3. Suppose a is identical to b.
4. Suppose φx is true of a, i.e. a satisfies φx.
5. Suppose φx expresses the property P.
6. a has P. [4,5,2]
7. b has P. [6,3,1]

So, assuming the following two uncontroversial principles

1. If φx expresses a property P, then for all y, y satisfies φx iff y has P. [fact about
    satisfaction]
2. If a is identical to b, then for any property P, a has P iff b has P. [indiscernibil-
    ity of identicals]

We have the following result:
    For any a and b,
IF
        a is identical to b
THEN
        for any formula φx,
        φx is true of a iff φx is true of b.

B) Argument for Restricted Substitution from the Satisfaction Principle

1. $\alpha=\beta$ is true iff the denotatum of α is identical to the denotatum of β.
2. For any x and y, if x is identical to y, then whatever is true of x is true of y.
3. Suppose, the occurrence of "x" in φx is purely designative.
4. Suppose $\alpha=\beta$ is true.
5. Suppose $\varphi\alpha$ is true.
6. Suppose the denotatum of α is a, and the denotatum of β is b.
7. The denotatum of α has the property expressed by φx, i.e. φx is true of the denotatum of α. [5,3]
8. φx is true of a. [7,6]
9. a is identical to b [4,1]
10. φx is true of b. [8,9,2]
11. φx is true of the denotatum of β, i.e. the denotatum of β has the property expressed by φx. [10,6]
12. $\varphi\beta$ is true. [11,3]

So, assuming the following two uncontroversial principles

1. For any s-terms α and β, $\alpha=\beta$ is true iff the denotatum of α is identical to the denotatum of β. [semantics of identity statements]
2. For any x and y, if x is identical to y, then whatever is true of x is true of y. [the true-of principle]

We have the following result:
For s-terms α, β, and formula φx,
IF
    The occurrence of "x" in φx is purely designative
THEN
    if $\alpha=\beta$ and $\varphi\alpha$ are true, then $\varphi\beta$ is true.

University of Florida, Gainesville

NOTES

[1] Note: nothing absolutely untoward follows directly for formally *de re* modal sentence constructions *per se*. Though we see that a) they are not genuinely or purely *de re*, b) they are similar in some way to quotational contexts, and c) there are certain limits on what can be inferred from them.

[2] Richard Cartwright (1971) muddies the waters by *seeming* to take up this position, and Genoveva Marti (1994) outright denies that the argument can be given. I will say a bit more about this in the discussion of Lemma 3 below.

[3] We will understand the satisfaction relation as obtaining just in case y is in the extension of φx. Thus, φx's being true of y already presupposes that φx expresses a property.

[4] Roughly, to say that an occurrence of a s-term is purely designative in a certain sentence is to say that, so far as truth conditions are concerned, the sentence may be thought of as of subject-predicate form with the s-term in question as subject term, and the remaining sentential context as (complex) predicate term.

[5] The principle in question is: if what are nominally two things are in fact one and the same, then whatever is true of one if true of the other.

[6] This is made more explicit in (Marti, 1994).

[7] Note that this implies that φx expresses a property. Thus, the supposition at line (7) may be discharged in the argument without additional assumption.

[8] Note that this, together with (1), implies that α and β each uniquely denotes something. Thus, the supposition at line (6) may be discharged in the argument without additional assumption.

[9] Note that the scope of the description is irrelevant to this assessment. So, we can construe the description as having that scope (namely narrow) which makes it the result of performing a substitution on the sentence mentioned in (2). The next remark presupposes that we are taking the description as having narrow scope. Notice that we are throughout respecting Quine's complaint to some who have favored Arthur Smullyan's criticism of this example. We respect it here by not supposing that it even makes sense to give a truth value assignment to the sentence which would give the description wide scope.

[10] Unless it be objected that (3) is unjustified, because the mentioned sentence is ambiguous as to the relative scope of the modal operator and the definite description. However, given that Quine has stipulated the sort of necessity involved, this is not a tenable line against the original argument.

[11] But perhaps, if problem there be, the one that Smullyan was pointing to has merely shown up in another place, for formulation-dependent reasons. I won't speculate here.

[12] I take Smullyan's line of objection and its ilk to be of this sort.

[13] I am taking for granted that the semantics of quantification is genuinely objectual. Alternative proposals to this assumption fostered interesting debate, of course, but I am not here interested in that discourse. I also happen to think that it is the better part of semantic innocence to construe ordinary universal and existential quantifiers objectually.

[14] Moreover, even for those who prefer a four-dimensionalist ontology, the corresponding pairs of four-dimensional objects would never be identical.

[15] Quine has not made much of these suggestions, and has not followed them out in any detail. Presumably, this is because it does rob the metaphysician's most cherished modal claims of much of their seeming-importance. Indeed, Quine is lately wont to say that modal claims just aren't of any great importance. The idea is that, once we understand what is going on with such claims as we make, we see that they do not play the important role in theorizing that we might have thought.

[16] Indeed, the view I favor may even *be* Quine's own view.

[17] If one could only rescue some kind of substitutional quantification for a modal context, then one would not have championed a de re modality.

[18] The modal sentences that I have in mind are those which 1) involve "necessarily", "possibly", "could have been", or "could not have been", and 2) are most naturally understood as *de re* and alethic.

[19] There are also important differences. In Lewis' theory what is pragmatically selected is the relevant counterpart relations for the subjects of a modal claim. Also, Lewis takes modal claims to be disguised quantifications over worlds which also involve disguised predications. The hidden predication involved is the one that says the subject of the claim stands in a certain counterpart relation with some possibilion. Thus, the particular counterpart relation gets into the semantics of the modal claim. So, on the one hand Lewis wants the counterpart relation to be picked out pragmatically, but wants the predication of that relation to be part of what is strictly and literally said.

[20] Naturally, when these sorts of pragmatic factors are at work and we do not recognize it, we may easily assign to sentences we *utter* truth values which properly belong only to sentences which literally express things which we, by so uttering, *implicate* instead.

[21] In (Lewis, 1983), David Lewis sketches how implicit restrictions may be determined and re-determined in a conversational context. See also (Nute, 1980) for a codification of some principles which govern this process, with special reference to our use of counterfactuals.

[22] On the other hand, if there is nothing qualitative about the object(s) in question that one needs to know in order to know the modal claim is true, this is evidence that the modal claim in question is not restricted.

[23] Actually, to take the implicitly restricting feature to just be *being a statue* is something of a simplification. The story we told will have led you to believe that Goliath is a statue of a certain kind. For example, you will naturally presuppose that it is not cleverly hinged so that when squeezed, it can collapse into the shape of a ball. Likewise, you have been given to

believe that Goliath is not a piece of modern or conceptual art whose shape is not constant or somehow otherwise not of critical importance.
[24] For convenience, I sometimes refer to *how one is regarding an object,* or *what one is thinking that object is,* or *one's way of conceiving of that object.* These locutions are casual, yet suggestive, identifiers for that feature of an object which restricts an utterance of a given modal operator. Please note that these same locutions are used sometimes to express views incompatible with mine. For example, Gareth Evans (1982) uses the idea of *a way of thinking about an object* as explicative of the notion of Fregean *sense,* but it is important to see that what I am attempting to illuminate here is a feature of the way we use modal locutions, not a feature of (the semantics of) singular terms. As I said, my view is non-Fregean with respect to names.
[25] Some of the dynamics of the evolution of implicit restrictions in a conversational context has been worked out in (Lewis, 1979) and (Nute, 1980), though these efforts are against the backdrop of Lewis' counterpart theory, and, in the latter case, specifically with regard to some conditionals.
[26] It can be proved that the "surface logic" of modal locutions is well-behaved in a certain way, on the current view. Places where this good behavior is predicted by the theory to break down are just the sorts of cases where puzzles like that of the statue and the clay arise. Cf. (Ray, 1992).

REFERENCES

Cartwright, Richard. "Identity and Substitutivity". *Identity and Individuation.* Ed. Milton K. Munitz. New York: New York University Press, 1971, pp. 119–133.

Evans, Gareth. *Varieties of Reference.* Ed. John McDowell. Oxford: Clarendon Press, 1982.

Gibbard, Allan. "Contingent Identity." *J Phil Logic* 4 (1975): 187–221.

Kratzer, Angelika. "What "Must" and "Can" Must and Can Mean." *Ling Phil* 1 (1977): 337–355.

Lewis, David. "Scorekeeping In a Language Game." *J Phil Log* 8 (1979): 339–359. Reprinted in *Philosophical Papers* 1. Oxford: Oxford University Press, 1983, pp. 233–249.

Lewis, David. *On the Plurality of Worlds.* New York: Basil Blackwell, 1986.

Lycan, William G. *Modality and Meaning.* Dordrecht: Kluwer Academic Publishers, 1994.

Marti, Genoveva. "Aboutness and Substitutivity". *Contemporary Perspectives in the Philosophy of Language II. Midwest Studies in Philosophy 14.* Ed. Peter A. French, Theodore E. Uehling, and Howard K. Wettstein. Notre Dame: University of Notre Dame Press, 1989, pp. 127–139.

Marti, Genoveva. "Is a Semantics for Modal Contexts Really Possible?" Typescript. University of California, Riverside, 1994.

Neale, Stephen. *Descriptions.* Cambridge: MIT Press, 1990.

Nute, Donald. "Conversational Scorekeeping and Conditionals." *J Phil Log* 9 (1980): 153–166.

Quine, Willard V.O. "Notes on Existence and Necessity." *J Phil* 40 (1943): 113–127. Reprinted in *Semantics and the Philosophy of Language.* Ed. Leonard Linsky. Urbana: University of Illinois Press, 1952, pp. 77–91.

Quine, Willard V.O. "Reference and Modality". *From a Logical Point of View.* 2nd, rev. ed. New York: Harper & Row, 1961. pp, 139–159.

Quine, Willard V.O. *Word and Object.* Cambridge: MIT Press, 1960.

Ray, Greg. "Modal Identities and De Re Necessity." Typescript, 1992.

Smullyan, Arthur F. "Modality and Description." *J Sym Log* 13 (1948): 31–37. Reprinted in *Reference and Modality.* Ed. Leonard Linsky. Oxford: Oxford University Press, 1971, pp. 35–43.

FRANÇOIS RECANATI

OPACITY AND THE ATTITUDES*

I. OPACITY

§1.1 Use vs. mention

When we mention an expression, do we use it? It depends on *how* we mention the expression in question. We can mention an expression A by using another expression B which names it. In such a case we are not using A, but its name ("heteronymous mention"). But we can also use A itself in "*suppositio materialis*", that is, autonymously. That is what is ordinarily called "mention" as opposed to "use". This traditional contrast between use and (autonymous) mention should not make us forget that in autonymous mention, the mentioned word itself is used, though deviantly.

I have just exploited the contrast between autonymous and heteronymous mention to lend credit to the idea that in autonymous mention (e.g. (1) below), the word is used. That is indeed the basis for the contrast; for the mentioned word does not occur at all in sentences such as (2) in which it is *heteronymously* mentioned.

(1) "Cat" is a three-letter word
(2) Wychnevetsky is a three-letter word[1]

Still, it may be argued that the word "cat" occurs only accidentally in (1), much as "nine" occurs in "canine" or "cat" in "cattle". Such a claim, made by Quine, would be highly implausible if taken at face value.[2] The occurrence of "cat" in "cattle" is indeed an accident; so much so that the word "cat" does not, *qua* word, occur in "cattle". A sequence of letters (or a sound) is not a word. To be sure, the individuation of words raises complex issues, but on any plausible account "cat" in "cattle" will not count as an occurrence of a word. In contrast, the occurrence of "cat" in (1) will count as an occurrence of the word "cat", rather than as an orthographic accident. We can go along with Quine and accept that the first word of (1) is not the word "cat", but a different expression formed from the word "cat" by appending quotation marks around it;[3] still, the occurrence of the word "cat" *within* the complex expression is no accident. The word "cat" is named by quoting *it*. That is how autonymous mention works. The mechanism of autonymous mention requires that we use *the word itself*, and put it within quotation marks. This is *toto mundo* different from a case of heteronymous mention (the word A is named by the word B) in which, by accident, B contains A in the manner in which "cattle" contains "cat". (Thus instead of "Wychnevetsky" in (2) we might have used another, no less arbitrary name, viz. "Wychnecatsky", in which by accident the orthographic sequence "cat" occurs.)

367

Alex Orenstein and Petr Kotatko (eds.), Knowledge, Language and Logic, 367–406.
© *2000 Kluwer Academic Publishers. Printed in Great Britain.*

The mentioned word is used, but, as I said, it is used *deviantly*. The word is not used according to its normal semantic function. Thus a word whose role is to name a certain object or to "make it the subject of discourse" (as Mill says) will be used to make *itself* the subject of discourse.

Deviant uses, in general, are far from uncommon, and come in many varieties. We may not only use the word "cat" autonymously, to denote that very word, but also to denote, say, a *representation* of a cat. Thus we can say:

(3)   In the middle of the piazza stood a gigantic cat, due to a local sculptor.

This is deviant because a stone cat is not a cat. So the word "cat", which means *cat*, can be used to mean many other things through the operation of various "primary pragmatic processes" – pragmatic processes involved in the determination of what is said (Recanati 1993). Autonymy is one such process; metonymy is another. Such processes generate *systematic ambiguities*. Whenever a word denotes a type of thing, we can use it alternatively to denote a representation of that type of thing; whatever a word denotes, we can use it to denote that very word. Such ambiguities are similar to those mentioned by Quine himself (1960: 130): the process/product ambiguity (e.g. "assignment" which can refer to the act of assigning or to the thing assigned), or the action/custom ambiguity ("skater", which can refer to someone who skates or to someone who is skating), or the type/token ambiguity.[4]

In natural language, such ambiguities flourish. It is through them that natural language gains its main virtue: its flexibility, which makes it fit to talk about anything. But what is a virtue from one point of view is a defect from another. From a *logical* point of view, ambiguity is to be avoided. That means that, instead of using the same word to mean different things, we should use distinct words.

When it comes to quotation and autonymous use, the reform is easy. It proceeds in three steps:

Step 1: Whenever a word is used autonymously, make that explicit by using e.g. quotation marks around the word. (Even logicians did not respect that precept in the earlier part of this century. Frege and Quine were pioneers in this regard.)
Step 2: Consider the complex expression – word plus quotation marks – as a new word which names the original expression.
Step 3: Ignore the occurrence of the original expression, as if it were accidental; treat it as a fragment of the new expression, as "cat" is a fragment of "Wychnecatsky".

I interpret Quine as urging this treatment of quotation as *desirable from the logical standpoint,* and as part of the "reform" which has to take place before we can subject natural language sentences to logical appraisal. But that does not mean that the treatment in question is descriptively correct (or thought by Quine to be correct), as an account of the way natural language works. Natural language does not fear ambiguities, it rather welcomes them. In particular there is no doubt that it allows using a word to refer to that very word.

§1.2 "Giorgione"

Is the autonymous word referential or not? It depends in what sense. If we accept that the word refers to itself, then it is referential after all.[5] Its referentiality can be checked using the Principle of Substitutivity. Replacing the autonymous word A, which refers to itself, by another, B, which also refers to A, preserves truth-value, as the possible transition from (1) to (2) shows. But the mentioned word is not referential in the *normal* sense: it does not refer to its normal referent. In what follows I will take "referential" to mean just that: referential in the normal sense. An occurrence of a word is referential, in that sense, if and only if it refers to the normal referent of the word.

A term's being referential does not guarantee that the word can be replaced *salva veritate* by an occurrence of another word referring to the same object. For that to be guaranteed, Quine says, the occurrence at issue must be *purely referential* – the term must be used "purely to specify its object" (Quine 1960: 142). This qualification is necessary because Quine thinks there is a continuum of cases from pure non-referentiality, as illustrated by (1), to pure referentiality. Quine gives the following example:

(4)   Giorgione was so-called because of his size

In such cases, Quine says, the word (here "Giorgione") has a dual role. It is both mentioned *and* used to refer. It is a mixture of autonymy and referentiality. It is because the word "Giorgione" is not used purely referentially that substitution of "Barbarelli" for "Giorgione" fails to preserve truth, despite the fact that Barbarelli and Giorgione are (were) one and the same person.

There is an apparent paradox in Quine's admitting such intermediate uses. For cases of autonymous mention are in principle eliminable in favour of heteronymous mention, in Quine's framework; but the occurrence of the word cannot be eliminated if, while mentioned, it keeps doing its normal referential work. Quine dispels the paradox by construing intermediate cases as involving two occurrences consolidated into a single one: a purely referential occurrence and an autonymous (hence eliminable) occurrence. A perspicuous paraphrase makes the duality explicit. Thus (4) is rendered as

(5)   Giorgione was called "Giorgione" because of his size

I think Quine's insight that there is a continuum of cases between pure autonymy and pure referentiality is correct and important. (See the Appendix.) But his classification of the "Giorgione" example in that category is misleading, for there is a sense in which the word "Giorgione" in (4) is used purely referentially. To be sure, the word "Giorgione" is mentioned in (4). But there is no inconsistency between holding that the word is used purely referentially, and holding that it is mentioned; for it is mentioned *heteronymously* in (4). Far from referring to itself, the word "Giorgione" is referred to *by means of a different expression*, viz. the demonstrative adverb "so" in "so-called". Hence the word "Giorgione" itself is not used in two ways (referentially and autonymously); it is used purely referentially.

In contrast to autonymous mention, heteronymous mention is compatible with purely referential use. This point can be driven home by splitting sentence (4) in two, as Kit Fine has suggested (1989: 253):

(8)
    A: Giorgione was Italian.
    B: Yes, and he was so-called because of his size

Who would deny that the occurrence of "Giorgione" in A's statement is purely referential? The fact that B's statement contains an expression demonstratively referring to the name "Giorgione" in no way conflicts with the purely referential character of the occurrence thus demonstrated.

Quine appeals to the failure of substitutivity as proof that the occurrence of "Giorgione" in (4) is not purely referential. For if it were, it would be substitutable. Now, even though Giorgione is Barbarelli, substitution of "Barbarelli" for "Giorgione" does not preserve truth. Substitution of "Barbarelli" for "Giorgione" in (4) yields (9), which is false:

(9)   Barbarelli was so-called because of his size

But this proof that the occurrence of "Giorgione" in (4) is not purely referential rests on an equivocation. The fallacy of equivocation is presented as follows in Quine's *Methods of Logic:*

The two conjunctions:
(10)   He went to Pawcatuck and I went along
(11)   He went to Saugatuck but I did not go along
may both be true; yet if we represent them as of the form "p&q" and "r&¬q", as seems superficially to fit the case, we come out with an inconsistent combination "p&q&r&¬q". Actually of course the "I went along" in (10) must be distinguished from the "I went along" whose negation appears in (11); the one is "I went along to Pawcatuck" and the other is "I went along to Saugatuck". When (10) and (11) are completed in this fashion they can no longer be represented as related in the manner of "p&q" and "r&¬q", but only in the manner of "p&q" and "r&¬s"; and the apparent inconsistency disappears. In general, *the trustworthiness of logical analysis and inference depends on our not giving one and the same expression different interpretations in the course of the reasoning. Violation of this principle was known traditionally as the fallacy of equivocation.* (…)
    The fallacy of equivocation arises… when the interpretation of an ambiguous expression is influenced in varying ways by immediate contexts, as in (10) and (11), so that the expression undergoes changes of meaning within the limits of the argument. In such cases *we have to rephrase before proceeding.* (Quine 1962: 42–43; notation and emphasis mine)

By the same reasoning, it can be shown that the alleged failure of substitutivity exhibited by the occurrence of "Giorgione" in (4) is merely apparent. Substitutivity fails, Quine says, because, although Giorgione was so-called because of his size, and Giorgione = Barbarelli, Barbarelli was *not* so-called because of his size. Paraphrasing Quine, however, we can respond as follows:

The two statements:
(4)   Giorgione was so-called because of his size
(12)   Barbarelli was not so-called because of his size

may both be true; yet if we represent them as of the form "Fa" and "¬Fb", as seems superficially to fit the case, we come out with an inconsistency, since a = b. Actually of course the "so-called" in (4) must be distinguished from the "so-called" which appears in (12); the one is "called *Giorgione*" and the other is "called *Barbarelli*". When (4) and (12) are rephrased in this fashion they can no longer be represented as related in the manner of "Fa" and "¬Fb", but only in the manner of "Fa" and "¬Gb"; and the apparent inconsistency disappears.

What this shows is that the substitution of "Barbarelli" for "Giorgione" does preserve truth after all. The appearance that it does not is caused by the fact that "the interpretation of an ambiguous expression is influenced in varying ways by immediate contexts, ... so that the expression undergoes changes of meaning within the limits of the argument." If, following Quine's advice, we "rephrase before proceeding" we must substitute "called *Giorgione*" for "so-called" in (4) *before* testing for substitutivity; and of course, if we do so, we see that substitutivity does not fail. From (5) and the identity "Giorgione = Barbarelli" we can legitimately infer (13):

(5)   Giorgione was called "Giorgione" because of his size
(13) Barbarelli was called "Giorgione" because of his size

I conclude that "Giorgione" in (4) is purely referential: substitution preserves truth, appearances notwithstanding. Yet the substitution which preserves truth is not any old substitution of coreferential singular terms, but substitution *under a uniform interpretation of whatever context-sensitive expression occurs elsewhere in the sentence* (Fine 1989: 221–5). This condition is crucial, for an apparent failure of substitutivity may be caused by the fact that the semantic value of some context-sensitive expression in the sentence changes as a result of the substitution itself. (That will be so in particular when, as in the "Giorgione" example, the sentence contains an expression demonstratively referring to the singular term which undergoes substitution). When that is the case, the failure of substitutivity is consistent with pure referentiality. Only a failure of substitutivity under conditions of uniform interpretation provides a reasonable criterion of non-purely referential use.

In his discussions of opacity Quine does not adhere to his own policy of "rephrasing before proceeding" when the sentence at issue is relevantly ambiguous or context-sensitive. Instead of using "substitutivity" in the sense of "substitutivity under conditions of uniformity", he uses it in the sense of "substitutivity *tout court*". In that sense the occurrence of "Giorgione" in (4) is indeed not substitutable. I will hereafter follow Quine and use "substitutable" in this way. My point concerning the "Giorgione" example can therefore be rephrased as follows: Pure referentiality does not entail substitutability; hence failure of substitutivity cannot be retained as a criterion of non-purely referential occurrence.

§1.4 Pure referentiality and transparency

An occurrence of a singular term is *purely referential,* Quine says, just in case the term, on that occurrence, is used "purely to specify its object". In other words, the term's semantic contribution, on that occurrence, is its (normal) referent, and nothing else. To be sure, a singular term will not only contribute its semantic value (its referent), it will also show or display whatever other properties it has: its form, its sense, its affective tone, its poetic qualities, and whatnot. But what matters from a semantic point of view is merely that which the term contributes to the truth-conditions of the whole.

What is meant exactly by a term's "semantic contribution", i.e. its contribution to the truth-conditions? There is an ambiguity here. On a broad reading, the semantic contribution of an expression is *the overall difference it makes to the truth-conditions of the sentence where it occurs*. In that sense, "Giorgione" in (4) does not make the same semantic contribution as "Barbarelli" in (9); for if they did, (4) and (9) would have the same truth-value. But there is a stricter reading, more relevant to semantic theory. From the standpoint of semantic theory, each expression has a semantic value, and the semantic value of the sentence depends upon the semantic values of its parts and the way they are put together. The semantic contribution of an expression, in the narrow sense, is its semantic value – that which, in part, determines the truth-value of the whole. Thus in the "Giorgione" example, what the word "Giorgione" contributes is the individual Giorgione, which it names. The name "Giorgione" serves also as referent for another expression, and affects the truth-conditions of the sentence in that respect too, but that is not part of the name's semantic contribution (in the narrow sense). Mentioning the name "Giorgione" is something which *another* expression does; hence it is the semantic contribution of that other expression – while the semantic contribution of the name "Giorgione" is the individual Giorgione, and nothing else.

As I am using it, the notion of a *purely referential occurrence* of a term is defined in terms of its narrow semantic contribution: a singular term is used purely referentially iff its semantic contribution is its referent, and nothing else. But there is room for a distinct notion, defined in terms of the "broader" type of contribution. Let me define a *transparent* occurrence of a singular term as an occurrence such that *the semantic value of the sentence depends only upon the referent of the term, not on its other qualities (its form, its sense, etc.)*. Thus an occurrence is transparent iff its contribution in the broad sense is its referent, and nothing else.

The distinction between the "broad" and the "narrow" semantic contribution of a term, and correlatively between pure referentiality and transparency, is important because it is possible for a term to be purely referential in a sentence, i.e. to contribute its referent and nothing else (in the narrow sense), without being transparent, i.e. such that the truth-value of the sentence does not depend upon any other quality of the term. For suppose that the sentence contains another singular term which demonstratively refers to the first one. Then, even if both terms are purely referential, the truth-value of the sentence will depend upon another property of the first term than merely its referent. That will not bar the first term from being purely referential since those aspects of the term, other than its referent, on which the truth-value of the sentence depends will not be part of the semantic contribution of *that term*, but part of the semantic contribution of the other term. That is exactly what happens in the "Giorgione" example, as we have seen: though purely referential the term "Giorgione" is not transparent; for the semantic value of the sentence depends not only upon the referent of the term, but also on its identity.

This analysis does not depend on my controversial construal of "so" in "so-called" as a demonstrative adverb. If we construe it as anaphorically linked to the name, the situation will be exactly the same: the semantic value of the sentence will depend upon the identity of the purely referential singular term *qua* antecedent of the anaphor. A striking example of that situation is provided by the following example, due to Kit Fine. He imagines a situation in which the man behind Fred is the man before Bill. Despite this identity we cannot infer (14) from (15):

(14) The man behind Fred saw him leave
(15) The man before Bill saw him leave

This does not show that the description "the man behind Fred" is not used purely referentially; only that the occurrence of the description is not "transparent", in the sense I have just defined.

To sum up, transparency entails pure referentiality, but not the other way round. There are *two* ways for an occurrence of a singular term not to be transparent.

- It can be non (purely) referential.
- The linguistic context in which the word occurs may be such that, even if it is purely referential, the truth-value of the sentence will depend upon other properties of the term than its referent. In this type of case I will say that the term occurs in a *reflecting context*; where a reflecting context is *a linguistic context containing an expression whose semantic value depends upon the identity of the term.*

In the second type of case, it's not the way the term is used but rather the context in which it is tokened that blocks substitutivity and generates opacity (the lack of transparency). Hence Quine's shift to talk of "positions" instead of "uses" or "occurrences". Quine defines a *position* as "non purely referential" just in case the term in that position is not substitutable; this may be because the term itself is not being used in a purely referential manner, *or* because the linguistic context contains some context-sensitive expression whose value depends upon the identity of the singular term. Quine's notion of a non-purely referential *position* thus corresponds to my notion of an *opaque* occurrence. If I am right in my interpretation, Quine's talk of "positions" was motivated by his realizing that opacity sometimes arises from the context rather than from the term itself. A term, in and of itself, may be as referential as is possible; if that term is demonstratively referred to by some other expression in the sentence, substitutivity will fail.[6]

§1.5 Transparency and substitutability

We have distinguished between a purely referential occurrence of a term, and a transparent occurrence (or, in Quine's terminology, an "occurrence in purely referential position"). Now I want to consider a third notion: that of a *substitutable* occurrence of a singular term, that is, an occurrence of a singular term which can be replaced by an occurrence of a coreferential singular term *salva veritate*.

We have seen that a purely referential occurrence may fail the substitutivity test if it is not transparent (if the "position" is not purely referential). At this point the question arises, whether we can equate substitutability and transparency.

The first thing we must note in this connection is that it is in fact possible for a purely referential term to be substitutable *without* being transparent. An example of that situation is provided by (16):

(16) Cicero is the person commonly referred to by means of the first word of this sentence.

There is no reason to deny that "Cicero" is purely referential in this sentence. Its semantic value is the individual Cicero, which it names. But the sentence's semantic value results from the contributions of all constituents, including the demonstrative phrase "this sentence". Now the referent of the demonstrative phrase, hence the semantic value of the sentence, depends upon the identity of the singular term occurring at the beginning of the sentence. If you change the singular term, you change the sentence, hence you change the referent of the phrase "this sentence", thereby possibly affecting the truth-value of the sentence. The singular term "Cicero" is therefore not transparent, because the truth-value of the sentence depends upon the form of the name, even though its semantic contribution is nothing other than its referent. The form of the name affects the truth-value of the sentence *via* the semantic value of another singular term in the sentence. Despite this lack of transparency, the singular term is substitutable: if we replace "Cicero" by "Tully", we change the truth-conditions, but the truth-value does not change.

In a case like that, the singular term is substitutable for quite extrinsic reasons. Indeed it can be replaced by any other personal name *salva veritate*, whether that name is coreferential with "Cicero" or not!

That a singular term can be substitutable without being transparent is not actually surprising. For a term can be substitutable without even being referential. Linsky (1967: 102) gives the following example:

(17) "Cicero" is a designation for Cicero

In this sentence the first occurrence of "Cicero" is (purely) autonymous, like the second occurrence of "Giorgione" in (5). Yet it is substitutable: replacement of "Cicero" by "Tully" or any other name of Cicero in (17) is truth-preserving.[7]

Let us grant that transparency cannot be equated with substitutability. Can we at least maintain, following Quine, that transparency *entails* substitutability? It seems that we should. Paraphrasing Quine (1960: 242), we can argue that

If an occurrence of a singular term in a true sentence is transparent, i.e. such that the truth-value of the sentence depends only upon the object which the term specifies, then certainly the sentence will stay true when any other singular term is substituted that designates the same object.

Yet even that has been (rightly) disputed. What I have in mind is Kaplan's insightful discussion of what he calls "Quine's alleged theorem" in "Opacity" (Kaplan 1986).

Kaplan argues that, technically, substitutability does not follow from transparency. But the same point can be made in a non-technical framework, by appealing to the same sort of observation which enabled us to draw a distinction between pure referentiality and transparency.

The crucial point, again, is that natural language sentences are context-sensitive to such a degree, that substituting a singular term for another one can affect the interpretation of other expressions in the same sentence. This may block substitutivity and generate opacity even if the terms at issue are purely referential. Now when a singular term is not only purely referential but *transparent*, it seems that no such thing can happen: for the context is (by definition) not reflecting; it does not contain expressions whose semantic values depend upon the identity of the term. How then can the substitution of coreferentials affect the interpretation of the rest of the sentence? It seems that it cannot, yet, I will argue, it can.

Let us imagine a purely referential occurrence of a term t in a sentence S(t), and let us assume that that occurrence is transparent in the sense that the truth-value of S(t) depends upon the referent of t but not on any other property of t. Since the occurrence of t is transparent, the context S() is not reflecting. Since it is not reflecting, it seems that if we replace t by a coreferential term t', and if the occurrence of t' also is purely referential, then t' can only be transparent. The truth-value of S(t') will therefore depend upon the referent of t' but not on any other property of t'. It follows that S(t') will have the same semantic value as S(t): t , therefore, is substitutable in S(t).

But there is a hidden assumption in the above argument, an assumption which is in fact questionable. It is this: that the linguistic context S() is "stable" in the sense that if it is non-reflecting in S(t), then it is also non-reflecting in S(t'). But suppose we lift that assumption; suppose we accept *unstable contexts*, that is, contexts whose interpretation can shift from non-reflecting to reflecting, depending on which singular term occurs in that context. Then we see that a transparent singular term may not be substitutable after all.

Let us, again, assume that the occurrence of t in S(t) is transparent. This entails that, on that occurrence, t is purely referential and S() is non-reflecting. Yet we cannot conclude that S() will remain non-reflecting after we have substituted t' for t. For an unstable context is a context which is ambiguous between a reflecting and a non-reflecting interpretation. If S() is unstable in this way, then it may be that S() is non-reflecting in S(t) but becomes reflecting in S(t'). Suppose that is the case; then t' is not transparent in S(t'): the truth-value of S(t') will not depend merely upon the referent of t' – it will depend on the identity of the term. The truth-conditions, hence possibly the truth-value, of S(t') will therefore be different from the truth-conditions of S(t). In such a case, therefore, t is not substitutable: replacing it by a purely referential occurrence of a coreferential term t' may result in a change of truth-value!

That is not a purely theoretical possibility. There are reasons to believe that attitude contexts are unstable. A belief sentence like "John believes that Cicero is bald" has two readings: a purely relational reading in which it says of John and Cicero that the former believes the latter to be bald, without specifying how (under which "mode of presentation') John thinks of Cicero; and a non-purely relational reading in which it is further understood that John thinks of Cicero as "Cicero". According to several authors, who use the "Giorgione" example as paradigm, "John believes that ... is bald" is a reflecting context on the non-purely relational reading;[8] that is, the sentence somehow involves a "logophoric" or demonstrative reference to the singular term which occurs in the context. Even if the term in question is construed as purely referential, the truth-value of the sentence depends not only on the referent of the term but also on its identity, on the non-purely relational reading. In contrast, the context is non-reflecting on the purely relational reading. If I say

John, who confuses me with my grandfather Frank Recanati, believes that I died twenty years ago

the truth-value of the sentence depends only upon the referent of "I".

If belief contexts are ambiguous and unstable in this manner, which particular singular term occurs in the sentence may affect its interpretation. This blocks substitutivity: even if the occurrence of the singular term t in "John believes that t is F" is not only purely referential but also transparent, substituting a purely referential occurrence of a coreferential singular term t' for t may shift the interpretation of "John believes that... is F" to its reflecting reading, thereby making the occurrence of t' opaque. That is what apparently happens if we replace "I" by "François Recanati" in (18):

(18) John believes that I died twenty years ago
(19) John believes that François Recanati died twenty years ago

In both cases John is said to have a belief concerning François Recanati, to the effect that he died twenty years ago; but in the second case there arguably is a logophoric or demonstrative reference to the singular term. (19) can be paraphrased as:

(19*) John *so-believes* that François Recanati died twenty years ago

That interpretation of the ambiguous "believes" is natural when the singular term is the proper name "François Recanati", while the pronoun "I" rules out this interpretation for pragmatic reasons (McKay 1981; Recanati 1993: 399–401).

I am not presently defending this analysis of belief sentences; I will do so in the third part of this paper. That brief anticipation was only meant to illustrate the notion of an unstable context, that is, a context ambiguous between a reflecting and a non-reflecting reading. In the same way in which a purely referential occurrence may not be transparent if it occurs in a reflecting context, a transparent occurrence may not be substitutable if it occurs in an unstable context. Thus in (18) the singular term "I" is not substitutable even though it is transparent, because the context is unstable.

To be sure, if, following Quine's general methodological recommendations, we get rid of context-sensitivity by suitably rephrasing the sentences we subject to logical treatment, then we automaticaly get rid of both reflecting and unstable contexts. It then becomes possible to equate (as Quine does) pure referentiality, transparency and substitutability. But, as we saw, Quine himself does not follow his own recommendations: he treats "Giorgione" as non-purely referential and non-substitutable in (4), something which is possible only if we take the context-sensitive sentences "as they come" (Quine 1960: 158), without prior rephrasal. It is this policy which enables him to put in the same basket non referential (autonymous) occurrences of terms and referential occurrences in reflecting contexts. I have shown that if we take this line, then we should draw a principled distinction between pure referentiality, transparency, and substitutability.

II. BELIEF SENTENCES

§2.1 Singular and general beliefs

In his classic paper "Quantifiers and propositional attitudes" (1956), Quine made a distinction between "two senses of believing", as he then put it: the notional and the relational sense. That is both a distinction between two readings of belief sentences, and a distinction between two types of belief. The distinction is very intuitive, but it faces difficulties. In later writings Quine expressed skepticism toward the distinction, and more or less gave it up (Quine 1977: 10). Contrary to Quine I think the distinction can be saved. What follows is my reconstruction of it.

Let us start with the distinction between two types of belief. Some beliefs are purely general, others are singular and involve particular objects. As an example of a general belief, we have the belief that there are spies, or the belief that all swans are black. As Frege put it, those beliefs are about concepts, if they are about anything at all: the first is the belief that the concept "spy" is satisfied by at least one object, the second is the belief that whatever satisfies the concept "swan" satisfies the concept "black". But the belief that Quine was a student of Carnap is a belief about two individual objects: Quine and Carnap. Of this belief we can say: There is an x and there is a y such that the belief is true iff x was a student of y. We cannot say anything similar concerning the belief that there are spies: there is no individual object x such that that belief is true iff x satisfies a given predicate.

A singular belief is relational in the sense that the believer believes something *about* some individual. The relation of "believing about" descends from more basic, informational relations such as the relations of perceiving, of remembering or of hearing about. All those relations are genuine *relations*. If John perceives, remembers, or hears about the table, there is something which he sees, remembers or hears about. Similarly, if John believes something about Peter, there is someone his belief is about.

Singular belief is based on, or grounded in, the basic informational relations from which it inherits its relational character. To have a thought about a particular object, one must be *"en rapport* with" the thing through perception, memory or communication. Pure thinking does not suffice. Thus inferring that there is a shortest spy does not put one in a position to entertain a singular belief about the shortest spy, in the relevant sense.

In terms of this distinction between singular and general beliefs, well-documented and elaborated in the philosophy of mind (see e.g. Evans 1982), I suggest that we define a relational belief report as one that reports the having of a singular belief; and a notional belief report as one that reports the having of a general belief.

How do we know whether a given sentence reports a singular or a general belief? Can we tell from the form of the sentence, or is each belief sentence ambiguous between the two readings? Before dealing with this important question (§2.2–3), we must pause to consider Quine's likely attitude toward the distinction between singular and general belief.

As my examples reveal, singular beliefs are typically expressed by means of singular statements such as "Quine was a student of Carnap"; and general beliefs by means of quantified statements such as "There are spies" or "Every swan is black". But Quine notoriously downplays the difference between the two types of statement. Singular statements, he holds, can be rephrased as general statements (Quine 1960: 178ff). Thus "Cicero is bald" says no more and no less than: "There is an x such that x is Cicero and x is bald". The difference between the two statements is purely rhetorical, Quine says.

Quine's elimination of singular terms in favour of general terms is not intended as a wholesale elimination of singularity, however, but as a displacement of it. If there is some distinguishing feature which singular statements possess, that feature will automatically be transmitted to the "general" statements into which singular statements are rephrased in canonical notation. Quine insists that nothing is lost in the manoeuver – the elimination of singular terms concerns only superficial grammar:

> It is felt... that the names differ from the predicates in their connotation of uniqueness, though predicates may just happen to apply uniquely. It is felt also that proper names lack connotation while predicates connote. Now these are traits of names that I simply transfer to the predicates, however unaccustomed the new setting. This is why I spoke of *reparsing*: the names can keep all their old traits except grammatical position. (Quine 1980: 173)

The difference between "Cicero is bald" and "There are spies", then, is not essentially structural (both, according to Quine, are best seen as quantified statements), but lies in the nature of the predicates involved: the first but not the second type of statement involves what we might call a "singular predicate", viz. "is Cicero".

Corresponding to the original distinction between singular and general statements, we now have a distinction between singular and general predicates. Singular predicates are those predicates which inherit, or otherwise possess, the distinguishing features of singular terms. If, as I have suggested, relationality is the distinguishing feature of genuine singular terms, then singular predicates will possess that feature as well. That means that one cannot believe that the predicate "is Cicero" is instantiated without being suitably related to (*en rapport* with) Cicero. From this point of view, the singular predicate "is Cicero" is very different from a truly general predicate like "is a spy" or "is called *Cicero*".

Earlier I used the following criterion of singularity:

Criterion C:
A belief (or a statement) is singular iff:
There is an x such that the belief (or the statement) is true iff ...x...

Can we still use that criterion, in Quine's framework? I think so. If we rephrase "Cicero is bald" as

(1)   There is an x such that x is Cicero and x is bald

that is still singular by criterion C. For *there is a y such that* (1) is true iff there is an x such that x = y and x is bald. The same thing cannot be said of a fully general statement such as "There are spies".

Quine would certainly object to the recurring of the proper name "Cicero" within the predicate "is Cicero", however. In order to complete the elimination of singular terms, the singular predicate must be construed as "notationally atomic" (Quine 1980: 173). Still, Quine says, the predicate will inherit the traits of the eliminated name, including – presumably – its essentially relational character. If that is so, then we can still use criterion C. We can define a singular predicate as follows:

A predicate F is singular iff:
there is a y such that the belief that there is an x such that x is F is true iff ...y...

Any belief to the effect that such a predicate is instantiated will count as singular by virtue of criterion C.

I conclude that Quine's elimination of singular terms other than variables does not threaten the distinction between singular and general beliefs (or between singular and general statements). Even if it did, however, we would not be forced to choose between the singular/general distinction and Quine's policy of letting only variables refer. For there is an alternative to Quine's way of eliminating singular terms other than variables – an alternative which, far from undermining the singular/general distinction, captures it in a rather elegant and straightforward manner.

Part of Quine's motivation for his regimentation is his belief that "names, like predicates, serve to characterize the thing referred to" (1980: 172). Thus when I refer to Cicero as "Cicero", I characterize him as (being) Cicero. That claim is not very convincing to someone who holds that proper names are non-connotative. Be that as it may, Quine himself accepts that a pronoun such as "he" does not do much by way of characterizing its referent. The pronoun, he says, is "purely referential [and] utterly uninformative"; it "connotes only the sex and scarcely that" (1980: 165). So there is a certain convergence between Quine and theorists of singular reference as far as pronouns such as "he" are concerned: both parties accept that pronouns are vehicles of pure reference. Quine accepts that because he sees pronouns as the natural language counterparts of variables (1974: 93–101); the theorist of singular reference because he sees *demonstrative* pronouns as paradigmatic singular terms (Kaplan 1989). I think this convergence can be exploited to eliminate singular terms other than variables in a way which is more congenial to the theorist of singular reference.

A pronoun, in general, is very much like a variable. Some pronouns are like bound variables: "Every man believes that *he* is brave". A demonstrative pronoun is more like a *free* variable, under a contextual assignment of value. The suggestion, then, is this. When we say "He is brave", pointing to some man, the sentence which we utter is neither true nor false: it is an open sentence. By asserting it, however, we present it as *true of* the object we are demonstrating. The assertion is true *tout court* iff that is indeed the case, that is, iff the demonstrated object satisfies "x is brave". In such a case *there is an x such that our assertion is true iff x is bald.*

On the other hand, when the sentence is a closed sentence such as "There are spies", there is no x such that the assertion is true iff ...x...

This treatment of demonstrative pronouns can be extended to all genuine singular terms, including proper names. A proper name such as "Cicero" can also be considered as a free variable. I refer the reader to Dever (forthcoming) for an elaboration of this view. In that framework, I think, the elimination of singular terms is conducted in a way that enhances the distinction between singular and general statements.

§2.2 Scope ambiguities in attitude contexts

How do we tell whether a given belief sentence reports the having of a general belief or the having of a singular belief? Quine thinks that a standard belief sentence like "Ralph believes that Ortcutt is a spy" is ambiguous between the relational and the notional reading; and that we can force the relational reading by "exporting" the singular term: "Ralph believes *of Ortcutt* that he is a spy". When exportation is thus possible, existential generalization is also possible: if Ralph believes that Ortcutt is a spy, in the "exportable" sense (that is, if he believes of Ortcutt that he is a spy), then there is someone Ralph's belief is about.

Even though Quine's claim concerning the ambiguity of belief sentences between the relational and the notional reading has been very popular, I think that it rests, in part, on a confusion; a confusion which is, again in part, responsible for Quine's despair of the distinction. In the next section I will argue that standard belief sentences such as "x Vs that p", where the embedded sentence contains a singular term, are not ambiguous between the relational and the notional reading. That ambiguity arises only when the embedded sentence contains a quantified or descriptive phrase.

The distinction between genuine singular terms and descriptive or quantified phrases such as "some man", "a man", "no man" or "the man" goes back to Russell (1905). While Russell wanted to restrict the class of "logically proper names" (as he called genuine singular terms) to only a couple of natural language devices, contemporary semanticists consider ordinary proper names and demonstratives, in general, as genuine singular terms. *Qua* genuine singular terms, they are purely referential, in the sense of §1.3. Definite and even indefinite descriptions can also be *used* purely referentially, according to some authors at least (Donnellan 1966; Chastain 1975); but the purely referential use of descriptions is not their normal semantic function, while it is the normal semantic function of genuine singular terms.

There is a good deal of controversy over the referential use of definite descriptions. Many people believe that it is irrelevant to semantics. I disagree, but we need not be concerned with this issue here. If, as I believe, definite descriptions have a non-deviant referential use,[9] then, when so used, they behave like genuine singular terms: they are purely referential and their semantic value (on that use) is their referent. What I have to say about the behaviour of genuine singular terms in belief contexts will therefore automatically apply to definite descriptions on their referential use. So I will put referential descriptions aside and consider only what Evans called the "pure" uses of definite descriptions, that is, their attributive uses.

As Russell pointed out in the above-mentioned paper, definite descriptions are very much like quantified phrases. Like them, they serve to make general statements. If John believes or asserts "The winner will be rich", we cannot say that there is an object x such that John's belief or statement is true iff x satisfies F, whichever predicate we put in place of the schematic letter "F". In particular, we cannot say that a certain person, namely the winner, is such that John's belief is true iff *she* will be rich; the condition "being the winner" must also be satisfied by her. Nor can we say that a certain person is such that the belief is true iff she is both rich and the winner. *Any* person's being rich and the winner can make the belief true.

Definite descriptions are similar to quantified phrases in another respect: like them, they induce scope ambiguities in complex sentences containing an intensional operator. Thus there are two readings for sentences such as (2) or (3):

(2) Someone will be in danger
(3) The President will be in danger

(2) says either that someone is such that she will be in danger, or that it will be the case that someone is in danger. The two readings can be represented as follows:

(2a) $(\exists x)$ (it will be the case that (x is in danger))
(2b) It will be the case that $((\exists x)$ (x is in danger))

The same duality of readings can be discerned in the case of (3). (3) says either that the President is such that he will be in danger, or that it will be the case that: the President is in danger. On the second reading it is the fate of a future president which is at issue, while on the first reading the sentence concerns the present president. Again, the two readings can be represented in terms of relative scope:

(3a) $(\iota x$ President x) (it will be the case that (x is in danger))
(3b) It will be the case that $((\iota x$ President x) (x is in danger))

In (2a) and (3a), the quantifier or descriptive phrase is given wide scope; thus it seems to reach into the intensional context created by he operator "it will be the case that". But, as Kaplan (1968, 1986) and Quine (1977) pointed out, (3a) and (2a) need not be construed as actually violating Quine's prohibition of quantification into intensional contexts. The intensional operator "it will be the case that", or "will-be" for short, can be thought of as *multigrade* (Quine 1956, 1977). Its taking narrow scope vis-a-vis the descriptive or quantified phrase in (2a) and (3a) means that it governs only the predicate "in danger", while it governs the whole sentence "someone is in danger" or "the President is in danger" when it is given wide scope, as in (2b) and (3b). That can be made notationally explicit in the manner of Quine 1977:

(2a′) $(\exists x)$ (will-be(in-danger) x)
(2b′) Will-be $((\exists x)$ (in-danger x))

(3a′) (ιx President x) (will-be(in-danger) x)
(3b′) Will-be ((ιx President x) (in-danger x))

In (2a′) and (3a′) the multigrade "will-be" is understood as a predicate functor making a new predicate, "will be in danger", out of the original predicate "in danger". The quantified variable thus falls outside the scope of the intensional operator. When the operator is given wide scope, as in (2b') and (3b'), it is understood as governing the whole sentence (including the quantifier and the variable). The quantified variable now falls within the scope of the operator, but, as Quine says, the sentence "exhibits only a quantification *within* the "believes that' context, not a quantification *into* it" (1956: 188).

Before proceeding, let us note that genuine singular terms give rise to no such scope ambiguities: they are, as Geach once put it, "essentially scopeless" (Geach 1972: 117). Thus sentence (4) is not ambiguous, contrary to (2) or (3); there is no truth-conditional difference between (4a) and (4b), as there was between (2a) and (2b) or between (3a) and (3b):

(4)   Cicero will be in danger
(4a)   Will-be (in-danger Cicero)
(4b)   Will-be(in-danger) Cicero

It is time to introduce belief sentences. Belief sentences with descriptive or quantified phrases are ambiguous in a way that exactly parallels the ambiguities we have just observed in temporal sentences with descriptive or quantified phrases. Thus (5) is ambiguous like (2), and (6) is ambiguous like (3):

(5)   John believes that someone is F
(5a)   Someone is such that John believes him to be F
        $(\exists x)\ (B_j(F)\ x)$
(5b)   John believes that: someone is F
        $B_j\ ((\exists x)\ (Fx))$
(6)   John believes that the President is in danger
(6a)   The President is such that John believes him to be in danger
        $(ιx\ President\ x)\ (B_j(in\text{-}danger)\ x)$
(6b)   John believes that: the President is in danger
        $B_j\ ((ιx\ President\ x)\ (in\text{-}danger\ x))$

The quantification is endorsed by the speaker in (5a), while it is ascribed to the believer in (5b). Similarly, the description is endorsed by the speaker in (6a), while it is ascribed to the believer in (6b).

Note that in (6a) the description can be read attributively even though it takes wide scope (Kripke 1977: 258). The speaker says that the President, whoever he is, is such that John believes him to be in danger. The description does not behave like a singular term here; it does not contribute an object. Still the *ascribed* belief is singular: the speaker says that there is a particular object such that the believer believes something of that object.

To sum up, when the quantified phrase or the description takes wide scope, belief reports like (5) and (6) have their relational reading: the belief they report is singular, even though the object the belief is about is only described in general terms.[10] In contrast, when the descriptive or quantified phrase takes narrow scope, the belief report is understood notionally. The believer is said to believe that there is an object x with such and such properties; that does not entail that there actually is an object y such that the believer believes that of y. Whatever quantification there is is strictly internal to the ascribed content; it is not endorsed by the speaker.

§2.3 Singular terms in belief sentences

So far, Quine's claim concerning the ambiguity of belief sentences has been vindicated. But quantified phrases and definite descriptions are not genuine singular terms (Neale 1990). As soon as what occurs in the embedded sentence is a genuine singular term (or a referential description), the scope ambiguity vanishes, along with the distinction between the notional and the relational reading of the belief report.

Since a singular term is purely referential (unless it is used deviantly), a statement in which it occurs is bound to be singular. That is true not only of a simple statement such as "Cicero is in danger", but also of a complex statement such as "John believes that Cicero is in danger". The former is about the individual Cicero; the latter is about two individuals, John and Cicero. It follows that exportation is always licensed when the embedded sentence contains a genuine singular term.[11] From:

(7) John believes that t is F

we can always go to

(8) John believes of t that it is F

and, through existential generalization, to

(9) $(\exists x) (B_j(F)x)$

That means that the ascribed belief is always singular, when the belief report contains a singular term. "Notional" readings are thus ruled out: only relational readings are available.

What I have just said, of course, presupposes that genuine singular terms are used normally (non-deviantly) in attitude contexts. That is, I am assuming what Davidson (1968) and Barwise and Perry (1981) call "semantic innocence" (see §3.2 below); and correlatively rejecting the notion that singular terms in attitude contexts refer to something different from their usual referent (Frege) or behave somewhat deviantly, as they do when they occur autonymously (Quine). I take singular terms to be purely referential, in the sense of §1.3, in all their non-deviant occurrences; and I assume that their occurrences in attitude contexts are non-deviant.

The picture I am advocating is highly controversial, of course; but at least it is neat. It is organized around two main distinctions:

(i) The embedded sentence in a belief report contains either a singular term, or a quantified/descriptive phrase.

(ii) A quantified/descriptive phrase can be given either wide scope or narrow scope vis-a-vis the epistemic operator.

Thus there are three possibilities: what occurs in the embedded sentence can be a singular term, a quantified/descriptive phrase with narrow scope, or a quantified/descriptive phrase with wide scope. The belief report counts as *relational* if, and only if, the embedded sentence contains either a singular term or a quantified/descriptive phrase with wide scope. Note that there remains a difference between the two types of case. When using a singular term, the speaker himself makes a singular statement about the individual object the belief is about. When using a descriptive/quantified phrase with wide scope, the speaker ascribes a singular belief, but she does not herself express a singular belief, or make a singular statement, about the individual object the ascribed belief is about.

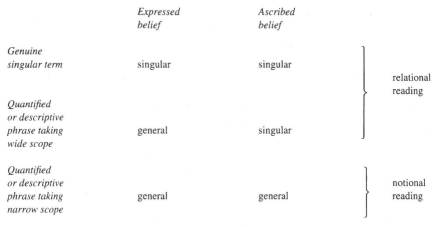

	Expressed belief	Ascribed belief	
Genuine singular term	singular	singular	relational reading
Quantified or descriptive phrase taking wide scope	general	singular	
Quantified or descriptive phrase taking narrow scope	general	general	notional reading

Table 1

At this point two main objections spring to mind:

- If the above theory was correct, it would be always be possible to infer from "John believes that t is F" that there is an x John believes to be F. But what about statements like (10)?

(10) My three-year old son believes that Santa Claus will come tonight

Since Santa Claus does not exist, there is no individual to whom my son is related in the manner required for singular belief. Hence from (10) we cannot infer "There is an x such that my son believes that x will come tonight". That is a counter-example to the theory.

- I claim that belief sentences with singular terms are not ambiguous, in contrast to belief sentences with quantifiers. But they are: the name can be either endorsed by the speaker as his own way of referring to whatever the belief is about, or ascribed to the believer. That is the same old *de re/de dicto* ambiguity which we have observed in the case of belief sentences with quantifiers.

The second objection is especially important; it is the main obstacle on the road to accepting the view I have just sketched. In the next section, I will argue that it rests on a confusion. Belief sentences with singular terms are indeed ambiguous between a "transparent' and an "opaque" reading, but that ambiguity is *distinct from*, indeed orthogonal to, the relational/notional ambiguity we have been considering so far. When the two ambiguities are confused under a singular heading (the so-called *"de re/de dicto"* distinction), the situation becomes intractable and leads one to despair. Once the ambiguities are kept apart, however, the apparently intractable problems disappear.

As for the first objection, it can be rebutted as follows. The reason why we can't infer "(∃x) (my son believes that x will come tonight)" from "My son believes that Santa Claus will come tonight' is the same reason why we can't infer (12) from (11).

(11) Santa Claus lives in the sky
(12) (∃x) (x lives in the sky)

So the objection is not a specific objection to the view that genuine singular terms behave as such in belief reports; rather, it is an objection to the view that fictional names such as "Santa Claus" *are* genuine singular terms, subject to ordinary logical principles. Since that problem is a general problem, it is not incumbent on the attitude theorist to solve it.

There is, however, an important difference between a fictional statement like (11) and a statement like "My son believes that Santa Claus will come tonight" or "In the story Santa Claus lives in the sky" ("metafictional" statements, as Currie [1990] aptly calls them). The author of a fictional statement does not really make assertions, but only pretends to do so. Thus in (11) she only pretends to say of a certain person that he lives in the sky.[12] Since that it so, the failure of existential generalization is unproblematic. (12) cannot really be inferred, because (11) was not really asserted. (Within the pretense, however, the inference goes through: the speaker pretends to be committed to (12), by pretending to assert (11).) In contrast, it seems that metafictional statements are serious and evaluable as true or false (Lewis 1978). Hence it is not obvious that the failure of existential generalization has the same source in both cases.

Despite what I have just said, it can be maintained that the author of a metafictional statement such as "In the story, Santa Claus lives in the sky" is also pretending: she pretends to assert of someone that the story says he lives in the sky. Similarly for (10): the speaker pretends to assert of a given individual that her son believes he will come tonight. In neither case does the speaker really make that assertion, as there is no individual the story (or the child's belief) is about. By pretending to do so, however, the speaker communicates something true about the story or about the child's belief – something which could be communicated literally only by means of a lengthy and cumbersome paraphrase (Walton 1990: 396ff; Crimmins forthcoming; see also Forbes 1996 for discussion of related isues).

A lot more needs to be said to flesh out this proposal. One must detail the mechanism of "semantic pretense" through which one can, in a more or less conventional manner, convey true things by pretending to say other things. One must also show how fictional statements like (11) can be distinguished from metafictional statements in which, intuitively at least, it seems that a genuine (and true) assertion is made. If pretense is involved in both cases, it is not quite the same sort of pretense; the theory owes an account of how the two kinds connect up with each other (Recanati 1998, 2000). I do not intend to go into those complex issues here, even though Quine turns out to have been a pioneer in this area too.[13] It is sufficient to have pointed out that a research programme exists to solve precisely the sort of problem that (10) raises, in a way which is consistent with the theory I have expounded concerning the behaviour of singular terms in attitude contexts.[14]

§2.4 The ambiguity of the de re/de dicto distinction

In §2.2 I glossed the relational/notional distinction in terms of the points of view involved. I said that the description (or the quantification) is "endorsed by the speaker" in relational readings, while it is "ascribed to the believer" in the notional reading. Now it seems that – contrary to what I claimed – exactly the same distinction can be made with respect to belief sentences containing singular terms instead of descriptions or quantifiers. Thus (13) can be understood in two ways.

(13) Ralph believes that Cicero denounced Catiline

On the transparent interpretation, Ralph is said to have a belief concerning the individual Cicero. Since Cicero is Tully, (13) can be rephrased as (14):

(14) Ralph believes that Tully denounced Catiline

The transparent reading of sentences like (13) is often rendered by appealing to the exported form, as in (15):

(15) Ralph believes of Cicero that he denounced Catiline

But there is another interpretation of (13) and (14), an interpretation in which they are not equivalent and cannot be rendered as (15). This is the "opaque" interpretation. On that interpretation, Ralph is said by (13) to have a belief such that he would assent to "Cicero denounced Catiline", but not necessarily to "Tully denounced Catiline". On the opaque interpretation, the use of the name "Cicero" (rather than "Tully") to refer to Cicero is *ascribed to the believer*. On the transparent reading, the choice of the name is up to the speaker and does not reflect the believer's usage; that is why replacement of "Cicero" by "Tully" in (13) on the transparent interpretation does not induce a change in the ascribed belief.

Quine and many philosophers and linguists after him have jumped to the conclusion that a single distinction applies to belief sentences whether they contain singular terms or descriptive/quantified phrases. They have equated the relational/notional distinction talked about in previous sections and the transparent/opaque distinction I have just introduced for belief sentences with singular terms. Both are viewed as instances of the so-called *"de re/de dicto"* distinction. The exported form (15) is the mark of the *de re*. Belief sentences on the *de dicto* (opaque, notional) reading resist exportation, because the epistemic operator takes wide scope – it governs the embedded sentence in its entirety. On the *de re* reading, the epistemic operator takes narrow scope and governs only the predicate: the subject expression, be it quantificational or referential, is endorsed by the speaker without being ascribed to the believer. That is the confused doctrine whose untenability led Quine and others to despair of the original relational/notional distinction.

In fact, there is a clear difference between the two distinctions – the relational/notional distinction, and the transparent/opaque distinction. Consider the notional reading of a belief sentence. In such a case the believer is said to believe that there is an object x with such and such properties; that does not entail that there actually is an object y such that the believer believes that of y. Whatever quantification there is is strictly internal to the ascribed belief; it is not endorsed by the speaker. But *even on the opaque reading of a belief sentence in which a singular term occurs, reference is made to some particular individual* (Loar 1972). Thus the speaker who utters (13) on its opaque reading is committed to there being an individual x, such that Ralph's belief concerns x and is true iff ...x... To be sure, the belief which is ascribed to Ralph on the opaque reading of (13) is not merely the belief that that individual denounced Catiline; that would correspond to the transparent reading of (13). On the opaque reading, Ralph is ascribed the belief that: *Cicero* denounced Catiline. Cicero is thought of by Ralph not only as having denounced Catiline, but also *as Cicero*. Yet that feature of opacity is compatible with the relational character of the belief report, that is, with the fact that the speaker himself refers to Cicero as the object the ascribed belief is about. We can represent the opaque reading of (13) as follows:

(16) Ralph believes of Cicero, thought of as "Cicero", that he denounced Catiline

The apposition "thought of as *Cicero*" is sufficient to distinguish the opaque reading from the transparent reading. Both readings are relational: in both cases Ralph believes something of Cicero, and the speaker himself refers to Cicero as what Ralph's belief is about. In the opaque reading, however, the name has a dual role: it serves not only to refer to the object the ascribed belief is about, but also tells us something about how the believer thinks of that object. As Brian Loar pointed out, this dual role is reminiscent of that of "Giorgione" in Quine's famous example (Loar 1972: 51).

The non-equivalence of (13) and (14) on their opaque readings is clearly compatible with the relational character of those readings. In the same way in which (13), on its opaque reading, is rendered as (16), the opaque reading of (14) can be rendered as (17):

(17) Ralph believes of Tully, thought of as "Tully", that he denounced Catiline

The name "Tully" in (14) refers to Cicero even on the opaque reading. The speaker is therefore committed to there being an individual, namely Cicero (= Tully), such that Ralph believes of that individual, thought of as "Tully", that he denounced Catiline. There is no such existential implication when a belief report (with a descriptive or quantified phrase) is understood notionally.

As we can see, the contrast between cases in which something is ascribed to the believer and cases in which it is endorsed by the speaker is not drawn in quite the same way for the two distinctions. On the notional reading of a belief sentence with a descriptive/quantified phrase, the quantification is ascribed to the believer *without* being endorsed by the speaker; but the reference to the object of belief, and the existential commitment that goes with it, is *both* ascribed to the believer *and* endorsed by the speaker on the opaque reading of a singular belief sentence. The relational/notional distinction articulates a simple contrast between the point of view of the speaker and the point of view of the believer; while the transparent/opaque distinction articulates a quite different contrast, between the point of view of the sole speaker and the point of view of *both* the speaker and the believer. As far as the respective points of view of the speaker and the believer are concerned, opaque readings are thus essentially "cumulative".

Far from being identical to the relational/notional distinction, the transparent/opaque distinction turns out to be a distinction between two sorts of *relational* reading. Hence there is no incompatibility between the claim that belief sentences with singular terms can only be understood relationally, and the observation that they have both a transparent and an opaque reading. Yet, precisely because belief reports with genuine singular terms cannot be interpreted notionally, but only relationally, it has seemed to many that a single distinction applies indifferently to all belief sentences: just as belief sentences with descriptive/quantified phrases can be interpreted relationally or notionally, belief sentences with singular terms can be interpreted transparently or opaquely. To dispell that illusion, one has only to notice that belief sentences with descriptive/quantified phrases are subject to *both* ambiguities. They can be interpreted notionally or relationally; and when relational, they can be interpreted transparently or opaquely. Loar gives the following example of a belief sentence with a quantified phrase which is naturally given a relational yet opaque interpretation:

(18) Ralph believes that a certain cabinet member is a spy

This does not mean that Ralph has a general belief to the effect that some cabinet member or other is a spy. As the phrase "a certain" is meant to indicate, there is a particular cabinet member Ralph's belief is about. The belief report, therefore, is relational. However, Loar (1972: 54) points out that (18)

will often be taken to imply more than
(19) $(\exists y)$ (y is a cabinet member & B (Ralph, "x is a spy", y)
Ralph, we may suppose, believes it of the fellow under a certain description; that is,
(20) $(\exists y)$ (y is a cabinet member & B (Ralph, "x is a cabinet member and x is a spy", y))

Loar's rendition of (18) as (20) nicely captures the cumulative aspect of opaque readings. Both the speaker and the believer view the person the belief is about as a cabinet member. As Loar pointed out (1972: 54), in a framework such as Quine's, in which the two distinctions are conflated under a single heading, one cannot account for belief reports which, like (18), are both relational and opaque. "Relational" entails "transparent", for Quine and his followers. For that reason also, examples like (13) and (14), on their "opaque" interpretation (corresponding to [16] and [17]), will have to be considered "notional", while they are clearly relational. Given the extreme confusion that results, it is only natural that Quine eventually gave up the distinction as hopeless. It *is* hopeless, considered as a single distinction covering all the cases.

III. OPACITY IN BELIEF SENTENCES

§3.1 "That"-clauses as complex demonstratives

According to Brian Loar, the singular term in an opaque belief report has a dual role. It refers to the object the belief is about, but also determines an aspect of the ascribed belief concerning that object. The ascribed belief is conjunctive, and the first conjunct depends upon the identity of the singular term.

This theory can be understood in two ways. On one interpretation a singular term behaves deviantly in belief contexts. Instead of merely referring to some object, as singular terms normally do, it refers to an object *and* contributes a "mode of presentation" to the content of the ascribed belief. That theory gives up semantic innocence, even if it does so in a less extreme manner than Frege's. It construes the singular term in a belief report as referential, but not as *purely* referential.[15]

There is another option, though. It consists in preserving semantic innocence and holding that the singular term in an opaque belief report is purely referential, in accordance with its normal semantic function. The opacity of the occurrence can then be explained by construing the context as *reflecting*, in analogy with the above analysis of the "Giorgione" example (§1.2–3).

A context for a singular term is reflecting if and only it contains a *dependent expression*, that is, an expression whose semantic value depends upon the identity of the singular term occurring in that context. In the "Giorgione" example, the dependent expression was the adverb "so" in "so-called", which we can construe either as demonstrative or as anaphoric. When we replace "Giorgione" by a coreferential term, e.g. "Barbarelli", the semantic value of the dependent expression changes. That accounts for the sentence's change in truth-value.

In the "Giorgione" example, the dependent expression ("so-called") is part of the *frame* in which the singular term occurs ("... is so-called because of his size").[16] The dependent expression is therefore disjoint (separable) from the singular term itself. But that need not be the case: for a context to be reflecting, it is not necessary that the dependent expression occur as part of the frame, in disjunction from the singular term itself. There are cases in which the singular term itself will be a constituent of the dependent expression. Let me give an example involving, not a singular but a general term.

Consider the demonstrative phrase "that nag". The semantic value of "nag" is the same as that of "horse"; the difference, as Frege would say, is one of "colouring" rather than a properly semantic (truth-conditional) difference. Despite their semantic equivalence, "nag" in "that nag" cannot be replaced by "horse", because the reference of a demonstrative phrase is linguistically underdetermined and crucially depends upon the referential intentions of the speaker, as revealed by the context. Now one aspect of the context which may be relevant to the determination of the speaker's referential intentions is the word which the speaker uses. If he uses a word such as "nag", that provides some evidence that he does not intend to refer to his beloved and much respected horse Pablo, who happens to be otherwise salient in the context, but rather to the deprecated Pedro. If the word "horse" was used, however, sheer salience would presumably promote Pablo to the status of referent. Substituting "horse" for "nag" can therefore change the likely interpretation, hence possibly the truth-value, of the sentence, by affecting the semantic value of the demonstrative phrase.

In general, whenever the semantic value of a phrase is linguistically underdetermined, and depends upon the intentions of the speaker, that phrase is a reflecting context for its constituents. A "part" of the global phrase cannot be replaced by a semantically equivalent expression without possibly affecting the semantic value of the whole, because any aspect of the context, including the actual words which are used, may be relevant to determining that semantic value.

Let us now go back to belief reports. If singular terms in belief sentences fulfill their ordinary function and are purely referential, substitutivity failures must be accounted for by appealing to the notion of a reflecting context. That means that we must find a dependent expression in the belief report – an expression whose semantic value depends upon the identity of the singular term.

One possible candidate is the "that"-clause itself. A "that"-clause is commonly taken to be a referring expression. Let us call what a "that"-clause allegedly refers to a "proposition", without going into the issue of what propositions exactly are. (I will construe them, heuristically, as $entences in the sense of Kaplan 1986, since propositions in that sense seem to have been found palatable by Quine.) The reason why "that"-clauses are generally considered as singular terms is that this enables us to account for inferences like the following:

John says that grass is green
Everything John says is true
Therefore, it is true that grass is green

If we rephrase "It is true that grass is green" as "That grass is green is true", as we are certainly entitled to do, the inference can easily be accounted for on the assumption that "that grass is green" is a singular term. The pattern is:

a is F	(That grass is green is said by John)
Every F is G	(Everything said by John is true)
Therefore, *a* is G	(That grass is green is true)

Most philosophers consider the reference of a "that"-clause as fixed by the following rule: a "that"-clause refers to the proposition expressed by the embedded sentence. In my book *Direct Reference* (Recanati 1993) I put forward an alternative proposal, in order to account for the well-known context-sensitivity of "that"-clauses. I claimed that a "that"-clause can, but need not, refer to the proposition expressed by the embedded sentence. It can also refer to a proposition obtained by contextually *enriching* the expressed proposition.

The relevant notion of contextual enrichment is that needed to account for examples like the following:

(1) She took out her key and opened the door

In that example, analysed in Carston (1985), the fact that the door was opened with the key is not linguistically specified, yet it is certainly part of what we undertand when we hear that sentence. It is an aspect of the meaning or content of the utterance which is provided through "contextual enrichment". John Perry calls that an "unarticulated constituent" of what is said (Perry 1986); and he and Crimmins hold that modes of presentation of the objects of belief are unarticulated constituents of the proposition expressed by opaque belief reports (Crimmins and Perry 1989).[17] I agree with the spirit, if not the details, of that analysis.

In my book I took a "that"-clause to be a demonstrative phrase whose reference is constrained, but not determined, by the proposition which the embedded sentence expresses – much like the reference of the demonstrative phrase "that horse" is constrained, but not determined, by the general term it contains. In other words, I took the reference of "that"-clauses to be linguistically *underdetermined*. Underdetermination is to be distinguished from mere context-dependence. The reference of words like "I" or "today" is context-dependent, but it does not exhibit the relevant feature of underdetermination. In a given situation, the meaning of a pure indexical like "I" or "today" fully determines what the reference is. Not so with demonstratives. The reference of "he" or "that" is not determined by any rigid rule; it is determined by answering questions such as, Who or what can the speaker plausibly be taken to be referring to, in that context? The same thing holds, I assumed, for "that"-clauses. A "that"-clause refers to a proposition which *resembles* the proposition expressed by the embedded sentence, but need not be identical with it; it can be an enrichment of it. What the reference of a given "that"-clause actually is will depend upon the speaker's intentions as manifested in the context.

On that theory, when a belief report such as "Ralph believes that Cicero is a Roman orator" is understood opaquely, the reference of the "that"-clause "that Cicero is a Roman orator" is distinct from what it is on the transparent interpretation. On the transparent interpretation the reference is, arguably, the "singular proposition" (or valuated formula) which the embedded sentence expresses, viz. a sequence whose first member is the individual Cicero, and whose second member is the predicate "Roman orator". On the opaque interpretation, the reference is a "quasi-singular" proposition, that is, the same thing except that the first member of the sequence is itself an ordered pair, consisting of the individual Cicero *and another predicate serving as "mode of presentation"* (Recanati 1993). The quasi-singular proposition is an enrichment of the expressed singular proposition. The extra constituent provided by the context is the mode of presentation (the predicate) under which the reference of the singular term is assumed to be thought of by the believer.

The "that"-clause thus turns out to be a dependent expression, whose semantic value is susceptible to change if a singular term occurring in the "that"-clause is replaced by a coreferential term. For the reference of the "that"-clause ultimately depends upon the speaker's communicative intentions as revealed by the context; and any aspect of the context, including the words which the speaker actually uses to report the ascribee's beliefs, may be relevant in figuring out the speaker's referential intention. In some contexts, the speaker's use of the name "Cicero" will suggest that the believer thinks of Cicero as "Cicero". That is no more than a contextual suggestion, accountable perhaps in Gricean terms (McKay 1981; Salmon 1986); yet it may influence the assignment of a particular semantic value to the "that"-clause, thereby affecting the truth-conditions of the belief report. That will be so whenever the belief report is understood opaquely: the "that"-clause will then refer to a quasi-singular proposition involving not only the individual Cicero and the predicate "Roman orator", but also a further predicate such as "called *Cicero*".

§3.2 Semantic innocence

Substitutivity problems have led many philosophers to give up semantic innocence in connection with attitude contexts. Both Frege and Quine thus appeal to the thesis of Semantic Deviance, according to which the extension of an expression is affected when it is embedded within a "that"-clause. For Frege, the extension of a sentence systematically shifts in such circumstances. Once embedded, a sentence no longer denotes its truth-value but it comes to denote its truth-*condition*. Quine does not accept that view, but he sticks to the thesis of Semantic Deviance. "That"-clauses, he says, are similar to quotation contexts: when we put a sentence in quotes, it no longer represents what it ordinarily represents. The sentence is mentioned rather than used.

In the case of quotation the thesis of Semantic Deviance is indeed very plausible. Do the first word of (2) and (3) below have e.g. the same extension? No, the first one refers to cats, the second one refers to the word "cats".

(2) Cats are nice
(3) "Cats" is a four-letter word

But what about belief sentences and "that"-clauses in general? Consider (4)

(4) John believes that grass is green

Is it credible to say that the words "Grass is green" do not represent what they normally represent? Does "grass" in the embedded sentence refer to anything else than grass? Does it do anything else than refer to grass? As Davidson emphasized,

If we could recover our pre-Fregean semantic innocence, I think it would seem to us plainly incredible that the words "The earth moves", uttered after the words "Galileo said that", mean anything different, or refer to anything else, than is their wont when they come in different environments. (Davidson 1968: 144)

I fully agree with Davidson that we should at least *try* to "recover our pre-Fregean innocence", that is, to do without the thesis of Semantic Deviance. In an "innocent" framework, the semantic value of an expression in the embedded part of a belief report is construed as its *normal* semantic value (whatever that may be).

I have shown how it is possible to preserve semantic innocence by construing the singular terms in belief contexts as purely referential (as they normally are), and accounting for failures of substitutivity in terms of reflecting context. Yet the theory elaborated in Recanati (1993), and summarized in §3.1, is not *thoroughly* innocent. It is innocent as far as singular terms are concerned, but when it comes to the complete embedded sentence, innocence is eventually abandoned.

Let us call the view that "that"-clauses are complex singular terms the "standard account". On that view, the sentential complement *names* a proposition. But that is not what the complement sentence does when it is not embedded. Unembedded, the sentence expresses a proposition, it doesn't name one. Hence, by construing "that"-clauses as names, it seems that the standard account violates semantic innocence.

Faced with that objection, the usual strategy consists in drawing a distinction between the embedded sentence and the complete "that"-clause. The embedded sentence, it is said, expresses a proposition, and it is that proposition which the "that"-clause names. In this way innocence is allegedly saved: the sentence does the same thing – it expresses a certain proposition – whether it is embedded or not; it never names a proposition, since that is a job for the complete "that"-clause.

I do not think this strategy works, however. First, the distinction between the embedded sentence and the complete "that"-clause has no obvious equivalent when we turn to non-standard belief sentences like "In John's mind, grass is green" or "According to John, grass is green". There is no "that"-clause in such examples – only the sentence "Grass is green". Second, even when the distinction makes syntactic sense, it is unclear that it enables us to preserve semantic innocence. I will show that by considering, once again, the case of quotation.

Faced with an instance of quotation such as (3), we have two options. We can say that the word "cats" in this context does something different from what it normally does: it is used "autonymously" (self-referentially). Or we can say that it is the complex expression consisting of the word "cats" *and the quotes* which denotes the word "cats". If, by taking the second route, we refrain from ascribing the word "cats" a deviant function in quotation contexts, we will be led to deny that the word "cats" really occurs; rather, with Tarski and Quine, we will say that it occurs there only as a "fragment" of the longer expression, much as "cat" occurs in "cattle". From the semantic point of view, the relevant unit is indeed the complete quotation; the word "cats" itself thus disappears from the picture. *In this way innocence is lost as surely as it is when we take the first option.* A truly innocent account is one that would *both* acknowledge the occurrence of the expression at issue in the special context under consideration *and* ascribe it, in that context, its normal semantic function. (Of course, there is no reason to expect an account of quotation to be semantically innocent in that sense.)

Similarly, we have two options with regard to attitude reports, in the standard framework. If we say that the complement sentence, once embedded, names the proposition which it would normally express, we give up semantic innocence: we accept that the embedded sentence does not do what it normally does. On the other hand, if, in order to protect innocence, we draw a sharp distinction between the embedded sentence (which expresses a proposition) and the "that"-clause (which names it), *we run the risk of making the former disappear from the logical scene.* For the relevant semantic unit is the complete "that"-clause. At the level of logical form the sentence "John believes that S" has the form aRb – it consists of a two-place predicate and two singular terms. The embedded sentence plays a role only via the "that"-clause in which it occurs. Which role? Arguably a *pre-semantic* role analogous to that of the demonstration which accompanies a demonstrative. If that is right, then semantically the complexity of the "that"-clause matters no more than the pragmatic complexity of a demonstrative-*cum*-demonstration or the "pictorial" complexity of a quotation.[18]

For that reason, I conclude that any theory which construes "that"-clauses as singular terms is bound to give up semantic innocence at some point or other. On the view presented in §3.1, we protect the innocence of singular terms, but not that of the embedded sentence. The theory we end up with is thus not very different from Quine's. In Quine's framework, there is a sense in which the singular term "Cicero" is purely referential in the sentence "Cicero is bald" which we find embedded in the belief report "John believes that Cicero is bald". But the belief context into which the sentence is embedded is said by Quine to be "opaque", like a quotation context. The situation is similar to that of:

(5) "Cicero is bald" is held by John

The singular term "Cicero" occurs purely referentially in the sentence "Cicero is bald", but that sentence is quoted, hence insulated from the outer context by the opaque barrier of quotation. The pure referentiality of "Cicero" thus becomes "strictly an internal affair" (Quine 1995: 356). At the outer level the insulated singular term no longer counts as purely referential. That is indeed Quine's definition of an opaque context: a context is referentially opaque "when, by putting a statement Φ into that context, we can cause a purely referential occurrence in Φ to be not purely referential in the whole context" (Quine 1953: 160).

A truly innocent account must give up the view that "that"-clauses are singular terms. That is, the embedded sentence must be treated as, logically, a sentence; it must not be converted into a term. That means that "Grass is green" has the same status in "John believes that grass is green" which it has in "The sky is blue and grass is green" (extensionality aside). The statement that grass is green is not mentioned, but simply occurs as a part of a longer statement. (On the distinction between mentioning and compounding, see Quine 1962: 38.)

There are two theories on the market which satisfy this requirement of semantic innocence.[19] One is Davidson's paratactic theory (Davidson 1968); the other is

Quine's sentential operator analysis, pursued and elaborated by Arthur Prior in a number of writings (Quine 1960; Prior 1963, 1971; Orenstein forthcoming). On this view, "believes that" is an "attitudinative" (Quine) or "connecticate" (Prior) which forms a sentence from a singular term and a sentence. When its first argument place is filled by a singular term, it yields a sentence-forming operator, e.g. "John believes that..."

The "John believes that" operator can be viewed as a *world-shifting operator* (Recanati, 2000). It presents the sentence which follows it as true in John's belief world, rather than in the actual world. Whichever expression is responsible for such a world-shift is taken by Quine to constitute "an opaque interface between two ontologies, two worlds: that of the attitudinist, however benighted, and that of our responsible ascriber" (Quine 1995: 356). If Quine is right, then it does not matter whether we treat the embedded statement as mentioned, or as falling in the scope of a sentence-forming operator. In both cases the context in which the sentence is embedded is opaque. But is Quine right? If what I said concerning singular terms in belief contexts is correct, then, *pace* Quine, the ontology remains that of the ascriber all along, even though the "world" which is described is that of the attitudinist: the objects the ascribee's belief is said to be about are picked out in the speaker's world, that is, in the actual world. If that view is tenable, there are two sorts of world-shift. One can use the singular terms with their normal references to describe counterfactual possibilities – worlds other than the actual world; let us call that an *innocuous* world-shift. Or one can "shift ontologies" and use the terms with deviant references or at least without their normal references. In attitude contexts we have the first type of world-shift, but it is controversial that we have the more radical sort as well. The truly innocent theory I would like to see developed is one in which "believes that" is an attitudinative, and the sentential operators built from it are innocuous world-shifters.

§3.3 The attitudinatives as dependent expressions

According to Hintikka (1962: 138–141), failures of substitutivity in belief contexts show that two co-referential singular terms, though they pick out the same individual in the actual world, may refer to different objects in the ascribee's belief world. That option is ruled out in the present framework; for we want the ontology to remain that of the ascriber all along: we want the singular terms to refer to the same objects, whether we are talking about the actual world or about the ascribee's belief world. That is the price to pay for semantic innocence. How, then, can we account for substitutivity failures?

Once again, we must appeal to the notion of a reflecting context. Consider the following inference:

(6) John believes that Emile Ajar wrote *La Vie devant soi*
(7) Emile Ajar = Romain Gary
(8) John believes that Romain Gary wrote *La Vie devant soi*

Despite the identity stated in (7), we cannot infer (8) from (6). For it is possible that (6) is true and (8) false. That entails that (6) and (8) have different truth-conditions.

Now, by virtue of (7) and the semantics of singular terms, the embedded sentence in (6) and (8) make the same (narrow) contribution to the truth-conditions of the global belief report. Hence it must be the interpretation of the prefix, "John believes that", that is, *its* contribution to the truth-conditions of the global belief report, which changes from (6) to (8). If the prefix was given the same interpretation in (6) and (8), there could be no difference of truth-value between (6) and (8), since the embedded sentences express the same proposition. The prefix, therefore, must be a "dependent expression" whose semantic value shifts as a result of the substitution.

To emphasize the similarity with Quine's "Giorgione" example, and borrowing an idea from Graeme Forbes (1990), I suggest that we rephrase (6) and (8) respectively as

(6′) John so-believes that Emile Ajar wrote *La Vie devant soi*

and

(8′) John so-believes that Romain Gary wrote *La Vie devant soi*

In general, I suggest that whenever an attitude sentence, "*a* Ψs that p", is interpreted opaquely, we render it as "*a* so-Ψs that p", where "so" is a demonstrative adverb referring to some manner of Ψ-ing instantiated in the context. Slightly more colloquially, we might use the phrase "*a* Ψs that p *thus*", or "*a* Ψs that p *in that manner*". For example, "*a* says that p", opaquely understood, will be interpreted as tacitly referring to some manner of saying that p, as if the speaker had said: "*a* said that p thus". Similarly for "*a* believes that p" and the other attitude verbs.

What is a manner of Ψ-ing? Consider the case of "saying that". Someone can say that I am ill by uttering the sentence "He is ill" (while pointing to me) or by uttering "Recanati is not well". Those are two ways of saying that I am ill. Similarly, there are different ways or manners of believing that I am ill: by mentally entertaining the thought "That guy is ill" or by entertaining the thought "Recanati is not well". As is well known, the distinction between a sentence and what it says extends to thought, and the corresponding distinction between *what* is believed and *how* that is believed provides a key to the puzzles of cognitive significance (Perry 1992).

Let us assume that the speaker utters

(9) John said that Recanati is not well

and that this is understood opaquely, as somehow reporting (some of) the words which John himself used. I analyse (9) as

(9′) John so-said that Recanati is not well
      = John said that Recanati is not well *thus*

where the demonstrative adverb, "so" or "thus", refers to some manner of saying that I am ill. Which manner of saying that? *The manner of saying which is instantiated by the speaker's utterance of the embedded sentence.*

In that framework the same prefix "John believes that" makes different semantic contributions in (6) and (8), because the semantic value of the implicit demonstrative shifts when we substitute "Romain Gary" for "Emile Ajar". The difference can be made explicit as follows:

(6″) John believes that Ajar/Gary wrote *La Vie devant soi* in that manner: "Emile Ajar wrote *La Vie devant soi*'

(8″) John believes that Ajar/Gary wrote *La Vie devant soi* in that manner: "Romain Gary wrote *La Vie devant soi*'

In (6′) and (8'), the adverbial "in that manner" must of course be interpreted as modifying the main verb "believes". That verb itself must be given the "transparent" interpretation: in (6′) and (8′) "believes" is *not* equivalent to "so-believes".

As it stands the analysis is not wholly satisfactory, for not all aspects of the embedded sentence need to play a role in the imputation of a particular manner of believing to the ascribee. To refine the analysis, we can appeal to Nunberg's useful distinction between the *index* and the *referent* of a given occurrence of a demonstrative (Nunberg 1993). The index is what Kaplan (1989) calls the "demonstratum" – that which is actually pointed to or attended at – but at least in cases of "deferred ostension" that is distinct from the referent: the referent is the intended object, identifiable in relation to the index. Thus if, pointing to a car key, I say "This is parked out back", the index (demonstratum) is the key, but the referent is the car. If we apply this distinction to our present case, we will say that the implicit demonstrative "so" or "thus" *demonstrates* the speaker's current utterance of the embedded sentence (= index), and thereby *refers to* a certain manner of Ψ-ing, namely, that manner of Ψ-ing which would be instantiated if one Ψ-ed by uttering/entertaining that sentence.[20] On this analysis not all aspects of the demonstrated utterance need to be relevant to the determination of the manner of Ψ-ing which the speaker ascribes to the believer.

We can achieve the same result without appealing to Nunberg's distinction, however. Instead of analysing "*a* so-believes that p" as "*a* believes that p *in that manner*", we can analyse it, more perspicuously perhaps, as: "*a* believes that p *like that*", where the demonstrative "that" refers to the utterance of the embedded sentence. The manner of Ψ-ing denoted by the whole adverbial phrase "like that" will then depend upon the dimensions of similarity which are contextually relevant.

Whichever method we choose, the prefix turns out to be context-sensitive in two distinct ways, on the opaque interpretation. First, its semantic value depends upon the embedded sentence which follows it; for the demonstratum *d* (the index, or the referent of the constituent demonstrative "that') automatically changes when we substitute an expression for another in the embedded sentence. That, in itself, is sufficient to account for failures of substitutivity in attitude contexts. Second, the manner of Ψ-ing *m* which the demonstrated utterance is taken to instantiate will itself depend upon the aspects of the demonstrated utterance which are considered relevant. Even if we fix the demonstrated utterance, it will still be possible, by changing the context, to change the manner of Ψ-ing ascribed to the Ψ-er, thereby affecting the semantic value of the prefix.

There is, of course, an even more basic dimension of contextual variation: the belief report can be understood as transparent or opaque in the first place. The opaque reading I take to be a contextual enrichment of the transparent reading.

Much as "She opened the door" in (1) is contextually enriched into "She opened the door *with the key*", "John believes that p" is enriched into "John believes that p *in such and such manner*". The transparent/opaque ambiguity for belief reports is therefore an ambiguity between the minimal reading and a contextually enriched reading of the sentence.[21] Here as elsewhere, the enriched reading entails the minimal reading. (That is a general property of enrichment. See Recanati 1993.)[22]

Table 2 summarizes the three dimensions of contextual variation we have discerned in belief sentences.

I conclude, first, that Quine was quite right to stress the extreme context-sensitivity of attitude reports; second, that the content of the embedded sentence need not be considered as affected by the contextual variation. All the shifts in interpretation talked about in this section can construed as changes in the semantic value of the prefix "a Ψs that". It is the prefix which can be interpreted minimally or in an enriched, opaque manner ('a so-Ψs that'), depending on the context;[23] and it is the semantic value of the prefix which, on the opaque interpretation, varies according to the two further sorts of contextual change I have described.

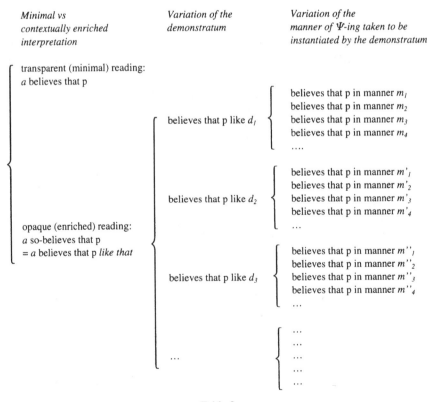

Minimal vs contextually enriched interpretation	Variation of the demonstratum	Variation of the manner of Ψ-ing taken to be instantiated by the demonstratum
transparent (minimal) reading: a believes that p		
	believes that p like d_1	believes that p in manner m_1
		believes that p in manner m_2
		believes that p in manner m_3
		believes that p in manner m_4
		....
	believes that p like d_2	believes that p in manner m'_1
		believes that p in manner m'_2
		believes that p in manner m'_3
		believes that p in manner m'_4
		...
opaque (enriched) reading: a so-believes that p = a believes that p *like that*	believes that p like d_3	believes that p in manner m''_1
		believes that p in manner m''_2
		believes that p in manner m''_3
		believes that p in manner m''_4
		...
	...	...

Table 2

§3.4 Opacity, substitution, and quantification

The prefix "John believes that" is a dependent expression only on the opaque reading, i.e., when it is interpreted as "John so-believes that". On the transparent reading it is not a dependent expression. Since the occurrences of singular terms in the embedded sentences are uniformly treated as purely referential, in accordance with their normal semantic function, they come out *transparent*, by the definitions given in part I, whenever the prefix itself is given the transparent reading: for (i) they are purely referential, and (ii) the context in which they occur is not reflecting (since the prefix is not a dependent expression, on the transparent reading). The truth-value of a transparent belief report therefore depends only upon the reference of the term, not on its identity. That strongly suggests that singular terms in transparent belief contexts should be substitutable, that is, freely replaceable by coreferential singular terms. Yet, I shall argue, they are not.

The reason why occurrences of singular terms in transparent belief reports are not substitutable, even though they are transparent, is very simple. Since (i) the prefix can be given an opaque (enriched) as well as a transparent (minimal) interpretation, depending on the context; and (ii) substituting an expression for another in the embedded sentence changes the context in which the prefix is tokened; it follows that the substitution can shift the interpretation of the prefix from transparent to opaque, by making it more likely that the speaker, using *those* words, intends to capture the believer's own way of thinking of the matter. In other words, belief contexts are *unstable* (§1.5). Only if we somehow fix (stabilize) the transparent interpretation of the prefix will substitution of coreferentials be a legitimate move.

Our findings so far can be summarized as follows:

- An occurrence of a singular term in the embedded portion of a belief sentence is purely referential, but not necessarily transparent: it is transparent only if the belief sentence is given a minimal interpretation ('transparent" reading). When the belief sentence is given an enriched interpretation ('opaque" reading), the occurrence of the singular term is not transparent, because the context in which it occurs is reflecting.
- Whether transparent or not, an occurrence of a singular term in the embedded portion of a belief sentence is not substitutable. It is non-substitutable either because the context *is* reflecting (opaque reading) or because the substitution can *make it* reflecting (transparent reading).

A last feature of singular belief reports must now be considered. As we saw in the first part of this paper, Quine tends to equate pure referentiality, transparency and substitutability. There is a fourth, no less important property on Quine's list: existential generalizability. When a singular term is purely referential (transparent, substitutable), Quine says, existential generalization is possible. When substitutivity fails because of opacity, existential generalization likewise fails. Thus we cannot go from

(10) Giorgione was so-called because of his size

to

(11) (∃x) (x was so-called because of his size)

Contrary to Quine, who holds that transparency entails substitutability, I emphasized that even transparent occurrences of singular terms in (transparent) belief reports are not substitutable – unless of course we stabilize the context by fixing the interpretation of all the other expressions in the sentence while we make the substitution. A first question that arises, therefore, is this: Is a transparent occurrence of a singular term in a belief context open to existential generalization? If the answer is, as I claim, "yes", then, *pace* Quine, substitution and existential generalization do not go hand in hand. I will go much further than that: I will argue that even opaque occurrences of singular terms in belief contexts are open to existential generalization. On the picture I am advocating (Table 3), substitution is *never* possible (even if the occurrence of the singular term at issue is transparent); while existential generalization is *always* possible (even if the occurrence of the singular term at issue is opaque).

Occurrence of singular term	in	'opaque' belief report	'transparent' belief report
purely referential?		yes	yes
transparent?		no	yes
substitutable?		no	no
open to existential generalization?		yes	yes

Table 3

The unstability of the context accounts for the (surprising) failure of substitutivity in transparent belief reports. Substitutivity fails because the substitution can, by changing the context, shift the interpretation of the prefix from transparent to opaque, thereby affecting the truth-conditions of the belief report. It is also the unstability of the context which accounts for the (no less surprising) possibility of existential generalization in opaque belief reports.

Normally, opacity blocks existential generalization. For the truth-value of a sentence containing an opaque occurrence of a singular term depends upon the identity of the term, not merely on its reference. When that term is eliminated through existential generalization, the statement is left incomplete and unevaluable: a reflecting context with nothing to reflect. Thus, Quine observes, (11) "is clearly meaningless, there being no longer any suitable antecedent for "so-called" " (Quine 1961: 145). There is, however, a crucial difference between an opaque belief sentence and a sentence like (10) – a difference which accounts for the success of existential generalization in opaque belief sentences.

(10) is a reflecting context for the singular term "Giorgione", and it is so in a stable manner: the context *remains* reflecting under operations such as substitution of coreferentials or existential generalization. But a belief sentence is a reflecting context for the singular terms occurring in the embedded sentence only when it is given the opaque ('so-believes') interpretation; and that interpretation is a highly context-sensitive hence *unstable* feature of the sentence. As I have repeatedly stressed, replacing a transparent occurrence of a singular term by an occurrence of

a coreferential singular term may change the truth-value of the report by shifting the prefix from the transparent to the opaque interpretation. In the other direction, replacing an opaque occurrence of a singular term, that is, an occurrence of a singular term in the embedded portion of an opaquely interpreted belief report, by a variable, automatically shifts the interpretation of the prefix from the opaque ('so-believes') to the transparent interpretation; for it is only on the transparent reading that the quantified statement makes sense. If the context remained reflecting, the statement would become meaningless once the singular term is eliminated. By virtue of this compensatory mechanism, we can go from "Tom believes that Cicero denounced Catiline", even on the opaque interpretation, to "Someone is such that Tom believes *he* denounced Catiline". The opacity of the original sentence is pragmatically filtered out in the very process of existential generalization.

At this point one might argue that, surely, the inference is illegitimate. We can go by existential generalization from "Fa" to "$(\exists x)$ (Fx)", but not from "Fa" to "$(\exists x)$ (Gx)". But in the type of inference I have just described, an expression (viz. the attitudinative) is interpreted differently in the premiss (the opaque belief sentence we start with) and the conclusion (the quantified statement). Logically, therefore, the inference does not take us from "Fa" to "$(\exists x)$ (Fx)", but from "Fa" to "$(\exists x)$ (Gx)". That is an instance of the fallacy of equivocation mentioned in §1.3.

But I think existential generalization from opaque belief reports with singular terms *is* a valid move. Since the opaque reading is an enrichment of the transparent reading, it entails the transparent reading. It is therefore legitimate to go from the opaque belief report "Tom believes$_o$ that Cicero denounced Catiline" to the meaningful quantified statement "There is someone of whom Tom believes$_t$ he denounced Catiline'; for the latter is entailed by the transparent belief report "Tom believes$_t$ that Cicero denounced Catiline", and that transparent belief report itself is entailed by the opaque belief report.

APPENDIX

Even though I consider "Giorgione" as purely referential in Quine's famous example, I accept Quine's point that there are intermediate cases between pure autonymy and pure referentiality. A good example is

(1) A "robin" is a thrush in American English, but not in British English.

Though it is quoted, the word "robin" here keeps its normal semantic value: it denotes a type of bird. It is a type of bird, not a word, which is said to be a thrush. But (1) also says something about the word "robin". For it is the word, not the bird, whose properties vary from one dialect of English to the next. As Austin pointed out, this mixture of mention and use is typical of semantic discourse:

Although we may sensibly ask "Do we *ride* the word "elephant" or the animal?" and equally sensibly "Do we *write* the word or the animal?" it is nonsense to ask "Do we *define* the word or the animal?" For defining an elephant (supposing we ever do this) is a compendious description of an operation involving both word and animal (do we focus the image or the battleship?). (Austin 1971: 124)

"Echoes" provide another example of mixed use. Often one uses a word while at the same time implicitly ascribing that use to some other person (or group of persons) whose usage one is blatantly echoing or mimicking. Thus one might say:

(2)   That boy is really "smart"

In such examples one is quoting, but at the same time using the words with their normal semantic values.

In (2) the fact that a word is quoted while being used does not affect the truth-conditions of the utterance. But sometimes it does. Thus I can refer to some object, A, using the name of another object, B, in quotes, providing the person I am mimicking uses the name for B as a name for A. I may well say

(3)   "Quine" has not finished writing his paper

and refer, by the name "Quine" in quotes, not to Quine but to that person whom our friend James mistakenly identified as Quine the other day. Any word can, by being quoted in this echoic manner, be ascribed a semantic value which is not its normal semantic value, but rather what some other person takes to be its semantic value.[24]

Many instances of mixed use lend themselves to a paraphrase where the expression in quotes is replaced by a descriptive phrase in which that expression occurs autonymously. Thus (1) could be rephrased as

(1*) The bird called "robin" is a thrush in American English, but not in British English[25]

The description "the bird called "robin"" describes a bird, but it does so by mentioning a word. We might, similarly, paraphrase the "echoic" examples above using such metalinguistic descriptions.[26]

CREA, CNRS/Ecole Polytechnique, Paris

NOTES

[*]   My thanks to Brian Loar for helpful comments on this paper.
[1]   I am using "Wychnevetsky" as an arbitrary name for the word "cat".
[2]   To the extent that he makes that claim, Quine does not take it at face value. He does not say that the occurrence is accidental *simpliciter*, but that it counts as accidental *from a logical point of view*. For a gloss, see the end of this section.
[3]   Indeed the expression in question may differ from the enclosed word in gender or number. Thus in:

"Cats" is a four-letter word

the first word of the sentence is singular, while the mentioned word ("cats") is plural.
[4]   The three examples I have just given contrast with the previous examples (autonymy, metonymy, etc.) in the following respect: In autonymy or metonymy there is a distinction between the normal semantic value and the semantic values which deviate from the norm. In the type/token or process/product ambiguity, arguably there is no such asymmetry. (See Nunberg 1979 for an investigation of those issues.) Despite this difference, all the examples belong to the family of "systematic ambiguities", whose study has proved important and fruitful in contemporary semantics (see e.g. Pustejovsky 1995).

⁵   The autonymous word need not be taken as self-referential, of course. Instead of having the word refer to itself, we can insist that it is the complex expression (word plus quotation marks) or the pair of quotation marks (construed as a demonstrative, in the manner of both Prior [1971: 60–61] and Davidson [1979]) which refers to the quoted word.

⁶   In Quine's framework, the first type of case (the case in which it is the singular term itself which is used non-purely referentially) can be reduced to the second type of case (the case in which it is the context or the position that generates opacity). For Quine takes an autonymous word to be a word that occurs in a special linguistic context, viz. "within quotation marks". Quine can thus get rid of non-purely referential occurrences or uses altogether and handle opacity entirely in terms of positions. Opaque or non-purely referential positions are linguistic contexts (e.g. quotation contexts or reflecting contexts) such that a singular term in that context is not subject to the Principle of Substitutivity. In this way a uniform treatment is provided for

"..." is a three-letter word

and

... is so-called because of his size

Even so, there remains a big difference between the two types of case. In the first context the singular term is not used referentially (it is deviant), while it is used with its normal referential function in the second context.

⁷   As Kit Fine has shown, we can make any non-referential occurrence of a singular term similarly substitutable merely by forming the disjunction of the sentence where it occurs with "$2 + 2 = 4$" (Fine 1989: 218).

⁸   See Forbes 1990; Crimmins 1992; Recanati 1993. The first after Quine himself to have drawn attention to the analogy between belief sentences and "Giorgione" sentences was Brian Loar in his 1972 article.

⁹   To say that they have a non-deviant referential use is compatible with saying that their referential use is not their normal semantic function. If their referential use was their normal semantic function, their non-referential uses would themselves be deviant. In Recanati (1993) I claimed that attributive and referential uses of descriptions are *both* non-deviant. (In contrast, only purely referential uses of singular terms are non-deviant.)

¹⁰   "If we say that there is someone of whom Othello believes that she is unfaithful, while we do not thereby put *ourselves* into any relation with anyone except Othello, we do thereby say that there is someone with whom *he* stands in the relation of believing her unfaithful" (Prior 1971: 135).

¹¹   In 1956 Quine said that exportation – the step from "*a* believes that t is F" to "*a* believes of t that it is F" – "should doubtless be viewed in general as implicative" (Quine 1956: 190). Afterwards he was moved by the Sleigh/Kaplan example of the shortest spy (Sleigh 1968; Kaplan 1968): if exportation is valid, then we can go from "John believes that the shortest spy is a spy" to "($\exists$x) (John believes x is a spy)", via "John believes of the shortest spy that he is a spy"; but if that is accepted, an obviously notional belief report is treated as if it were relational. As Quine concludes, "we must find against exportation" (Quine 1977: 9). Indeed, insofar as exportation opens the way for existential quantification, it is clear that exportability must be restricted to those cases in which the relational reading is intuitively appropriate. It cannot be treated as generally permissible.

    Still, I think Quine was right in the first place: exportation *is* generally valid, provided t is a genuine singular term. In the Sleigh/Kaplan example, it isn't. Of course, exportation also works when t, though not a singular term, is given scope over the epistemic operator. But that is not the case in the Sleigh/Kaplan example either. In the Sleigh/Kaplan example Ralph is said to believe "The shortest spy is a spy"; he is not said to believe, of some particular individual known to him (and described by the speaker as "the shortest spy"), that he is a spy.

¹²   (11) can also be interpreted as short for "In the story, Santa Claus lives in the sky" (Lewis 1978). On that interpretation (11) is a metafictional statement, like (10).

¹³   See Quine's seminal and oft-quoted remarks on the role of empathy and pretense in belief ascriptions (Quine 1960: 219).

¹⁴   The theory in question should not be considered as dependent upon the success of the Walton-Crimmins research programme, however. Should the latter fail, it would still be

possible to argue that metafictional uses (e.g. the use of "Santa Claus" in (10)) are somehow deviant, like the non-purely referential uses mentioned in the Appendix.

[15]   To flesh out this interpretation, one might construe the occurrence of a singular term in an opaque belief context as "echoic", along the lines of

Ralph believes of "Cicero" that he denounced Catiline

where the quotes around the singular term "Cicero" indicate that the occurrence is partly autonymous. On echoic uses, see the Appendix.

[16]   Brian Loar also uses the word "frame" in his 1972 article, but in a different sense.

[17]   That is, in effect, the "hidden indexical theory" of belief reports, the first formulations of which can be found in Linsky (1967: 113) and Schiffer (1977: 32–33).

[18]   "From the standpoint of logical analysis each whole quotation must be regarded as a single word or sign, whose parts count no more than serifs or syllables. A quotation is not a *description*, but a *hieroglyph*; it designates its object not by describing it in terms of other objects, but by picturing it. The meaning of the whole does not depend upon the meanings of the constituent words." (Quine 1951: 26)

[19]   Three if we count the theory put forward in Panaccio 1996. According to Panaccio, "John believes that S" can be analysed as "John believes something which is true *tantumsi* S", where "*tantumsi*" is a special, indexical connective.

[20]   This counterfactual circumlocution is necessary because the speaker herself need not be Ψ-ing in uttering the embedded sentence.

[21]   Some will hasten to conclude that only the transparent reading is relevant to semantics, since semantics is supposed to deal with literal meanings and "minimal" interpretations, without considering "speaker's meaning". (See e.g. Salmon 1986.) I disagree. If we want our theory to keep in touch with our intuitions, we cannot disregard those aspects of meaning resulting from enrichment. Moreover, when we embed an utterance, we see that the aspects of meaning which result from contextual enrichment often become part of the minimal interpretation of the complex utterance, thereby making the intended segregation of the semantic from the pragmatic untenable. For more on those matters, see Recanati 1993, part II.

[22]   At this point a suggestion must be considered: can we not reduce the third form of context-sensitivity to the second one? Remember that in the opaque reading, there are two dimensions of variation. (i) Any change in the wording of the embedded sentence can, by changing the demonstratum, affect the semantic value of the sentential operator; (ii) even if the demonstratum is fixed, the manner of Ψ-ing which is tacitly referred to can vary depending on the aspects of the index that are considered relevant in the context at hand. Now we could consider the transparent reading as the special case in which the relevant aspect of the demonstrated utterance is nothing other than its semantic content – its truth-conditions – to the exclusion of any other properties of the embedded sentence.

I have two objections to this move:

(i) Instead of considering the opaque reading as an enrichment of the transparent reading, it construes the transparent reading as a limiting case of the opaque. But in analysing the opaque reading in terms of "so-believing" or "believing thus", I implicitly accepted the primacy of the transparent sense. If we use subscripts to distinguish the transparent sense from the opaque, then, on my analysis, to believe$_0$ that p is to believe$_t$ that p in such and such a manner. That is hardly consistent with analysing believing$_t$ that p as a variety of believing$_0$ that p.

(ii) If we construe "that"-clauses as singular terms, it is indeed tempting to view transparency as a limiting case of opacity. Thus we can say that a "that"-clause always refers to some enrichment of the proposition expressed, where that proposition itself counts as the minimal or "zero" enrichment. But such a move is much less tempting when, giving up the view that "that"-clauses are singular terms, we construe the embedded sentence as a *bona fide* sentence. On that construal the semantic content of the embedded sentence (the proposition it expresses) is its contribution to the content of the global attitude report: the embedded sentence expresses a proposition, and the sentential operator maps that proposition on a truth-value (truth, whenever the proposition in question holds in the ascribee's belief world). The semantic content of the embedded sentence is thus given prior to and independent of whatever tacit reference to some manner of Ψ-ing the sentential operator may additionally convey.

[23]   The "context" in which the prefix is tokened includes the words it is prefixed to. See §3.4 for some consequences of that type of context-sensitivity.

24 In Recanati 1987: 63 I offered an example involving a definite description:

Hey, "your sister" is coming over

In that example the person who is coming over is not actually the addressee's sister, but is thought to be the addressee's sister by some person whom the speaker is ironically mimicking.

25 Note that the description has to be construed as an "incomplete" definite description, that is, as a definite description whose domain is contextually variable, on the pattern of: "The window is open in the kitchen, but not in the living room." (See Recanati 1996 for a treatment of incomplete definite descriptions in terms of contextually variable domains.)

26 There is a clear difference, in all cases, between the original sentence and the explicit paraphrase; a difference which should prevent one from saying that one is "elliptical for" the other. See Walton 1990: 222–224.

REFERENCES

Austin, J.L. (1971), *Philosophical Papers*, 2nd ed. Oxford: Clarendon Press.

Barwise, J. and J. Perry (1981), Semantic Innocence and Uncompromising Situations. *Midwest Studies in Philosophy* 6: 387–403.

Carston, R. (1988), Implicature, Explicature, and Truth-Theoretic Semantics. In R. Kempson (ed.), *Mental Representations: The Interface Between Language and Reality*, Cambridge: Cambridge University Press, p. 155–81.

Chastain, C. (1975), Reference and Context. In K. Gunderson (ed.), *Language, Mind, and Knowledge*, Minneapolis: University of Minnesota Press, p. 194–269.

Crimmins, M. (1992), *Talk about Belief*. Cambridge, Mass: MIT Press/ Bradford Books.

Crimmins, M. (forthcoming), Hesperus and Phosphorus.

Crimmins, M. and J. Perry (1989), The Prince and the Phone Booth. *Journal of Philosophy* 86: 685–711.

Currie, G. (1990), *The Nature of Fiction*. Cambridge: Cambridge University Press.

Davidson, D. (1968), On Saying That. *Synthese* 19: 130–146.

Davidson, D. (1979), Quotation. *Theory and Decision* 11: 27–40.

Dever, J. (forthcoming), Strangers on a Train: Engineering a Kripke-Frege Reunion.

Donnellan, K. (1966), Reference and Definite Descriptions. *Philosophical Review* 75: 281–304.

Evans, G. (1982), *The Varieties of Reference*, ed. by J. McDowell. Oxford: Clarendon Press.

Fine, K. (1989), The Problem of De Re Modality. In J. Almog, H. Wettstein and J. Perry (eds.), *Themes from Kaplan*, New York: Oxford University Press, p. 197–272.

Forbes, G. (1990), The Indispensability of *Sinn*. *Philosophical Review* 99: 535–563.

Forbes, G. (1996), Substitutivity and the Coherence of Quantifying In. *Philosophical Review* 105: 337–372.

Geach, P.T. (1972), *Logic Matters*. Oxford: Basil Blackwell.

Hintikka, J. (1962), *Knowledge and Belief*. Ithaca: Cornell University Press.

Hornsby, J. (1977), Singular Terms in Contexts of Propositional Attitude. *Mind* 86: 31–48.

Kaplan, D. (1968), Quantifying In. *Synthese* 19: 178–214.

Kaplan, D. (1986), Opacity. In L. Hahn and P. A. Schilpp (eds.) *The Philosophy of W.V. Quine*, La Salle, Illinois: Open Court, p. 229–89.

Kaplan, David (1989), Demonstratives. In J. Almog, H. Wettstein and J. Perry (eds.) *Themes from Kaplan*, New York: Oxford University Press, p. 481–563.

Kripke, S. (1977), Speaker's Reference and Semantic Reference. *Midwest Studies in Philosophy* 2: 255–276.

Lewis, D. (1978), Truth in Fiction. *American Philosophical Quarterly* 15: 37–46.

Linsky, L. (1967), *Referring*. London: Routledge & Kegan Paul.

Loar, B. (1972), Reference and Propositional Attitudes. *Philosophical Review* 81: 43–62.

McKay, T. (1981), On Proper Names in Belief Ascriptions. *Philosophical Studies* 39: 287–303.

Neale, S. (1990), *Descriptions*. Cambridge, Mass: MIT Press/Bradford Books.

Nunberg, G. (1979), The Non-Uniqueness of Semantic Solutions: Polysemy. *Linguistics and Philosophy* 3: 143–84.

Nunberg, G. (1993), Indexicality and Deixis. *Linguistics and Philosophy* 16: 1–43.

Orenstein, A. (forthcoming), Propositional Attitudes without Propositions. In P. Kotatko and A. Grayling (eds.), *Meaning*, Oxford: Clarendon Press.

Panaccio, C. (1996), Belief-Sentences: Outline of a Nominalistic Approach. In M. Marion and R.S. Cohen (eds.), *Québec Studies in the Philosophy of Science II*, Dordrecht: Kluwer, p. 265–277.

Perry, J. (1986), Thought without Representation. *Proceedings of the Aristotelian Society, Supplementary Volume* 60: 137–51.

Perry, J. (1992), *The Problem of the Essential Indexical and Other Essays*. New York Oxford University Press.

Prior, A. (1963), Oratio Obliqua. *Proceedings of the Aristotelian Society, Supplementary Volume* 37: 115–126.

Prior, A. (1971), *Objects of Thought*, ed. by P. Geach and A. Kenny. Oxford: Clarendon Press.

Pustejovsky, J. (1995), *The Generative Lexicon*. Cambridge, Mass: MIT Press.

Quine, W.V.O. (1951), *Mathematical Logic*, 2nd ed. Cambridge, Mass: Harvard University Press.

Quine, W.V.O. (1953), Three Grades of Modal Involvement. Reprinted in Quine 1976, p. 158–176.

Quine, W.V.O. (1956), Quantifiers and Propositional Attitudes. Reprinted in Quine 1976, p. 185–196.

Quine, W.V.O. (1960), *Word and Object*. Cambridge, Mass: MIT Press.

Quine, W.V.O. (1961), *From a Logical Point of View*, 2nd ed. Cambridge, Mass: Harvard University Press.

Quine, W.V.O (1962), *Methods of Logic*, 2nd ed. London: Routledge & Kegan Paul.

Quine, W.V.O. (1974), *The Roots of Reference*. La Salle, Illinois: Open Court.

Quine, W.V.O. (1976), *The Ways of Paradox and Other Essays,* 2nd ed. Cambridge, Mass: Harvard University Press.

Quine, W.V.O. (1977), Intensions Revisited. *Midwest Studies in Philosophy* 2: 5–11.

Quine, W.V.O. (1980), The Variable and Its Place in Reference. In Z. van Straaten (ed), *Philosophical Subjects: Essays presented to P.F. Strawson*, Oxford: Clarendon Press, p. 164–173.

Quine, W.V.O. (1995), Reactions. In P. Leonardi and M. Santambrogio (eds.), *On Quine*, Cambridge: Cambridge University Press, p. 347–361.

Recanati, F. (1987), Contextual Dependence and Definite Descriptions. *Proceedings of the Aristotelian Society* 87: 57–73.

Recanati, F. (1993), *Direct Reference: From Language to Thought*. Oxford: Basil Blackwell.

Recanati, F. (1996), Domains of Discourse. *Linguistics and Philosophy* 19: 445–475.

Recanati, F. (1998), Talk about Fiction. *Lingua e Stile* 33: 547–58.

Recanati, F. (2000), *Obliquities*. Cambridge, Mass: MIT Press.

Russell, B. (1905), On Denoting. *Mind* 14: 479–493.

Salmon, N. (1986), *Frege's Puzzle*. Cambridge, Mass: MIT Press/Bradford Books.

Schiffer, S. (1977), Naming and Knowing. *Midwest Studies in Philosophy* 2: 28–41.

Sleigh, R.C. (1968), On a Proposed System of Epistemic Logic. *Noûs* 2: 391–398.

Walton, K.L. (1990), *Mimesis as Make-Believe*. Cambridge, Mass: Harvard University Press.

QUINE'S RESPONSES

RESPONSE TO SZUBKA

The distal scene, shared by the field linguist and the native, is the proper focus for the linguist in his enterprise of radical translation. The distal scene shared by mother and child is likewise the proper focus for the mother in helping the child with the language. Observation sentences are mostly reports on the distal scene. It is there, and not at the neural intake, that the action is. It is where minds meet.

To seek proximal rather than distal common ground is to court chaos. Darwin found that even simple insects from the same swarm have widely dissimilar nerve nets. Physiologically similar neural reactions in different observers are not to be expected. Yet similarity of verbal response to the scene, on the part of adult compatriots, is taken as a matter of course. How is this distal harmony across proximal heterogeneity to be explained? Why does it all come out right?

My explanation turns on a preestablished intersubjective harmony of subjective standards of what I call perceptual similarity, which I shall now clarify and account for. Instinct and natural selection are at the bottom of it.

I must begin by defining some terms. By an individual's *neural intake* on a given occasion I shall mean the temporally ordered set of all of his neural receptors that were triggered on that occasion. Each of us is born with subjective standards of perceptual similarity of neural intakes. Each intake is, for him, more similar to some than to others. His scale is private and subjective, but it can be probed objectively by the behavioral psychologist through the reinforcement and extinction of responses.

Perceptual similarity of an individual's neural intakes deviates widely from mere degree of overlap or quantity of shared receptors, what I call *re*ceptual similarity. The various figures that a cube projects on the retina, when seen from various angles, are geometrically dissimilar and make for receptually dissimilar neural intakes, but perceptually similar ones: the subject sees, he says, a cube. Similarly a tiger, seen from various angles and in various postures, induces perceptually similar intakes despite receptual diversity.

Perceptual similarity is private. Not only do we share no receptors; they are presumably not even homologous, and the nerve nets into which they lead are presumably far from homologous as well. Yet at a reasonable distance we evidently see eye to eye. Our vagrant neural intakes and our processing of them issue somehow in consonant perceptions. Such is the preestablished harmony of perceptual similarity. It applies not only to vision but to all the senses. The word "gavagai", as spoken by the native and the linguist, sounds enough alike to both, and likewise

Alex Orenstein and Petr Kotatko (eds.), Knowledge, Language and Logic, 407–430.

"Mama" as spoken by mother and child. But we still have to account for the harmony, and that will take a few more steps. Patience, please.

Perceptual similarity is essential, obviously, to conditioning. Hence, indeed, the use of conditioning in probing a subject's similarity standards. Since conditioning is essential to all learning, it follows that one's similarity standards cannot all have been learned. They are rooted in instinct, but change somewhat with experience.

Another instinct, of a piece with perceptual similarity, is the instinct of induction: the instinct to expect perceptually similar stimulations to have similar sequels. This instinct meshes with that of perceptual similarity, clearly, in conditioning.

Philosophers have marveled that expectation by induction, though fallible, is so much more successful than random guessing. This is explained by natural selection. Successful expectation is conducive to survival, as in eluding predators and catching prey; so natural selection down the ages has bolstered induction by warping our standards of perceptual similarity somewhat into conformity with trends in our environment. Witness my tiger example above.

Now we can account for the preestablished intersubjective harmony of our subjective standards of perceptual similarity. Natural selection molded our shared ancestors' standards into partial conformity with a shared environment.

RESPONSE TO GEORGE

George's title "Quine and Observation" prompted me first of all to scan his eleven pages of footnotes for references to my latest publications; for it is only in them that by a theory of preestablished harmony I dispelled, to my satisfaction, a stubborn thirty-year riddle of the meeting of minds in their perceptions of the shared world. Yes, George cites *From Stimulus to Science* and, still more to the point, "Progress on two fronts."

With a sigh of relief I settled into his essay. The fifth page brought shock and bewilderment. He quotes my statement of the harmony but evinces no sense of breakthrough and relief. It was as if to say "So, and what else is new?" His only evident point in bringing it in at all was to say that it did not offset an error in one of my definitions of "observation sentence". It had not been meant to.

The harmony has to do with our innate standards of perceptual similarity of stimulations. Picture two observers side by side observing some event, and later another event. If the events struck the one observer as similar, chances are that they struck the other as similar; such is the intersubjective harmony of our innate similarity standards. I have no neurological explanation of the harmony, but it would be a matter of a large but limited store of repeatable sensory features of environmental events, lodged deep in the brain of each of us as reusable modules of perception and accessed by each individual's idiosyncratic nerve net. For all our present ignorance of the neurological detail, natural selection has a clear role in inducing the harmony, jointly with the instinct of inductive expectation. See my adjoining response to Szubka.

What was the long-standing riddle? I had stressed already, in *Word and Object* (1960) and before, that observation sentences treat, for the most part, of the external world and not of private experience. The riddle, then, had concerned the gap between the privacy of our neural intake and the publicity of our testimony.

In *Word and Object* I had lamely hazarded a homology of receptors between one individual and another, but within five years ("Propositional objects," in *Ontological Relativity*) I was finding that implausible. Like Davidson, I was eventually doing without intersubjectively shared stimulations and shared stimulus meanings, but unlike him I retained subjective stimulus meanings and rested uneasily with the manifestly distal meeting of minds. This is the riddle that is now solved by preestablihed harmony.

In his section IV we find George worrying about what it means for two witnesses to perceive an event, since we cannot assume shared or homologous neural structures. This is typical business for the preestablished harmony that he passed over rough shod. Instead he proposes a retreat to mentalism.

He frequently mentions my "commitment to anomalous monism," so I must clarify that. The verbs of propositional attitude and various other mentalistic words contribute vitally to everyday explanation and are practically indispensable, today anyway, in the conduct of human affairs. Happily the grammatical idioms of propositional attitudes *de dicto* can by recourse to quotation be rendered extensional and thus reconciled with the logical grammar of regimented science. Their verbs with their faltering empirical criteria can then be accommodated as integral to a regimented language. But propositional attitudes *de re* end up rather alongside in the limbo of indexicals, along with the personal and demonstrative pronouns and the contrafactual conditionals.

The useful verbs of propositional attitude owe their intelligibility partly to visible manifestations of the subject's present emotions and partly to narrations of past events that induced the subject's attitudes. In this respect they are on the same footing as organic ailments whose aetiology has not yet been determined. Both are ultimately bodily, however ill understood, and equally deserving of a niche in our loose-knit global system of natural knowledge.

For all my tolerance of the verbs of propositional attitude *de dicto*, I find no hardship in dispensing with the corresponding mental entities. We can believe and perceive without trafficking in beliefs or percepts. Along with my treatment of the propositional attitudes *de dicto* by means of quotation, the role of beliefs and percepts is taken on by the quoted sentences themselves. These diverge from beliefs and percepts in respect of individuation, but then there has been no clear individuation of beliefs and percepts.

George devotes the second half of his essay to an intricate mentalistic dialectic in defense of meaning and a more forthright mentalism, sparked by my conjecture of the indeterminacy of translation. Unlike the indeterminacy of reference, which has its simple and conclusive proof in proxy functions, the indeterminacy of translation was always a conjecture, albeit a plausible one. It is a dismissal neither of translation nor of meaning. I have questioned the reification of meanings, plural, as abstract entities, and this not an on the score of their abstractness, but of their individuation; for there is no entity without identity. Seeing meaning as vested primarily in the sentence and only derivatively in the word, I sought in vain an operational line on sameness of sentential meaning by reflecting on the radical translation of sentences.

My conclusion was that the only overall test of a good manual of radical translation was fluent dialogue and successful negotiation, and my conjecture was that two manuals could pass muster and still conflict in translation of some sentences remote from observation and from social and commercial concerns. What was challenged was the philosophical notion of propositions, the meanings of sentences.

George sees my reservations about meanings as directed against meaning, and diagnoses them as rooted in my physicalism. I demur on both points. I just sought a definition of sentential synonymy that one could in general see how to apply.

Early and late I recognized empathy as the strategy in radical translation. My use of the word "empathy" is only recent and has been noticed, but I had already recognized the radical translator's approach as empathetic[1] in *Word and Object* and indeed nine years before. "The lexicographer," I wrote, "... depend[s] ... on a projection of himself, with his Indo-European *Weltanschauung*, into the sandals of his Kalaba informant."[2] It is by empathy that we estimate our interlocutor's perceptions. Their neural implementation is as may be.

Thus I already accept much of what George is arguing for. Where, he must now wonder, do I draw the line? I countenance mentalistic predicates when their applicability is outwardly observable enough for practical utility. Our materialistic predicates, after all, are likewise vague in varying degrees, and I would apply the same standards. But mental entities I dispense with, as explained, and extensionalism I insist on for applicability of our quantificational logic.

RESPONSE TO GRAYLING

The abdication of epistemology to psychology, in which I connive, is less abject than Grayling sees it. The pertinent motivations and aptitudes remain those of the analytic philosopher rather than the experimental psychologist. Analysis of reification was called for, which had been passed over by psychologist and philosopher alike. This analysis branched into settling on what counts as reification and what service it discharges in the structuring of science and our spatiotemporal conception of the world. An incidental question, germane to epistemology but not traditional, was as to what aspects of our ontology are essential to science and what ones are merely subjective. There is philosophical progress here for which we would not look to psychology.

Another dimension into which these speculations lead is subjective similarity of perceptions. This is recognizable as psychology, but I doubt that the intersubjective harmony of these subjective standards was looked into and accounted for until motivated by naturalized epistemology, where it is seen to underlie both communication and induction.

---

[1] Not "empathic", please. That, like "phonemic" for "phonematic", smacks of "little Latin and less Greek."

[2] "The problem of meaning in linguistics," presented at a linguistics conference in 1951 and published in *From a Logical Point of View* in 1953. The quotation is from page 63.

I see naturalized epistemology rather as enlivening than as superseding its eponym. Grayling is mistaken, however, in writing that "Quine insists that ... naturalized epistemology is truly epistemology." I have written at least twice (e.g. in *Pursuit of Truth*, p. 19) that I stretch the term perhaps unduly.

Not quite midway in his paper, Grayling alarmingly misreads these words in my response to a paper of Davidson's: "in my theory of evidence the term 'evidence' gets no explication and plays no role." Grayling takes this somehow as disavowing a theory of evidence. My chariness of the word had been due to Davidson's insistence that only a belief can be evidence for a belief; but evidence itself was for me a central concern. I treat of it, and even in Davidson's sense, when I treat of the testing of a theory by experiment. The scientist deduces from his hypotheses that a certain observable situation should bring about another observable situation; then he realizes the one situation and watches for the other. Evidence for or against his set of hypotheses ensues, however inconclusive.

There is scope for evidence of another sort when the scientist is thinking up hypotheses worth testing. Here considerations of simplicity seem pertinent, as well as logical links with his present theory. In both these domains of evidence I see no departure from the old epistemology.

I have written occasionally that I use the word "science" broadly, covering history no less than natural science. Grayling supposed the contrary. It is awkward that "science", unlike *scientia* and *Wissenschaft*, so strongly connotes natural science nowadays. But it is in natural science that methodological considerations stand forth most clearly.

In his later pages Grayling is at pains, in my behalf, to safeguard naturalism in case a scientific revolution supersedes physicalism. For my part, I find it ironical that naturalism, with its doctrine of scientific fallibility and no first philosophy, should exempt itself from these strictures. I see nothing sacrosanct about naturalism or physicalism; both are fallible, unless saved by vague edges.

If conclusive evidence of telepathy or even clairvoyance were forthcoming, I envision a scurry to the cyclotron, computers, and drawing boards to invent a new and more adequate theory, which would still be called physics. Even today the elementary particles are particles at all only thanks to a succession of ever more strained analogies. Physics can even be pursued as field theory, free of bodies, and no complaints. The name "physics" would survive the clairvoyance revolution too. Continuity of the enterprise is what matters.

In general I tend to be impatient with the quest for precision in the names for disciplines and schools of thought: in asking what really counts as naturalism, epistemology, physics. Like our everyday terms, these are at best helpful makeshifts, vague around the edges, and no matter.

RESPONSE TO LEHRER

Lehrer cites Grayling for a duality of epistemological objectives: the meaning-basing objective, as Lehrer puts it, and the inference-basing objective. Groping for a corresponding duality of agencies, I suggest stimulus meaning on the one hand

and induction on the other. Both depend on subjective standards of perceptual similarity, induced by natural selection.

Sense data were the traditional epistemological proving ground for natural science. Translation of science into a language of sense data was accordingly seen as the way to justify science. Hence Carnap's *Logischer Aufbau der Welt*. It provided translation up to a point and then proceeded with counsels for rational reconstruction short of translation. But the reduction of science to sense data fails if the translation cannot be completed: and one sees clearly from the *Aufbau* that it cannot. In asking at this point "Why not settle for psychology?" I did not mean, as Lehrer supposed, that psychology would advance the justification process. I meant "Let us just get clear on the psychology of what we are actually doing, and look elsewhere if at all for justification."

Where I do find justification of science and evidence of truth is rather in successful prediction of observations, and this evidence is conclusive only in varying degrees. Karl Popper argued that experiment can only refute hypotheses, not prove them. I hold that experiment is fallible both ways. I prefer Popper's analogy of science to an edifice supported only by a multitude of long piles driven deep down into a bottomless swamp.

So observation, however inconclusive, is in my view the locus of evidence. But coherence governs our prediction of the observations, since a substantial bundle of interlocking hypotheses is usually needed in order to predict a particular observation in particular observable circumstances.

Besides evidence in this strict sense, there is a weaker sort of something like evidence that underlies our production of hypotheses worth testing. It is evidence of the *promise* of a hypothesis, whereas predicted observation is evidence of the *truth* of a hypothesis. Simplicity, symmetry, economy are taken as evidence of promise. Just why is not altogether clear, but some considerations are marshaled in my paper "On simple theories of a complex world."[1]

RESPONSE TO BERGSTRÖM

Taking my naturalized epistemology as his point of departure, Lars Bergström has proposed an empiricist conception of truth. An observation sentence is true on just the occasions that would prompt the subject's assent to it if he were to observe the occasions. An observation categorical, then, being a universal conditional joining two observation sentences, is true if and only if its consequent is true on all occasions where its antecedent is true. A theoretical sentence is true, finally, if and only if logically implied by a "tight" theory that implies all true observation categoricals and no false ones. He goes on to sketch tightness.

As Bergström remarks, I have warned now and again that the authority of observation sentences is properly speaking a matter of degree. For simplicity and convenience, then, I have set this detail aside and proceeded much as if they were

---

[1] Reprinted in my *Ways of Paradox and Other Essays*, 1966; Harvard, 1976.

uniformly infallible. Bergström has followed me in this course. His paper, however, has prompted me now to focus rather on the gradations. I see a spectrum reaching from sense data to science. I shall develop this thought.

I continue to use my term "observation sentence" broadly: an occasion sentence is an observation sentence for a speaker if it has become keyed to a range of global neural inputs any one of which will prompt his immediate assent to it. That range of appropriate inputs will of course be vague along the edges; the speaker may hesitate over "It's raining" in a fine mist, and over "That's a swan" in the startling presence of a black swan. Vagueness of boundaries infests language at every turn, and I shall continue to take it in stride. Thus far, nothing new.

The gradations that are my new concern are degrees rather of susceptibility of unequivocal assent to unequivocal recantation. "It's raining", unhesitatingly affirmed in full view of a drenched window pane, is recanted when the water proves to have come from a hose. "It's a rabbit", affirmed in full view of the object in question, is recanted when the object proves to have been a toy. Such recantation reflects theoretical connections among observation sentences. The degree of susceptibility to recantation measures how theoretic the observation sentence is. It is its *degree of theoreticity*.

An observation sentence that is perhaps minimally theoretic is "This looks blue". I write "looks" here, rather than "is", to allow for the possibility that reflected light or environmental contrast may be affecting the color that the object would otherwise show. For the reference of "This" is still to an external patch or body, or is to become so with the flowering of reification.

Speakers vary in how they arrive at identical usage of an observation sentence. One speaker may have acquired the sentence in the primitive way, by direct holophrastic conditioning to global neural inputs in the appropriate range. Another speaker may have assembled the sentence rather from words learned in earlier contexts. In either case, assent to the sentence may be recanted in the light of subsequent evidence.

There are sentences that are learned only through theory, and that become observational only late in the specialist's career: thus "There was copper in it", said by the chemist after a glance at the solution, or "There goes a hyperthyroid", said by the physician after a glimpse of a stranger's face. Sophisticated observation sentences such as these are apt to be reducible to more primitive ones, delineating more directly sensory evidence, and these will tend also to be less theoretic by the stated criterion, that is, less susceptible to recantation. Some sophisticated ones, however, are not thus reducible. I think of the subtle traits that the wine expert learns to detect.

Such reduction, where possible, bolsters scientific theory; for the increased resistance to recantation of observation sentences increases the dependability of the corresponding observation categoricals. The categoricals implied by a theory are its checkpoints, and flabby categoricals are insensitive touchstones. Reducing theoreticity by buttressing or supplanting the more theoretic observation sentences by less theoretic ones then enhances the dependability of scientific theory. This reflection is quite in the spirit of traditional phenomenalistic epistemology.

A thoroughgoing reduction project of the kind, however, is surely a forlorn hope. It is utterly alien to what goes on and went on in the development of language and science in the child and in the race. Observation sentences, already theoretic to varying degrees, are learned outright and helter-skelter by direct holophrastic conditioning. Further ones are synthetized along the way from bits of those at hand. They vie with one another in a surging equilibrium of evidential claims. Such is the web of belief.

This vision of science is a step from Karl Popper toward Thomas Kuhn. The observation categoricals that are the checkpoints of a theory are built of observation sentences that are themselves irreducibly theoretic to various degrees, so an apparent counterinstance of such a categorical is strong evidence against the theory but not necessarily lethal. We are left weighing subjective probabilities, not only in confirming theories but in refuting them.

Back now to Bergström's paper, to which this has thus far been rather a reaction than a response. An observation sentence should be true, by his account, on just the occasions that would prompt the subject's assent to it. An observation sentence that perhaps meets this condition is "This looks blue". I see only such cases as fulfilling Bergström's empiricist conception of truth. Even here there is the vagueness of boundary of blue to reckon with, but then vagueness of predicates besets ascriptions of truth under any conception.

I like Bergström's injection of empiricism into the truth concept, if only at the very nadir of theoreticity: "This looks blue". Then man's theoretical creativity takes the lead. But we faithfully keep paying our dues to empiricism by deducing observation categoricals from theory and checking them for falsity. Observation is our empirical and sole objective check from first to last.

At the end of his paper Bergström credits his empirical conception of truth with suggesting why simplicity, generality, and other virtues of theories promote our pursuit of truth: they enhance the "tightness" of theory to the implied observation categoricals. But this reflection retains its interest without his empiricist conception of truth, since the implied categoricals are still the checkpoints.

It may be felt that rejection of a full empiricist conception of truth leaves the meaning of "true" again a mystery. This feeling is odd in view of the disquotational account of truth, for this of itself determines "true" uniquely; any two predicates fulfilling it are co-extensive. Indeed it over-determines truth, engendering paradox. And surely no one can gainsay disquotation, once we block the paradox. Granted, disquotation is language-bound; but we transcend those bounds by choosing our manual of translation.

RESPONSE TO GIBSON

True to form, Gibson has provided a masterly sketch of my epistemological position, even to my infrequently noted point about mutual containment. I shall just add some remarks on analyticity and my threadbare "Two Dogmas."

"If ... we conclude that moderate holism is true," Gibson writes, "... it is also very unlikely that there are analytic statements.... . As Quine has argued, any statement can be held true ... if we make drastic enough revisions to others ... " Here I would dissociate analyticity from incorrigibility. Even Carnap would give up an analytic statement. But doing so was for him a change of meaning rather than of substantive theory, and I was questioning that distinction.

Nor was I denying that there are analytic statements. In questioning the distinction I was seeking a definition. My very question was based on purported samples and a sketchy idea of what I wanted to see defined. "Truth by meaning" was the rough idea, and "No bachelor is married" was the paradigm. Further specimens, according to the literature, were the logical truths. Finally there was every layman's intuition that some truths, surely "Circles are round" and the bachelor example, are an empty matter of words.

In *Roots of Reference* decades later I even ventured a definition of analyticity meeting these conditions, but the boundary that it draws fades and dissolves once it broaches the sentences of scientific and literary theory. This disqualifies it for the philosophical use to which Carnap was putting analyticity, and Carnap was my concern in "Two Dogmas." My failure to set these matters forth was one of the shortcomings of that early paper.

Turning to Wittgenstein, Gibson finds him rejecting three of G.E. Moore's tenets: (1) that "I know" is used correctly in "I know there is an external world," (2) that knowing is a mental state, and (3) that Moore's "This is a hand" expressed a sensible proposition. For my part, I agree with Moore in accepting (1) and (3) but with Wittgenstein in rejecting (2). My objection to (2) is that knowing is a hybrid of warranted belief, which is mental, and truth, which is not.

Gibson takes up the distinction between relative and absolute foundationalism that Stroll draws in reporting on Wittgenstein's last work, *On Certainty*. The paragraphs that Gibson excerpts under each of these heads are of course free from the appeals to sense data that foundationalism used to connote, but I am surprised that there is no hint even of fallible occasion sentences conditioned to sensory stimulation. I see these as the links between science and reality.

RESPONSE TO MISCEVIC

When I represented mathematics as indispensable to science I was not alluding to its apodictic certainty. I meant the indispensability of mathematical entities and language to the formulations of natural science. Mathematics is indispensable equipment, whatever its epistemology.

I attributed the certainty of mathematical laws partly to our practice of favoring them over other tenets in the event of the experimental refutation of a theory. Of course, as Sober observes, this practice exploits mathematics in the implying of observation categoricals without any enhancement of its own credibility when the categorical is confirmed.

Proof is a luxury that the familiar mathematical truths enjoy, but they are not beyond empirical testing. I picture primitive man discovering by tests that an array of objects can be counted indifferently in any order.

On the other hand I see the logic of truth functions as mastered in learning the words that express them. Affirming a conjunction and denying a component of it is simply misuse of a word, "and" or "not", like calling a cat a dog. Correspondingly for modus ponens.

Quantification in its various guises in various languages fares perhaps like the truth functions: it is mastered in learning the words. But when we get beyond logic in the narrow sense and into the reification of classes, hence set theory, even mathematics reflects human fallibility. It was becoming clear around 1900 that all the concepts and known laws of classical mathematics could be expressed and proved strictly within pure logic and set theory. The basic law of set theory, intrinsic indeed to the very notion of a class, was that every membership condition we can formulate determines a class. But in 1901 the structure collapsed. Russell produced his paradox.

Various ways have since been devised of weakening the disastrous law so as to avert Russell's and related paradoxes, and we proceed confidently. But we were confident before, and we have had a salutary lesson in the frailty even of human mathematics.

Still Miscevic's effort in his later pages to make mathematics share the fallibility of natural science is in my view misguided: for the failure that he cites, unlike the paradoxes, turn up only when the mathematics is applied rather than pure. I have stressed the kinship of mathematics to natural science, but there is no denying the difference. Pure mathematics has the advantage of being deducible from first principles without sensory disruption.

RESPONSE TO GJELSVIK

Gjelsvik's paper concerns acrasia, the prevailing of present temptation over one's estimate of long-term benefits. It is a topic of decision theory, a domain in which I am unprepared. I will just note down some reactions from my disadvantaged vantage point.

In his early pages he draws contrasts that I fail to grasp. He contrasts "what we do is what we want most to do" with "what we want most to do is what we judge best to do." He then goes on to a second pair whose contrast is likewise unclear to me. He is contrasting, he writes, the forward connection of two concepts with their backward connection. Is it a matter of earlier and later? or cause and effect? or necessary and sufficient condition? I thought wryly of Brooklyn Bridge, which links Brooklyn with Manhattan on the one hand and Manhattan with Brooklyn on the other. A few pages later he collapses these and other such pairs as single "packages." In his place I might have begun with the packages. But I am missing something, for in footnote 5 he identifies Davidson with the backward and Bratman with the forward connection.

After a third of the paper Gjelsvik announces, to my relief, that he is switching from the mentalistic to the naturalistic mode. Murkiness clears appreciably, but I despair still of doing justice to the elaborate reasoning. Acrasia emerges as unrelated to compulsion and only remotely to rational choice, if at all, through extravagant discounting of future benefits. The extravagance of the discounting that might be required is brought out by his early example of the boring television versus the salutary run; for the discounting of future health would have to be extravagant indeed to put the television into the running with the run.

RESPONSE TO SEGAL

Segal sounds his first alert already in his eighth to tenth lines, where he writes that

in giving up on meaning it seems that we must give up also on truth and falsity. For if sentences do not have meaning, it would seem that they do not have truth conditions.

But we have Tarski's recursive or inductive definition of truth. Can anything remotely comparable be achieved for meaning? The words " ... along with other semantical phenomena" in his footnote suggest guilt by association: both truth and meaning are "semantical."

In footnote 3 he follows this up by correctly citing me as holding that two acceptable manuals of translation might translate a foreign sentence into English sentences that both translators recognize as opposite in truth value. He sees this as supporting his remark that "truth appears to follow meaning down the path to oblivion." No. The catch is that the two translations would be English sentences on whose truth values neither translator had an opinion except for agreeing that they must be opposite. Probably the foreigner was likewise open-minded about the truth value of his original sentence. Open-mindedness does not banish truth values.

Segal writes that a "fan of mentalistic semantics might [hold] that there is only one correct translation manual." He shouldn't, even by his own lights. A good manual will seldom state an integral translation for a sentence, but will support many by implication as acceptable paraphrases of one another. Two rival manuals will disagree on what set of translations of a foreign sentence they by implication support. This is where, by my lights, open-mindedness does give way to truth-valuelessness: there is no fact of the matter. Such, in partial answer to a question of Segal's, is indeterminacy as distinct from under-determination. But I anticipate.

Since a manual of translation normally supports a range of "equivalent" translations of a foreign sentence, one may wonder how to define and detect disagreement of manuals; for appeal to "equivalence" begs the whole question about meanings. In answer I have suggested applying the two manuals alternately sentence by sentence to a text that each manual separately makes coherent sense of, and seeing if the result visibly bewilders English listeners.

Segal chafes at my linguistic behaviorism. Let me then stress its limits. It disciplines data, not explanation. On the explanatory side my readers are familiar rather with my recourse to innate endowments. I cite instinct and hence natural selection to explain induction, and to explain also our innate subjective standards of perceptual similarity and their preestablished intersubjective harmony. All this is essential to language readiness. Behaviorism welcomes genetics, neurology, and innate endowments. It just excludes mentalistic explanation. It defines mentalistic concepts rather, if at all, by their observable manifestations in behavior.

Segal proposes a symmetrical stance toward physics and semantics. I agree. I see science in the broadest sense as an inclusive, loose-jointed theory of reality. Linguistics is part of it. The whole system becomes more closely knit, here and there, as science progresses. Our successes in prediction and technology assure us that we are on the right track on the whole, but some irreducibly different turn, deep in the fundamentals, might have fared as well; such is the conjecture of under-determination.

My conjecture of indeterminacy of translation is a different sort of thing. It is that in the general interlinguistic case the notion of sameness of meaning is an objectively indefinable matter of intuition. This implies that the notion of meanings as entities, however abstract, is untenable, there being no entity without identity. I reject introspection as an objective criterion, however invaluable heuristically.

RESPONSE TO ANTHONY

Her paper opens doors on another world and an alluring one: linguistics, or, as it was called in my college days, philology. In college I was torn between that domain and mathematics, in which I ended up majoring. I still idly ponder etymologies and check my speculations. I treasure translation, having lectured in six languages, and I gloat over the fifty-odd translations of books of mine into fifteen languages. So I can empathize in her tendency to miss the philosophical woods for the linguistic twigs and foliage.

She adduces her samples of current linguistics under the misconception that I, trammeled by behaviorism, underestimate the translator. She misinterprets my conjecture of the indeterminacy of translation. I postulate two ideal manuals of translation both of which translate the alien language impeccably, and I conjecture that they may, even so, sustain incompatible translations of some alien sentences on highly theoretical matters. Both manuals cover the ground to perfection, but in partly incompatible ways. They have missed nothing; the indeterminacy is objective. The point of my conjecture was a challenge to synonymy and hence to the reification of meanings, notably propositions.

In *Word and Object* I based the conjecture on the cantilever character of the scaffold of analytical hypotheses that relates the theoretical reaches of language to the linguist's evidence in verbal behavior. Conflict between the two manuals seems likely over one sentence or another on whose truth value the natives are open-minded. I see it not as a failure of translation, but as a commendable rounding out of translation beyond the bounds of actuality. It would be a case where there was no reality to uncover, but only a blank to fill.

So I have not been subjecting translation to a behaviorist onslaught. Anyway Anthony overestimates the austerity of my behaviorism. One could scarcely miss the central role that I ascribe to empathy, both in translation and in language learning. Radical translation begins with it in *Word and Object*: the linguist pictures himself in the native's place at the outset, in guessing at an observation sentence. The word "empathy" does not occur there, but it does in my later writings.

Anthony misinterprets my thought experiment in radical translation as an inquiry into the child's acquisition of language. That is quite another matter, and a fascinating one. But in my writings I have limited my concern with it to the minimum necessities of ontology, the structuring of science, and the meeting of minds regarding events in the external world: traditional concerns of philosophy.

She alights on "gavagai" as my prime example of indeterminacy of translation. But my conjecture of indeterminacy of translation concerned not terms like "gavagai" but sentences as wholes, for I follow Frege in deeming sentences the primary vehicles of meaning. The indeterminacy ascribed to "gavagai" comes under the head rather of indeterminacy of reference, or ontological relativity. This indeterminacy is proved, unlike my conjecture of the indeterminacy of holophrastic translation. Its proof is trivial and undebatable. In essence it comes down to the equivalence of "x is an F" to "the proxy of x is the proxy of an F". It does not imply the indeterminacy of holophrastic translation, because the indeterminacy of reference of a term can commonly be pinned down by the rest of the sentence.

If we take "gavagai" not as a term but as a one-word sentence, "Lo, a rabbit," it still does not illustrate the indeterminacy of holophrastic translation. It is an observation sentence, and hence, according to *Word and Object*, determinate in translation. "Lo, a rabbit," "Lo, undetached rabbit parts," and "Lo, rabbithood" are all equivalent.

RESPONSE TO HORWICH

My thought experiment in radical translation, in *Word and Object*, was meant as a challenge to the reality of propositions as meanings of cognitive sentences. Since there is no entity without identity, no reification without individuation, I needed only to challenge *sameness* of meaning of cognitive sentences. For pure sameness of meaning, unsullied by shared origins of words or mutual influences of cultures, where better to look than in radical translation?

A conclusion that Horwich draws in his last paragraph from his meticulous analysis is that it can indeed happen "that there is no determinate fact of the matter whether [some sentence] A's meaning is also possessed by A*." This sounds like me, but he sees it as disqualifying translation as a criterion of sameness of meaning rather than as challenging the reality of meanings. He feels that he has defended the reality of meanings. But subject then to what still conceivable sense of identity? There are no meanings without sameness of meaning.

The misunderstanding surfaces also where he has me assuming that the only route to meaning is via translation. This was not the idea.

Horwich seems to visualize my fiction of a manual of translation somewhat in the image of a bilingual dictionary, with stress on referential words. I picture it rather as an exhaustive account in the home language of the vocabulary and grammar of the foreign language. His conception is already hinted in his use of the definite singular, "*the* translation of", instead of "*a* translation of". The manual should afford, by implication, many equivalent translations of a sentence.

His conception is further reflected in the statements R, M, S, I, and A (not the A of my second paragraph above) that Horwich sets down for consideration in the course of his argument. Condensed, they run thus:

R: Adequate translation of a word preserves its reference.
M: Adequate translation of a word preserves its meaning.
S: Synonymous English terms are correferential.
I: Two good manuals may translate a term non-correferentially.
A: A manual that preserves assertability is adequate.

All but A, we see, are directed at terms rather than sentences.

In my work the inscrutability of reference was one thing and the indeterminacy of holophrastic translation was another. The one admitted of conclusive and trivial proof by proxy functions, hence model theory, while the other remained a plausible conjecture. Midway in his paper Horwich makes this point, and presents the argument from proxy functions in minute detail. But then he resumes holophrastic indeterminacy and R, M, S, and I and the correlation of foreign words with English words and phrases.

He claims to disprove A. He constructs a situation where the foreigner purportedly disagrees with us on an inferential relation between unassertables. Though agreeing on all assertables, he disagrees on an implication. I don't see how this could be known. He nowhere infers a falsehood from a truth by our lights, for we agree on truth values. We must just disagree on some sentence of the form "ζ implies η", where my Greek letters stand for names of sentences. But then, waiving the question how we might have learned to translate "implies", we have in the disagreement over the whole sentence "ζ implies η" itself a violation of the hypothesis that we agree on assertables. The whole sentence is assertable.

RESPONSE TO PAGIN

It gradually dawns on me in his first few pages that Pagin is not taking my profession of linguistic behaviorism seriously, try as he will. Consider, to begin with, his Basic Publicness: "What a speaker means by his words can be known by others." I agree, but what I am accepting is no more than this: "What paraphrases the speaker would be prepared to accept, in describable circumstances, can be known by others." This is behaviorally acceptable, and my intention in mentioning meaning runs no deeper.

The subtlety of the matter emerges at the foot of his third typewritten page, where he asks rhetorically whether I view the learning of language as acquisition of speech dispositions without necessarily any knowledge of what they mean. He thinks that I do not. I agree that I do not, but simply because knowing what expressions mean *consists*, for me, in being disposed to use them on appropriate occasions. Pagin is harboring a distinction that I in my behaviorism reject.

He actually goes on to quote, from my *Roots of Reference*, a passage to the effect that learning occasion sentences "amounts to learning what occasions warrant assent to the sentences, or dissent." But he seems to persist in crediting me with a mentalistic notion of meaning over and above, or under and below, my sketchy behavioral substitute.

A juxtaposition at the top of his fifth typewritten page is worth noting. Pagin quotes my "no sort of equivalence, however loose". My deprecatory "no sort" and "however loose" were by way of apology for my undefined term "equivalence", solely in need of definition. In *Pursuit of Truth* I liquidated that promissory note in terms of coherence of compared manuals. But in Pagin's next quotation, where I appeal to the intuitive notions of coherence and interchangeability, I sensed nothing to apologize for. These are behaviorally recognizable in the observable reactions of native listeners to the translations.

He quotes me as writing that "there is nothing in linguistic meaning beyond what is to be gleaned from overt behavior," but he will not take me at my word. He writes that I am, as he understands me, "speaking of ... meaning *insofar* as can be gleaned from overt behavior." He is too kind. He goes on: "There would not be much point to the statement if Quine's view was simply that there is nothing in linguistic meaning at all." He does his level best to save me from myself. Let us not forget that we still have *dispositions* to observable behavior to work from.

Despite his basic misconception of my attitude toward meaning, I find Pagin's statement of my indeterminacy thesis satisfactory. This cheers me, for it suggests that my behavioral line on meaning is, insofar, a serviceable substitute for his mentalistic line.

But the cheer does not endure. We next find Pagin eliciting and appraising an argument from the intersubjectivity of linguistic meaning to linguistic behaviorism. I should have expected the reverse: linguistic behaviorism can accommodate only intersubjective meaning. In support of linguistic behaviorism itself I expect no deductive argument. The doctrine rests only on our observation of language acquisition and the empirical implausibility of supplementary channels such as telepathy.

My difficulty in following his reasoning in the rest of the paper is due, I think, to his lingering tacit assumption that of course we are not questioning our familiar intuitive notion of meaning, and that we are only differing on whether and to what extent it can be realized in behavioral terms.

RESPONSE TO STOUTLAND

I am uncomfortable with the concept of norm and normativity that pervades Stoutland's thoughtful paper. He sees language as normative and stimulus meanings not. What is he telling us?

Perhaps the keynote to normativity is rule-governed behavior. Our behavior in a second language, learned in school in the traditional way, is indeed rule-governed. But the acquisition of a first language in a primitive community without formal teaching is apt not to be by stated rules. The language conforms to regularities that can be compiled, but so in large part does nature. Normativity must be more than conformity to laws.

The regularities in language have for the most part evolved without anyone's deliberate intervention, and most speakers conform without noticing the regularity. Schoolteachers and statutes are atypical in the long run, however welcome.

What I see as emphatically normative is the rule book itself. The handbook of English grammar emerges as a guide to "good" English, the dialect of cultivated speakers who take pride in it and even promote it. Here we have normativity *par excellence*, but still the normativity is in the handbook and the promotors and not in the good English itself, a dialect among dialects.

Stoutland finds stimulus meaning subjective, or individual, and linguistic meaning social. We can cut matters finer. The perceptual similarity that binds the subject's stimulations into a stimulus meaning is subjective, but the segregation of that particular bundle of similars under an observation sentence, and indeed that particular bundle of similars under an observation sentence, and indeed that particular observation sentence, is socially imposed. I do not agree that "any disposition to respond to stimuli is as correct as any other." Not in the language game.

He speculates on why I call stimulus meaning. It was just that I saw it as the naturalistic analogue of sensory meaning. He goes on to say that "all sentences have stimulus meanings." I applied the notion only to observation sentences, and I see no application beyond occasion sentences.

Stoutland rightly notes the vagueness of boundaries of a language, or linguistic community, and the doubtfulness of full linguistic homogeneity short of an idiolect. Vagueness permeates language and language about language in varying degrees, and demands treatment where it matters. Vagueness of the limits of the intended linguistic community does matter to the notion of observation sentence, and I deal with it by taking the community as a parameter, varying from a broad society to a clutch of specialists according to our purpose. Vagueness is emphatic and deliberate in my standard of adequacy of translation: just smooth dialogue and successful negotiation.

RESPONSE TO ORENSTEIN

I can sympathize somewhat with Orenstein's wish to narrow the gap between formalized logic and ordinary language. I have seen elaborate resort to symbolic logic in marine biology where ordinary language would have been more perspicuous and equally brief. I have heard an arch reference to model theory when all that was in point was a correlation. See my *Quiddities*, under "Mathematosis."

But values bi- or trifurcate, and can be cultivated separately. My elimination of singular terms, which Orenstein deplores, shows that old perplexities over irreferential singular terms are the product of idioms that are in principle eliminable. This is philosophical progress. Also there was aesthetic progress in the resulting simplification of logical theory, and a metamathematical gain in simplifying proof theory.

But just as the reduction of classical mathematics to set theory does not enable us in practice to dispense with numerals, differential operators, and other mathematical hieroglyphs, so the elimination of singular terms does not enable us in practice to dispense with singular terms. It was never meant to, for that would have been unthinkable. Computation and research in mathematics would be paralyzed without singular terms – numerals, to begin with – to substitute for variables. All results of substitution could in principle be got via the contextual definitions, but too circuitously for practical purposes.

So we have here two conflicting interests: elegant simplicity on the one hand, utility on the other. Definition is the solvent, affording us the best of both worlds. Let us not see the two as a dilemma; we can live it up in both. Even a third or fourth is not excluded. Predicate-functor logic does without singular terms even to the extent of variables; but it and the familiar quantificational logic are intertranslatable. The latter fits our intuitions better, but the other is sufficiently unlike to afford a philosophically interesting perspective, particularly on the nature and function of variables themselves.

In its relation to traditional reasoning and ordinary language, moreover, modern logic must be seen as much more than a servile formalization. In its penetration of polyadic predicates it raised logic to a metamathematical level, and in its quantifiers and variables it both afforded an explicit standard of ontic commitment and revealed the utility of reification in the structuring of science. Its utility hinges on the deductive strength of the universally quantified conditional as over against the underlying truth-functional "if-then".

Orenstein questions the quantificational criterion of reification and claims rather that something exists just in case a singular term names it. But this disqualifies the mathematics of real numbers that is so crucial to natural science. Most of the irrational reals are unnamed and indeed individually unspecifiable, as Cantor made clear. Yet much of science would be immobilized without them. Natural science is deeply committed to abstract objects, named and nameless, though only extensional ones. Plato's beard is trimmed but otherwise intact.

A difficulty that Orenstein finds in quantification as attesting to existence is that generalization of a quantification from an irreferential singular term or instantiation by such a term issues in paradox. He is venturing on free logic without heeding its rules: no generalization or instantiation without existence, proved or premissed.

Writing of the negation "¬x is human", Orenstein puzzlingly represents Tarski as according it existential import. Do open sentences have existential import? Of course variables do, in the sense that all their values exist, even in free logic.

Then he proceeds to contrast the "Lesniewskian" or "Terminist" view, which "accords existential commitment to atomic sentences but not to their negations." Thus "Alex is human" implies the existence of Alex, and "¬Alex is human" implies no existence. But I see no disagreement with Tarski. Orenstein has switched in midstream from open sentences to closed. Orenstein's ensuing contrast between two schools of thought does not speak to me.

RESPONSE TO TERENCE PARSONS

Our problem is where to accommodate the ever-present vagueness of empirical terms. I limit vagueness to language, leaving reality unsullied. Some of us acquiesce rather in a fuzzy world, subject perhaps to a fuzzy logic. Parsons surrenders the last bastion of rigid objectivity, identity itself. Awed by the gulf that yawns between us, I shall begin with an independent account of my position.

I am wedded to classical first-order predicate logic, couched in truth functions and quantification. It is linked to general language by its schematic letters for predicates, and it is linked to reality by its variables, which take all objects, specifiable and unspecifiable, as their values.

As a foundation for classical mathematics all that is needed, atop this logic, is the two-place set-theoretic predicate "$\in$" of class membership. The identity predicate "$=$" is then at hand as well: we define "$x = y$" in the familiar fashion "$\forall z(x \in z \cdot\equiv\cdot y \in z)$". Other cognitive needs, beyond mathematics, are met by adding further predicates without preassigned limit. No special provision is needed for proper names, nor for term functors such as "plus", "cosine of", "altitude of", "father of", for these are all reducible to predicates through singular description. They are indeed indispensable as a convenience, but they are dispensable in principle.

The limitless fund of extra-mathematical predicates is acquired largely by ostension. Others, such as "electron",[1] are learned merely through explanations that instill adequate understanding without either ostension or strict definition. Sometimes we fit a new predicate into our vocabulary by mere imitation and guesswork. It would be quixotic to try to eliminate vagueness from the resulting language, but I find it gratuitous to project it beyond language. Maximal simplicity compatible with observation is a precept of science, and it is counter to that precept to let the vagueness of language seep into its ontology.

Though uneliminable, the ubiquitous vagueness of our predicates for commonplace physical objects is sporadically reducible in various degrees. Meanwhile it can be accounted for and accommodated without projecting it into our logic and ontology. Parsons has illustrated what it would do to our logic.

My ontology includes all three of the spatiotemporal objects that Parsons calls ships in his example, and it includes all the innumerable microscopically diverse approximations to Mrs. Parsons. My way around his paradoxes is vagueness not of ships or wives, but of reference into a tight-packed and mostly nameless ontology of intrinsically precise entities. "Ship" denotes only one of Parsons's so-called ships, and "Mrs. Parsons" designates just an unspecifiable one of her continuum of mutual approximations. Parsons is confirmed in his monogamy. We merely have not decided which of the three quasi-ships to call a ship, and we could never hope to pick Mrs. Parsons out from among her approximations. This vagueness of reference is the place of Parsons's truth-value gaps, and welcome. But this is semantics, the arena of vagueness, and not natural science, the arena of elusive and recalcitrant but clean-cut reality. For science I assume as values of its variables indiscriminately the ship and the quasi-ships, also Mrs. Parsons and her indiscriminable deviants, indeed all occupants of space-time, specifiable and unspecifiable, and all classes thereof, classes of classes, and so on up. Each is uniquely identical to itself.

---

[1] I apply the word "predicate" to common nouns as well as to verbs and adjectives, for our predicate letters represent them indiscriminately.

On the semantic side we tolerate the vagueness until it matters, and then we make an *ad hoc* ruling or even a lasting one at just the crucial point. An example can be imagined in the case of the vague term "person" in the course of the controversy over abortion. The supreme court might slightly reduce the vagueness by ruling that the fetus becomes a person at the end of its fourth month. We enhance precision where it matters.

A further species of semantic indeterminacy, independent of vagueness, sprouted in my writings in 1968 under the head of indeterminacy or inscrutability of reference or ontological relativity. I put it in terms of proxy functions. Putnam has since put it in terms of models. What it shows is that any ontology in an expressible one-to-one correlation with our own is equally supported by sensory input. Thanks to ostension, this freedom of reference is no hindrance to communication. Nor does it destabilize science, thanks to the isomorphism; for the global shift to proxies is obliterated by a correspondingly global reinterpretation of our words.

I turn in conclusion to a few technical details in Parson's paper. At the very end he explains, to my pleasant surprise, that by properties he simply means classes, identical when coextensive. But then seven pages earlier he need not have written three times "the property's extension." For him the property *is* its extension.

In describing his set D of "ontons" he writes that "an image i pictures its object as having a property p if every onton in i is in ... p." (My dots supplant a redundant "extension of".) But then, if I follow him, he is not distinguishing between having a property (being a member of a class) and being included in the property (being a subclass of the class).

A few pages later he formulates the condition under which "a formula will be capable of expressing a property." He renders the condition as the negation of a quantification of a five-clause conjunction. The conjunction is doubly redundant. Its third and fifth clauses are implied respectively by its second and fourth. What the whole condition boils down to is that if "$x = y$" lacks truth value and $\Phi(x)$ then $\Phi(y)$.

He presents seventeen rules for his three-valued logic of truth, falsity, and non-falsity. I have trouble with his rule for conjoining two non-falsities. It prescribes non-falsity of the conjunction, but I wonder about this where the non-falsities are mutual negations.

RESPONSE TO WOODRUFF

In my response to Parsons I expressed myself on his and Woodruff's joint project, a theory of indeterminacy of identity and a logic to accommodate it. Against it I launched what Woodruff calls the Basic Objection to three-valued logic and encapsulates in a word: "messiness." But I still want to respond to some points in Woodruff's paper.

He sees some properties as having sets as their extensions, and other properties as not, on pain of Russell's and other paradoxes. He evidently does not appreciate that properties face paradoxes parallel to the paradoxes of sets. Just as there can be no set of all and only the sets that are not members of themselves, so there can be no property of not being a property of self. Properties offer no escape from paradox. Their only difference from sets is an infirmity, namely failure of extensionality and consequent dimness of individuation. Hence my repudiation of them.

Woodruff's definition of transparency puzzled me. He defines ϕ as transparent if, when "$\phi(a)$" is definitely true and "$a = b$" is not definitely false, "$\phi(b)$" is not definitely false. I fail to picture failure of such transparency. He offers a counterexample, but I still founder. My intuitions are halting in three-valued logic. I am more pleased than surprised that two-valued logic has reigned so long.

Woodruff argues, in support of his and Parsons' thesis of indeterminacy of identity, that vagueness of properties implies indeterminacy of identity of properties. I agree, and I find the point important. It speaks for my policy of allocating vagueness to predicates and other expressions, not to objects, concrete or abstract; to language, not to reality.

Granted, I have rejected properties anyway, but let us not confuse the two issues; for "properties" read "classes".

I came in seeing the thesis of indeterminacy of identity as the wanton fogging up of a crystal-clear concept. I go out prizing it as a further support, by *reductio ad absurdum*, of my conception of language as vague discourse about clean-cut reality.

RESPONSE TO NEALE

As Neale brings out in his admirable analysis, *although Arthur Smullyan's observations did not exonerate modal logic, ultimately they cleared it* * of any suspicion of inconsistency raised by my slingshot argument. This applies also to my variant of 1960 in terms of Kronecker's "characteristic function." I would blush if it were not for my prestigious fellow slingshooters Church and Gödel.

But my initial and enduring disdain of modal logic is due rather to the obscurities and laborious complexities that arise from its intensionality. In part, as Neale brings out, these complexities come of an impairment of contextual definition. We are compelled now and again to expand the context into primitive notation and check whether a rule applicable to primitive singular terms applies to the fac- or fake-simile.

In justice I must point out, however, that Russell's contextual definition of singular description raises a contextual problem even in extensional logic. It has to do with Russell's cumbersome scope-marking prefixes and his convention of leaving them tacit when the scope is minimal. Trouble can arise when a contextual definition of something else shortens a context of a singular description. To avoid fallacy on this score we must make sure that what is prima facie the shortest context of a description is still shortest in primitive notation. This imposes on us the same sort of burden that I ascribed to modal logic in my preceding paragraph.

The problem deepens when we reflect that Russell's theory of descriptions is meant not just for formal systems with a fixed primitive lexicon, but for the

---

* The italicized words were added in proofs by the editors as Professor Quine was not available for clarification. The original words were: "Arthur Smullyan cleared modal logic".

philosophical clarification of ordinary usage that does not recognise any fixed hierarchy of definitions on fixed primitive foundations. Prima facie "The king of Ohio is innocent" comes out false, since the scope is minimal and there is no such king. But if we rephrase it as "The king of Ohio is not guilty" it comes out true, being the negation of the minimal scope "The king of Ohio is guilty".

The moral of this digression is that we must drop Russell's convention and use his scope markers when we apply his theory of descriptions beyond fully formalized systems. Still, they are cumbersome. We are tempted back with Hilbert to the original treatment of "$(\iota x)Fx$" by Peano, namely as primitive and devoid of laws except contingently on a lemma or premise of unique existence.

It is with a sigh of relief that one moves from first-order logic into set theory, where singular description can be defined directly rather than contextually, and with absurd simplicity: the sole member of a class α, if any, is Uα, which is to say the union of α, which is to say the class of all members of members of α. This bewilderingly brief definition "Uα" depends on my identification of an individual with its unit class, or singleton. The result is that Uα is the member of α if there is only one, and in other cases Uα has its usual uses.

Another digression is prompted by Neale's mention of my elimination of singular terms other than variables by construing them as singular descriptions and then defining them away in Russell's way. The elimination brings out a startling contrast between the theory of a formalism and the use of it. For the theory we prize simplicity, and this simplification bypasses considerable apparatus. In practice, however, the elimination of singular terms would paralyze the algorithms of mathematics, whose very essence is the substitution of complex singular terms for variables. Such, then, is the boon of contextual definition, in reconciling theory and practice.

In closing, I return to my disdain for modal logic. There is no place in my philosophy for metaphysical or physical necessity in general, but there is indeed logical truth, and there are the provisional and transitory analogues in the way of deducibility within one or another system. Insofar, then, I find modal logic applicable. My complaint is rather its departure from extensionalism, whose clarity and simplicity I so highly prize.

Another domain that is notoriously intensional, and one that is indispensable, is that of the propositional attitudes. My recourse both in the propositional attitudes and in logical necessity has been to what I call semantic ascent. I shift from the use of crucial sentences to discourse about them, as Neale and my other readers are well aware.

RESPONSE TO RAY

We have here a meticulous inquiry into vague and elusive idioms of necessity, possibility, and contrafactual conditions. In particular I applaud Ray's concluding remarks on necessity. Instead of speciating it into physical necessity, conceptual necessity, analytic necessity, and whatever else, he treats it as a single idiom but an indexical or token-reflexive one like the personal and demonstrative pronouns: each occurrence of it depends for its interpretation on its textual or circumstantial context. The same is conspicuously true of contrafactual conditionals. In *From Stimulus to Science* I likewise declared necessity indexical, along with propositional attitudes *de re*.

In discussing examples of purported referential opacity, mine and others, Ray notes that some are disqualified by Smullyan's appeal to distinctions of scope of singular descriptions. But an example that he singles out as withstanding Smullyan is his Example One, in which "$H = P$" is true but not necessarily. I find no gloss for "H" or "P", but surely they are "Hesperus" and "Phosphorus", for "Evening Star" and "Morning Star". As simple proper names they circumvent Smullyan's appeal to singular descriptions, and nothing could be more clearly a matter of contingent empirical discovery than "$H = P$". I am with Ray in wondering at the insistence by some modal logicians that true identities are necessarily true.

I find more fault with Gibbard's example of Goliath and the lump of clay than Ray seems to find, though he finds some. Goliath is a clay statue, and Gibbard distinguishes it from its constituent clay on the grounds that the clay would retain its identity if compressed to a ball, while Goliath would not. To ponder this topic we must come to terms, some terms or other, with time, tense, and objectitude. If we construe physical objects in the way best suited to quantificational logic, namely as spatiotemporally four-dimensional, then the statue and the lump of clay are *sub specie aeternitatis* not identical. The statue is a temporally proper part of the lump of clay. The lump is older than the statue and, if eventually compressed, it outlasts the statue. If on the other hand we cleave to our vernacular of tense, we can still say no better than that the statue and the lump of clay are now identical but did not use to be, and would cease to if compressed. No paradox, no puzzle, no failure of pure referentiality.

RESPONSE TO RECANATI

Recanati's sprightly first page recalled Carnap, in whose writing I first encountered "autonomous" for the use of a word to designate itself. The practice is vivid and convenient, but we do well to signal it, as Recanati remarks, with a pair of quotation marks; for confusion of the use of expressions with the mention of them can run deep. To take the extreme case, nobody prey to that confusion could follow the proof of Gödel's great incompleteness theorem.

In unformalized usage there can be gradations from use to mention due to what Recanati calls reflections from context. His perceptive account of these matters is crowned with this gem, nearly enough: "The last word of this sentence designates Cicero."

In his §1, 4 he applies my notion of the purely referential occurrence of a singular term, and along with this notion he uses a purportedly narrower one of *transparency* that is meant to take the reflections from context into account. It is a distinction that recurs throughout his essay, but his definitions are not readily distinguished:

[A] singular term is used purely referentially iff its semantic contribution is its referent, and nothing else ... [An] occurrence is transparent iff its contribution in the broad sense is its referent, and nothing else.

He cites an example from Kit Fine, but it does not help me.

Moving to propositional attitudes, Recanati takes up a purported distinction between *relational* and *notional* belief that I propounded in 1956. I repudiated it in the light of Robert Sleigh's argument, twelve years later, featuring the shortest spy. But Recanati thinks I erred in repudiating it.

What was at issue, back then, was the contrast between just believing there are spies and believing of someone that he is a spy. The natural answer is that in the second case we can specify a spy. But, Sleigh protests, we believe there is a shortest spy (dismissing the unlikely case of a tie), so we can specify him by that description. If the answer does not satisfy us, what is missing? Name, address, photograph, social-security number—one or another of these might serve, depending on the circumstances. Merely knowing something peculiar to him does not suffice, for we already know that he is *the* shortest spy. Knowing who or what someone or something is makes practical sense only relative to background circumstances, and accordingly should be banished from regimented science to the limbo of indexicals along with the demonstratives and personal pronouns. Consequently I have cast thither the propositional attitudes *de re*, and retained only the *de dicto* as self-contained discourse.

Recanati's argument to the contrary, defending my original position, occupies his eloquent five-page §2.1. I grant that the crucial notion of knowing who or what someone or something is is a natural one. I rested placidly with it for those intervening twelve years, occupied with other matters. But in §2.1 his one gesture toward an explicit clarification of the challenged concept is his cryptic Criterion C, which appears twice:

There is an x such that the belief is true iff ... x ...

If this is just meant to require that the belief contain some singular term, represented here by "x", then "the shortest spy" is not excluded. If it is meant in some other way, I can't think what.

Recanati's progress through the rest of the essay depends heavily on his restoration of the pre-Sleigh innocence that I no longer enjoy, so there is little point in my pursuing his further developments along this line. Still there are occasional offshoots that catch my eye.

One of them has to do with Mill's characterization of proper names as non-connotative. I have no quarrel with that. My regimentation of proper names as predicates then just widens the category of predicates to include some non-connotative cases.

He credits me with the word "multigrade" in §2.2, but seems to misconstrue it. I meant "-grade" to suggest *degree* of a relation: dyadic, triadic, etc. Perhaps "among", seen as a relative term, expresses a multigrade relation.

He then moves to a novel combination of quantification and tense, in which an existential quantification holds true only for the duration of its instance. He accommodates this departure with an auxiliary phrase "it will be the case that" or "is" or "was the case that". I respect this as a linguist's convenience in treating of natural language, but let me stress the difference. In the all-purpose interpretation of the logic of quantification the variables range over all things, some of which long since ceased to exist and some of which will some day happen to exist though not as yet identified or even causally determined. Temporal distinctions are left to the extralogical vocabulary of application along with details of color, mass, and the rest. Time is thus treated on a par with place. For general purposes the resulting simplicity is richly rewarding, quite apart from relativity physics.

INDEX OF NAMES

Boston Studies in the Philosophy of Science

Editor: Robert S. Cohen, *Boston University*

1. M.W. Wartofsky (ed.): *Proceedings of the Boston Colloquium for the Philosophy of Science, 1961/1962.* [Synthese Library 6] 1963   ISBN 90-277-0021-4
2. R.S. Cohen and M.W. Wartofsky (eds.): *Proceedings of the Boston Colloquium for the Philosophy of Science, 1962/1964.* In Honor of P. Frank. [Synthese Library 10] 1965
    ISBN 90-277-9004-0
3. R.S. Cohen and M.W. Wartofsky (eds.): *Proceedings of the Boston Colloquium for the Philosophy of Science, 1964/1966.* In Memory of Norwood Russell Hanson. [Synthese Library 14] 1967   ISBN 90-277-0013-3
4. R.S. Cohen and M.W. Wartofsky (eds.): *Proceedings of the Boston Colloquium for the Philosophy of Science, 1966/1968.* [Synthese Library 18] 1969   ISBN 90-277-0014-1
5. R.S. Cohen and M.W. Wartofsky (eds.): *Proceedings of the Boston Colloquium for the Philosophy of Science, 1966/1968.* [Synthese Library 19] 1969   ISBN 90-277-0015-X
6. R.S. Cohen and R.J. Seeger (eds.): *Ernst Mach, Physicist and Philosopher.* [Synthese Library 27] 1970   ISBN 90-277-0016-8
7. M. Čapek: *Bergson and Modern Physics.* A Reinterpretation and Re-evaluation. [Synthese Library 37] 1971   ISBN 90-277-0186-5
8. R.C. Buck and R.S. Cohen (eds.): *PSA 1970.* Proceedings of the 2nd Biennial Meeting of the Philosophy and Science Association (Boston, Fall 1970). In Memory of Rudolf Carnap. [Synthese Library 39] 1971   ISBN 90-277-0187-3; Pb 90-277-0309-4
9. A.A. Zinov'ev: *Foundations of the Logical Theory of Scientific Knowledge (Complex Logic).* Translated from Russian. Revised and enlarged English Edition, with an Appendix by G.A. Smirnov, E.A. Sidorenko, A.M. Fedina and L.A. Bobrova. [Synthese Library 46] 1973
    ISBN 90-277-0193-8; Pb 90-277-0324-8
10. L. Tondl: *Scientific Procedures.* A Contribution Concerning the Methodological Problems of Scientific Concepts and Scientific Explanation.Translated from Czech. [Synthese Library 47] 1973   ISBN 90-277-0147-4; Pb 90-277-0323-X
11. R.J. Seeger and R.S. Cohen (eds.): *Philosophical Foundations of Science.* Proceedings of Section L, 1969, American Association for the Advancement of Science. [Synthese Library 58] 1974   ISBN 90-277-0390-6; Pb 90-277-0376-0
12. A. Gr~nbaum: *Philosophical Problems of Space and Times.* 2nd enlarged ed. [Synthese Library 55] 1973   ISBN 90-277-0357-4; Pb 90-277-0358-2
13. R.S. Cohen and M.W. Wartofsky (eds.): *Logical and Epistemological Studies in Contemporary Physics.* Proceedings of the Boston Colloquium for the Philosophy of Science, 1969/72, Part I. [Synthese Library 59] 1974   ISBN 90-277-0391-4; Pb 90-277-0377-9
14. R.S. Cohen and M.W. Wartofsky (eds.): *Methodological and Historical Essays in the Natural and Social Sciences.* Proceedings of the Boston Colloquium for the Philosophy of Science, 1969/72, Part II. [Synthese Library 60] 1974   ISBN 90-277-0392-2; Pb 90-277-0378-7
15. R.S. Cohen, J.J. Stachel and M.W. Wartofsky (eds.): *For Dirk Struik.* Scientific, Historical and Political Essays in Honor of Dirk J. Struik. [Synthese Library 61] 1974
    ISBN 90-277-0393-0; Pb 90-277-0379-5
16. N. Geschwind: *Selected Papers on Language and the Brains.* [Synthese Library 68] 1974
    ISBN 90-277-0262-4; Pb 90-277-0263-2
17. B.G. Kuznetsov: *Reason and Being.* Translated from Russian. Edited by C.R. Fawcett and R.S. Cohen. 1987   ISBN 90-277-2181-5

Boston Studies in the Philosophy of Science

18. P. Mittelstaedt: *Philosophical Problems of Modern Physics.* Translated from the revised 4th German edition by W. Riemer and edited by R.S. Cohen. [Synthese Library 95] 1976
ISBN 90-277-0285-3; Pb 90-277-0506-2

19. H. Mehlberg: *Time, Causality, and the Quantum Theory.* Studies in the Philosophy of Science. Vol. I: *Essay on the Causal Theory of Time.* Vol. II: *Time in a Quantized Universe.* Translated from French. Edited by R.S. Cohen. 1980   Vol. I: ISBN 90-277-0721-9; Pb 90-277-1074-0
Vol. II: ISBN 90-277-1075-9; Pb 90-277-1076-7

20. K.F. Schaffner and R.S. Cohen (eds.): *PSA 1972.* Proceedings of the 3rd Biennial Meeting of the Philosophy of Science Association (Lansing, Michigan, Fall 1972). [Synthese Library 64] 1974   ISBN 90-277-0408-2; Pb 90-277-0409-0

21. R.S. Cohen and J.J. Stachel (eds.): *Selected Papers of LÇon Rosenfeld.* [Synthese Library 100] 1979   ISBN 90-277-0651-4; Pb 90-277-0652-2

22. M. Čapek (ed.): *The Concepts of Space and Time.* Their Structure and Their Development. [Synthese Library 74] 1976   ISBN 90-277-0355-8; Pb 90-277-0375-2

23. M. Grene: *The Understanding of Nature.* Essays in the Philosophy of Biology. [Synthese Library 66] 1974   ISBN 90-277-0462-7; Pb 90-277-0463-5

24. D. Ihde: *Technics and Praxis.* A Philosophy of Technology. [Synthese Library 130] 1979
ISBN 90-277-0953-X; Pb 90-277-0954-8

25. J. Hintikka and U. Remes: *The Method of Analysis.* Its Geometrical Origin and Its General Significance. [Synthese Library 75] 1974   ISBN 90-277-0532-1; Pb 90-277-0543-7

26. J.E. Murdoch and E.D. Sylla (eds.): *The Cultural Context of Medieval Learning.* Proceedings of the First International Colloquium on Philosophy, Science, and Theology in the Middle Ages, 1973. [Synthese Library 76] 1975   ISBN 90-277-0560-7; Pb 90-277-0587-9

27. M. Grene and E. Mendelsohn (eds.): *Topics in the Philosophy of Biology.* [Synthese Library 84] 1976   ISBN 90-277-0595-X; Pb 90-277-0596-8

28. J. Agassi: *Science in Flux.* [Synthese Library 80] 1975
ISBN 90-277-0584-4; Pb 90-277-0612-3

29. J.J. Wiatr (ed.): *Polish Essays in the Methodology of the Social Sciences.* [Synthese Library 131] 1979   ISBN 90-277-0723-5; Pb 90-277-0956-4

30. P. Janich: *Protophysics of Time.* Constructive Foundation and History of Time Measurement. Translated from German. 1985   ISBN 90-277-0724-3

31. R.S. Cohen and M.W. Wartofsky (eds.): *Language, Logic, and Method.* 1983
ISBN 90-277-0725-1

32. R.S. Cohen, C.A. Hooker, A.C. Michalos and J.W. van Evra (eds.): *PSA 1974.* Proceedings of the 4th Biennial Meeting of the Philosophy of Science Association. [Synthese Library 101] 1976   ISBN 90-277-0647-6; Pb 90-277-0648-4

33. G. Holton and W.A. Blanpied (eds.): *Science and Its Public.* The Changing Relationship. [Synthese Library 96] 1976   ISBN 90-277-0657-3; Pb 90-277-0658-1

34. M.D. Grmek, R.S. Cohen and G. Cimino (eds.): *On Scientific Discovery.* The 1977 Erice Lectures. 1981   ISBN 90-277-1122-4; Pb 90-277-1123-2

35. S. Amsterdamski: *Between Experience and Metaphysics.* Philosophical Problems of the Evolution of Science. Translated from Polish. [Synthese Library 77] 1975
ISBN 90-277-0568-2; Pb 90-277-0580-1

36. M. Marković and G. Petrović (eds.): *Praxis.* Yugoslav Essays in the Philosophy and Methodology of the Social Sciences. [Synthese Library 134] 1979
ISBN 90-277-0727-8; Pb 90-277-0968-8

Boston Studies in the Philosophy of Science

37. H. von Helmholtz: *Epistemological Writings*. The Paul Hertz / Moritz Schlick Centenary Edition of 1921. Translated from German by M.F. Lowe. Edited with an Introduction and Bibliography by R.S. Cohen and Y. Elkana. [Synthese Library 79] 1977
ISBN 90-277-0290-X; Pb 90-277-0582-8
38. R.M. Martin: *Pragmatics, Truth and Language*. 1979
ISBN 90-277-0992-0; Pb 90-277-0993-9
39. R.S. Cohen, P.K. Feyerabend and M.W. Wartofsky (eds.): *Essays in Memory of Imre Lakatos*. [Synthese Library 99] 1976   ISBN 90-277-0654-9; Pb 90-277-0655-7
40. Not published.
41. Not published.
42. H.R. Maturana and F.J. Varela: *Autopoiesis and Cognition*. The Realization of the Living. With a Preface to "Autopoiesis' by S. Beer. 1980   ISBN 90-277-1015-5; Pb 90-277-1016-3
43. A. Kasher (ed.): *Language in Focus: Foundations, Methods and Systems*. Essays in Memory of Yehoshua Bar-Hillel. [Synthese Library 89] 1976
ISBN 90-277-0644-1; Pb 90-277-0645-X
44. T.D. Thao: *Investigations into the Origin of Language and Consciousness*. 1984
ISBN 90-277-0827-4
45. F.G.-I. Nagasaka (ed.): *Japanese Studies in the Philosophy of Science*. 1997
ISBN 0-7923-4781-1
46. P.L. Kapitza: *Experiment, Theory, Practice*. Articles and Addresses. Edited by R.S. Cohen. 1980   ISBN 90-277-1061-9; Pb 90-277-1062-7
47. M.L. Dalla Chiara (ed.): *Italian Studies in the Philosophy of Science*. 1981
ISBN 90-277-0735-9; Pb 90-277-1073-2
48. M.W. Wartofsky: *Models*. Representation and the Scientific Understanding. [Synthese Library 129] 1979   ISBN 90-277-0736-7; Pb 90-277-0947-5
49. T.D. Thao: *Phenomenology and Dialectical Materialism*. Edited by R.S. Cohen. 1986
ISBN 90-277-0737-5
50. Y. Fried and J. Agassi: *Paranoia*. A Study in Diagnosis. [Synthese Library 102] 1976
ISBN 90-277-0704-9; Pb 90-277-0705-7
51. K.H. Wolff: *Surrender and Cath*. Experience and Inquiry Today. [Synthese Library 105] 1976
ISBN 90-277-0758-8; Pb 90-277-0765-0
52. K. Kosík: *Dialectics of the Concrete*. A Study on Problems of Man and World. 1976
ISBN 90-277-0761-8; Pb 90-277-0764-2
53. N. Goodman: *The Structure of Appearance*. [Synthese Library 107] 1977
ISBN 90-277-0773-1; Pb 90-277-0774-X
54. H.A. Simon: *Models of Discovery* and Other Topics in the Methods of Science. [Synthese Library 114] 1977   ISBN 90-277-0812-6; Pb 90-277-0858-4
55. M. Lazerowitz: *The Language of Philosophy*. Freud and Wittgenstein. [Synthese Library 117] 1977   ISBN 90-277-0826-6; Pb 90-277-0862-2
56. T. Nickles (ed.): *Scientific Discovery, Logic, and Rationality*. 1980
ISBN 90-277-1069-4; Pb 90-277-1070-8
57. J. Margolis: *Persons and Mind*. The Prospects of Nonreductive Materialism. [Synthese Library 121] 1978   ISBN 90-277-0854-1; Pb 90-277-0863-0
58. G. Radnitzky and G. Andersson (eds.): *Progress and Rationality in Science*. [Synthese Library 125] 1978   ISBN 90-277-0921-1; Pb 90-277-0922-X
59. G. Radnitzky and G. Andersson (eds.): *The Structure and Development of Science*. [Synthese Library 136] 1979   ISBN 90-277-0994-7; Pb 90-277-0995-5

Boston Studies in the Philosophy of Science

60. T. Nickles (ed.): *Scientific Discovery*. Case Studies. 1980
ISBN 90-277-1092-9; Pb 90-277-1093-7

61. M.A. Finocchiaro: *Galileo and the Art of Reasoning*. Rhetorical Foundation of Logic and Scientific Method. 1980   ISBN 90-277-1094-5; Pb 90-277-1095-3

62. W.A. Wallace: *Prelude to Galileo*. Essays on Medieval and 16th-Century Sources of Galileo's Thought. 1981   ISBN 90-277-1215-8; Pb 90-277-1216-6

63. F. Rapp: *Analytical Philosophy of Technology*. Translated from German. 1981
ISBN 90-277-1221-2; Pb 90-277-1222-0

64. R.S. Cohen and M.W. Wartofsky (eds.): *Hegel and the Sciences*. 1984   ISBN 90-277-0726-X

65. J. Agassi: *Science and Society*. Studies in the Sociology of Science. 1981
ISBN 90-277-1244-1; Pb 90-277-1245-X

66. L. Tondl: *Problems of Semantics*. A Contribution to the Analysis of the Language of Science. Translated from Czech. 1981   ISBN 90-277-0148-2; Pb 90-277-0316-7

67. J. Agassi and R.S. Cohen (eds.): *Scientific Philosophy Today*. Essays in Honor of Mario Bunge. 1982   ISBN 90-277-1262-X; Pb 90-277-1263-8

68. W. Krajewski (ed.): *Polish Essays in the Philosophy of the Natural Sciences*. Translated from Polish and edited by R.S. Cohen and C.R. Fawcett. 1982
ISBN 90-277-1286-7; Pb 90-277-1287-5

69. J.H. Fetzer: *Scientific Knowledge*. Causation, Explanation and Corroboration. 1981
ISBN 90-277-1335-9; Pb 90-277-1336-7

70. S. Grossberg: *Studies of Mind and Brain*. Neural Principles of Learning, Perception, Development, Cognition, and Motor Control. 1982   ISBN 90-277-1359-6; Pb 90-277-1360-X

71. R.S. Cohen and M.W. Wartofsky (eds.): *Epistemology, Methodology, and the Social Sciences*. 1983.   ISBN 90-277-1454-1

72. K. Berka: *Measurement*. Its Concepts, Theories and Problems. Translated from Czech. 1983
ISBN 90-277-1416-9

73. G.L. Pandit: *The Structure and Growth of Scientific Knowledge*. A Study in the Methodology of Epistemic Appraisal. 1983   ISBN 90-277-1434-7

74. A.A. Zinov'ev: *Logical Physics*. Translated from Russian. Edited by R.S. Cohen. 1983
[*see also* Volume 9]   ISBN 90-277-0734-0

75. G-G. Granger: *Formal Thought and the Sciences of Man*. Translated from French. With and Introduction by A. Rosenberg. 1983   ISBN 90-277-1524-6

76. R.S. Cohen and L. Laudan (eds.): *Physics, Philosophy and Psychoanalysis*. Essays in Honor of Adolf Gr~nbaum. 1983   ISBN 90-277-1533-5

77. G. Bhme, W. van den Daele, R. Hohlfeld, W. Krohn and W. Schîfer: *Finalization in Science*. The Social Orientation of Scientific Progress. Translated from German. Edited by W. Schîfer. 1983   ISBN 90-277-1549-1

78. D. Shapere: *Reason and the Search for Knowledge*. Investigations in the Philosophy of Science. 1984   ISBN 90-277-1551-3; Pb 90-277-1641-2

79. G. Andersson (ed.): *Rationality in Science and Politics*. Translated from German. 1984
ISBN 90-277-1575-0; Pb 90-277-1953-5

80. P.T. Durbin and F. Rapp (eds.): *Philosophy and Technology*. [*Also* Philosophy and Technology Series, Vol. 1] 1983   ISBN 90-277-1576-9

81. M. Marković: *Dialectical Theory of Meaning*. Translated from Serbo-Croat. 1984
ISBN 90-277-1596-3

82. R.S. Cohen and M.W. Wartofsky (eds.): *Physical Sciences and History of Physics*. 1984.
ISBN 90-277-1615-3

Boston Studies in the Philosophy of Science

Boston Studies in the Philosophy of Science

Boston Studies in the Philosophy of Science

Boston Studies in the Philosophy of Science

Boston Studies in the Philosophy of Science

171. M.A. Grodin (ed.): *Meta Medical Ethics*: The Philosophical Foundations of Bioethics. 1995
ISBN 0-7923-3344-6

172. S. Ramirez and R.S. Cohen (eds.): *Mexican Studies in the History and Philosophy of Science.*
1995 ISBN 0-7923-3462-0

173. C. Dilworth: *The Metaphysics of Science.* An Account of Modern Science in Terms of Principles, Laws and Theories. 1995 ISBN 0-7923-3693-3

174. J. Blackmore: *Ludwig Boltzmann, His Later Life and Philosophy, 1900–1906* Book Two: The Philosopher. 1995 ISBN 0-7923-3464-7

175. P. Damerow: *Abstraction and Representation.* Essays on the Cultural Evolution of Thinking.
1996 ISBN 0-7923-3816-2

176. M.S. Macrakis: *Scarcity's Ways: The Origins of Capital.* A Critical Essay on Thermodynamics, Statistical Mechanics and Economics. 1997 ISBN 0-7923-4760-9

177. M. Marion and R.S. Cohen (eds.): *QuÇbec Studies in the Philosophy of Science.* Part I: Logic, Mathematics, Physics and History of Science. Essays in Honor of Hugues Leblanc. 1995
ISBN 0-7923-3559-7

178. M. Marion and R.S. Cohen (eds.): *QuÇbec Studies in the Philosophy of Science.* Part II: Biology, Psychology, Cognitive Science and Economics. Essays in Honor of Hugues Leblanc.
1996

ISBN 0-7923-3560-0
Set (177–178) ISBN 0-7923-3561-9

179. Fan Dainian and R.S. Cohen (eds.): *Chinese Studies in the History and Philosophy of Science and Technology.* 1996 ISBN 0-7923-3463-9

180. P. Forman and J.M. Snchez-Ron (eds.): *National Military Establishments and the Advancement of Science and Technology.* Studies in 20th Century History. 1996

ISBN 0-7923-3541-4

181. E.J. Post: *Quantum Reprogramming.* Ensembles and Single Systems: A Two-Tier Approach to Quantum Mechanics. 1995 ISBN 0-7923-3565-1

182. A.I. Tauber (ed.): *The Elusive Synthesis: Aesthetics and Science.* 1996 ISBN 0-7923-3904-5

183. S. Sarkar (ed.): *The Philosophy and History of Molecular Biology: New Perspectives.* 1996
ISBN 0-7923-3947-9

184. J.T. Cushing, A. Fine and S. Goldstein (eds.): *Bohmian Mechanics and Quantum Theory: An Appraisal.* 1996 ISBN 0-7923-4028-0

185. K. Michalski: *Logic and Time.* An Essay on Husserl's Theory of Meaning. 1996
ISBN 0-7923-4082-5

186. G. MunÇvar (ed.): *Spanish Studies in the Philosophy of Science.* 1996 ISBN 0-7923-4147-3

187. G. Schubring (ed.): *Hermann G~nther Graßmann (1809–1877): Visionary Mathematician, Scientist and Neohumanist Scholar.* Papers from a Sesquicentennial Conference. 1996
ISBN 0-7923-4261-5

188. M. Bitbol: *Schrdinger's Philosophy of Quantum Mechanics.* 1996 ISBN 0-7923-4266-6

189. J. Faye, U. Scheffler and M. Urchs (eds.): *Perspectives on Time.* 1997 ISBN 0-7923-4330-1

190. K. Lehrer and J.C. Marek (eds.): *Austrian Philosophy Past and Present.* Essays in Honor of Rudolf Haller. 1996 ISBN 0-7923-4347-6

191. J.L. Lagrange: *Analytical Mechanics.* Translated and edited by Auguste Boissonade and Victor N. Vagliente. Translated from the *MÇcanique Analytique, novelle Çdition* of 1811. 1997
ISBN 0-7923-4349-2

192. D. Ginev and R.S. Cohen (eds.): *Issues and Images in the Philosophy of Science.* Scientific and Philosophical Essays in Honour of Azarya Polikarov. 1997 ISBN 0-7923-4444-8

Boston Studies in the Philosophy of Science

193. R.S. Cohen, M. Horne and J. Stachel (eds.): *Experimental Metaphysics*. Quantum Mechanical Studies for Abner Shimony, Volume One. 1997   ISBN 0-7923-4452-9

194. R.S. Cohen, M. Horne and J. Stachel (eds.): *Potentiality, Entanglement and Passion-at-a-Distance*. Quantum Mechanical Studies for Abner Shimony, Volume Two. 1997
                    ISBN 0-7923-4453-7; Set 0-7923-4454-5

195. R.S. Cohen and A.I. Tauber (eds.): *Philosophies of Nature: The Human Dimension*. 1997
                    ISBN 0-7923-4579-7

196. M. Otte and M. Panza (eds.): *Analysis and Synthesis in Mathematics*. History and Philosophy. 1997   ISBN 0-7923-4570-3

197. A. Denkel: *The Natural Background of Meaning*. 1999   ISBN 0-7923-5331-5

198. D. Baird, R.I.G. Hughes and A. Nordmann (eds.): *Heinrich Hertz: Classical Physicist, Modern Philosopher*. 1999   ISBN 0-7923-4653-X

199. A. Franklin: *Can That be Right?* Essays on Experiment, Evidence, and Science. 1999
                    ISBN 0-7923-5464-8

200. Reserved

201. Reserved

202. Reserved

203. B. Babich and R.S. Cohen (eds.): *Nietzsche, Theories of Knowledge, and Critical Theory*. Nietzsche and the Sciences I. 1999   ISBN 0-7923-5742-6

204. B. Babich and R.S. Cohen (eds.): *Nietzsche, Epistemology, and Philosophy of Science*. Nietzsche and the Science II. 1999   ISBN 0-7923-5743-4

205. R. Hooykaas: *Fact, Faith and Fiction in the Development of Science*. The Gifford Lectures given in the University of St Andrews 1976. 1999   ISBN 0-7923-5774-4

206. M. Fehér, O. Kiss and L. Ropolyi (eds.): *Hermeneutics and Science*. 1999 ISBN 0-7923-5798-1

207. R.M. MacLeod (ed.): *Science and the Pacific War*. Science and Survival in the Pacific, 1939-1945. 1999   ISBN 0-7923-5851-1

208. I. Hanzel: *The Concept of Scientific Law in the Philosophy of Science and Epistemology*. A Study of Theoretical Reason. 1999   ISBN 0-7923-5852-X

209. G. Helm; R.J. Deltete (ed./transl.): *The Historical Development of Energetics*. 1999
                    ISBN 0-7923-5874-0

210. A. Orenstein and P. Kotatko (eds.): *Knowledge, Language and Logic*. Questions for Quine. 1999   ISBN 0-7923-5986-0

211. R.S. Cohen and H. Levine (eds.): *Maimonides and the Sciences*. 2000   ISBN 0-7923-6053-2

212. H. Gourko, D.I. Williamson and A.I. Tauber (eds.): *The Evolutionary Biology Papers of Elie Metchnikoff*. 2000   ISBN 0-7923-6067-2

213. S. D'Agostino: *A History of the Ideas of Theoretical Physics*. Essays on the Nineteenth and Twentieth Century Physics. 2000   ISBN 0-7923-6094-X

Also of interest:
R.S. Cohen and M.W. Wartofsky (eds.): *A Portrait of Twenty-Five Years Boston Colloquia for the Philosophy of Science, 1960-1985*. 1985   ISBN Pb 90-277-1971-3

Previous volumes are still available.

KLUWER ACADEMIC PUBLISHERS – DORDRECHT / BOSTON / LONDON